Lecture Notes in Computer Science 2450

Edited by G. Goos, J. Hartmanis, and J. van Leeuwen

Springer

Berlin
Heidelberg
New York
Barcelona
Hong Kong
London
Milan
Paris
Tokyo

Masami Ito   Masafumi Toyama (Eds.)

Developments in Language Theory

6th International Conference, DLT 2002
Kyoto, Japan, September 18-21, 2002
Revised Papers

 Springer

Series Editors

Gerhard Goos, Karlsruhe University, Germany
Juris Hartmanis, Cornell University, NY, USA
Jan van Leeuwen, Utrecht University, The Netherlands

Volume Editors

Masami Ito
Masafumi Toyama
Kyoto Sangyo University
Faculty of Science and Faculty of Engineering
Kyoto 603-8555, Japan
E-mail: {ito,toyama}@ksuvx0.kyoto-su.ac.jp

Cataloging-in-Publication Data applied for

Bibliographic information published by Die Deutsche Bibliothek
Die Deutsche Bibliothek lists this publication in the Deutsche Nationalbibliografie;
detailed bibliographic data is available in the Internet at <http://dnb.ddb.de>.

CR Subject Classification (1998): F.4.3, F.4.2, F.4, F.3, F.1, G.2

ISSN 0302-9743
ISBN 3-540-40431-7 Springer-Verlag Berlin Heidelberg New York

Springer-Verlag Berlin Heidelberg New York
a member of BertelsmannSpringer Science+Business Media GmbH

http://www.springer.de

© Springer-Verlag Berlin Heidelberg 2003

Typesetting: Camera-ready by author, data conversion by Olgun Computergrafik
Printed on acid-free paper   SPIN: 10871102   06/3142   5 4 3 2 1 0

Preface

DLT 2002 was the 6th Conference on Developments in Language Theory. The topics dealt with at the conference were: grammars and acceptors for strings, graphs, arrays, etc.; efficient algorithms for languages; combinatorial and algebraic properties of languages; decision problems; and relations to complexity theory, logic, picture description and analysis, DNA computing, cryptography, concurrency, quantum computing and algebraic systems.

DLT 2002 was held at Kyoto Sangyo University from September 18 to September 21, 2002. It was the first DLT conference held in Asia. The organizing committee consisted of Teruo Imaoka (Matsue), Bálazs Imreh (Szeged), Masami Ito (Kyoto, Co-chair), Yuji Kobayashi (Funabashi), Yoshiyuki Kunimochi (Fukuroi) and Masafumi Toyama (Kyoto, Co-chair). The Program Committee invited Prof. Grzegorz Rozenberg as a special guest and the conference was dedicated to his 60th birthday.

Previous DLTs were held in Turku (1993), Magdeburg (1995), Thessaloníki (1997), Aachen (1999) and Vienna (2001). These will be followed by the next DLT in Szeged (2003).

The Program Committee consisted of Jorge Almeida (Porto), Jean Berstel (Marne-la-Valée), Christian Calude (Auckland), Zoltán Fülöp (Szeged), Julaj Hromkovic (Aachen), Masami Ito (Kyoto, Chair), Juhani Karhumäki (Turku), Werner Kuich (Vienna), Alexandru Mateescu (Bucharest), Kenichi Morita (Hiroshima), Gheorghe Păun (Bucharest), Antonio Restivo (Palermo), Kai Salomaa (Kingston, Canada), Takashi Yokomori (Tokyo) and Sheng Yu (London, Canada). The program committee selected 28 papers from 63 submitted papers. The papers came from the following countries: Australia, Austria, Canada, China, Denmark, Finland, France, Germany, Hungary, India, Italy, Japan, Czech Republic, The Netherlands, The Philippines, The Ukraine, USA, Romania, Russia, Spain, Vietnam, and Yugoslavia. Each submitted paper was evaluated by at least four members of the Program Committee. All 28 selected papers are contained in this volume together with 9 invited presentations. In this volume, invited presentations are printed first, followed by contributed papers according to the order of presentation at DLT 2002.

The editors thank the members of the Program Committee for the evaluation of the papers, and the many referees who assisted the program committee members in this process. We are also grateful to the contributors of DLT 2002, in particular to the invited speakers for the realization of a very successful conference. The conference was supported by Kyoto Sangyo University, the Inoue Foundation for Science, the Kayamori Science Foundation and the Asahi Glass Foundation, and the conference was also under the auspices of EATCS. We are grateful to these organizations.

Finally, we would like to express our thanks to the following people for their assistance during the conference: Ms. Miyuki Endo, Ms. Tomomi Hirai, Mr. Tetsuya Hirose, Mr. Fumiaki Ishigure, Ms. Keiko Kobayashi, Mr. Shoichi Ohno, Mr. Hiroki Onoda, Mr. Kei Saito, Mr. Nobuaki Tanaka and Ms. Chikage Totsuka.

We would like also to thank Prof. Manfred Kudlek for his report on the conference in the bulletin of EATCS. We appreciate Mr. Yoshiyuki Kunimochi's assistance in editing this volume.

March 2003   Masami Ito
                                                              Masafumi Toyama
                                                                       Editors

Table of Contents

Invited Presentations

Contributions

Computational Processes in Living Cells: Gene Assembly in Ciliates

Tero Harju[1] and Grzegorz Rozenberg[2]

[1] Department of Mathematics, University of Turku, FIN-20014 Turku, Finland
harju@utu.fi
[2] Leiden Institute for Advanced Computer Science, Leiden University
Niels Bohrweg 1, 2333 CA Leiden, the Netherlands, and
Department of Computer Science, University of Colorado at Boulder
Boulder, Co 80309-0347, USA
rozenber@liacs.nl

Abstract. One of the most complex DNA processing in nature known to us is carried out by ciliates during the sexual reproduction when their micronuclear genome is transformed to the macronuclear genome. This process of gene assembly is intriguing and captivating also from the computational point of view. We investigate here three intramolecular molecular operations (*ld, hi,* and *dlad*) postulated to accomplish gene assembly. The formal models for these operations are formulated on three different abstraction levels: MDS descriptors, legals strings and overlap graphs. In general both legal strings and overlap graphs contain strings and graphs that do not model any micronuclear gene. After a short survey of gene assembly we study the problem of recognizing whether a general legal string or a general overlap graph is a formalization of a micronuclear gene.

> "One of the oldest forms of life on Earth has been revealed as a natural born computer programmer."
> BBC 10th September 2001

1 Introduction

Ciliates are complex single-cell organisms; around 8,000 species are known by today, but it is generally believed that the actual number of species is much higher than this. Ciliates are about two billion years old, and they live in almost every environment containing water including oceans, lakes, ponds, soils, and rivers. They are characterized by their cilia (Latin *cilium*, eyelash) – tiny hairs used for moving around as well as for moving food (e.g., bacteria or algae) towards mouth opening. Ciliates possess a unique feature of *nuclear dualism*: they have two, functionally different, types of nucleus – the *macronucleus*, which provides for the RNA transcripts needed for the cell to function, and the *micronucleus*, which is mostly dormant – activated only during sexual reproduction.

In *stichotrichs ciliates*, that are considered in this paper, micronucleus and macronucleus differ from each other drastically both on global and local levels.

M. Ito and M. Toyama (Eds.): DLT 2002, LNCS 2450, pp. 1–20, 2003.

A micronucleus has about 100 chromosomes; each of them containing very long, hundreds of thousands base pairs (bp), DNA molecule. Micronuclear genes are scattered along these molecules with long stretches of spacer DNA separating them. On the other hand, the macronuclear genome consists of short chromosomes (on average about 2000 base pairs) with the shortest of these about 200 bps – shorter than any other known DNA molecule occurring in nature. As for the local level, the micronuclear genes are not functional and each of them they consists of segments of the macronuclear version of this gene (these segments are called *macronuclear destined segments* or *MDSs*) separated by noncoding segments of DNA (called *internal eliminated segments* or *IESs*).

During the sexual reproduction process a macronucleus develops from a micronucleus. This transformation is the most involved DNA processing known in living cells. During the whole process of gene assembly about 100,000 IESs are excised, and MDSs are ligated to form competent genes. One of the amazing computational features of gene assembly is the fact that ciliates implement this this process using one of the standard data structures of computer science – *linked lists*.

Computational aspects of gene assembly were was first investigated by Landweber and Kari [10,11]. The focus of their studies is the computational power (in the sense of theoretical computer science) of molecular operations used in gene assembly. A different model was proposed by Ehrenfeucht, Prescott and Rozenberg [9,19,20]. Their system of operations for gene assembly is based on three molecular operations: *ld, hi,* and *dlad* (see [19]). It was proved by Ehrenfeucht, Petre, Prescott and Rozenberg [7] that every micronuclear gene can be assembled using the three postulated molecular operations. The major difference between the two models for molecular operations is that the model of Landweber and Kari is based on *intermolecular* operations while the other model is based on *intramolecular* operations.

Three abstraction levels of formalizing the three molecular operations *ld, hi* and *dlad* were considered in the literature: MDS descriptors, legal strings and overlap graphs. While the MDS descriptors give a rather faithful description of the micronuclear genes, the legal strings and the overlap graphs are more general in nature, that is, there exist legal strings and overlap graphs that do not correspond to any micronuclear gene. In this paper, after surveying the basic notions and results for the operations of gene assembly, we shall study the problem of recognizing when a general legal string, or a general overlap graph, is a formalization of a micronuclear gene.

For problems and results concerning formal aspects of gene assembly we refer to [4] (for characterizations of micronuclear MDS/IES patterns that can be assembled using various subsets of the three molecular operations), [5] (for general graph theoretic method for modelling and analysing gene assembly), and [8] (for invariant properties of the assembly process). For background on DNA molecules, we refer the reader to [12] and [13].

2   Legal Strings and Overlap Graphs

Let Σ be a (finite or infinite) alphabet, and let $\overline{\Sigma} = \{\overline{a} \mid a \in \Sigma\}$ be a copy of Σ such that $\Sigma \cap \overline{\Sigma} = \emptyset$. We also write

$$\Sigma^{\overline{\Sigma}} = (\Sigma \cup \overline{\Sigma})^*$$

for the set of all strings over $\Sigma \cup \overline{\Sigma}$. Each $v \in \Sigma^{\overline{\Sigma}}$ is a *signed string* over Σ. The signing $a \mapsto \overline{a}$ can be extended as follows. For each $a \in \Sigma$, let $\overline{\overline{a}} = a$, and for a nonempty signed string $u = a_1 a_2 \ldots a_n \in \Sigma^{\overline{\Sigma}}$, where $a_i \in \Sigma \cup \overline{\Sigma}$ for each i, let the *inversion* of u be $\overline{u} = \overline{a}_n \overline{a}_{n-1} \ldots \overline{a}_1 \in \Sigma^{\overline{\Sigma}}$. Let also $u^R = a_n a_{n-1} \ldots a_1$ and $u^C = \overline{a}_1 \overline{a}_2 \ldots \overline{a}_n$ be the *reversal* and the *complementation* of u, respectively. It is then clear that $\overline{u} = (u^R)^C = (u^C)^R$.

A letter $a \in \Sigma \cup \overline{\Sigma}$ *occurs* in v, if either a or $\overline{a}$ is a substring of v. Let $\mathrm{dom}(v) = \{a \in \Sigma \mid a \text{ occurs in } v\}$ be the *domain* of v.

Let Σ and Γ be two alphabets. A function $\tau \colon \Sigma^{\overline{\Sigma}} \to \Gamma^{\overline{\Sigma}}$ is a *morphism*, if $\tau(uv) = \tau(u)\tau(v)$ for all $u, v \in \Sigma^{\overline{\Sigma}}$, and τ is a *substitution*, if, moreover, $\tau(\overline{u}) = \overline{\tau(u)}$ for all $u, v \in \Sigma^{\overline{\Sigma}}$. Note that the images $\tau(a)$ for the letters $a \in \Sigma$ determine the substitution τ. Two strings $u \in \Sigma^{\overline{\Sigma}}$ and $v \in \Gamma^{\overline{\Sigma}}$ are *isomorphic*, if $\tau(u) = v$ for an injective substitution $\tau \colon \Sigma^{\overline{\Sigma}} \to \Gamma^{\overline{\Sigma}}$ such that $\tau(\Sigma) \subseteq \Gamma$. Let ξ be the morphism such that for all $a \in \Sigma$,

$$\xi(a) = a = \xi(\overline{a}).$$

A signed string $v \in \Sigma^{\overline{\Sigma}}$ is a *signing* of a string $u \in \Sigma^*$, if $\xi(v) = u$.

Example 1. The signed string $u = 43\overline{3}5\overline{4}5 \in \{3, 4, 5\}^{\overline{\Sigma}}$ is isomorphic to $v = 23\overline{3}424$. The isomorphism τ is defined in this example by $\tau(4) = 2$, $\tau(3) = 3$ and $\tau(5) = 4$. Also, u is a signing of the string 433545.   □

Let $v = a_1 a_2 \ldots a_n \in \Sigma^*$. Then a string $u \in \Sigma^*$ is a *permutation* of v, if there exists a permutation $(i_1 i_2 \ldots i_n)$ of $\{1, 2, \ldots, n\}$ such that $u = a_{i_1} a_{i_2} \ldots a_{i_n}$. Moreover, a signing of a permutation of v is said to be a *signed permutation* of v. A string $v \in \Sigma^*$ is a *double occurrence string*, if every letter $a \in \mathrm{dom}(v)$ occurs exactly twice in v. A signing of a nonempty double occurrence string is called a *legal string*. If a legal string $u \in \Sigma^{\overline{\Sigma}}$ contains one occurrence of $a \in \Sigma$ and one occurrence of $\overline{a}$, then a is said to be *positive* in u; otherwise, a is *negative* in u.

Example 2. In the legal string $u = 243\overline{2}\,\overline{5}\,345$ letters 2 and 5 are positive while 3 and 4 are negative. The string $w = 243\overline{2}\,\overline{5}\,35$ is not legal, since it has only one occurrence of 4.   □

Let $u = a_1 a_2 \ldots a_n \in \Sigma^{\overline{\Sigma}}$ be a legal string over Σ, where $a_i \in \Sigma \cup \overline{\Sigma}$ for each i. Then for each $a \in \mathrm{dom}(u)$, there are indices i and j with $1 \leq i < j \leq n$ such that $\xi(a_i) = a = \xi(a_j)$. The substring

$$u_{(a)} = a_i a_{i+1} \ldots a_j$$

is called the a-*interval* of u. Two different letters $a, b \in \Sigma$ are said to *overlap in* u, if the a-interval and the b-interval of u overlap: if $u_{(a)} = a_{i_1} \dots a_{j_1}$ and $u_{(b)} = a_{i_2} \dots a_{j_2}$, then either $i_1 < i_2 < j_1 < j_2$ or $i_2 < i_1 < j_2 < j_1$. Moreover, for each letter a, we denote by $O_u(a)$ ($O_u^+(a)$, $O_u^-(a)$, resp.) the set of letters (positive, negative, resp.) overlapping with a in u. For technical reasons, it is convenient to include a in $O_u(a)$: if a is positive in u, then $a \in O_u^+(a)$, and if a is negative in u, then $a \in O_u^-(a)$.

Example 3. The 2-interval $u_{(2)} = 2\,4\,3\,5\,3\,\overline{2}$ of the string $u = 2\,4\,3\,5\,3\,\overline{2}\,\overline{6}\,5\,7\,4\,6\,7$ contains only one occurrence of 4 and 5, but two or no occurrences of 3, 6 and 7. Therefore 2 overlaps with 4 and 5 but not with 3, 6 and 7, and so $O_u(2) = \{2, 4, 5\}$, $O_u^+(2) = \{2, 5\}$ and $O_u^-(2) = \{4\}$. Similarly,

$$u_{(3)} = 3\,5\,3 \qquad\qquad \text{and } 3 \text{ overlaps with } 5\,,$$

$$u_{(4)} = 4\,3\,5\,3\,\overline{2}\,\overline{6}\,5\,7\,4 \qquad \text{and } 4 \text{ overlaps with } 2, 6, 7\,,$$

$$u_{(5)} = 5\,3\,\overline{2}\,\overline{6}\,5 \qquad\qquad \text{and } 5 \text{ overlaps with } 2, 3, 6\,,$$

$$u_{(6)} = \overline{6}\,\overline{5}\,7\,4\,6 \qquad\qquad \text{and } 6 \text{ overlaps with } 4, 5, 7\,,$$

$$u_{(7)} = 7\,4\,6\,7 \qquad\qquad \text{and } 7 \text{ overlaps with } 4, 6\,.$$

□

For each letter $a \in \Sigma$, let $\mathbf{a} = \{a, \overline{a}\}$, and for each legal string $v \in \Sigma^{\Phi}$, let $\mathbf{P}_v = \{\mathbf{a} \mid a \in \mathrm{dom}(v)\}$. Define the *overlap graph* $\gamma_v = (\mathbf{P}_v, E, \sigma)$ of v as a graph on the set $\mathbf{P}_v$ of vertices together with the labelling σ of the vertices defined by

$$\sigma(\mathbf{a}) = \begin{cases} + , & \text{if } a \in \Sigma \text{ is positive in } v\,, \\ - , & \text{if } a \in \Sigma \text{ is negative in } v\,, \end{cases}$$

and

$$\{\mathbf{a}, \mathbf{b}\} \in E \iff a \text{ and } b \text{ overlap in } v\,.$$

The label of a vertex $\mathbf{x}$ in an overlap graph is usually called the *sign* of $\mathbf{x}$. Overlap graphs of double occurrence strings are also known as *circle graphs* (see [1,2]).

Example 4. The overlap graph of the legal string u of Example 3 is given in Fig. 1, where the sign of each vertex is given as a superscript.   □

The mapping $w \mapsto \gamma_w$ of legal strings to overlap graphs is not injective: for each legal string $w = w_1 w_2$, we have

$$\gamma_{w_1 w_2} = \gamma_{w_2 w_1} \quad \text{and} \quad \gamma_w = \gamma_{\varsigma(w)}\,,$$

where ς is any mapping that chooses an element $\varsigma(\mathbf{a})$ from $\mathbf{a} = \{a, \overline{a}\}$ for each a. In particular, all conjugates of a legal string w have the same overlap graph. Also, the reversal w^R and the complementation w^C of a legal string w define the same overlap graph as w does.

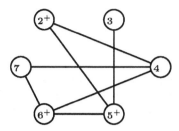

Fig.1. The overlap graph of $u = 2\,4\,3\,5\,3\,\overline{2}\,\overline{6}\,\overline{5}\,7\,4\,6\,7$

3 Gene Assembly in Ciliates

The genes in a micronuclear chromosome in ciliates consist of MDSs separated by IESs. Moreover, some of the MDSs may have been inverted. From the viewpoint of gene assembly, the micronuclear gene can be seen as a sequence of MDSs and IESs forming the gene.

Example 5. The actin I gene in *Sterkiella nova* has the following MDS/IES micronuclear structure:

$$M_3 I_1 M_4 I_2 M_6 I_3 M_5 I_4 M_7 I_5 \overline{M}_2 I_7 M_1 I_8 M_8 \,. \tag{1}$$

This structure is illustrated in Fig. 2, where each MDS is drawn as a rectangle and the interspacing IESs are represented by lines. Note that the MDS M_2 is inverted. □

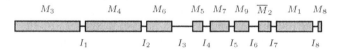

Fig.2. The micronuclear version of the *actin I* gene in *Sterkiella nova*

In the process of gene assembly each micronuclear gene is translated into a macronuclear gene by excising all IESs and by splicing the MDSs in the orthodox order $M_1, M_2, \ldots, M_\kappa$ (see, [13,14,18]). Each MDS M_i has the form

$$M_i = (\pi_i, \mu_i, \pi_{i+1}), \quad \text{where } \pi_1 = \frac{b}{b}, \ \pi_i = \frac{p_i}{p_i} \text{(for } 1 < i < \kappa\text{), and } \pi_{\kappa+1} = \frac{e}{e} \,.$$

Here π_i, μ_i and π_{i+1} correspond to double stranded molecules; π_i and π_{i+1} are the (*incoming* and *outgoing*) *pointers* and μ_i is the *body* of the MDS M_i. The *markers* b and e (and their inversions $\overline{b}$ and $\overline{e}$) designate the locations where the macronuclear DNA gene is to be excised. In the final gene in the macronucleus, the MDSs $M_1, M_2, \ldots, M_\kappa$ are spliced together by "gluing" M_j and M_{j+1} on the common pointer π_{j+1} for each j.

We now describe the operations *ld*, *hi* and *dlad* introduced in [9,19].

1. The operation *(loop, direct repeat)-excision*, or *ld*, for short, is applied to a molecule with a direct repeat pattern $(- \pi - \pi -)$ of a pointer: either the two occurrences of π are separated by one IES or they are at the two opposite ends of the molecule. The molecule is folded into a loop in such a way that the pointers are aligned and then, the operation proceeds as shown in Fig. 3. The *ld* operation yields two molecules: a linear and a circular one. The circular molecule either contains the whole gene or it contains one IES only.

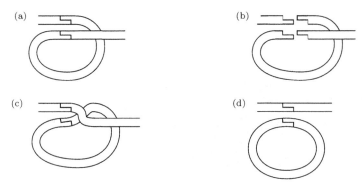

F ig.3. The *ld* operation

2. The operation *(hairpin, inverted repeat)-excision/reinsertion*, or *hi*, for short, is applied to a molecule with an *inverted repeat pattern* $(- \pi - \overline{\pi} -)$ of a pointer. The molecule is folded into a hairpin in such a way that the pointers are aligned and the operation proceeds as in Fig. 4. The operation *hi* yields only one molecule.

3. The operation *(double loop, alternating direct repeat)-excision/reinsertion*, or *dlad*, for short, is applied to a molecule with an *alternating direct repeat pattern* $(- \pi - \pi' - \pi - \pi'-)$ for two pointers π and π'. The molecule is folded into a double loop in such a way that the pointers π and the pointers π' are aligned and then the operation proceeds as in Fig. 5. The operation *dlad* yields one molecule.

4   MDS Descriptors

The process of gene assembly can be seen as a process of assembling MDSs through splicing so that finally the macronuclear gene is obtained (recall that the macronuclear gene is obtained by gluing together, on common pointers, the

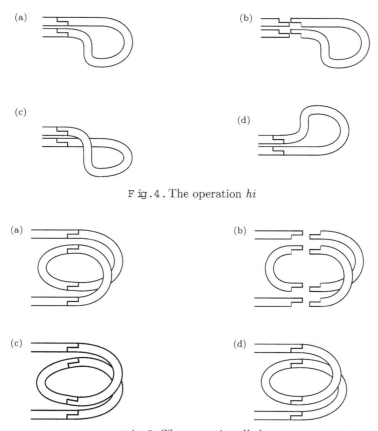

Fig.4. The operation hi

Fig.5. The operation $dlad$

MDSs arranged in the orthodox order $M_1, M_2, \ldots, M_\kappa$). Therefore the structural information about the micronuclear gene or an intermediate precursor of a macronuclear gene can be given by the sequence of MDSs of the gene. Thus one can represent a macronuclear gene, as well as its micronuclear or an intermediate precursor, by its sequence of MDSs only.

Example 6. The representation of the actin I gene in *Sterkiella nova* from Example 5 has the following simplified representation: $\alpha = M_3 M_4 M_6 M_5 M_7 \overline{M}_2 M_1 M_8$ (all the IESs were removed). From the point of view of our considerations has the same information as the structure in (1). □

We shall use the alphabets $\Theta_\kappa = \{ M_{i,j} \mid 1 \le i \le j \le \kappa \}$ to denote the MDSs for all $\kappa \ge 1$. Also, we let

$$\Theta = \bigcup_{\kappa \ge 1} \Theta_\kappa .$$

The signed strings in $\Theta^{\maltese}$ are *(MDS) arrangements*. Elements $M_{i,i}$ (that denote micronuclear MDSs) are called *elementary MDSs*, and they are often written simply as M_i. Letters $M_{i,j}$ with $j > i$ denote *composite MDSs* formed during the assembly process by splicing the MDSs $M_i, M_{i+1}, \ldots, M_j$.

We say that an arrangement $\alpha \in \Theta_\kappa{}^{\maltese}$ is *orthodox*, if it is of the form

$$\alpha = M_{1,i_2-1} M_{i_2,i_3-1} \ldots M_{i_n,\kappa} \,. \tag{2}$$

Note that an orthodox arrangement does not contain any inverted MDSs. A signed permutation of an orthodox arrangement ($M_1 M_2 \ldots M_\kappa$, resp.) is a *realistic arrangement* (*micronuclear arrangement*, resp.) of size κ.

Each orthodox arrangement α such as (2) can be mapped onto the orthodox arrangement $M_1 M_2 \ldots M_{i_n}$ by mapping $M_{i_r,i_{r+1}-1}$ to M_r (with $1 = i_1$). Hence every realistic arrangement is an image of a micronuclear arrangement by such a mapping.

Example 7. The arrangement $M_{1,1} M_{2,5} M_{6,9}$ is orthodox. It is not a micronuclear arrangement, since it contains composite MDSs $M_{2,5}$ and $M_{6,9}$. □

In an assembled gene no pointers are present, because the gene has no IESs. On the other hand, the micronuclear form of a gene has (possibly many) pointers. Thus the gene assembly process can be analysed by

1. representing the micronuclear and each of the intermediate genes by the pattern of pointers present in this gene, and then
2. representing the process by a sequence of such patterns, where each next pattern results by the application of molecular operations to the previous one.

Consequently, we can simplify the formal framework by denoting each MDS by the ordered pair of its pointers and markers only, i.e., $M_{i,j} = (\pi_i, \mu, \pi_j)$ is represented as (p_i, p_j), and its inversion $\overline{M}_{i,j} = (\overline{\pi}_j, \overline{\mu}, \overline{\pi}_i)$ as $(\overline{p}_j, \overline{p}_i)$ for all $i \le j$.

More formally, let $\Psi = \{b, e, \overline{b}, \overline{e}\}$ denote the set of the *markers* (b stands for "beginning", and e for "end"). For $\kappa \ge 2$, let

$$\Pi_\kappa = \Delta_\kappa \cup \overline{\Delta}_\kappa, \ \Pi_{ex,\kappa} = \Pi_\kappa \cup \Psi \ \text{ where } \Delta_\kappa = \{2, \ldots, \kappa\} \text{ and } \overline{\Delta}_\kappa = \{\overline{2}, \ldots, \overline{\kappa}\} \,.$$

Also, let $\Delta = \{2, 3, \ldots\}$ and $\Pi = \Delta \cup \overline{\Delta}$. We do not use 1, since it represents a begin marker in the encoding defined in a latter section. The letters in Π are called *pointers*, and for each $p \in \Pi$, the pair $\mathbf{p} = \{p, \overline{p}\}$ is the *pointer set* of p (and of $\overline{p}$). Whenever we deal with a specific gene (in any specific species) the number of elementary MDSs in the micronuclear determines the index κ. In order to avoid too involved formalism we shall assume that, unless explicitly stated otherwise, the index κ is clear from the context. Let

$$\Gamma_\kappa = \{(b, e), (\overline{e}, \overline{b})\} \cup \{(i, j), (\overline{j}, \overline{i}) \mid 2 \le i < j \le \kappa\} \cup$$
$$\cup \{(b, i), (\overline{i}, \overline{b}), (i, e), (\overline{e}, \overline{i}) \mid 2 \le i \le \kappa\} \,.$$

A string over Γ_κ is called an *MDS descriptor*.

We define the morphism $\psi_\kappa : (\Theta_\kappa)^{\maltese} \to \Gamma_\kappa^*$ as follows:

$$\psi_\kappa(M_{1,\kappa}) = (b,e) \quad\quad \text{and} \quad \psi_\kappa(\overline{M}_{1,\kappa}) = (\overline{e},\overline{b}),$$
$$\psi_\kappa(M_{1,i}) = (b,i+1) \quad \text{and} \quad \psi_\kappa(\overline{M}_{1,i}) = (\overline{i+1},\overline{b}) \quad (1 \le i < \kappa),$$
$$\psi_\kappa(M_{i,\kappa}) = (i,e) \quad\quad \text{and} \quad \psi_\kappa(\overline{M}_{i,\kappa}) = (\overline{e},\overline{i}) \quad\quad (1 < i \le \kappa),$$
$$\psi_\kappa(M_{i,j}) = (i,j+1) \quad \text{and} \quad \psi_\kappa(\overline{M}_{i,j}) = (\overline{j+1},\overline{i}) \quad (1 < i \le j < \kappa).$$

The mapping ψ_κ is bijective, and therefore, for formal purposes, an MDS arrangement α and its image MDS descriptor $\delta = \psi_\kappa(\alpha)$ are equivalent – that is, the structure of these two strings is the same. In particular, the mapping ψ_κ is invertible: if $\delta = \psi_\kappa(\alpha)$ then $\alpha = \psi_\kappa^{-1}(\delta)$ is well defined.

Example 8. For the realistic arrangement $\alpha = M_{3,5}\overline{M}_{9,11}\overline{M}_{1,2}M_{12}\overline{M}_{6,8}$, we obtain the MDS descriptor $\psi_{12}(\alpha) = (3,6)(\overline{12},\overline{9})(\overline{3},\overline{b})(12,e)(\overline{9},\overline{6})$. □

Let $\delta = (x_1,x_2)\ldots(x_{2n-1},x_{2n})$ and $\delta' = (y_1,y_2)\ldots(y_{2n-1},y_{2n})$ be two MDS descriptors. They are *isomorphic*, if

(1) $x_i \in \Delta \iff y_i \in \Delta$ and $x_i \in \Psi \implies y_i = x_i$,
(2) $\xi(x_i) < \xi(x_j) \iff \xi(y_i) < \xi(y_j)$ for $x_i, x_j \notin \Psi$.

Example 9. The MDS descriptors $(4,5)(\overline{8},\overline{6})(b,4)$ and $(2,3)(\overline{5},\overline{4})(b,2)$ satisfy the above requirements, and thus they are isomorphic.

An MDS descriptor δ is said to be *realistic*, if δ is isomorphic with an MDS descriptor $\psi_\kappa(\alpha)$ for some κ and a micronuclear arrangement α. The following lemma is clear from the definition of the mapping ψ.

Lemma 1. *An MDS descriptor δ is realistic if and only if $\delta = \psi_\kappa(\alpha)$ for some κ and a realistic MDS arrangement α of size κ.*

Let $\delta = (x_1,x_2)(x_3,x_4)\ldots(x_{2n-1},x_{2n})$ be a realistic MDS descriptor. Each pointer $p \in \Delta$ either does not occur in δ or it occurs exactly twice in δ. If there are two occurrences, let these be $x_i, x_j \in \mathbf{p} = \{p,\overline{p}\}$ for $1 \le i < j \le 2n$. The *p-interval* of δ is then defined to be the set

$$\delta_{(p)} = \{x_i, x_{i+1}, \ldots, x_j\}.$$

If $x_i = x_j$, then p is *negative* in δ; otherwise (i.e., if $x_i = \overline{x}_j$) p is *positive* in δ.

5   The Assembly Operations

We now formalize the gene assembly operations through formal operations on realistic MDS descriptors. Corresponding to the three molecular operations *ld, hi,* and *dlad*, we have three operations ld, hi, and dlad on MDS descriptors.

1. For each $p \in \Pi_\kappa$, the ld-*rule* for p is defined as follows:

$$\mathsf{ld}_p(\delta_1(q,p)(p,r)\delta_2) = \delta_1(q,r)\delta_2 \tag{l1}$$

$$\mathsf{ld}_p((p,q)\delta_1(r,p)) = (r,q)\delta_1 \tag{l2}$$

where $q, r \in \Pi_{ex,\kappa}$ and $\delta_1, \delta_2 \in (\Gamma_\kappa)^*$.

The case (l1) is called a *simple* ld-*rule*, and it applies to two adjacent occurrences of p separated by one IES only. The case (l2) is called a *boundary* hi-*rule*, and it applies to two occurrences of p at the two ends of the molecule. Both of these cases are illustrated in Fig. 6, where rectangles denote MDSs with their pointers indicated, the zigzag line denotes a segment of a molecule that may contain both MDSs and IESs; a simple straight line represents an IES.

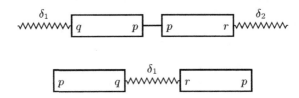

Fig.6. The MDS/IES structure to which ld_p is applicable

2. For each $p \in \Pi_\kappa$, the hi-*rule* for p is defined as follows:

$$\mathsf{hi}_p(\delta_1(p,q)\delta_2(\overline{p},\overline{r})\delta_3) = \delta_1\overline{\delta}_2(\overline{q},\overline{r})\delta_3 \tag{h1}$$

$$\mathsf{hi}_p(\delta_1(q,p)\delta_2(\overline{r},\overline{p})\delta_3) = \delta_1(q,r)\overline{\delta}_2\delta_3 \tag{h2}$$

where $q, r \in \Pi_{ex,\kappa}$ and $\delta_i \in (\Gamma_\kappa)^*$ for each $i = 1, 2, 3$.

Here one occurrence of p is the incoming pointer and the other occurrence is the outgoing pointer. These cases are illustrated in Fig. 7.

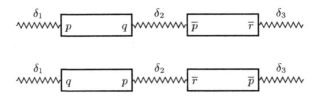

Fig.7. The MDS/IES structure to which hi_p is applicable

3. For each $p, q \in \Pi_\kappa$, $p \neq q$, /the dlad-*rule* for p and q is defined as follows:

$$\mathsf{dlad}_{p,q}(\delta_1(p,r_1)\delta_2(q,r_2)\delta_3(r_3,p)\delta_4(r_4,q)\delta_5) = \delta_1\delta_4(r_4,r_2)\delta_3(r_3,r_1)\delta_2\delta_5 \quad (4a)$$

$$\mathsf{dlad}_{p,q}(\delta_1(p,r_1)\delta_2(r_2,q)\delta_3(r_3,p)\delta_4(q,r_4)\delta_5) = \delta_1\delta_4\delta_3(r_3,r_1)\delta_2(r_2,r_4)\delta_5 \quad (4b)$$

$$\mathsf{dlad}_{p,q}(\delta_1(r_1,p)\delta_2(q,r_2)\delta_3(p,r_3)\delta_4(r_4,q)\delta_5) = \delta_1(r_1,r_3)\delta_4(r_4,r_2)\delta_3\delta_2\delta_5 \quad (4c)$$

$$\mathsf{dlad}_{p,q}(\delta_1(r_1,p)\delta_2(r_2,q)\delta_3(p,r_3)\delta_4(q,r_4)\delta_5) = \delta_1(r_1,r_3)\delta_4\delta_3\delta_2(r_2,r_4)\delta_5 \quad (4d)$$

$$\mathsf{dlad}_{p,q}(\delta_1(p,r_1)\delta_2(q,p)\delta_4(r_4,q)\delta_5) = \delta_1\delta_4(r_4,r_1)\delta_2\delta_5 \quad (3a)$$

$$\mathsf{dlad}_{p,q}(\delta_1(p,q)\delta_3(r_3,p)\delta_4(q,r_4)\delta_5) = \delta_1\delta_4\delta_3(r_3,r_4)\delta_5 \quad (3b)$$

$$\mathsf{dlad}_{p,q}(\delta_1(r_1,p)\delta_2(q,r_2)\delta_3(p,q)\delta_5) = \delta_1(r_1,r_2)\delta_3\delta_2\delta_5 \quad (3c)$$

where $r_i \in \Pi_{ex,\kappa}$ and $\delta_i \in (\Gamma_\kappa)^*$ for each i.

In each of the above instances, the pointer p overlaps with the pointer q. The cases (4a) – (4d) are illustrated in Fig. 8, and the 'short' cases (3a) – (3c) in Fig. 9.

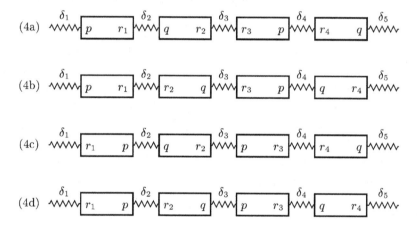

Fig. 8. The main MDS/IES structures to which $\mathsf{dlad}_{p,q}$ is applicable

The following lemma is clear by the form of the operations and the definition of isomorphism of MDS descriptors.

Lemma 2. *Let δ be a realistic MDS descriptor, p and q be pointers in δ, and $\varphi \in \{\mathsf{ld}_p, \mathsf{hi}_p, \mathsf{dlad}_{p,q}\}$. If φ is applicable to δ, then also $\varphi(\delta)$ is realistic.*

Let a composition $\varphi = \varphi_k \ldots \varphi_1$ of operations $\varphi_i \in \{\mathsf{ld}_p, \mathsf{hi}_p, \mathsf{dlad}_{p,q} \mid p,q \in \Pi\}$ be applicable to an MDS descriptor δ. In this case, we also say that φ is a *reduction* of δ. Moreover, φ is *successful* for δ, if either $\varphi(\delta) = (b,e)$ or $\varphi(\delta) = (\bar{e}, \bar{b})$.

Example 10. Let $\delta = (4,5)(\bar{2}, \bar{b})(5,e)(\bar{4}, \bar{3})(\bar{3}, \bar{2})$. Now δ is a realistic MDS descriptor, since $\delta = \psi(\alpha)$ for the micronuclear arrangement $\alpha = M_4 \overline{M}_1 M_5 M_3 \overline{M}_2$.

$$\text{(3a)} \qquad \overset{\delta_1}{\wedge\!\wedge\!\wedge} \boxed{\quad p \qquad r_1 \quad} \overset{\delta_2}{\wedge\!\wedge\!\wedge} \boxed{\quad q \qquad p \quad} \overset{\delta_4}{\wedge\!\wedge\!\wedge} \boxed{\quad r_4 \qquad q \quad} \overset{\delta_5}{\wedge\!\wedge\!\wedge}$$

$$\text{(3b)} \qquad \overset{\delta_1}{\wedge\!\wedge\!\wedge} \boxed{\quad p \qquad q \quad} \overset{\delta_3}{\wedge\!\wedge\!\wedge} \boxed{\quad r_3 \qquad q \quad} \overset{\delta_4}{\wedge\!\wedge\!\wedge} \boxed{\quad q \qquad r_4 \quad} \overset{\delta_5}{\wedge\!\wedge\!\wedge}$$

$$\text{(3c)} \qquad \overset{\delta_1}{\wedge\!\wedge\!\wedge} \boxed{\quad r_1 \qquad p \quad} \overset{\delta_2}{\wedge\!\wedge\!\wedge} \boxed{\quad q \qquad r_2 \quad} \overset{\delta_3}{\wedge\!\wedge\!\wedge} \boxed{\quad p \qquad q \quad} \overset{\delta_5}{\wedge\!\wedge\!\wedge}$$

F ig . 9 . The special MDS/IES structures to which $\text{dlad}_{p,q}$ is applicable

The operations $\text{ld}_{\overline{3}}$, hi_4, and $\text{dlad}_{5,\overline{2}}$ are applicable to δ:

$$\text{ld}_{\overline{3}}(\delta) = (4,5)(\overline{2},\overline{b})(5,e)(\overline{4},\overline{2}),$$
$$\text{hi}_4(\delta) = (\overline{e},\overline{5})(b,2)(\overline{5},\overline{3})(\overline{3},\overline{2}),$$
$$\text{dlad}_{5,\overline{2}}(\delta) = (4,e)(\overline{4},\overline{3})(\overline{3},\overline{b}).$$

We also have that $\text{hi}_4\,\text{dlad}_{5,\overline{2}}(\delta) = (\overline{e},\overline{3})(\overline{3},\overline{b})$ and $\text{ld}_{\overline{3}}\,\text{hi}_4\,\text{dlad}_{5,\overline{2}}(\delta) = (\overline{e},\overline{b})$, and therefore $\text{ld}_{\overline{3}}\,\text{hi}_4\,\text{dlad}_{5,\overline{2}}$ is successful for δ.   □

A single application of the three operations ld, hi, and dlad shortens the MDS descriptor. The following theorem was first proved in [6]. It shows that the set of our three operations on MDS descriptors, ld, hi, and dlad, is *universal*, i.e., any realistic MDS descriptor δ has a successful reduction for it.

Theorem 1. *Each realistic MDS descriptor δ has a successful reduction.*

Proof. It is sufficient to prove that for every realistic MDS descriptor δ at least one of the three operations is applicable to δ, since a realistic MDS descriptor yields again a realistic MDS descriptor by any one of these operations.

Let then δ be such a descriptor, and assume that neither ld nor hi is applicable for δ. Since hi is not applicable, all pointers in δ must be negative, and since ld is not applicable, δ has no simple direct repeat pattern. Consequently, if $p \in \Pi$ occurs in δ, then the p-interval $\delta_{(p)}$ must contain at least one other pointer.

Let p then be a pointer in δ such that the number of pointers within the p-interval is minimal (no other pointer from δ has less pointers in its interval). Let q be a pointer that has an occurrence within the p-interval. Since all pointers in δ are negative, δ contains two occurrences of q. The other occurrence of q must be outside the p-interval, as otherwise the minimality of p is contradicted. But now either $\text{dlad}_{p,q}$ or $\text{dlad}_{q,p}$ is applicable to δ.   □

6   Realizable Legal Strings

The model of MDS descriptors was greatly simplified in [3] and [7] by considering legal strings instead of MDS descriptors. In order to discuss the reduction of the model of MDS descriptors to the model using legals strings, we need the following definitions.

Let μ be the morphism that removes the parenthesis and the markers from each MDS descriptor. To be more precise, let $\nu\colon (\Delta_{ex,\kappa})^{\maltese} \to \Delta^{\maltese}$ be a substitution defined by $\nu(x) = \Lambda$ if $x \in \Psi$ and $\nu(x) = x$, otherwise. Then let $\mu(x,y) = \nu(x)\nu(y)$ for all $x, y \in \Pi_{ex,\kappa}$. For the MDS arrangements, we define a substitution $\varrho_\kappa \colon (\Theta_\kappa)^{\maltese} \to (\Delta_\kappa)^{\maltese}$ by $\varrho_\kappa = \mu\psi_\kappa$. In particular, for the elementary MDSs, we have

$$\varrho_\kappa(M_1) = 2, \quad \varrho_\kappa(M_\kappa) = \kappa, \quad \varrho_\kappa(M_i) = i\,i+1 \quad \text{for } 2 < i < \kappa,$$

and $\varrho_\kappa(\overline{M}_i) = \overline{\varrho_\kappa(M_i)}$ for $1 \leq i \leq \kappa$.

The operations ld, hi, and dlad carry over to legal strings in simplified form. Indeed, as seen from the definition given below instead of the 11 cases for the MDS descriptors, we have only four cases for legal strings.

Let $p, q \in \Pi$.

1. The string operations corresponding to ld_p are the following:

$$\mathsf{ld}_p(uppv) = uv \quad \text{and} \quad \mathsf{ld}_p(pup) = u.$$

2. The string operation corresponding to hi_p is the following:

$$\mathsf{hi}_p(upw\overline{p}v) = u\overline{w}v.$$

3. The string operation corresponding to $\mathsf{ld}_{p,q}$ is the following:

$$\mathsf{dlad}_{p,q}(u_1pv_1qu_2pv_2qu_3) = u_1v_2u_3v_1u_3.$$

For a more detailed discussion concerning the correspondence of the operations for MDS descriptors and legal strings, we refer to [3,6]

We say that a legal string u is *realistic* if there exists a realistic MDS descriptor δ such that $u = \mu(\delta)$, or equivalently $u = \varrho(\alpha)$ for a signed permutation of an orthodox MDS arrangement α (see Lemma 1).

Example 11. (1) The MDS arrangement $\alpha = M_3M_4M_6M_5M_7M_9\overline{M}_2M_1M_8$ in Example 6 gives the realistic legal string $\varrho(\alpha) = 3\,4\,4\,5\,6\,7\,5\,6\,7\,8\,9\,\overline{3}\,\overline{2}\,2\,8\,9$.

(2) The string $u = 2\,3\,4\,3\,2\,4$ is legal and $\mathrm{dom}(u) = \Delta_\kappa$ for $\kappa = 4$. However, u is not realistic. (It is easy to see that u has no 'realistic parsing'.) $\square$

A legal string $u \in \Sigma^{\maltese}$ is *realizable*, if u is isomorphic to a realistic legal string, that is, if there exists an injective substitution $\tau\colon \Sigma^{\maltese} \to \Delta^{\maltese}$ such that $\tau(\Sigma) \subseteq \Delta$ and $\tau(u)$ is realistic.

Example 12. The legal string $u = 5\,4\,\overline{3}\,2\,5\,\overline{5}\,2\,4\,3 \in (\Delta_5)^{\maltese}$ is not realistic, since $4\,\overline{3}\,2$ can never be a substring of a realistic legal string. However, u is realizable: the substitution τ is defined by $\tau(2) = 2$, $\tau(3) = 5$, $\tau(4) = 4$, and $\tau(5) = 3$, and then $\tau(u) = 3\,4\,\overline{5}\,2\,3\,\overline{3}\,2\,4\,5$. Now $\tau(u) = \varrho(\alpha)$ for the micronuclear arrangement $\alpha = M_3\overline{M}_5M_1\overline{M}_2M_4$, and thus $\tau(u)$ is realistic.

The following example shows that there are legal strings that are not realizable.

Example 13. The legal string $u = 2\,2\,3\,4\,4\,3\,5\,5$ is not realizable. To see this, assume to the contrary: let v be a realistic legal string and τ an isomorphism such that $u = \tau(v)$. Since the string pp is never an image $\varrho(M_i)$ of any MDS M_i, either one of the following cases (i) or (ii) holds.

(i) $\tau(2) = 2$ and $\tau(5) = 5$. Now also $\tau(3) = 3$, since u begins with $2(23)$. This yields a contradiction, since now 35 is a substring of v.

(ii) $\tau(2) = 5$ and $\tau(5) = 2$. Now, v should begin with 55, but 55 is never a prefix of any realistic legal string (although $\overline{5}\overline{5}$ can be). □

We relax now the conditions of realizability by considering realizable signings. Let Σ be an alphabet. Recall that a signed string v over Σ is a signing of a string $u \in \Sigma^*$, if $\xi(v) = u$, where ξ removes the bars from the string. Moreover, if v is realizable, it is a *realizable signing* of u.

Example 14. The legal strings $v_1 = 2\,3\,\overline{2}\,\overline{4}\,\overline{3}\,5\,4\,5$ and $v_2 = \overline{2}\,3\,2\,\overline{4}\,3\,5\,4\,5$ are both signings of $u = 2\,3\,2\,4\,3\,5\,4\,5$. Note that v_1 is realistic ($v_1 = \varrho(M_2\overline{M}_1\overline{M}_3 M_5 M_4)$), but u is not. Also, it is easy to see that the signing v_2 of u is not realistic. □

By definition every signing of a double occurrence string is legal. We shall now study the problem which double occurrence strings have realizable signings.

For a double occurrence string $w = a_1 a_2 \ldots a_{2\kappa}$, we define an edge-coloured and vertex-labelled (multi)graph A_w on the set $\{1, 2, \ldots, 2\kappa\}$ of vertices as follows: the (undirected) edges e together with their colours $c(e)$ are

$$e_{ij} = \{i, j\} \text{ and } c(e_{ij}) = 1, \text{ if } |i - j| = 1,$$
$$e'_{ij} = \{i, j\} \text{ and } c(e'_{ij}) = 0, \text{ if } a_i = a_j.$$

The vertex-labelling is defined by

$$\ell(i) = a_i \text{ for } i = 1, 2, \ldots, 2\kappa.$$

Note that if both $|i - j| = 1$ and $a_i = a_j$, then there will be two edges between i and j, one edge of each colour. A path $e_1 e_2 \ldots e_k$ in the graph A_w is said to be *alternating*, if $c(e_{2r+1}) = 0$ and $c(e_{2(r+1)}) = 1$ for each r. An alternating path is *alternating hamiltonian*, if it visits every vertex of the graph exactly once.

Each edge $e = \{y, z\}$ of the graph A_w has two *orientations*: (y, z) and (z, y). We also denote by

$$y \xrightarrow{c(e)} z$$

the orientation (y, z) together with the colour of the edge. In a drawing of the graph A_w, see Fig. 10, a solid line represents colour 1 and a dashed line colour 0. We also write the values $\ell(x)$ of the vertices beneath them.

Example 15. Let $w = 2\,3\,5\,4\,6\,4\,6\,3\,2\,5$ be a double occurrence string over Δ_6. Then the graph A_w is drawn in Fig. 10. The path

$$1 \xrightarrow{0} 9 \xrightarrow{1} 10 \xrightarrow{0} 3 \xrightarrow{1} 4 \xrightarrow{0} 6 \xrightarrow{1} 5 \xrightarrow{0} 7 \xrightarrow{1} 8 \xrightarrow{0} 2$$

is an alternating hamiltonian path of A_w.   □

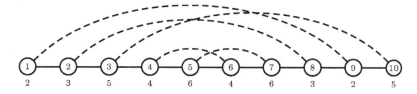

Fig.10. The graph A_w in Example 15

Theorem 2. *A double occurrence string w has a realizable signing if and only if the graph A_w has an alternating hamiltonian path.*

Proof. (1) We show first by induction on the length of the strings that every double occurrence string w that has a realizable signing w' has an alternating hamiltonian path that starts from the vertex $\varrho(M_1)$ corresponding to the beginning marker and ends in the vertex $\varrho(M_\kappa)$ corresponding to the end marker of the realistic legal string $\tau(w')$, where τ is an isomorphism.

Let w be a double occurrence string of length $2m$ over Σ, and assume that w has a realizable signing $w' \in \Sigma^{\maltese}$. Consequently, there exists a substitution $\tau \colon \Sigma^{\maltese} \to (\Delta_{m+1})^{\maltese}$ such that $\tau(w')$ is realistic. Clearly, the original string w and its image $\tau(w)$ under the isomorphism τ have the same graph, $A_w = A_{\tau(w)}$, except for the labelling of the vertices. Hence, without loss of generality, we can assume that $w = \tau(w)$. In particular, w is over Δ_{m+1}, and w' is a realistic string (and a signing of w).

For each $i \in \Delta_{m+1}$, let $x_{i1}, x_{i2} \in \{1, 2, \ldots, 2m\}$ be the vertices of A_w such that $\ell(x_{i1}) = i = \ell(x_{i2})$ and $x_{i1} < x_{i2}$.

Now $w = w_1(m+1)w_2(m+1)w_3$ for some substrings $w_1, w_2, w_3 \in \Delta_m^*$. The string $v = w_1w_2w_3$ is a signing of the realistic string v', which is obtained from the signing w' of w by removing the two occurrences of the letters from $\{m+1, \overline{m+1}\}$. By the induction hypothesis, v' has an alternating hamiltonian path from x_{2t} to x_{mr}, where $t, r \in \{1, 2\}$, x_{2t} is the position of the beginning marker $\varrho(M_1)$ and x_{mr} is the position of the end marker $\varrho(M_m)$ (of v'). Since w' is realistic, the occurrence of this $m = \varrho(M_m)$ is a part of the substring $m(m+1)$ or $(m+1)\overline{m}$ in w'. It is now obvious that we have an alternating hamiltonian path

$$x_{2t} \xrightarrow{0} \ldots \xrightarrow{0} x_{mr} \xrightarrow{1} x_{(m+1)r'} \xrightarrow{0} x_{(m+1)r''}$$

in A_w, where $\{r', r''\} = \{1, 2\}$ and $x_{(m+1)r''}$ is the position of the end marker $m+1 = \varrho(M_{m+1})$ in w'. This proves the claim in one direction.

(2) In the other direction, suppose that w is a double occurrence string such that the graph A_w does have an alternating hamiltonian path. Let the hamiltonian path be

$$x_{11} \xrightarrow{0} x_{12} \xrightarrow{1} \ldots \xrightarrow{1} x_{i1} \xrightarrow{0} x_{i2} \xrightarrow{1} \cdots \to x_{m1} \xrightarrow{0} x_{m2},$$

where again $\ell(x_{i1}) = \ell(x_{i2})$, for each $i = 1, 2, \ldots, m$, so that the edge $\{x_{i1}, x_{i2}\}$ has colour 0. Here $\{x_{i1}, x_{i2} \mid i = 1, 2, \ldots, m\} = \{1, 2, \ldots, 2m\}$. Define the substitution τ by $\tau(\ell(x_{i1})) = i + 1$, and then define the signing as follows: if $x_{i2} > x_{(i+1)1}$ (and so $x_{i2} = x_{(i+1)1} + 1$) in the edge $\{x_{i2}, x_{(i+1)1}\}$ of colour 1, then sign both $\tau(\ell(x_{i2}))$ and $\tau(\ell(x_{(i+1)1}))$; otherwise $x_{i2} = x_{(i+1)1} - 1$, and in this case, $\tau(\ell(x_{i2}))$ and $\tau(\ell(x_{(i+1)1}))$ are left unsigned. By the construction, the so obtained string is realistic, and the claim follows from this. □

Example 16. We illustrate the previous theorem and its proof by continuing Example 15. In Fig. 11 we have drawn the alternating hamiltonian path

$$1 \xrightarrow{0} 9 \xrightarrow{1} 10 \xrightarrow{0} 3 \xrightarrow{1} 4 \xrightarrow{0} 6 \xrightarrow{1} 5 \xrightarrow{0} 7 \xrightarrow{1} 8 \xrightarrow{0} 2$$

of A_w. The labels beneath the vertices are now obtained by following the hamiltonian path as in the proof above. The signing given by the proof produces the realistic legal string $2\,6\,3\,4\,\overline{5}\,\overline{4}\,5\,6\,2\,3$. □

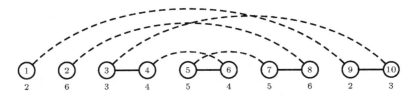

F ig .11 . Alternating hamiltonian path of Example 16

From the proof of Theorem 2 we have the following procedure to determine whether a given legal string is realizable. Let v be a legal string of length $2m$ over an alphabet Σ. We consider the graph $A_v = A_{v'}$. Let x be a vertex of A_v. Define an *induced alternating path* $\mathrm{alt}_v(x)$ *of v from x* as follows:

(1) Start with the edge $x_1 \xrightarrow{0} x_2$, where $x_1 = x$ and $\ell(x_2) = \ell(x_1)$. Then $\mathrm{alt}(w)$ is the maximum path obtained by iterating the step (2):

(2) Suppose the alternating path $x_1 \xrightarrow{0} x_2 \xrightarrow{1} \ldots \xrightarrow{1} x_{i-1} \xrightarrow{0} x_i$ has been constructed so that the colour of the edge $\{x_{i-1}, x_i\}$ is 0. If the label of x_i is signed in w, then the next edges are $\{x_i, x_i - 1\}$ (of colour 1) and $\{x_i - 1, x_{i+1}\}$ (of colour 0), where $\ell(x_{i+1}) = \ell(x_i)$.

Notice that after the vertex x is chosen, the induced alternating path $\mathrm{alt}_v(x)$ is well defined, that is, in every step the next edge is uniquely determined.

The following result is a corollary to Theorem 2.

Theorem 3. *Let v be a legal string. Then v is realizable if and only if there exists a vertex x of A_v such that the induced alternating path $\mathrm{alt}(v)$ is an alternating hamiltonian path of A_v.*

7   Nonrealizable Strings and Graphs

The model of MDS descriptors and (realistic) legal strings can be further abstracted by considering overlap graphs of legal strings. This line of research was initiated in [3] and [7].

By Example 13, there are legal strings that are not realizable. We prove a stronger result in this section: there exist overlap graphs that cannot be 'realized' by any micronuclear arrangement.

For the next two proofs we adopt the following notation: let $A = \{x_1, x_2, x_3\}$ be an alphabet, and let $[x_i] = x_{i1}x_{i2}x_{i3}x_{i0}x_{i1}x_{i2}x_{i3}$.

Lemma 3. *The cyclic conjugates of the double occurrence string*

$$w' = x_{10}x_{20}x_{30}[x_1][x_2][x_3]$$

do not have realizable signings.

Proof. We let the letters y and y_i be variables, and let

$$[y] = y_1y_2y_3y_0y_1y_2y_3 . \tag{3}$$

Let u be a realizable double occurrence string such that either $u = v_1[y]v_2$ or $u = w_2vw_1$, where $[y] = w_1w_2$. The specified substring $[y]$ or the scattered cyclic conjugate w_2w_1 of $[y]$ in the above is called a $[y]$-*block* of u. By Theorem 2, there exists an alternating hamiltonian path of the graph A_u. Let H be any such path.

It is straightforward to show by case analysis that if the path H does not start at a position inside the $[y]$-block, then H ends at a position inside the $[y]$-block, and symmetrically, if H does not end at a position inside the $[y]$-block, then H begins at a position inside the $[y]$-block. In Fig. 12 we have illustrated one possibility to travel along the substring $[y]$ in H for the case $u = v_1[y]v_2$. (In the figure, the vertices of A_u are simply represented by their labels.) There the path H visits outside the string $[y]$ between the last y_3 and the next $z = y_0$. The path ends in y_0 in the middle of $[y]$.

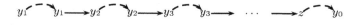

F ig.12. A part of the path H

Therefore H either starts or ends at a vertex corresponding to an occurrence of a pointer in the $[y]$-block. It follows that any cyclic conjugate of a realizable double occurrence string can contain at most two different blocks, a $[y]$-block and a $[y']$-block, of the form (3) such that these do not share letters. In particular, the string w' of the claim does not have conjugates with realizable signings.   □

As we have seen the mapping $w \mapsto \gamma_w$ from the legal strings to the overlap graphs is not injective. Therefore Lemma 3 leaves unanswered the question whether there are overlap graphs γ_w that are not 'realizable'. In the following theorem we answer this question by constructing an (unsigned) overlap graph that is not realizable.

Theorem 4. *There exists an overlap graph of a double occurrence string that has no signing of the vertices such that the result is an overlap graph of a realistic legal string.*

Proof. Let $w' = x_{10}x_{20}x_{30}[x_1][x_2][x_3]$ be as stated in Lemma 4. The overlap graph $\gamma = \gamma_{w'}$ of w' is given in Fig. 13. We show now that this graph is different from γ_w for all realizable double occurrence strings w.

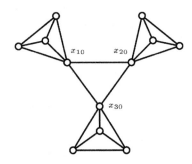

F i g .13 . The overlap graph from the proof of Theorem 4

Assume to the contrary that there exists a realizable double occurrence string w such that $\gamma_w = \gamma$. In particular, w has the same pairs of overlapping letters as w'. For each i, the vertices corresponding to x_{i1}, x_{i2} and x_{i3} are adjacent to each other in γ, and they have the same neighbourhood in the rest of γ (the vertex corresponding to x_{i0}). Hence, we can assume that they occur in w in the given order: $w = v_{i0}x_{i1}v_{i1}x_{i2}v_{i2}x_{i3}v_{i3}x_{i1}v_{i4}x_{i2}v_{i5}x_{i3}v_{i6}$ for some substrings v_{ij} for each i.

Since the letters $x_{j0}, \ldots, x_{j3}$ overlap with each other in w and they do not overlap with x_{i1}, x_{i2}, x_{i3} for $i \neq j$, we conclude that all occurrences of x_{jk} for each $0 \leq k \leq 3$ are either (a) in one substring v_{it} for some $1 \leq t \leq 5$, or (b) they are all in $v_{i0}v_{i6}$. We show now that there are indices i and j such that the case (a) holds. Indeed, there is an index i such that $[x_i]$ is not a substring of w, since otherwise there would be three disjoint substrings in w of the form (3), which is not possible by the proof of Lemma 3. Now if the occurrences of $x_{j0}, x_{j1}, x_{j2}, x_{j3}$, for both indices j with $j \neq i$, are in $v_{i0}v_{i6}$, then an occurrence of x_{i0} must also be in $v_{i0}v_{i6}$ (since x_{i0} overlaps with x_{j0}), and therefore the second occurrence of x_{i0} must be in v_{i3} (since x_{i0} overlaps with x_{i1}, x_{i2} and x_{i3}). This means, however, that $[x_i]$ is a substring of w; a contradiction.

Moreover, exactly one occurrence of x_{i0} is in v_{it}, since x_{i0} overlaps with x_{j0}. This occurrence must be in $v_{j0}v_{j6}$, since the other occurrence of x_{i0} is not in v_{it}.

Now, by the overlapping properties of x_{j0}, we have for $m = j$ that

$$[x_m]x_{i0}x_{m0} \quad \text{or} \quad x_{m0}x_{i0}[x_m] \text{ is scattered substring in } v_{it}. \tag{4}$$

Since for the remaining index $k \notin \{i, j\}$, x_{k0} overlaps with x_{j0}, an occurrence of x_{k0} lies in the interval of the letter x_{j0}, and thus it is in v_{it}. Similarly to the above, we can show that (4) also holds for $m = k$.

By the above, x_{k0} does not occur in v_{jr} for any $1 \le r \le 5$, because otherwise both occurrences of x_{i0} are in v_{jr}. Therefore the index i is unique with respect to the property (a), and then it follows that $[x_m]$ is a substring of v_{it} for both m with $m \ne i$. This implies that a conjugate of w has three disjoint substrings of the form $[y]$, which yields, by the proof of Lemma 3, a contradiction.   □

Example 17. The simplest form of the counter example discussed in the above proof is the string   2 3 4 5 6 7 2 5 6 7 8 9 10 3 8 9 10 11 12 13 4 11 12 13.   □

Acknowledgements

G. Rozenberg gratefully acknowledges partial support by NSF grant 0121422.

References

1. Bouchet, A., Circle graphs. *Combinatorica* 7 (1987), 243 – 254.
2. Bouchet, A., Circle graph obstructions. *J. Combin. Theory Ser B* 60 (1994), 107 – 144.
3. Ehrenfeucht, A., T. Harju, I. Petre, D. M. Prescott, and G. Rozenberg, Formal systems for gene assembly in ciliates. *Theoret. Comput. Sci.* 292 (2003), 199 – 219.
4. Ehrenfeucht, A., T. Harju, I. Petre, and G. Rozenberg, Characterizing the micronuclear gene patterns in ciliates. *Theory of Computation and Systems* 35 (2002), 501 – 519.
5. Ehrenfeucht, A., T. Harju, and G. Rozenberg, Gene assembly through cyclic graph decomposition. *Theoretic Comput. Syst.* 281 (2002), 325 – 349.
6. Ehrenfeucht, A., I. Petre, D. M. Prescott, and G. Rozenberg, Universal and simple operations for gene assembly in ciliates. In *Words, Sequences, Languages: Where computer science, biology and linguistics come across*, V. Mitrana, C. Martin-Vide (eds.), Kluwer Academic Publishers, Dortrecht/Boston, 329 – 342, (2001).
7. Ehrenfeucht, A., I. Petre, D. M. Prescott, and G. Rozenberg, String and graph reduction systems for gene assembly in ciliates. *Math. Structures Comput. Sci.* 12 (2001), 113 – 134.
8. Ehrenfeucht, A., I. Petre, D. M. Prescott, and G. Rozenberg, Circularity and other invariants of gene assembly in cliates. In *Words, semigroups, and transductions*, M. Ito, Gh. Păun, S. Yu (eds.), World Scientific, Singapore, 81 – 97, 2001.
9. Ehrenfeucht, A., D. M. Prescott, and G. Rozenberg, Computational aspects of gene (un)scrambling in ciliates. In *Evolution as Computation*, L. Landweber, E. Winfree (eds.), 45 – 86, Springer-Verlag, Berlin, Heidelberg, 2001.

10. Landweber, L. F., and L. Kari, The evolution of cellular computing: nature's solution to a computational problem. In *Proceedings of the 4th DIMACS meeting on DNA based computers*, Philadelphia, PA, 3 – 15 (1998).

11. Landweber, L. F., and L. Kari, Universal molecular computation in ciliates. In *Evolution as Computation*, L. Landweber, E. Winfree (eds.), Springer-Verlag, Berlin, Heidelberg, 2002.

12. Păun, Gh., G. Rozenberg, and A. Salomaa, *DNA Computing*. Springer-Verlag, Berlin, Heidelberg, 1998.

13. Prescott, D. M., Cutting, splicing, reordering, and elimination of DNA sequences in Hypotrichous ciliates. *BioEssays* 14 (1992), 317 – 324.

14. Prescott, D. M., The unusual organization and processing of genomic DNA in Hypotrichous ciliates. *Trends in Genet.* 8 (1992), 439 – 445.

15. Prescott, D. M., The DNA of ciliated protozoa. *Microbiol Rev.* 58(2) (1994), 233 – 267.

16. Prescott, D. M., The evolutionary scrambling and developmental unscabling of germlike genes in hypotrichous ciliates. *Nucl. Acids Res.* 27 (1999), 1243 – 1250.

17. Prescott, D. M., Genome gymnastics: unique modes of DNA evolution and processing in ciliates. *Nat Rev Genet.* 1(3) (2000), 191 – 198.

18. Prescott, D. M., and M. DuBois, Internal eliminated segments (IESs) of Oxytrichidae. *J. Eukariot. Microbiol.* 43 (1996), 432 – 441.

19. Prescott, D. M., A. Ehrenfeucht, and G. Rozenberg, Molecular operations for DNA processing in hypotrichous ciliates. *European Journal of Protistology* 37 (2001), 241 – 260.

20. Prescott, D. M., and G. Rozenberg, How ciliates manipulate their own DNA – A splendid example of natural computing. *Natural Computing* 1 (2002), 165 – 183.

Experimental Quantum Computation with Molecules

Masahiro Kitagawa[1,2]

[1] Graduate School of Engineering Science, Osaka University
Toyonaka, Osaka 560–8531 Japan
kitagawa@ee.es.osaka-u.ac.jp
[2] CREST, Japan Science and Technology Corporation

1 Introduction

Quantum computation was born from a gap between the law of computation (Church–Turing thesis) and the law of physics (quantum mechanics). When Alan Turing came up with his computation model which is now widely used and referred to as a Turing machine in 1936, he emulated human calculation process. At that time, quantum mechanics was already born and rapidly gaining the position of the most fundamental law of physics. However human thoughts were and still are strongly bound by the experiences in daily life, which are overwhelmingly classical. The gap was not noticed until Paul Benioff and Richard Feynman studied in early 1980's and David Deutsch formulated quantum Turing machine in 1985 [1].

The gap was not extensively exploited until Peter Shor discovered his efficient quantum algorithm for factorization in 1994[2]. The discovery was sensational because it could potentially break the public key crypto system widely used on the Internet. Unfortunately or fortunately, quantum computers which could factor hundreds of digits have been and still are far beyond the reach of our technology. However physicists are rather enthusiastically working on it. Why? Because a quantum computer is strongly conjectured (although not proved yet) to be much faster than classical computer from computational complexity viewpoint unlike other physically new computers such as optical or superconducting computers.

In this *Paper*, I will introduce the experimental developments of quantum computation using molecules, obstacles, and perspectives.

2 NMR Quantum Computation

Among the many physical systems studied as a candidate for quantum computing device, the nuclear spins in solution molecules have favorable characteristics at least as a small-scale quantum computer or abacus[3,4]. Under the strong static magnetic field, these nuclear spins act as a tiny magnet and precess around the static field like a spinning top. They can be rotated by resonant radio frequency magnetic pulses. This is the Nuclear Magnetic Resonance (NMR).

M. Ito and M. Toyama (Eds.): DLT 2002, LNCS 2450, pp. 21–27, 2003.

Any single qubit operations including NOT and H (Hadamard transformation) can be performed as rotations. The nuclear spins in the same molecule are also coupled together via electron spins in the chemical bonds. These J-couplings enable two-qubit operation called *controlled-rotation* which is very similar to controlled-NOT. The rotations and controlled-rotations together form a universal set of quantum operations which enable any unitary operation of any number of qubits.

While physicists are still struggling to make and control a qubit in other physical systems such as semiconductors or superconductors, molecular NMR has achieved various quantum computation experiments such as Deutsch-Jozsa algorithm, Grover's quantum search, and Shor's factoring. The group at IBM has experimentally demonstrated the factoring of $15 \rightarrow 5 \times 3$ by Shor's quantum algorithm using 7-qubit molecules with NMR[5].

3   Controversies on NMR Quantum Computation

However, there are also criticisms and controversies[6,7,8] over NMR quantum computation. The major criticism is that the present NMR experiments consume exponentially huge physical or computational resources[8] such as the number of molecules[3] or the number of experiments[9] and therefore retain no quantum merit. The problem is associated with the thermal equilibrium state at room temperature which has near maximum entropy.

4   Initialization of NMR Quantum Computers

The simplest answer to the above problem would be cooling all the spins down to $T=0$K. Then the states of all molecular quantum computers would be initialized to $|0\rangle^{\otimes n}$. However it is not that simple. First of all, $T=0$K is never reached. We need the scenario which works not only at 0K. That is, we still need some scheme to extract the pure state signal out of the mixed states. We will start with lowering temperature in existing initialization schemes. Secondly, if the material is simply cooled, it is frozen to the solid in which the dipolar interactions make the computational basis including $|0\rangle^{\otimes n}$ non-stationary. We have to decouple dipolar interactions by magic angle spinning or applying special pulse sequences. Alternatively we may seek the possibility of refrigerating spins without actually refrigerating materials[10].

4.1   Logical Labeling with Typical Sequences

The exhaustive averaging[9] which requires 2^n-1 experiments is out of question from the computational complexity viewpoint and will never be used for the real quantum computation. By real quantum computation, we mean the exponential speed-up over classical computation without requiring exponentially huge physical resources other than the Hilbert space.

The logical labeling[3] also suffers from the exponentially small population of effective pure state at room temperature. However if the origin of the problem is the extremely small polarization at room temperature, it is natural to seek the possibility of improving it by lowering the temperature. We have generalized the logical labeling[3] to arbitrary temperature by using typical sequences $\{|0\rangle^{\otimes np}|1\rangle^{\otimes nq}\}$ as the background (np is treated as an integer for simplicity) and the most probable state $|0\rangle^{\otimes n}$ as the signal state[11].

Under the effective Hamiltonian of homonuclear solution NMR, the thermal equilibrium states can be well-approximated by the density operator $\rho^{\otimes n}$ with $\rho=p|0\rangle\langle0|+q|1\rangle\langle1|$ (*i.i.d.*) where $p=(1+\delta)/2$, $q=1-p$, $\delta=\tanh\alpha$, and $\alpha=\hbar\omega/2k_BT$. Unitary logical labeling transforms it into

$$\rho' = P(0)\underbrace{|0\rangle\langle0|}_{n-m}\otimes[(1-\epsilon)\underbrace{I/2^m}_{m}+\epsilon\underbrace{|0\rangle\langle0|]}_{m}+\sum_{a\neq0}P(a)\underbrace{|a\rangle\langle a|}_{n-m}\otimes\underbrace{\rho_a}_{m}.$$

The first term is the conditional effective pure state and its population $P(0) > 1/\sqrt{2\pi n}$ scales very well.

The entanglement is restored if p is large enough. However, the population ϵ of the effective pure state still decreases exponentially with n. This is the limit of using single signal state since any state is subject to exponentially small population for increasing n at finite temperature.

Our generalization has revealed that the number of qubits available is given by $m\approx nH(p)$ where $H(p)=-p\log p - (1-p)\log(1-p)$ is the entropy. We lose qubits by lowering the temperature. The counter-intuitive nature of the logical labeling is the consequence of using the most typical states as the signal-less background.

4.2 Typical Sequences as Signal States

The last two considerations naturally lead to the conclusions; (a) we should use as many signal states as possible, (b) all signal states should give the same result, and (c) the typical states should be used as signal state rather than background[11].

Each of the typical states $\mathcal{A} = \{|0\rangle^{\otimes np}|1\rangle^{\otimes nq}\}$ has the probability of $2^{-nH(p)}$ and the number of the typical states is $|\mathcal{A}| = \binom{n}{np} \approx 2^{nH(p)}/\sqrt{2\pi npq}$. If all the typical states give the same result, the signal strength is in proportion to the population of the signal states $P(x \in \mathcal{A}) = 2^{-nH(p)}|\mathcal{A}| \approx 1/\sqrt{2\pi npq}$, which scales very well.

One way of doing this is the unitary Schumacher compression[12] S which transforms the thermal equilibrium state into

$$\rho' = \sum_{x=0}^{|\mathcal{A}|-1} 2^{-nH(p)}\underbrace{|0\rangle\langle0|}_{n[1-H(p)]}\otimes\underbrace{|x\rangle\langle x|}_{nH(p)}+\sum_{y\notin\mathcal{A}}P(y)S\underbrace{|y\rangle\langle y|}_{n}S^\dagger.$$

The number of qubits available is $m\approx n[1-H(p)]$, which is very reasonable since we gain qubits by lowering the temperature and all qubits are available in the limit of $T=0$.

The notion of initialization by data compression has been first presented by Schulman and Vazirani in their scalable initialization scheme[13]. However their algorithm relies on Monte Carlo simulation and therefore has not been explicitly given.

The notion itself is very natural from information theoretical viewpoint as shown in Fig. 1. Such a compression can be regarded as a one-to-one transformation from 2^{nH} typical states to compressed states.

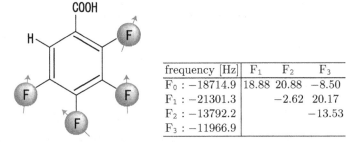

Fig.1. Initialization by compression: the typical states (left) are transformed into the compressed states (right) with common initialized substate.

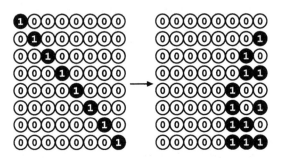

frequency [Hz]	F_1	F_2	F_3
F_0 : -18714.9	18.88	20.88	-8.50
F_1 : -21301.3		-2.62	20.17
F_2 : -13792.2			-13.53
F_3 : -11966.9			

Fig.2. 4-qubit molecule for initialization experiment

4.3   Experimental Demonstration

We have experimentally demonstrated the initialization by compression[14] using the molecules shown in Fig. 2. We have prepared the pseudo low temperature state of $n=4$ and $p=3/4$, the equal-weight mixture of the pseudo typical states, $|0001\rangle$, $|0010\rangle$, $|0100\rangle$ and $|1000\rangle$ as shown in Fig. 3 (a) by exhaustive averaging. Application of the simple compression circuit shown in Fig. 4 has transformed it into the equal mixture of $|0000\rangle$, $|0001\rangle$, $|0010\rangle$ and $|0011\rangle$ as shown in Fig. 3 (b) in which the leading two qubits are successfully initialized to $|00\rangle$. However, such a compression circuit is not necessarily efficient. In general, the brute-force implementations may take $O(2^{nH})$ rather than $poly(n)$.

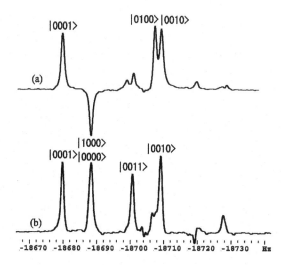

F ig.3. Initialization by compression experiment: (a) pseudo typical state, (b) compressed state

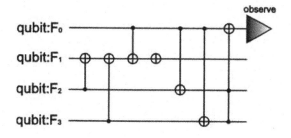

F ig.4. brute-force compression circuit

4.4   Efficient Initialization Scheme

The data compression circuit for the initialization of NMR quantum computers is subject to the special requirements: a) no clean qubit is initially available, b) no state reduction by measurement is available, and c) the total computational complexity must be $poly(n)$. The first requirement makes the application of existing compression algorithms very difficult since they use clean workspace for granted. For example, Schumacher compression circuit of Cleve and DiVincenzo [15] requires $2\sqrt{n} + \log n$ clean qubits and therefore cannot be used. The second requirement is specific to the NMR.

The third requirement rejects the brute-force implementations which take $O(2^n)$ steps. However, if it is applied to $\log n$ qubits instead of n, it completes in $poly(n)$ steps. Therefore $\log n$ clean qubits may be available by pre-compression of $\log n/(1 - H)$ qubits in $O(n^{cH/(1-H)})$ steps.

By replacing additions and comparisons which require $2\sqrt{n}$ work qubits in the enumerative coding circuit[15] with Draper's in-place additions[16] based on QFT (Quantum Fourier Transformation), we have developed efficient initialization circuit which requires only $O(n^4)$ steps and $\log n$ clean ancilla qubits[17]. The clean $\log n$ qubits can be supplied in $poly(n)$ steps even with brute-force method as discussed above. The overall scheme can be further improved to operate in less steps. However it is enough to prove that the NMR quantum computation can be done in polynomial steps including the initialization phase and therefore eligible for real quantum computation.

In summary, to solve the initialization problem, the nuclear spins must be first cooled down to very low temperature and then further logical cooling must be performed to make zero temperature subsystem within each molecule[13,17]. We are working on the physical[18] and logical initializations[17] to make NMR quantum computation a genuine quantum computation which enables exponential speed up without requiring exponential resources other than the size of Hilbert space itself.

5   Molecular Architecture

Other scalability considerations include molecular architecture. Obviously the larger molecule has the sparser network of J-couplings and the narrower separations of resonant frequencies. These problems may be solved by unidirectional periodic structure such as the one proposed by Seth Lloyd in 1993[19].

6   Quantum Languages

Currently we know only one quantum algorithm exponentially faster than classical counterpart – Shor's algorithm[2]. Why? Is this the only one which exists? Or haven't we just spent enough time and effort on it? In quantum computation, theorists speak quantum Turing machine language (QTM)[1] and experimentalists speak quantum circuit language (QC). Probably neither is enough and we need new language (Q) which enhances our ability to exploit quantum interference. Because our thoughts are too much dependent on our classical experiences.

Acknowledgments

This work has been supported by CREST of JST (Japan Science and Technology Corporation). The author would like to acknowledge A. Kataoka for stimulating collaborations on the initialization problems and T. Nishimura for devoted experiment on initialization by compression. The author would like to also thank Prof. A. Fujiwara for enlightening discussions on information theoretical aspects, K. Takeda for useful discussions on physical aspects and A. Saitoh for helpful discussions on Ref. [13].

References

1. D. Deutsch, *Proc. Roy. Soc. Lond. A* 400, pp. 97–114 (1985).
2. P. W. Shor, *SIAM J. Comp.* 26, pp. 1484–1509 (1997).
3. N. Gershenfeld and I. L. Chuang, *Science* 275, pp. 350–356 (1997).
4. D. G. Cory, A. F. Fahmy and T. F. Havel, *Proc. Natl. Acad. Sci. USA,* 94, pp. 1634–1639 (1997).
5. L. M. K. Vandersypen, M. Steffen, G. Breyta, C. S. Yannoni, M. H. Sherwood, and I. L. Chuang, *Nature* 414, pp. 883–887 (2001).
6. S. L. Braunstein, C. M. Caves, R. Jozsa, N. Linden, S. Popescu and R. Schack, *Phys. Rev. Lett.* 83, pp. 1054–1057 (1999).
7. R. Schack and C. M. Caves, *Phys. Rev. A* 60, pp. 4354–4362 (1999).
8. N. Linden and S. Popescu, *Phys. Rev. Lett.* 87, 047901 (2001).
9. E. Knill, I. Chuang and R. Laflamme, *Phys. Rev., A* 57, pp. 3348–3363 (1998).
10. M. Iinuma, Y. Takahashi, I. Shaké, M. Oda, A. Masaike, T. Yabuzaki and H. M. Shimizu, *Phys. Rev. Lett.* 84, pp. 171–174 (2000).
11. M. Kitagawa, "Initialization, Entanglement and Scalability in Bulk-Ensemble NMR Quantum Computation," *EQIS'01* (2001).
12. B. Schumacher, *Phys. Rev. A* 51, pp. 2738–2747 (1995).
13. L. J. Schulman and U. Vazirani, *Proc. 31st STOC*, pp. 322–329 (1999).
14. T. Nishimura, A. Kagawa, A. Kataoka and M. Kitagawa, "Initialization of NMR quantum computer at low temperature using data compression process," *ISQC'02.*
15. R. Cleve and D. P. DiVincenzo, *Phys. Rev. A* 54, pp. 2636–2650 (1996).
16. T. G. Draper, quant-ph/0008033 (2000).
17. M. Kitagawa and A. Kataoka, "Efficient Initialization Scheme for Real Quantum Computation using NMR," *EQIS'02* (2002).
18. K. Takeda, T. Takegoshi and T. Terao, *Chem. Phys. Lett.* 345, pp. 166–170 (2001).
19. S. Lloyd, Science 261, pp. 1569–1571 (1993).

Efficient Transformations
from Regular Expressions to Finite Automata

Sebastian Seibert

Lehrstuhl für Informatik I (Algorithms and Complexity)
RWTH Aachen, 52056 Aachen, Germany
seibert@cs.rwth-aachen.de

Abstract. We examine several methods to obtain nondeterministic finite automata without ε-transitions (NFA) from regular expressions. The focus is on the size (number of transitions) of the resulting automata, and on the time complexity of the transformation.

We show how recent developments [9,6] have improved the size of the resulting automaton from $\mathcal{O}(n^2)$ to $\mathcal{O}(n(\log n)^2)$, and even $\mathcal{O}(n \log n)$ for bounded alphabet size (where n is the size of the regular expression). A lower bound [11] shows this to be close to optimal, and also one of those constructions can be computed in optimal time [8].

1 Introduction

Comparing different devices to describe formal languages not only with respect to their expressive power but also with respect to their efficiency is a long standing topic of automata and formal language theory. Still, there are many open questions, even when restricting to devices describing regular languages only.

Regular expressions and finite automata without ε-transitions (NFA) are two of the best studied formalisms. It is known that converting NFA into regular expressions may cause an exponential blow-up [5]. For the converse direction, a quadratic blow-up was thought best possible for a long time. Only recently, it was discovered that a poly-logarithmic blow-up is sufficient and necessary at the same time, though there remains a small (i.e. double logarithmic) gap.

We will try to summarize the development here, present the new transformation methods and how they relate to the well-known methods. Also, we look at the lower bound and the time complexity needed for executing the different transformation methods. Our focus is on giving rather complete descriptions of the constructions involved. Additionally we sketch the central ideas of the proofs of the claimed complexity results whereas for correctness of the methods we refer to the cited references.

In the next section, we will, after giving the necessary definitions, describe three essentially different methods to transform regular expressions into NFA of quadratic size, i.e. of quadratic number of transitions. First, we look at what one could consider the "textbook method" or better the ε-NFA method (see e.g. [10]). Here, a regular expression is converted first into a nondeterministic finite

M. Ito and M. Toyama (Eds.): DLT 2002, LNCS 2450, pp. 28–42, 2003.
© Springer-Verlag Berlin Heidelberg 2003

automaton *with* ε-transitions, and the ε-transitions are substituted in a second step.

The second method is the construction of the so called position automaton [12,7,2]. The idea is to have basically one state for each occurrence of a letter in the regular expression, called a position. Transitions are obtained from the *follow sets* that describe which positions may "follow each other" in the generation of a word from the expression.

Partial derivatives are a generalization of the derivatives of Brzozowski [3], marking in some sense the step from determinism to non-determinism. They were introduced in [1] as a means to get another method for transforming regular expressions into NFA.

Section 3 is devoted to the two recently introduced refinements of these methods, resulting in essential improvements of complexity results. We will demonstrate here how both methods rely essentially on a tool which we call balanced tree decomposition. We will also point out the differences between the two methods.

In [9], the *common follow sets automaton* construction was invented. It refines the position automaton construction in that the follow set of each position is divided into several parts such that different positions use large parts of those sets in common. If the given regular expression has n occurrences of symbols from an Alphabet Σ, the common follow sets automaton has $\mathcal{O}(n)$ states and $\mathcal{O}\left(n(\log n)^2\right)$ transitions. This construction was proven to be implementable in essentially optimal time [8].

Finally, in [6] a refinement of the ε-NFA method was introduced that also produces automata of size $\mathcal{O}\left(n(\log n)^2\right)$. Moreover, it improves the conversion for limited alphabet size to $\mathcal{O}\left(|\Sigma|n\log n\right)$ transitions.

In Section 4, we will have a short glance at the lower bound. It was shown ([9], see [8] for a simple proof) that there is no linear conversion. This was improved in [11] by proving that no construction can produce automata with less than $\Omega(n(\log n)^2/\log\log n)$ transitions. Since all these results use linear alphabet size (in n), it remains open whether bounded alphabets allow a linear conversion.

2 Definitions and Basic Constructions

A non-deterministic finite automaton (NFA) will be considered here as a quintuple $\mathcal{A} = (Q, \Sigma, q_I, \Delta, Q_F)$, consisting as usual of a finite state set Q, a finite alphabet Σ, an initial state $q_I \in Q$, a set of final states $Q_F \subseteq Q$, and a transition function $\Delta \subseteq Q \times \Sigma \times Q$. We require this automata type to always read an alphabet symbol in each step, as opposed to ε-NFA, where we allow $\Delta \subseteq Q \times (\Sigma \cup \{\varepsilon\}) \times Q$. Transitions labeled with the the empty word ε are called ε-*transitions*, and *alphabet transitions* are those that are labeled with an alphabet letter. For the sake of readability, occasionally we denote a transition $x = (p, a, q)$ by $p \xrightarrow{a} q$, and we let $p \xrightarrow{\varepsilon^*} q$ denote the existence of a sequence of ε-transitions, leading from p to q.

Clearly, the size of the transition function is dominating the size of the description of an automaton. Therefore, it is reasonable to take the number of transitions as the main measure of an automaton, that is, we define $size(\mathcal{A}) = |\Delta|$.

For regular expressions over alphabet Σ, we apply the usual inductive definition. That is, ε and each single letter a for $a \in \Sigma$ are regular expressions. And with F and G, also $F^*, (F+G), (F \cdot G)$ are regular expressions. We have omitted $\emptyset$ since it is needed only to describe the empty language itself, for which a trivial NFA exists. Also, we will always assume that an expression is *reduced* in that it contains no iteration of stars (like F^{**}), and that it contains no concatenation with ε and no iterated addition of ε. We use precedence rules "$*$ before $\cdot$ before $+$" and omit superfluous brackets. The Language $L(E)$, described by expression E, is inductively defined in the usual way.

We define as size of a regular expression its number of *occurrences of alphabet symbols*. Looking at $E_n = (a_1 + \varepsilon)(a_2 + \varepsilon) \ldots (a_n + \varepsilon)$, the example we will use throughout the paper, we obtain $size(E_n) = n$. This definition is easily justified by the fact that in a reduced regular expression, the number of symbols it contains is linear in $size(E)$.

As we will see later, an important tool in obtaining the more efficient transformation methods is what we call a balanced tree decomposition. To this end, we have to define how we will associate regular expressions to binary trees.

We define the *tree representation* t_E of a regular expression as follows. If $E \in \{\varepsilon\} \cup \{a | a \in \Sigma\}$, t_E consists only of the root (being a leaf at the same time) labeled with E. If $E \in \{F^*, G + F, G \cdot F\}$, t_E has a root labeled $*, +, \cdot$, respectively, connected to one (case F^*) respectively two (cases $G + F, G \cdot F$) sons being the roots of the trees t_G (left son) and t_F (right, respectively only son).

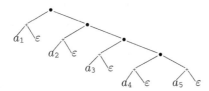

F ig. 1. A tree representation of regular expressions $E_5 = (a_1 + \varepsilon)(a_2 + \varepsilon)(a_3 + \varepsilon)(a_4 + \varepsilon)(a_5 + \varepsilon)$

Clearly, for expressions like E_5 from our running example, one could also choose a more balanced representation than that in Figure 1. But since this not always possible, we just fix one arbitrary tree representations for each given regular expression. The constructions described below will work in any case.

2.1   The ε-NFA Method

Let us look now at what we will call the ε-NFA method of converting regular expressions into NFA (see e.g. [10]). It uses ε-NFA as an intermediate step.

First, by a simple recursive construction, the regular expression E is converted into an ε-NFA $\mathcal{A}(E)$. There are different variants of this method to be found in the literature. Important for the correctness is that either no transition leads into the initial state or no transition leaves the final state (or both). Here we have chosen the first, "start separated" variant and at the same time assured that only one final state exists. The reason is that the refined method, which we will describe below, builds on this version.

Also needed for the refinement only is that the regular expressions are brought into a special normal form first. The essential aspect of this is first to hide the occurrence of ε in a new unary operator $F^\diamond := F + \varepsilon$, and then to connect every unary operator either to a single letter or to a binary operator[1].

Proposition 1 ([6]). *Every regular expression over an alphabet Σ can efficiently be transformed into an equivalent one that is made up from only*

(a) elementary expressions $a, a^*, a^\diamond$, for $a \in \Sigma$;
(b) binary operators $F + G$, $F \cdot G$, $(F \cdot G)^*$, $(F \cdot G)^\diamond$.

Now, the construction of the ε-NFA is done as depicted in Figure 2. There, for the binary operator cases, the used sub-automata are shown as ellipse with their initial state on the left hand side and their unique final state on the right hand side. Some of these states are merged in the new automaton, and a dashed circle around the final state of a sub-automaton marks the fact that it is no longer final in the resulting automaton.

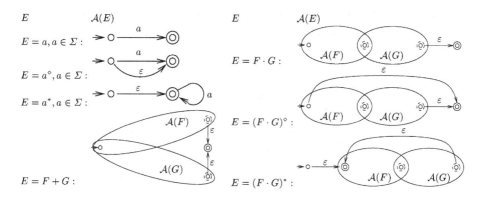

Fig.2. ε-NFA for regular expressions in special normal form

The second step of the ε-NFA method consists of the conversion from ε-NFA into NFA. That is, from ε-NFA $\mathcal{A} = (Q, \Sigma, q_I, \Delta, Q_F)$ we obtain NFA $\mathcal{A}' = (Q, \Sigma, q_I, \Delta', Q_F')$ by defining

---

[1] Note that we need no combination of unary operators with $+$ since $L((F + G)^*) = L((F^*G^*)^*)$, and $L((F + G)^\diamond) = L(F + G^\diamond)$.

$$\Delta' = \{(p, a, q) \in Q \times \Sigma \times Q \mid p \xrightarrow{\varepsilon^*} p' \xrightarrow{a} q \text{ in } \mathcal{A}\}, \text{ and}$$
$$Q'_F = Q_F \cup \{q \in Q \mid q \xrightarrow{\varepsilon^*} Q_F\}.$$

Proposition 2. *If $n = size(E)$ for a given regular expression, the automaton produced by the ε-NFA method has $2n$ states and $\mathcal{O}(n^2)$ transitions.*

Here, as in the following, the given bounds are asymptotically sharp. That is, for each construction, the sequence of automata obtained for the sequence of regular expressions E_n grows in the described order of magnitude.

2.2   The Position Automaton

We will describe now what we call the position automaton construction, first introduced by R. Book, Sh. Even, Sh. Greibach, and G. Ott [2], based on ideas developed by R. McNaughton and H. Yamada [12] and V. M. Glushkov [7].

We call *position* any occurrence of a letter in E, usually numbered $1, 2, \ldots,$ $size(E)$. Let $pos(E)$ be the set of all positions in E. The basis of the position automaton construction is the computation of two sets, $first(E)$, and $last(E)$, for E, and for each $x \in pos(E)$ one computes the set $follow(E, x)$.

Intuitively speaking, $first(E)$ is the set of positions in E that can generate a first letter of a word in $L(E)$, and $last(E)$ is the set of positions that can generate a last letter of a word in $L(E)$. For $x \in pos(E)$, $follow(E, x)$ gives the set of positions that may follow position x in the generation of a word in $L(E)$.

The following inductive definition formalizes the intuitive notion and gives a way to compute those sets at the same time.

$$first(\varepsilon) = \emptyset, \qquad first(a) = pos(a) \text{ for } a \in \Sigma,$$
$$first(F + G) = first(F) \cup first(G), \qquad first(F^*) = first(F), \text{ and}$$
$$first(FG) = \begin{cases} first(F) & \text{if } \varepsilon \notin L(F), \\ first(F) \cup first(G) & \text{if } \varepsilon \in L(F). \end{cases} \tag{1}$$

In order to obtain rules for $last(E)$, we just have to substitute "last" for "first" and replace (1) by:

$$last(FG) = \begin{cases} last(G) & \text{if } \varepsilon \notin L(G), \\ last(F) \cup last(G) & \text{if } \varepsilon \in L(G). \end{cases}$$

Finally, the follow sets are obtained by the following rules.

$$follow(a, x) = \emptyset \qquad \text{for } a \in \Sigma,$$
$$follow(F + G, x) = \begin{cases} follow(F, x) & \text{if } x \in pos(F), \\ follow(G, x) & \text{if } x \in pos(G), \end{cases}$$
$$follow(FG, x) = \begin{cases} follow(F, x) & \text{if } x \in pos(F) \setminus last(F), \\ follow(F, x) \cup first(G) & \text{if } x \in last(F), \\ follow(G, x) & \text{if } x \in pos(G), \end{cases}$$
$$follow(F^*, x) = \begin{cases} follow(F, x) & \text{if } x \in pos(F) \setminus last(F), \\ follow(F, x) \cup first(F) & \text{if } x \in last(F). \end{cases}$$

Now the idea of the position automaton is to take the positions of E as states. Then, after reading a letter from the input, the automaton should be in a position which could have generated that letter. To this end, we notate by $l(x)$ he letter at position x.

Definition 1. *The position automaton for E is $(Q, \Sigma, q_I, \Delta, Q_F)$ where, for some $q_I \notin \mathrm{pos}(E)$,*

$$Q = \{q_I\} \cup \mathrm{pos}(E), \qquad Q_F = \begin{cases} \{q_I\} \cup \mathrm{last}(E) & \text{if } \varepsilon \in L(E), \\ \mathrm{last}(E) & \text{otherwise,} \end{cases}$$

$$\Delta = \{(q_I, l(x), x) \mid x \in \mathrm{first}(E)\} \cup \{(x, l(y), y) \mid y \in \mathrm{follow}(E, x)\}.$$

For our example, $E_5 = (a_1 + \varepsilon)(a_2 + \varepsilon)(a_3 + \varepsilon)(a_4 + \varepsilon)(a_5 + \varepsilon)$, we have $\mathrm{pos}(E_5) = \mathrm{last}(E_5) = \mathrm{first}(E_5) = \{1, 2, 3, 4, 5\}$, and the follow sets are shown in Figure 3, together with the resulting position automaton.

position x	follow(E, x)
1	$\{2, 3, 4, 5\}$
2	$\{3, 4, 5\}$
3	$\{4, 5\}$
4	$\{5\}$
5	$\emptyset$

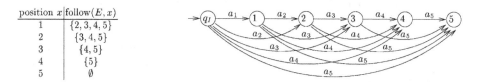

Fig. 3. The position automaton for E_5

Proposition 3. *Given a regular expression E of size n, the position automaton $\mathcal{A}$ for E has $n + 1$ states, and $\mathrm{size}(\mathcal{A}) \leq n^2$. It can be computed in time $\mathcal{O}(n^3)$.*

2.3 The Partial Derivatives Method

In [1], V. Antimirov introduced yet another basically different approach to obtain an NFA from a given regular expression, based on algebraic ideas. Since here, we are interested in transformation algorithms, we give a glance on the algebraic side only and rather look at it as an instruction how to construct an automaton.

The starting point is the notion of word derivatives as introduced by Brzozowski [3]. Those can be looked at as "left quotients of regular expressions", transferring the Myhill-Nerode congruence from languages to regular expressions. Also, they can be seen as another way to construct a deterministic automaton.

Now, the partial derivatives are in a way a non-deterministic version of those word derivatives.

Definition 2. *For a given regular expression, a non-deterministic linear form $lf(E)$ is a finite set of pairs $(a, E) \in \Sigma \times re(\Sigma)$ (where $re(\Sigma)$ is the set of all regular expressions over Σ), defined inductively as follows.*

$$lf(\varepsilon) \quad = \emptyset, \qquad\qquad lf(F+G) = lf(F) \cup lf(G),$$
$$lf(a) \quad = \{(a,\varepsilon)\}, \qquad\qquad lf(F^*) = lf(F) \odot F^*, \; and$$
$$lf(F \cdot G) = \begin{cases} lf(F) \odot G, & if \; \varepsilon \notin L(F) \\ lf(F) \odot G \cup lf(G), & if \; \varepsilon \in L(F) \end{cases}$$

Here, $\{(a_1, E_1), \ldots, (a_k, E_k)\} \odot F = \{(a_1, E_1 \cdot F), \ldots, (a_k, E_k \cdot F)\}.$

Now, the partial derivatives of E are obtained in the following way. First, we define what a derivative with respect to a single letter is. Then we extend this to non-empty words and languages in the natural way.

$\partial_a(E) = \{F \in re(\Sigma) \mid (a, F) \in lf(E)\}$ *for* $a \in \Sigma.$

$\partial_w(E) = \bigcup_{F \in \partial_v(E)} \partial_a(F)$ *for a word* $w = va,\; |v| \geq 1.$

$\partial_L(E) = \bigcup_{w \in L} \partial_w(E)$ *for a language* $L \subseteq \Sigma^+.$

We denote by $PD(E) = \partial_{\Sigma^+}(E)$ *the set of all partial derivatives of E.*

Note that $lf(E)$ can be interpreted as taking a left quotient of E and splitting up sums into separate elements, thus reflecting the nondeterministic choice.

Theorem 1 (Antimirov [1]). *For every regular expression E,* $|PD(E)| \leq size(E).$

Definition 3. *The partial derivatives automaton for E is* $\mathcal{A} = (Q, \Sigma, q_I, \Delta, Q_F)$ *where* $Q = PD(E)$, $q_I = E$, $Q_F = \{G \in PD(E) \mid \varepsilon \in L(G)\}$, *and*
$$\Delta = \{(F, a, G) \in PD(E) \times \Sigma \times PD(E) \mid G \in \partial_a(F)\}$$

Proposition 4. *For a regular expression E of size n the partial derivatives automaton $\mathcal{A}$ for E has $\leq n+1$ states, and* $size(\mathcal{A}) \in \mathcal{O}(n^2).$

Theorem 2 (J.M. Champarnaud, D. Ziadi [4]). *The partial derivatives automaton can be computed in time* $\mathcal{O}(n^2).$

3   Refinements

In this section, we describe refinements of the basic constructions mentioned above that allow to construct NFA with less than quadratically many transitions.

A central tool for obtaining the desired complexity improvement is the balanced tree decomposition. It is used explicitly in the common follow sets method and implicitly also in [6] to refine the ε-NFA method. At the moment, it is not clear whether something similar can be done for the partial derivatives method.

Let us remind that the positions of E appear as leafs in t_E.

Definition 4. *Let t_E be a tree representation of E. A balanced tree decomposition of t_E is obtained as follows.*

1. *Cut out a subtree t_1 (i.e. one vertex and all its successors) that has $\geq \frac{1}{3}$ and $\leq \frac{2}{3}$ of the positions of t. Let t_2 be the rest of t after removing t_1, see Figure 4.*

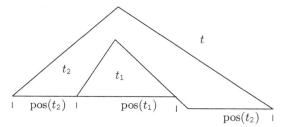

Fig.4. Balanced tree decomposition

2. *Apply the same procedure to each of t_1, t_2 that contains more than one position of E.*

It is easy to see that, for a binary tree, such a decomposition always exists, and that the "height" of such a decomposition, i.e. the number of subtrees occurring in that process to which each position belongs, is at most $\log_{\frac{3}{2}} n$, if n is the size of E.

3.1   The Common Follow Sets Automaton

A transformation with only poly-logarithmic blow-up was achieved for the first time in [9] by the so called "common follow sets automaton" construction. It is a refinement of the position automaton construction. The basic idea is to split up follow sets of states (and thus the states themselves) and to use large parts of follow sets in common.

Definition 5. *Let* $\mathrm{first}(E), \mathrm{last}(E)$, *and* $\mathrm{follow}(E, x)$ *be defined as above. A decomposition of follow sets consists of a set* $\mathrm{dec}(x) \subseteq 2^{\mathrm{pos}(E)}$ *for every* $x \in \mathrm{pos}(E)$ *such that*

$$\mathrm{follow}(E, x) = \bigcup_{C \in \mathrm{dec}(x)} C.$$

The resulting family of common follow sets is

$$\mathcal{C} = \{\mathrm{first}(E)\} \cup \bigcup_{x \in \mathrm{pos}(E)} \mathrm{dec}(x).$$

The common follow sets automaton associated with $\mathcal{C}$ *is* $(Q, \Sigma, q_I, \Delta, Q_F)$ *with*

$$Q = \mathcal{C} \times \{0, 1\}, \qquad Q_F = \mathcal{C} \times \{1\}, \qquad q_I = \begin{cases} (\mathrm{first}(E), 1) & \text{if } \varepsilon \in L(E), \\ (\mathrm{first}(E), 0) & \text{otherwise}, \end{cases}$$

$$\Delta = \{((C, f), l(x), (C', f_x)) \mid x \in C, f \in \{0, 1\}, C' \in \mathrm{dec}(x)\},$$

where $f_x = 1$ *if* $x \in \mathrm{last}(E)$ *and* $f_x = 0$ *if* $x \notin \mathrm{last}(E)$, *for* $x \in \mathrm{pos}(E)$.

As shown in [9], we obtain an NFA recognizing $L(E)$ for every valid decomposition. But the complexity very much depends on choosing or rather constructing a clever one. Here, we use the balanced tree decomposition to obtain such a decomposition of follow sets. In order to facilitate the manner-of-speaking, we will refer to the resulting automaton as *the* common follow sets automaton henceforth.

The main point is that for each subtree to which a position x belongs, at most one set is included into $\text{dec}(x)$ which at the same time is not larger than the mentioned subtree. Since a priori it is not clear that such a set exists, it is the crucial point of the whole construction to get such a set, even constructively.

Now to define those sets, we analyze how the follow sets can be looked at with respect to t_E. To this end, we define a sort of pointer between subtrees or rather between the subexpressions they represent. We use $\bullet$ as a place-holder by defining $\text{first}(\bullet) = \emptyset$.

$$\text{next}(F) = \begin{cases} F & \text{if } F \text{ is a son of } F^* \text{ in } t_E, \\ G & \text{if } F \text{ is a son of } FG \text{ in } t_E, \\ \bullet & \text{otherwise.} \end{cases}$$

Lemma 1 ([9]). *For every regular expression E and $x \in \text{pos}(E)$,*

$$\text{follow}(E, x) = \bigcup_{\substack{H \succ E \\ x \in \text{last}(H)}} \text{first}(\text{next}(H))$$

where $H \succ E$ means that H is a subexpression of E.

Now $\text{dec}(x)$ is obtained by defining recursively sets $\text{dec}(x, t)$ for the mentioned subtrees and all positions x occurring in them. Then, we let $\text{dec}(x) = \text{dec}(x, t_E)$.

When x is the only position in t, $\text{dec}(x, t)$ is either $\emptyset$ or $\{\{x\}\}$, depending on whether $x \in \text{follow}(E, x)$ or not.

Next, consider a tree t with $size(t) > 1$. Assume F_1 is a subexpression of E, and t_1 is what of the subtree t_{F_1} of t_E remains in t. (Some parts of t_{F_1} might be cut out already.) Let t_2 be the rest of t after cutting out t_1. Let $x \in \text{pos}(t)$. Then we can set

$$\text{dec}(x, t) = \begin{cases} \text{dec}(x, t_1) \cup \{C_1\} & \text{if } x \in \text{last}(F_1), \\ \text{dec}(x, t_1) & \text{otherwise,} \end{cases}$$

where

$$C_1 = \text{pos}(t) \cap \bigcup_{\substack{F \prec G \preceq F_1 \\ \text{last}(G) \cap \text{pos}(F_1) = \text{last}(F_1)}} \text{first}(\text{next}(G)).$$

Similarly, if $x \in \text{pos}(t_2)$, then we can set

$$\text{dec}(x, t) = \begin{cases} \text{dec}(x, t_2) \cup \{C_2\} & \text{if } \text{first}(F_1) \subseteq \text{follow}(E, x), \\ \text{dec}(x, t_2) & \text{otherwise,} \end{cases}$$

where

$$C_2 = \text{pos}(t) \cap \text{first}(F_1).$$

We remark (without proof) that C_1 and C_2 are independent of the particular position x.

In Figure 5, the corresponding calculation is shown for the tree representation we have chosen for E_5. In each step, it is shown how the remaining trees are cut, and which are the sets C_1, C_2 calculated for the respective cut. Also, under each position x it is shown which common follow set, if any, is added to $dec(x)$ according to the rules given above. In the end, the tabular shows the resulting decompositions, and we see the automaton calculated from this. (The empty set is trivially omitted from each $dec(x)$ containing other sets too.)

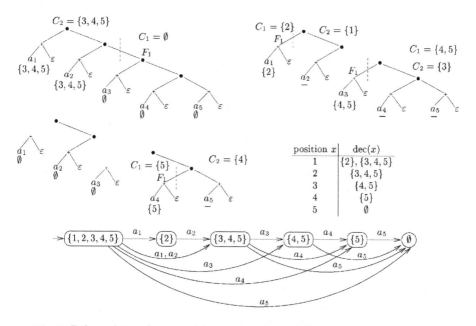

F ig.5. Balanced tree decomposition and common follow sets automaton for E_5

Theorem 3 ([9]). *For a regular expression E of size n, the common follow sets automaton $\mathcal{A}$ for E has $\leq 2n - 1$ states, and*

$$size(\mathcal{A}) \leq \tfrac{4}{(\log_2 3/2)^2} n(\log_2 n)^2.$$

As mentioned above, one $\log n$ factor comes from the "height" of the tree decomposition. It implies that the follow set of each position is decomposed into at most $\mathcal{O}(\log n)$ common follow sets. That C_1, C_2 are always subsets of $pos(t)$ implies that the summed up size of all common follow sets is in $\mathcal{O}(n \log n)$. Multiplied, those two components give the size of $\mathcal{A}$. We have to add that in order to get the small constants stated in Theorem 3, one has to do a bit of "fine tuning", described in detail in [9].

In addition to the small size of the common follow sets automaton, it has the advantage of being computable in essentially optimal time, as shown by Ch. Hagenah and A. Muscholl[2].

Theorem 4 ([8]). *Let E be a regular expression of size n, and let $\mathcal{A}$ be the common follow sets automaton for it.*

(a) There is a sequential algorithm that computes $\mathcal{A}$ in time $\mathcal{O}(n \log n + \mathrm{size}(\mathcal{A}))$;
(b) There is a parallel algorithm that computes $\mathcal{A}$ in time $\mathcal{O}(\log n)$ on a CREW PRAM using $\mathcal{O}(n + \mathrm{size}(\mathcal{A})/\log n)$ processors.

We will shortly sketch the idea behind the algorithm. Its central tool is the data structure `firstdata`. It allows the efficient handling of unions and intersections of first sets from which the common follow sets are made up.

Consider the front of t_E without ε-leaves. It simply consists of the positions of E, ordered as they occur in it. Each first set is a subsequence of this front but, unfortunately, not a contiguous one. This situation is remedied now by rearranging the front, and the result will be `firstdata`. We observe that in the inductive definition of $\mathrm{first}(E)$, there is only one case where the first set of a subexpression is not included in the first set of the larger expression, described in rule (1).

Therefore, we take t_E and look at each inner node labeled with the concatenation dot. Let the subtree below this node represent $G \cdot F$. Whenever $\varepsilon \notin L(G)$, we cut off the right son, i.e. the subtree representing F. Thus, we obtain a forest of "first trees", and `firstdata` is just the concatenation of the tree fronts.

The consequence is that each set $\mathrm{first}(H)$, for any subexpression of E, is a contiguous subsequence, i.e. a substring, of `firstdata`.

3.2   Combining Tree Decomposition and ε-NFA

In [6], V. Geffert refined the ε-NFA method in order to get automata of worst case size $\mathcal{O}(n(\log n)^2)$ as obtained by the previous method. Moreover, for a constant size alphabet Σ, the resulting automata can be bound by $\mathcal{O}(|\Sigma|n \log n)$, too. We will describe this method here in terms of the balanced tree decomposition.

The quadratic size in the ε-NFA method is generated when for each alphabet transition and *every* sequence of ε-transitions preceding it, a new alphabet transition is introduced. In general, this is wasted, because we need only so many new transitions that, whenever there were in the ε-NFA two alphabet transitions with a sequence of ε-transitions in between, we get two alphabet transitions in the new automaton that can follow each other immediately. What we need is to single out states where "many transitions can meet". Then, for each alphabet transition in the ε-automaton, we need only a few copies in the new automaton, originating from, respectively ending in those *separator states* as they are called in [6].

So the goal is to find a separator state $sep_{x,y}$ for each two alphabet transitions x and y connected by a chain of ε-transitions. In Figure 6, for a transition y, separator states and the new transition to be constructed in the NFA are shown.

---

[2] Note that in [8], slightly different sets C_1 and C_2 are used from those defined above.

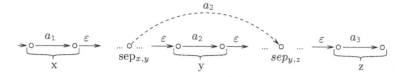

Fig.6. New transitions between separator states

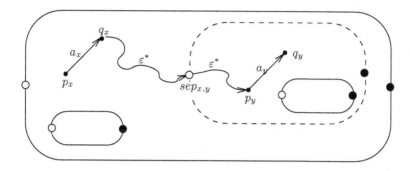

Fig.7. Decomposition of ε-automata

The computation of those separator states is based essentially on a tree decomposition of the given regular expression. The basic observation is that by the recursive construction of the ε-NFA, each sub-automaton $\mathcal{A}_F$ for a subexpression F of E is still present in the final automaton $\mathcal{A}_E$ in such a way that the sub-automaton can be entered only via a single state, and it can also be left only via a single state. In Figure 7, sub-automata are drawn as ovals, and the entering and leaving states are white respectively black circles.

Now the balanced tree decomposition of t_E induces a decomposition of $\mathcal{A}_E$. When two transitions x and y, connected by a sequence of ε-transitions in $\mathcal{A}_E$, are separated for the first time in the decomposition by cutting out a sub-automaton (dashed line in Figure 7), the entering or leaving state of that automaton where the sequence of ε-transitions passes through becomes the separator state $sep_{x,y}$.

Once the separator states are known, the computation of an NFA $\mathcal{A}' = (Q, \Sigma, q_I, \Delta', Q'_F)$ from a given ε-NFA $\mathcal{A} = (Q, \Sigma, q_I, \Delta, Q_F)$ is rather straightforward. Let $p_x \xrightarrow{a_x} q_x$ denote the transition $x = (p_x, a_x, q_x) \in \Delta$. Then we obtain the new components Δ' and Q'_F of $\mathcal{A}'$ as follows.

$$\Delta' = \{(sep_{x,y}, a_y, sep_{y,z}) \in Q \times \Sigma \times Q \mid p_x \xrightarrow{a_x} q_x \xrightarrow{\varepsilon^*} p_y \xrightarrow{a_y} q_y \xrightarrow{\varepsilon^*} p_z \xrightarrow{a_z} q_z\}$$
$$\cup \{(q_I, a_y, sep_{y,z}) \in Q \times \Sigma \times Q \mid q_I \xrightarrow{\varepsilon^*} p_y \xrightarrow{a_y} q_y \xrightarrow{\varepsilon^*} p_z \xrightarrow{a_z} q_z\}$$
$$\cup \{(sep_{x,y}, a_y, q_f) \in Q \times \Sigma \times Q \mid p_x \xrightarrow{a_x} q_x \xrightarrow{\varepsilon^*} p_y \xrightarrow{a_y} q_y \xrightarrow{\varepsilon^*} q_f \in Q_F\}$$
$$\cup \{(q_I, a_y, q_f) \in Q \times \Sigma \times Q \mid q_I \xrightarrow{\varepsilon^*} p_y \xrightarrow{a_y} q_y \xrightarrow{\varepsilon^*} q_f \in Q_F\}$$
$$Q'_F := Q_F \cup \{q_I \mid q_I \xrightarrow{\varepsilon^*} q_f \in Q_F\}$$

Theorem 5 ([6]). *For a regular expression of size n the equivalent NFA $\mathcal{A}$, computed via ε-NFA using separator states and tree decomposition, has $\leq 2n$ states, and*

$$size(\mathcal{A}) \leq \frac{2}{(\log_2 3/2)^2} n (\log_2 n)^2 + \frac{1}{(\log_2 3/2)} n \log_2 n,$$

respectively, for a fixed alphabet Σ, $size(\mathcal{A}) \leq |\Sigma| \frac{6}{(\log_2 3/2)} n \log_2 n$.

Here, the fact that the "height" of the tree decomposition is in $\mathcal{O}(\log n)$ is used to bound the number of separator states on both sides of each transition. Thus, one gets $\mathcal{O}(\log n)^2$ copies of the n alphabet transitions of the ε-NFA. For constant alphabet size, the estimate is calculated differently. It relies on some additional technical notions, and we refer the interested reader to [6].

A rough calculation shows the above construction can be performed in time $\mathcal{O}(n^3)$. It remains open to find a more efficient implementation.

Let us illustrate the construction for our running example E_5. In in the upper part of Figure 8, the ε-NFA generated in the first step is shown. The first cut in the tree decomposition (cf. Figure 5) corresponds to the dashed line. This lets s be the separator state $sep_{x,y}$ for all alphabet transitions x labeled $a_i, i \leq 2$ and y labeled $a_j, j \geq 3$. The resulting automaton is drawn in the lower half of Figure 8.

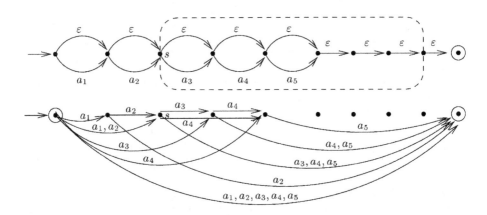

F ig . 8 . NFA from ε-NFA using separator states

The transitions drawn straight are of the form $(sep_{x,y}, a_y, sep_{y,z})$. All transitions drawn curved are obtained by the special rules for q_I and q_F, respectively.

4   A Lower Bound

To show that the refined transformation methods are nearly optimal in the order of magnitude of the resulting automaton, we look again at the sequence (E_n) of expressions $E_n = (a_1 + \varepsilon)(a_2 + \varepsilon) \dots (a_n + \varepsilon)$.

Theorem 6 ([9]). *Each NFA $\mathcal{A}$ with $L(\mathcal{A}) = L(E_n)$ has at least $n + 1$ states, and*

$$\text{size}(\mathcal{A}) \in \Omega(n \log n).$$

A simpler proof was given in [8] which we will shortly sketch. We assume that the NFA $\mathcal{A} = (Q, \Sigma, q_I, \Delta, Q_F)$ recognizes $L(E_n)$.

First, by unifying states, we may assume $Q = \{1, 2, \ldots, n + 1\}$, where for each state $i \leq n$ we have $\min\{j \mid (i, a_j, k) \in \Delta\} = i$. It is easy to see that each state is needed for accepting $a_1 a_2 \ldots a_n$.

Next, we cut $\mathcal{A}$ in the middle and observe that the resulting sub-automata recognize copies of $L(E_{n/2})$. To show that there are $\geq n/2$ transitions across the cut we look at all words $a_l a_m \in L(E_n), l \leq n/2 < m$. Either for every $l \in \{1, \ldots, n/2\}$, there is a transition across the cut labeled a_l, or for an l where there is no such transition, there need to exist a transition labeled a_m for every $m \in \{n/2 + 1, \ldots, n\}$ in order to recognize $a_l a_m$. By induction, this gives the desired bound.

Recently, Y. Lifshits has improved this.

Theorem 7 ([11]). *For each NFA $\mathcal{A}$ with $L(\mathcal{A}) = L(E_n)$*

$$\text{size}(\mathcal{A}) \in \Omega\left(\frac{n(\log n)^2}{\log \log n}\right).$$

The proof method is basically the same as above. But to show that there are at least $\frac{(\log n)^2}{\log \log n}$ many transitions across the cut one needs a much more intricate analysis. Instrumental to this is to split up each transition into two parts, introducing new states carrying the edge label in the middle.

5   Conclusion

Let us sum up what we know about the considered transformation methods in the following table.

method	states	size	time
via ε-NFA	$2n$	$\mathcal{O}(n^2)$	$\mathcal{O}(n^2)$
position automaton	$n + 1$	$\leq n^2$	$\mathcal{O}(n^3)$
partial derivatives	$\leq n + 1$	$\mathcal{O}(n^2)$	$\mathcal{O}(n^2)$
common follow sets	$\leq 2n - 1$	$\frac{4}{(\log_2 3/2)^2} n(\log_2 n)^2$	$\mathcal{O}(n(\log_n)^2)$
via ε-NFA, refined	$\leq 2n$	$\frac{2}{(\log_2 3/2)^2} n(\log_2 n)^2 + \frac{1}{(\log_2 3/2)} n \log_2 n$	$\mathcal{O}(n^3)$
(resp. for alphabet size s)		$s \frac{6}{(\log_2 3/2)} n \log_2 n$	

We see that the two refined methods improve substantially over the corresponding basic methods for the price of only a small increase in the number of states. In view of the lower bound, they are nearly optimal in case of general alphabets. For bounded alphabet size, this remains to be seen, though it seems to be unlikely to get a linear transformation here.

Comparing the two methods, one notes a constant factor of two in favor of the refined ε-NFA method. But this might be just resulting from a sharper estimate. So far the examples which show that the estimates are sharp in the order of magnitude are not sharp with the respect to the constants.

More important is the difference with respect to constant size alphabets. So far, we haven't succeeded in improving the common follow sets method for this case. It should be noted that this may come from the fact that both methods make somewhat different use of the tree decomposition.

On the other hand, looking at the algorithmic performance, the refined ε-NFA method needs some improvement. The efficient implementation from [8] seems not so easily to be transferred. Of course, we would like to see the advantages of both methods combined into one.

Another question is whether the partial derivatives method can be improved similarly. It would be interesting, especially if we consider the fact that the partial derivatives automaton can occasionally have much less states than the other basic ones, depending on the regular expression.

References

1. V. Antimirov, Partial derivatives of regular expressions and finite automaton constructions. *Theoret. Comp. Sci.* 155 (1996), 291–319.
2. R. Book, Sh. Even, Sh. Greibach, and G. Ott. Ambiguity in graphs and expressions. *IEEE Trans. Comput.* 20 (1971), 149–153.
3. J. Brzozowski, Derivatives of regular expressions. *J. ACM* 11 (1964), 481–494.
4. J.M. Champarnaud, D. Ziadi, From C-continuations to new quadratic algorithms for automaton synthesis. *Intl. J. of Algebra and Comput.* 11 (2001), 707–735.
5. A. Ehrenfeucht, P. Zeiger. Complexity measures for regular expressions. *J. Comput. System Sci.* 12 (1976), 134–146.
6. V. Geffert, Translation of Binary Regular Expressions into Nondeterministic ε-free Automata with $\mathcal{O}(n \log n)$ Transitions. To appear in *J. Comp. Syst. Sci.*.
7. V. M. Glushkov. The abstract theory of automata. *Russian Math. Surveys*, 16 (1961), 1–53. Translation from Usp. Mat. Naut. 16 (1961), 3–41, by J. M. Jackson.
8. Chr. Hagenah and A. Muscholl. Computing ε-Free NFA from Regular Expressions in $\mathcal{O}(n \log^2(n))$ Time. *RAIRO Theoret. Informatics and Applications* 34 (2000), 257–277.
9. J. Hromkovič, S. Seibert, and Th. Wilke. Translating regular expression into small ε-free nondeterministic automata. In R. Reischuk, M. Morvan (eds.), *STACS 97*, LNCS 1200, 55–66, Springer-Verlag 1997. Full version in *J. Comp. Syst. Sci.* 62 (2001), 565–588.
10. J. E. Hopcroft and J. D. Ullman. *Introduction to Automata Theory, Languages and Computation.* Addison-Wesley, Reading, Mass., 1979.
11. Y. Lifshits. A Lower Bound on the Size of ε-free NFA Corresponding to a Regular Expression, Manuscript, St. Petersburg State Univ. 2002.
12. R. F. McNaughton and H. Yamada. Regular expressions and state graphs for automata. *IRE Trans. Electron. Comput.* EC-9(1), (1960), 39–47.

Extended Temporal Logic on Finite Words and Wreath Product of Monoids with Distinguished Generators

Zoltán Ésik*

Dept. of Computer Science, University of Szeged
P.O.B. 652, 6701 Szeged, Hungary

Abstract. We associate a modal operator with each language belonging to a given class of regular languages and use the (reverse) wreath product of monoids with distinguished generators to characterize the expressive power of the resulting logic.

1 Introduction

The wreath product and its variants have been very useful and powerful tools in the characterization of the expressive power of several logical systems over finite and infinite words, including first-order logic and its extension with modular counting, cf. Straubing [11] and Straubing, Therien, Thomas [12], temporal logic and the until hierarchy, Cohen, Pin, Perrin [3] and Therien, Thomas [13], and modular temporal logic, Bazirambawo, McKenzie, Therien [2]. In this paper, we associate a modal operator with each language in a given subclass of regular languages, and use the reverse wreath product (of monoids with distinguished generators) to provide an algebraic characterization of the expressive power of (future) temporal logic on finite words endowed with these modal operators. Our logic is closely related to that proposed in Wolper [14], and our methods are related to those of Cohen, Pin, Perrin [3] and Bazirambawo, McKenzie, Therien [2] and Ésik, Larsen [7]. Moreover, our methods and results extend to ω-words, (countable) ordinal words and, more generally, to all discrete words. These extensions will be treated in subsequent papers.

Some Notation. An *alphabet* is a finite nonempty set. *We assume that each alphabet is equipped with a fixed linear order* $<$. For an alphabet Σ, we denote by Σ^* the free monoid of all finite *words* over Σ including the *empty word* ϵ. The *length* of a word u is denoted $|u|$. The notation $u = u_1 \cdots u_n$ for a word $u \in \Sigma^*$ means that u is a word of length n whose letters are $u_1, \ldots, u_n$. A subset of Σ^* is called a *language* over Σ. The *boolean* and *regular operations* on languages, and *regular languages* are defined as usual. When $L \subseteq \Sigma^*$ and $u \in \Sigma^*$, the *left*

* Supported in part by a grant from the National Foundation of Hungary for Scientific Research and by the Japan Society for the Promotion of Science.

M. Ito and M. Toyama (Eds.): DLT 2002, LNCS 2450, pp. 43–58, 2003.

quotient $u^{-1}L$ and *right quotient* Σu^{-1} of L with respect to u are respectively given by

$$u^{-1}L = \{v : uv \in L\}$$
$$Lu^{-1} = \{v : vu \in L\}.$$

A *class of (regular) languages* $\mathcal{L}$ consists of a set of (regular) languages for each alphabet Σ. If n is a nonnegative integer, we let $[n]$ denote the set $\{1,\ldots,n\}$. Thus, $[0]$ is another name for the empty set.

2　Extended Temporal Logic

Syntax. For an alphabet Σ, the set of formulas over Σ is the least set containing the letters p_σ, for all $\sigma \in \Sigma$, closed with respect to the boolean connectives $\vee$ (disjunction) and $\neg$ (negation), as well as the following construct. Suppose that $L \subseteq \Delta^*$ and that for each $\delta \in \Delta$, φ_δ is a formula over Σ. Then

$$L(\delta \mapsto \varphi_\delta)_{\delta \in \Delta} \tag{1}$$

is a formula over Σ. The notion of *subformula* of a formula is defined as usual.

　　Semantics. Suppose that φ is a formula over Σ and $u \in \Sigma^*$. We say that u *satisfies* φ, in notation $u \models \varphi$, if

- $\varphi = p_\sigma$, for some $\sigma \in \Sigma$, and u is a nonempty word whose first letter is σ, i.e., $u = \sigma u'$ for some $u' \in \Sigma^*$, or
- $\varphi = \varphi' \vee \varphi''$ and $u \models \varphi'$ or $u \models \varphi''$, or
- $\varphi = \neg\varphi'$ and it is not the case that $u \models \varphi'$, or
- $\varphi = L(\delta \mapsto \varphi_\delta)_{\delta \in \Delta}$, $u = u_1 \cdots u_n$, where each u_i is a letter, and the *characteristic word* $\delta_1 \cdots \delta_n$ belongs to L, where for each $i \in [n]$, δ_i is the first letter of Δ with respect to the linear order on Δ with $u_i \cdots u_n \models \varphi_{\delta_i}$, if such a letter exists, and δ_i is the last letter of Δ, otherwise.

For any formula φ of over Σ, we let L_φ denote the *language defined by* φ:

$$L_\varphi = \{u \in \Sigma^* : u \models \varphi\}.$$

We say that formulas φ and ψ over Σ are *equivalent* if $L_\varphi = L_\psi$. Throughout the paper we will use the boolean connective $\wedge$ (conjunction) as an abbreviation. Moreover, for any alphabet Σ, we define $\mathbb{t} = p_\sigma \vee \neg p_\sigma$ and $\mathbb{f} = \neg\mathbb{t}$, where σ is a letter in Σ.

　　We will consider subsets of formulas associated with a class $\mathcal{L}$ of (regular) languages. We let FTL($\mathcal{L}$) denote the collection of formulas all of whose subformulas of the form (1) above satisfy that L belongs to $\mathcal{L}$. We define **FTL**($\mathcal{L}$) to be the class of all languages definable by formulas in FTL($\mathcal{L}$). It is clear that for each formula $L(\delta \mapsto \varphi_\delta)_{\delta \in \Delta}$ in FTL($\mathcal{L}$) over an alphabet Σ there is an equivalent formula $L(\delta \mapsto \varphi'_\delta)_{\delta \in \Delta}$ in FTL($\mathcal{L}$) such that the subformulas φ'_δ are *pairwise inconsistent*: There exists no $u \in \Sigma^*$ and distinct letters $\delta, \delta' \in \Delta$ such that $u \models \varphi'_\delta \wedge \varphi'_{\delta'}$. Indeed, when the given linear order on Δ is $\delta_1 < \cdots < \delta_k$, then we define

$$\varphi'_{\delta_i} = \varphi_{\delta_i} \wedge \bigwedge_{j<i} \neg\varphi_{\delta_j},$$

for all $i \in [k]$. Alternatively, we may define φ'_{δ_i} for all $i < k$ as above, and

$$\varphi'_{\delta_k} = \bigwedge_{j<k} \neg\varphi_{\delta_j}.$$

Thus, the modal formulas in FTL($\mathcal{L}$) over Σ associated with a language $L \subseteq \Delta^*$ in $\mathcal{L}$ may equivalently be written as $L(\delta \mapsto \varphi_\delta)_{\delta \in \Delta}$, where the φ_δ are pairwise inconsistent and $\bigvee_{\delta \in \Delta} \varphi_\delta$ is equivalent to $\mathbb{t}$, so that each word in Σ^* satisfies exactly one φ_δ. Below we will call such families φ_δ, $\delta \in \Delta$ *deterministic*. Moreover, we will sometimes write modal formulas over Σ as $L(\delta \mapsto \varphi_\delta)_{\delta \in \Delta}$, where φ_δ, $\delta \in \Delta$ is a deterministic family of formulas over Σ. When φ_δ, $\delta \in \Delta$ is a deterministic family, we have

$$u \models L(\delta \mapsto \varphi_\delta)_{\delta \in \Delta} \Leftrightarrow \exists \delta_1 \cdots \delta_n \in L \, \forall i \in [n] \, u_i \cdots u_n \models \varphi_{\delta_i},$$

for all $u = u_1 \cdots u_n \in \Sigma^*$. We call a formula φ deterministic if for every subformula of φ of the form $L(\delta \mapsto \varphi_\delta)_{\delta \in \Delta}$, the family φ_δ, $\delta \in \Delta$ is deterministic. As shown above, for each $\varphi \in$ FTL($\mathcal{L}$) there is a deterministic formula in FTL($\mathcal{L}$) which is equivalent to φ.

We end this section with the definition of *formula substitution*. Suppose that φ is a formula over an alphabet Σ, and suppose that for each $\sigma \in \Sigma$ we are given a formula ψ_σ over Σ'. Then the formula over Σ',

$$\tau = \varphi[p_\sigma \mapsto \psi_\sigma],$$

is obtained from φ by replacing, for each letter $\sigma \in \Sigma$, each occurrence of the symbol p_σ by the formula ψ_σ. Formally, we define

- $\tau = \psi_\sigma$ if $\varphi = p_\sigma$,
- $\tau = \varphi_1[p_\sigma \mapsto \psi_\sigma] \vee \varphi_2[p_\sigma \mapsto \psi_\sigma]$ if $\varphi = \varphi_1 \vee \varphi_2$,
- $\tau = \neg(\varphi_1[p_\sigma \mapsto \psi_\sigma])$ if $\varphi = \neg\varphi_1$,
- $\tau = L(\delta \mapsto \varphi_\delta[p_\sigma \mapsto \psi_\sigma])_{\delta \in \Delta}$ if $\varphi = L(\delta \mapsto \varphi_\delta)_{\delta \in \Delta}$.

Note that when φ and ψ_σ are in FTL($\mathcal{L}$), for all $\sigma \in \Sigma$, then $\varphi[p_\sigma \mapsto \psi_\sigma]$ belongs to FTL($\mathcal{L}$).

3   Some Elementary Properties

In this section, we establish some elementary properties of the classes **FTL**($\mathcal{L}$), where $\mathcal{L}$ denotes a class of languages. We also study conditions on $\mathcal{L}$ and $\mathcal{L}'$ under which **FTL**($\mathcal{L}$) = **FTL**($\mathcal{L}'$).

Suppose that Δ and Δ' are alphabets. A *literal homomorphism* $\Delta^* \to \Delta'^*$ is a homomorphism $h : \Delta^* \to \Delta'^*$ such that $h(\Delta) \subseteq \Delta'$. Note that a homomorphism $h : \Delta^* \to \Delta'^*$ is a literal homomorphism iff it is *length preserving*, i.e., when $|h(u)| = |u|$, for all $u \in \Sigma^*$.

Proposition 1. *For each $\mathcal{L}$, the class of languages $\mathbf{FTL}(\mathcal{L})$ contains $\mathcal{L}$ and is closed with respect to the boolean operations and inverse literal homomorphisms.*

Proof. It is obvious that $\mathbf{FTL}(\mathcal{L})$ is closed under the boolean operations. Moreover, each language $L \subseteq \Sigma^*$ in $\mathcal{L}$ is definable by the formula $L(\sigma \mapsto p_\sigma)_{\sigma \in \Sigma}$ in $\mathbf{FTL}(\mathcal{L})$. Assume now that $h : \Sigma'^* \to \Sigma^*$ is a literal homomorphism. We argue by induction on the structure of the formula φ over Σ in $\mathbf{FTL}(\mathcal{L})$ to show that $h^{-1}(L_\varphi)$ is definable by some formula ψ in $\mathbf{FTL}(\mathcal{L})$. When $\varphi = p_\sigma$, for some letter σ, then we define $\psi = \bigvee_{h(\sigma')=\sigma} p_{\sigma'}$. It is clear that $L_\psi = h^{-1}(L_\varphi)$. Suppose now that $\varphi = \varphi_1 \vee \varphi_2$ and that $L_{\psi_i} = h^{-1}(L_{\varphi_i})$, $i = 1, 2$. Then we define $\psi = \psi_1 \vee \psi_2$. When $\varphi = \neg\varphi_1$ and $L_{\psi_1} = h^{-1}(L_{\varphi_1})$, then let $\psi = \neg\psi_1$. In either case, we have $L_\psi = h^{-1}(L_\varphi)$. Finally, assume that $\psi = L(\delta \mapsto \varphi_\delta)_{\delta \in \Delta}$, and that for each δ there is a formula ψ_δ in $\mathbf{FTL}(\mathcal{L})$ with $L_{\psi_\delta} = h^{-1}(L_{\varphi_\delta})$. Then define $\psi = L(\delta \mapsto \psi_\delta)_{\delta \in \Delta}$. Let $u = u_1 \cdots u_n \in \Sigma'^*$. Since for all $\delta \in \Delta$ and $i \in [n]$,

$$u_i \cdots u_n \models \psi_\delta \Leftrightarrow h(u_i \cdots u_n) \models \varphi_\delta,$$

the characteristic word determined by u and the formulas ψ_δ is the same as that determined by $h(u)$ and the formulas φ_δ. It follows that $u \models \psi$ iff $h(u) \models \varphi$. $\square$

Next we show that $\mathbf{FTL}$ is a closure operator.

Proposition 2. *For any class $\mathcal{L}$ of languages, $\mathbf{FTL}(\mathbf{FTL}(\mathcal{L})) = \mathbf{FTL}(\mathcal{L})$.*

Proof. The inclusion from right to left follows from Proposition 1. To prove that $\mathbf{FTL}(\mathbf{FTL}(\mathcal{L})) \subseteq \mathbf{FTL}(\mathcal{L})$, we argue by induction on the structure of a formula φ over Δ in $\mathbf{FTL}(\mathcal{L})$ to show that for every deterministic family φ_δ, $\delta \in \Delta$ of formulas in $\mathbf{FTL}(\mathcal{L})$ over an alphabet A, the formula $L_\varphi(\delta \mapsto \varphi_\delta)_{\delta \in \Delta}$ is expressible in $\mathbf{FTL}(\mathcal{L})$, i.e., there exists a formula in $\mathbf{FTL}(\mathcal{L})$ which is equivalent to it. Assume first that $\varphi = p_{\delta_0}$, for some $\delta_0 \in \Delta$. Then $L_\varphi = \delta_0 \Delta^*$. It is clear that a word $u \in A^*$ satisfies $L_\varphi(\delta \mapsto \varphi_\delta)_{\delta \in \Delta}$ iff $|u| > 0$ and u satisfies φ_{δ_0}, so that $L_\varphi(\delta \mapsto \varphi_\delta)_{\delta \in \Delta}$ is equivalent to $\varphi_{\delta_0} \wedge \bigvee_{a \in A} p_a$. In the induction step, assume first that $\varphi = \varphi_1 \vee \varphi_2$. Then $L_\varphi = L_{\varphi_1} \cup L_{\varphi_2}$ and thus $L_\varphi(\delta \mapsto \varphi_\delta)_{\delta \in \Delta}$ is equivalent to $L_{\varphi_1}(\delta \mapsto \varphi_\delta)_{\delta \in \Delta} \vee L_{\varphi_2}(\delta \mapsto \varphi_\delta)_{\delta \in \Delta}$. By induction, there exist ψ_1 and ψ_2 in $\mathbf{FTL}(\mathcal{L})$ such that $L_{\varphi_i}(\delta \mapsto \varphi_\delta)_{\delta \in \Delta}$ is equivalent to ψ_i, $i = 1, 2$. It follows that $L_\varphi(\delta \mapsto \varphi_\delta)_{\delta \in \Delta}$ is equivalent to $\psi_1 \vee \psi_2$ which is in $\mathbf{FTL}(\mathcal{L})$. Suppose next that $\varphi = \neg\varphi_1$, so that $L_\varphi = \overline{L_{\varphi_1}}$. Then we have $L_\varphi(\delta \mapsto \varphi_\delta)_{\delta \in \Delta}$ is equivalent to $\neg(L_{\varphi_1}(\delta \mapsto \varphi_\delta)_{\delta \in \Delta})$. It follows from the induction hypothesis that $L_\varphi(\delta \mapsto \varphi_\delta)_{\delta \in \Delta}$ is equivalent to a formula in $\mathbf{FTL}(\mathcal{L})$. Assume finally that $\varphi = K(\sigma \mapsto \tau_\sigma)_{\sigma \in \Sigma}$, where the family τ_σ, $\sigma \in \Sigma$ is deterministic. Then for any word $u_1 \cdots u_n \in A^*$ and for any $i \in [n]$, let φ_{δ_i} denote the unique formula φ_δ with $u_i \cdots u_n \models \varphi_\delta$. Moreover, for each $i \in [n]$, let τ_{σ_i} denote the unique formula τ_σ with $\delta_i \cdots \delta_n \models \tau_\sigma$. Then we have:

$$u_1 \cdots u_n \models L_\varphi(\delta \mapsto \varphi_\delta)_{\delta \in \Delta} \Leftrightarrow \delta_1 \cdots \delta_n \in L_\varphi$$
$$\Leftrightarrow \delta_1 \cdots \delta_n \models \varphi$$
$$\Leftrightarrow \sigma_1 \cdots \sigma_n \in K.$$

But for every $i \in [n]$,

$$u_i \cdots u_n \models L_{\tau_{\sigma_i}}(\delta \mapsto \varphi_\delta)_{\delta \in \Delta},$$

since for every $j \geq i$, $u_j \cdots u_n \models \varphi_{\delta_j}$ and since $\delta_i \cdots \delta_n \in L_{\tau_{\sigma_i}}$. Moreover, the formulas $L_{\tau_\sigma}(\delta \mapsto \varphi_\delta)_{\delta \in \Delta}$, $\sigma \in \Sigma$ form a deterministic family. Thus,

$$u_1 \cdots u_n \models K(\sigma \mapsto L_{\tau_\sigma}(\delta \mapsto \varphi_\delta)_{\delta \in \Delta})_{\sigma \in \Sigma} \Leftrightarrow \sigma_1 \cdots \sigma_n \in K.$$

We have thus shown that φ is equivalent to $K(\sigma \mapsto L_{\tau_\sigma}(\delta \mapsto \varphi_\delta)_{\delta \in \Delta})_{\sigma \in \Sigma}$. By the induction hypothesis, for each σ there is a formula ψ_σ in FTL($\mathcal{L}$) which is equivalent to $L_{\tau_\sigma}(\delta \mapsto \varphi_\delta)_{\delta \in \Delta}$. Thus, φ is equivalent to $K(\sigma \mapsto \psi_\sigma)_{\sigma \in \Sigma}$. □

Since $\mathbf{FTL}(\mathcal{L}_1) \subseteq \mathbf{FTL}(\mathcal{L}_2)$ whenever $\mathcal{L}_1 \subseteq \mathcal{L}_2$ and since $\mathcal{L} \subseteq \mathbf{FTL}(\mathcal{L})$, for all $\mathcal{L}$, we have:

Corollary 1. FTL *is a closure operator.*

Proposition 3. *Suppose that for each formula $L(\delta \mapsto \varphi_\delta)_{\delta \in \Delta}$ in FTL($\mathcal{L}$), over any alphabet Σ, and for each $w \in \Delta^*$ there is a formula in FTL($\mathcal{L}$) which is equivalent to $(w^{-1}L)(\delta \mapsto \varphi_\delta)_{\delta \in \Delta}$. Then $\mathbf{FTL}(\mathcal{L})$ is closed with respect to left quotients.*

Proof. Suppose that φ is a formula over Σ in FTL($\mathcal{L}$) and σ is a letter in Σ. We show that $\sigma^{-1}L_\varphi$ belongs to $\mathbf{FTL}(\mathcal{L})$. The generalization to quotients $w^{-1}L_\varphi$, where w is a word, is left to the reader. When φ is p_σ, then $\sigma^{-1}L_\varphi$ is Σ^*, which is definable by the formula $\mathbf{tt}$. When φ is $p_{\sigma'}$, where $\sigma' \neq \sigma$, then $\sigma^{-1}L_\varphi$ is $\emptyset$, which is definable by the formula $\mathbf{ff}$. We continue by induction on the structure of φ. Suppose that $\varphi = \varphi_1 \vee \varphi_2$ or $\varphi = \neg\varphi_1$, and assume that $\sigma^{-1}L_{\varphi_i}$ is defined by $\widetilde{\varphi}_i$ in FTL($\mathcal{L}$), $i = 1, 2$. Then $\sigma^{-1}L_\varphi$ is defined by $\widetilde{\varphi}_1 \vee \widetilde{\varphi}_2$ or $\neg\widetilde{\varphi}_1$, respectively. Assume finally that φ is $L(\delta \mapsto \varphi_\delta)_{\delta \in \Delta}$, where φ_δ, $\delta \in \Delta$ is a deterministic family, and that for each δ, $\sigma^{-1}L_{\varphi_\delta}$ is defined by $\widetilde{\varphi}_\delta$ in FTL($\mathcal{L}$). Note that $\widetilde{\varphi}_\delta$, $\delta \in \Delta$ is also a deterministic family. By assumption, for each δ_0 in Δ there is a formula τ_{δ_0} in FTL($\mathcal{L}$) such that for all words $u \in \Sigma^*$,

$$u \models \tau_{\delta_0} \Leftrightarrow u \models (\delta_0^{-1}L)(\delta \mapsto \varphi_\delta)_{\delta \in \Delta}.$$

Then let

$$\widetilde{\varphi} = \bigvee_{\delta_0 \in \Delta} (\widetilde{\varphi}_{\delta_0} \wedge \tau_{\delta_0}).$$

We have, for all $u = u_1 \cdots u_n \in \Sigma^*$,

$$
\begin{aligned}
u \models \widetilde{\varphi} &\Leftrightarrow \exists \delta_0 \ u \models \widetilde{\varphi}_{\delta_0} \wedge u \models \tau_{\delta_0} \\
&\Leftrightarrow \exists \delta_0 \ \sigma u \models \varphi_{\delta_0} \wedge u \models (\delta_0^{-1}L)(\delta \mapsto \varphi_\delta)_{\delta \in \Delta} \\
&\Leftrightarrow \exists \delta_0 \ \sigma u \models \varphi_{\delta_0} \wedge \exists \delta_1 \cdots \delta_n \in \delta_0^{-1} L \ \forall i \in [n] \ u_i \cdots u_n \models \varphi_{\delta_i} \\
&\Leftrightarrow \exists \delta_0 \cdots \delta_n \in L \ \sigma u \models \varphi_{\delta_0} \wedge \forall i \in [n] \ u_i \cdots u_n \models \varphi_{\delta_i} \\
&\Leftrightarrow \sigma u \models L(\delta \mapsto \varphi_\delta)_{\delta \in \Delta} \\
&\Leftrightarrow \sigma u \models \varphi.
\end{aligned}
$$

This concludes the proof of Proposition 3. □

Proposition 4. *Suppose that for each formula $L(\delta \mapsto \varphi_\delta)_{\delta \in \Delta}$ in FTL($\mathcal{L}$), over any alphabet Σ, and for each $w \in \Delta^*$ there is a formula in FTL($\mathcal{L}$) which is equivalent to $(Lw^{-1})(\delta \mapsto \varphi_\delta)_{\delta \in \Delta}$. Then **FTL**($\mathcal{L}$) is closed with respect to right quotients.*

Proof. Suppose that φ is a formula over Σ in FTL($\mathcal{L}$), and let σ be a letter in Σ. We only show that $L_\varphi \sigma^{-1}$ belongs to **FTL**($\mathcal{L}$). When φ is p_σ, then $L_\varphi \sigma^{-1}$ is $\sigma \Sigma^* \cup \{\epsilon\}$, which is defined by the formula $p_\sigma \vee \bigwedge_{\sigma' \in \Sigma} \neg p_{\sigma'}$. When φ is $p_{\sigma'}$, where $\sigma' \neq \sigma$, then $\sigma^{-1} L_\varphi$ is $\sigma' \Sigma^*$, which is defined by the formula $p_{\sigma'}$. We proceed by induction on the structure of φ. The cases when $\varphi = \varphi_1 \vee \varphi_2$ or $\varphi = \neg \varphi_1$ can be handled as above. Assume finally that φ is $L(\delta \mapsto \varphi_\delta)_{\delta \in \Delta}$, and that for each δ, $L_{\varphi_\delta} \sigma^{-1}$ is defined by $\widetilde{\varphi}_\delta$ in FTL($\mathcal{L}$). By assumption, for each δ_0 in Δ there is a formula τ_{δ_0} such that for all words $u \in \Sigma^*$,

$$u \models \tau_{\delta_0} \Leftrightarrow u \models (L\delta_0^{-1})(\delta \mapsto \widetilde{\varphi}_\delta)_{\delta \in \Delta}.$$

We define

$$\widetilde{\varphi} = \bigvee_{\sigma \models \varphi_{\delta_0}} \tau_{\delta_0}.$$

Then, $u \in L_\varphi \sigma^{-1}$ iff $u\sigma \in L_\varphi$ iff the characteristic word determined by $u\sigma$ and the formulas φ_δ belongs to L iff there exists some δ_0 such that $\sigma \models \varphi_{\delta_0}$ and the characteristic word determined by u and the formulas $\widetilde{\varphi}_\delta$ belongs to $L\delta_0^{-1}$ iff $u \models \widetilde{\varphi}$. $\square$

Corollary 2. *1. For any class $\mathcal{L}$ of languages, **FTL**($\mathcal{L}$) = **FTL**($\mathcal{L}'$), where $\mathcal{L}'$ is the least class containing $\mathcal{L}$ closed with respect to the boolean operations and inverse literal morphisms.*

*2. For any class $\mathcal{L}$ of languages closed with respect to quotients, or such that the modal operators associated with the quotients of the languages in $\mathcal{L}$ are expressible in FTL($\mathcal{L}$) as in Propositions 3 and 4, **FTL**($\mathcal{L}$) = **FTL**($\mathcal{L}'$), where $\mathcal{L}'$ is the least class containing $\mathcal{L}$ closed with respect to the boolean operations, quotients, and inverse literal morphisms.*

4   Monoids with Distinguished Generators

Suppose that M is a monoid and A is a nonempty set of distinguished generators for M. Then we call the pair (M, A) a *monoid with distinguished generators*, or *mg-pair*, for short. When M is finite, the mg-pair (M, A) is also called finite.

Suppose that (M, A) and (N, B) are mg-pairs. A *homomorphism* $(M, A) \to (N, B)$ is a monoid homomorphism $h : M \to N$ such that $h(A) \subseteq B$. It is clear that mg-pairs equipped with these homomorphisms form a category. We call (M, A) a *sub mg-pair* of (N, B) if M is a submonoid of N and A is a subset of B. Moreover, we call (M, A) a *quotient* of (N, B) if there is a surjective homomorphism $(N, B) \to (M, A)$, i.e., a homomorphism of mg-pairs which maps B onto A. We say that (M, A) *divides*, or *is a divisor of* (N, B), denoted $(M, A) <$

(N, B), if (M, A) is a quotient of a sub mg-pair of (N, B). *We identify any monoid* M *with the mg-pair* (M, M).

Example 1. For every alphabet Σ, (Σ^*, Σ) is an mg-pair with the following property: For every mg-pair (M, A) and function $h : \Sigma \to A$ there is a unique homomorphism $h^\sharp : (\Sigma^*, \Sigma) \to (M, A)$ extending h. We call such mg-pairs *free*.

Let $L \subseteq \Sigma^*$. Recall that the *syntactic monoid* of L is the quotient $\Sigma^* / \sim_L$ of Σ^* with respect to the syntactic congruence $\sim_L$ defined on Σ^* by

$$u \sim_L v \Leftrightarrow \forall x, y \in \Sigma^* \ xuy \in L \Leftrightarrow xvy \in L.$$

The *syntactic mg-pair* of L is $Synt(L) = (\Sigma^* / \sim_L, \Sigma / \sim_L)$.

We call a language $L \subseteq \Sigma^*$ *recognizable by an mg-pair* (M, A) if there is a homomorphism $h : (\Sigma^*, \Sigma) \to (M, A)$ with $L = h^{-1}(h(L))$. It follows by standard arguments that a language L is recognizable by an mg-pair (M, A) iff $Synt(L) < (M, A)$. Moreover, a language is recognizable by a finite mg-pair iff it is regular.

For the definition of the *(reverse) semidirect product* of monoids we refer to Eilenberg [5], and for the extension of these notions to mg-pairs to Ésik and Larsen [7]. When (S, A) and (T, B) are mg-pairs equipped with a (monoidal) *right action* of T on S,

$$S \times T \to S$$
$$(s, t) \mapsto st$$

such that $st \in A$ whenever $s \in A$, we let $(S, A) \star_r (T, B)$ denote the reverse semidirect product of (S, A) and (T, B) determined by the right action. This is the mg-pair $(R, A \times B)$, where R is the submonoid of the ordinary reverse semidirect product $S \star_r T$ of the monoids S and T determined by the action. When the right action is trivial, i.e., $st = s$ for all $s \in S$ and $t \in T$, the reverse semidirect product $(S, A) \star_r (T, B)$ becomes the *direct product* $(S, A) \times (T, B)$, i.e., the mg-pair $(R, A \times B)$, where R is the submonoid generated by $A \times B$ in the usual direct product $S \times T$ of the monoids S and T.

In addition to the reverse semidirect product, we will also make use of the *reverse wreath product*. Suppose that (S, A) and (T, B) are mg-pairs. Then consider the direct power $(S, A)^T$ of (S, A), i.e., the mg-pair (R, A^T), where R is the submonoid of S^T generated by A^T. Define the right action of T on R by

$$(ft)(t') = f(tt'),$$

for all $f \in R$ and $t, t' \in T$. Then the reverse wreath product $(S, A) \circ_r (T, B)$ is the reverse semidirect product $(R, A^T) \star_r (T, B)$ determined by the above action.

In the sequel, except for free mg-pairs, we will only consider finite mg-pairs. We call a nonempty class of finite mg-pairs a *variety* if it is closed with respect to division and direct product. A *closed variety* is also closed with respect the reverse semidirect product (or reverse wreath product). For any class **K** of finite mg-pairs, we let $\widehat{\mathbf{K}}$ denote the least closed variety containing $\mathbf{K}$. An example of a closed variety is the class $\mathbf{D}^r$ of all *reverse definite mg-pairs*. We call a finite mg-pair (M, A) reverse definite if there exists an integer $n \geq 0$ such that $a_1 \cdots a_n = a_1 \cdots a_{n+1}$ for all $a_1, \ldots, a_{n+1}$ in A. For example, when M_n denotes the monoid of all words over the two-letter alphabet $\{a, b\}$ whose length is at most n equipped with the product operation $u \cdot v = w$ iff w is the maximal prefix of uv of length $\leq n$, then $E_n = (M_n, \{a, b\})$ is a reverse definite mg-pair. Below we will write E for E_2. Note that E_n is a quotient of E_{n+1}, for all n. Each E_n generates $\mathbf{D}^r$:

Proposition 5. *For each $n \geq 1$, $\mathbf{D}^r$ is the least closed variety containing E_n.*

This follows by adapting the proof of a well-known fact for definite semigroups, proved in Eilenberg [5]. Further examples of closed varieties will be introduced when needed. When $\mathbf{V}$ and $\mathbf{W}$ are closed varieties, we let $\mathbf{V} \vee \mathbf{W}$ denote the least *closed* variety containing $\mathbf{V} \cup \mathbf{W}$.

Suppose that $\mathbf{K}$ is a class of finite mg-pairs. We let $\mathcal{L}_{\mathbf{K}}$ denote the class of all regular languages recognizable by the mg-pairs in $\mathbf{K}$. By standard arguments, it follows that a language is in $\mathcal{L}_{\mathbf{K}}$ iff its syntactic mg-pair is in the variety generated by $\mathbf{K}$. Conversely, when $\mathcal{L}$ is a class of regular languages, let $\mathbf{K}_{\mathcal{L}}$ denote the class of all syntactic mg-pairs of the languages in $\mathcal{L}$.

For each class $\mathbf{K}$ of finite mg-pairs, we define $\mathrm{FTL}(\mathbf{K}) = \mathrm{FTL}(\mathcal{L}_{\mathbf{K}})$ and $\mathbf{FTL}(\mathbf{K}) = \mathbf{FTL}(\mathcal{L}_{\mathbf{K}})$.

Corollary 3. *Let $\mathcal{L}$ denote a class of regular languages. We have $\mathbf{FTL}(\mathcal{L}) = \mathbf{FTL}(\mathbf{K}_{\mathcal{L}})$ iff there exists some class $\mathbf{K}$ of finite mg-pairs with $\mathbf{FTL}(\mathcal{L}) = \mathbf{FTL}(\mathbf{K})$ iff for each $L \subseteq \Delta^*$ in $\mathcal{L}$ and for each $w \in \Delta^*$, the modal operators associated with $w^{-1}L$ and Lw^{-1} are expressible in $\mathrm{FTL}(\mathcal{L})$ as in Propositions 3 and 4.*

Remark 1. Given a class $\mathbf{K}$ of finite mg-pairs, let $\mathcal{L}$ denote the class of all regular languages $L \subseteq A^*$ such that there exists some mg-pair $(S, A) \in \mathbf{K}$ with $L = h^{-1}(h(L))$, where h denotes the homomorphism $(A^*, A) \to (S, A)$ which is the identity function on A. Then it follows from Corollary 2 and Corollary 1 that $\mathbf{FTL}(\mathbf{K}) = \mathbf{FTL}(\mathcal{L})$.

5   Main Results

We say that the *next modality is expressible in* $\mathrm{FTL}(\mathcal{L})$ if for each formula φ in $\mathrm{FTL}(\mathcal{L})$ over any alphabet Σ there is a formula $\mathsf{X}\varphi$ over Σ such that for all $u \in \Sigma^*$,

$$u \models \mathsf{X}\varphi \Leftrightarrow \exists \sigma \in \Sigma, v \in \Sigma^* \, u = \sigma v \wedge v \models \varphi.$$

Proposition 6. *The next modality is expressible in $\mathrm{FTL}(\mathcal{L})$ iff the two-letter regular language $(a+b)b(a+b)^*$ and the one-letter language a belong to $\mathbf{FTL}(\mathcal{L})$.*

Proof. Suppose first that $L_1 = (a + b)b(a + b)^*$ and $L_2 = a$ are in **FTL**($\mathcal{L}$). Let φ be any formula in FTL($\mathcal{L}$) over the alphabet Σ. If $\epsilon \not\models \varphi$, then $\mathsf{X}\varphi$ is expressible as $L_1(a \mapsto \neg\varphi, b \mapsto \varphi)$. If $\epsilon \models \varphi$, then $\mathsf{X}\varphi$ is expressible by $L_1(a \mapsto \neg\varphi, b \mapsto \varphi) \vee L_2(a \mapsto \mathsf{t})$. It follows from Corollary 1 that the next modality is expressible in FTL($\mathcal{L}$).

Suppose now that the next modality is expressible in FTL($\mathcal{L}$), so that $\mathsf{X}\varphi$ exists for each formula φ in FTL. Then $(a + b)b(a + b)^*$ is definable by $\mathsf{X}p_b$, and a is definable by $p_a \wedge \mathsf{X}\neg p_a$. $\square$

Note that the mg-pair E defined above is isomorphic to the syntactic mg-pair of the language $(a + b)b(a + b)^*$. Using this fact, we have:

Corollary 4. *For any class* **K** *of finite mg-pairs, the next modality is expressible in* FTL(**K**) *iff* $(a+b)b(a+b)^*$ *belongs to* **FTL**(**K**) *iff every language recognizable by* E *belongs to* **FTL**(**K**).

Proof. If $L = (a + b)b(a + b)^*$ belongs to **FTL**(**K**), then, by Corollary 2, so does a, regarded as a one-letter language, since it can be constructed from L by the boolean operations, left quotients, and inverse literal homomorphisms. The first equivalence in Corollary 4 now follows from Proposition 6. As for the second, it is clear that if every language recognizable by E belongs to **FTL**(**K**), then so does the language $(a + b)b(a + b)^*$, since it is recognizable by E. On the other hand, it can be shown by standard arguments that any language recognizable by the syntactic mg-pair of a regular language L is the inverse image under a literal homomorphism of a boolean combination of quotients of L. Thus, since E is the syntactic monoid of $(a + b)b(a + b)^*$, if $(a + b)b(a + b)^*$ is in **FTL**(**K**), then since **FTL**(**K**) is closed with respect to the above operations, it follows that every language recognizable by E is in **FTL**(**K**). $\square$

Corollary 5. *For any class* **K** *of finite mg-pairs, the next modality is expressible in* FTL(**K**) *iff* **FTL**(**K**) = **FTL**($\mathbf{K}_1$), *where* $\mathbf{K}_1 = \mathbf{K} \cup \{E\}$.

Proof. By Corollary 4 and Corollary 1. $\square$

Proposition 7. *Suppose that* (S, A) *and* (T, B) *are finite mg-pairs and* $(R, A \times B)$ *is a reverse semidirect product of* (S, A) *and* (T, B) *determined by a right action of* T *on* S. *If every language recognizable by* (S, A) *and* (T, B) *belongs to* **FTL**($\mathcal{L}$), *and if the next modality is expressible in* FTL($\mathcal{L}$), *then every language recognizable by* $(R, A \times B)$ *also belongs to* **FTL**($\mathcal{L}$).

Proof. Let h denote a homomorphism

$$(\Sigma^*, \Sigma) \to (R, A \times B)$$
$$u \mapsto h(u) = (h_\ell(u), h_r(u)).$$

It suffices to show that for each $(s, t) \in R$, the language $h^{-1}((s, t))$ belongs to **FTL**($\mathcal{L}$).

For each $u \in A^*$, let $\bar{u}$ denote the image of u under the homomorphism $(A^*, A) \to (S, A)$ which is the identity map on A. Moreover, let $\hat{h}$ denote the function $\Sigma^* \to A^*$ defined by

$$\hat{h}(\epsilon) = \epsilon$$
$$\hat{h}(\sigma u) = [h_\ell(\sigma)h_r(u)]\hat{h}(u),$$

for all $u \in \Sigma^*$ and $\sigma \in \Sigma$. Here, $h_\ell(\sigma)h_r(u)$ is the result of the right action of $h_r(u)$ on $h_\ell(\sigma)$. Note that $|\hat{h}(u)| = |u|$ and that $\hat{h}(u)$ is a suffix of $\hat{h}(v)$ whenever u is a suffix of v. Also, $\overline{\hat{h}(u)} = h_\ell(u)$, for all $u \in \Sigma^*$.

By assumption, for each $s \in S$ and $t \in T$ there exist a formula φ_s over A and a formula φ_t over Σ in FTL($\mathcal{L}$) such that

$$L_{\varphi_s} = \{w \in A^* : \overline{w} = s\}$$
$$L_{\varphi_t} = \{u \in \Sigma^* : h_r(u) = t\}.$$

Given a formula ψ over A in FTL($\mathcal{L}$), let

$$\psi' = \psi[p_a \mapsto \bigvee_{h_\ell(\sigma)t=a} (p_\sigma \wedge \mathsf{X}\varphi_t)].$$

CLAIM For all $u \in \Sigma^*$,

$$u \models \psi' \Leftrightarrow \hat{h}(u) \models \psi.$$

We prove this claim by induction on the structure of ψ. When ψ is p_a, for some $a \in A$, we have

$$u \models \psi' \Leftrightarrow \exists \sigma, t[u \models p_\sigma \wedge u \models \mathsf{X}\varphi_t \wedge h_\ell(\sigma)t = a]$$
$$\Leftrightarrow \exists \sigma, t, v[u = \sigma v \wedge v \models \varphi_t \wedge h_\ell(\sigma)t = a]$$
$$\Leftrightarrow \exists \sigma, t, v[u = \sigma v \wedge h_r(v) = t \wedge h_\ell(\sigma)t = a]$$
$$\Leftrightarrow \exists \sigma, v[u = \sigma v \wedge h_\ell(\sigma)h_r(v) = a]$$
$$\Leftrightarrow \hat{h}(u) \models p_a$$
$$\Leftrightarrow \hat{h}(u) \models \psi.$$

The induction step is obvious when ψ is the disjunction $\psi_1 \vee \psi_2$ or a negation $\neg\psi_1$. Suppose now that ψ is $L(\delta \mapsto \psi_\delta)_{\delta \in \Delta}$, where $L \subseteq \Delta^*$ is in $\mathcal{L}$ and each φ_δ is a formula in FTL($\mathcal{L}$) over Σ satisfying the induction assumption. Suppose that $u = u_1 \cdots u_n$, say. By the induction hypothesis we have that for all $i \in [n]$ and $\delta \in \Delta$,

$$u_i \cdots u_n \models \psi'_\delta \Leftrightarrow \hat{h}(u_i \cdots u_n) \models \psi_\delta.$$

Thus, since $\hat{h}$ preserves suffixes, the characteristic word determined by u and ψ' is the same as that determined by $\hat{h}(u)$ and ψ, proving that $u \models \psi'$ iff $\hat{h}(u) \models \psi$.

We now complete the proof of the proposition. For any $(s, t) \in R$ and $u \in \Sigma^*$,

$$h(u) = (s, t) \Leftrightarrow h_\ell(u) = s \wedge h_r(u) = t$$
$$\Leftrightarrow \overline{\hat{h}(u)} = s \wedge h_r(u) = t$$
$$\Leftrightarrow \hat{h}(u) \models \varphi_s \wedge u \models \varphi_t$$
$$\Leftrightarrow u \models \varphi'_s \wedge \varphi_t.$$

Thus, $\varphi'_s \wedge \varphi_t$ defines $h^{-1}((s,t))$.   □

Proposition 8. *Suppose that* $\varphi = K(\delta \mapsto \varphi_\delta)_{\delta \in \Delta}$ *is a formula over* Σ *in* FTL($\mathcal{L}$), *where* φ_δ, $\delta \in \Delta$ *is a deterministic family. Suppose that* K *is recognized by* $h_K : (\Delta^*, \Delta) \to (S, A)$ *and that each* L_{φ_δ} *is recognized by the morphism* $h : (\Sigma^*, \Sigma) \to (T, B)$. *Then* $L_\varphi \subseteq \Sigma^*$ *is recognizable by the reverse wreath product* $(S, A) \circ_r (T, B)$.

Proof. Without loss of generality we may assume that h is surjective. For each $\delta \in \Delta$, let F_δ denote the set $h(L_{\varphi_\delta})$. Note that the sets F_δ are pairwise disjoint by assumption, and that $\bigcup_{\delta \in \Delta} F_\delta = T$, since h is surjective. Define $k : (\Sigma^*, \Sigma) \to (S, A) \circ_r (T, B)$ by

$$k(\sigma) = (f_\sigma, h(\sigma)),$$

for all $\sigma \in \Sigma$, where for each $t \in T$, $f_\sigma(t) = h_K(\delta)$ for the unique δ with $h(\sigma)t \in F_\delta$. Let $u = u_1 \cdots u_n$ in Σ^*. We have

$$k(u) = (f, h(u)),$$

where

$$f(t) = h_K(\delta_1 \cdots \delta_n)$$

for the unique word $\delta_1 \cdots \delta_n$ with $h(u_i \cdots u_n)t \in F_{\delta_i}$ for each $i \in [n]$. In particular,

$$f(1) = h_K(\delta_1 \cdots \delta_n)$$

for the unique word $\delta_1 \cdots \delta_n$ with $h(u_i \cdots u_n) \in F_{\delta_i}$, i.e.,

$$u_i \cdots u_n \models \varphi_{\delta_i}$$

for each $i \in [n]$. Thus,

$$f(1) \in h_K(K) \Leftrightarrow \delta_1 \cdots \delta_n \in K$$
$$\Leftrightarrow u \models \varphi.$$

It follows that

$$L_\varphi = \{u \in \Sigma^* : f(1) \in h_K(K)\}.$$

This proves that L_φ is recognizable by $(S, A) \circ_r (T, B)$.   □

Recall that E_1 denotes the mg-pair $(\{1, a, b\}, \{a, b\})$, where 1 is the identity element and a, b are both left-zeroes. Note that E_1 is just the syntactic mg-pair of the language $a(a + b)^*$.

Theorem 1. *For any class* **K** *of finite mg-pairs, every language in* **FTL(K)** *is recognizable by some mg-pair in* $\hat{\mathbf{K}} \vee \mathbf{D}^r$.

Proof. Let φ denote a deterministic formula over Σ in $\mathrm{FTL}(\mathbf{K})$. When φ is p_σ, for some $\sigma \in \Sigma$, then $L_\varphi = \sigma\Sigma^*$, which is recognizable by $E_1 \in \mathbf{D}^r$. We continue by induction on the structure of φ. Assume that $\varphi = \varphi_1 \vee \varphi_2$ such that L_{φ_i} is recognizable by (S_i, A_i) in $\widehat{\mathbf{K}} \vee \mathbf{D}^r$, $i = 1, 2$. Then L_φ is recognizable by the direct product $(S_1, A_1) \times (S_2, A_2)$ which is also in $\widehat{\mathbf{K}} \vee \mathbf{D}^r$. When $\varphi = \neg\varphi_1$, where L_{φ_1} is recognizable by (S_1, A_1) above, then L_φ is recognizable by the same mg-pair (S_1, A_1). Finally, when $\varphi = L(\delta \mapsto \varphi_\delta)_{\delta \in \Delta}$ and each L_{φ_δ} is recognizable by some mg-pair in $\widehat{\mathbf{K}} \vee \mathbf{D}^r$, then it follows by Proposition 8 that L_φ is recognizable by some mg-pair in $\widehat{\mathbf{K}} \vee \mathbf{D}^r$. (Note that since $\widehat{\mathbf{K}} \vee \mathbf{D}^r$ is closed with respect to the direct product, we may assume without loss of generality that each L_{φ_δ} is recognizable by the same mg-pair (M, A) in $\widehat{\mathbf{K}} \vee \mathbf{D}^r$, and in fact by the same morphism $(\Sigma^*, \Sigma) \to (M, A)$.) □

Theorem 2. *Suppose that the next modality is expressible in* $\mathrm{FTL}(\mathbf{K})$, *where* $\mathbf{K}$ *is a class of finite mg-pairs. Then a language L belongs to* $\mathbf{FTL}(\mathbf{K})$ *iff* $Synt(L)$ *belongs to* $\widehat{\mathbf{K}} \vee \mathbf{D}^r$.

Proof. First, by Corollary 5, we have $\mathbf{FTL}(\mathbf{K}) = \mathbf{FTL}(\mathbf{K}_1)$, where $\mathbf{K}_1 = \mathbf{K} \cup \{E\}$, so that $\widehat{\mathbf{K}}_1 = \widehat{\mathbf{K}} \vee \mathbf{D}^r$. Let us define the *rank* of $(S, A) \in \widehat{\mathbf{K}} \vee \mathbf{D}^r$ to be the smallest number of reverse semidirect product and division operations needed to generate (S, A) from $\mathbf{K}_1$. We prove by induction on the rank of (S, A) that every language recognizable by (S, A) is in $\mathbf{FTL}(\mathbf{K}_1)$. When the rank is 0 we have $(S, A) \in \mathbf{K}_1$. Thus the result follows from Proposition 1. When the rank of (S, A) is positive, then (S, A) either divides an mg-pair (T, B) in $\widehat{\mathbf{K}} \vee \mathbf{D}^r$ of smaller rank, or (S, A) is the reverse semidirect product of some mg-pairs in $\widehat{\mathbf{K}} \vee \mathbf{D}^r$ of smaller rank. In the first case, every language recognizable by (S, A) is recognizable by (T, B). In the second case, the result follows from Proposition 7.

To necessity part of Theorem 2 follows from Theorem 1. □

Corollary 6. *For each class $\mathcal{L}$ of regular languages,* $\mathbf{FTL}(\mathcal{L})$ *consists of regular languages.*

Call a nonempty class of regular languages $\mathcal{L}$ *closed* if $\mathbf{FTL}(\mathcal{L}) \subseteq \mathcal{L}$ and if $\mathcal{L}$ is closed with respect to quotients. By Propositions 1, 3 and 4, every closed class is a *literal variety*, i.e., it is closed with respect to the boolean operations, quotients, and inverse literal homomorphisms. Moreover, by Corollaries 1, 2 and 3, $\mathcal{L}$ is closed iff $\mathcal{L} = \mathbf{FTL}(\mathcal{L}')$ for a class $\mathcal{L}'$ of regular languages closed with respect to quotients iff $\mathcal{L} = \mathbf{FTL}(\mathbf{K})$ for a class $\mathbf{K}$ of finite mg-pairs.

The assignment $\mathbf{V} \mapsto \mathcal{L}_{\mathbf{V}}$ defines an order isomorphism between varieties $\mathbf{V}$ of finite mg-pairs and literal varieties of regular languages, cf. Ésik, Larsen [7]. The inverse assignment maps a literal variety $\mathcal{L}$ to the class of those finite mg-pairs (M, A) such that every language recognizable by (M, A) belongs to $\mathcal{L}$.

Theorem 3. *The assignment* $\mathbf{V} \mapsto \mathcal{L}_{\mathbf{V}} = \mathrm{FTL}(\mathbf{V})$ *defines an order isomorphism between closed varieties $\mathbf{V}$ of finite mg-pairs containing $\mathbf{D}^r$ and closed classes $\mathcal{L}$ of regular languages containing $(a + b)b(a + b)^*$.*

Proof. If **V** is a closed variety containing $\mathbf{D}^r$, then by Theorem 1, $\mathcal{L}_{\mathbf{V}}$ is a closed class of regular languages containing $(a+b)b(a+b)^*$. Since $\mathbf{FTL}(\mathbf{V})$ is the least closed class containing $\mathcal{L}_{\mathbf{V}}$, it follows that $\mathcal{L}_{\mathbf{V}} = \mathbf{FTL}(\mathbf{V})$. As mentioned above, we have $\mathbf{V}_1 \subseteq \mathbf{V}_2$ iff $\mathcal{L}_{\mathbf{V}_1} \subseteq \mathcal{L}_{\mathbf{V}_2}$. Finally, the map is surjective, for if $\mathcal{L}$ is a closed class of regular languages containing $(a+b)b(a+b)^*$, then $\mathcal{L} = \mathcal{L}_{\mathbf{V}}$ for some variety **V** of finite mg-pairs containing E. By Proposition 7, **V** is closed with respect to the reverse semidirect product. Since **V** contains E, by Proposition 5 it also contains all the reverse definite mg-pairs. $\qquad\square$

We refer to Almeida [1] for a detailed study of varieties of finite semigroups closed with respect to the semidirect product. Any such variety gives rise to a closed variety of finite mg-pairs.

Example 2. The closed class of regular languages corresponding to $\mathbf{D}^r$ is the class of reverse definite languages, where a language $L \subseteq \Sigma^*$ is termed reverse definite iff there is some $n \geq 0$ such that the membership of a word u in L depends only on the maximal prefix of u of length $\leq n$. (This condition is equivalent to requiring that the language is recognizable by some E_n.)

6   Some Applications

Propositional (future) temporal logic (FTL) was introduced in Pnueli [10]. The formulas of FTL over an alphabet Σ are constructed from the letters p_σ, $\sigma \in \Sigma$ by the boolean connectives $\vee$ and $\neg$ and the *next* and *until* modalities, denoted $\mathsf{X}\varphi$ and $\varphi\mathsf{U}\psi$. The semantics of FTL are defined similarly to that of $\mathbf{FTL}(\mathcal{L})$. In particular, when $u = u_1 \cdots u_n \in \Sigma^*$ and φ and ψ are formulas over Σ,

1. $u \models \mathsf{X}\varphi$ iff $n \geq 1$ and $u_2 \cdots u_n \models \varphi$,
2. $u \models \varphi\mathsf{U}\psi$ iff there exists some $i \in [n]$ such that $u_i \cdots u_n \models \psi$ and $u_j \cdots u_n \models \varphi$ for all $j < i$.

We let **FTL** denote the class of languages definable by the formulas in FTL.

Let U^r denote the monoid component of E_1, i.e., the monoid $\{1, a, b\}$, where a, b are left-zero elements.

Proposition 9. FTL $= \mathbf{FTL}(\{U^r, E\})$.

Proof. The inclusion $\mathbf{FTL} \subseteq \mathbf{FTL}(\{U^r, E\})$ follows from Corollary 4 and the fact that $\varphi\mathsf{U}\psi$ is expressible as $L_{\mathsf{U}}(a \mapsto \varphi \wedge \neg\psi,\ b \mapsto \psi,\ c \mapsto \neg\varphi \vee \neg\psi)$, where L_{U} denotes the language $a^*b(a+b+c)^*$ over the three-letter alphabet $\{a, b, c\}$ which can be recognized by U^r. For the reverse inclusion, one can show easily that every language recognizable by U^r or E is definable in **FTL**. For example, L_{U} is defined by $p_a\mathsf{U}p_b$. It then follows that the modal operator corresponding to any language recognized by U^r or E is expressible in **FTL**. $\qquad\square$

We call a finite mg-pair *aperiodic* if its monoid component is aperiodic, cf. Eilenberg [5]. For example, U^r and E are aperiodic. It follows from (the dual of) the Krohn-Rhodes Decomposition Theorem [5] that the aperiodic mg-pairs form a closed variety, namely the closed variety generated by U^r. Let **A** denote the closed variety of aperiodic mg-pairs. For *first-order definable* languages, we refer to [9].

Theorem 4. Mc Naughton-Papert [9] *A language is first-order definable iff* $Synt(L) \in \mathbf{A}$.

Theorem 5. *Let* $\mathbf{K}$ *denote a class of finite mg-pairs such that the next modality is expressible in* FTL($\mathbf{K}$). *Then* **FTL**($\mathbf{K}$) *contains the first-order definable languages iff* $\mathbf{A} \subseteq \widehat{\mathbf{K}} \vee \mathbf{D}^r$. *Moreover,* **FTL**($\mathbf{K}$) *is the class of all first-order definable languages iff* $\mathbf{A} = \widehat{\mathbf{K}} \vee \mathbf{D}^r$.

Proof. Immediate from Theorem 2, Theorem 3, the Krohn-Rhodes Decomposition Theorem and the fact that E is aperiodic. □

From Theorem 5 and Proposition 9 we immediately have:

Corollary 7. Kamp [8] *A language is first-order definable iff it is in* **FTL**.

We now consider cyclic counting. For any formula φ over Σ and integers $k \geq 1$ and $0 \leq r < k$, let $C_k^r \varphi$ be a formula with the following semantics. For any $u = u_1 \cdots u_n$ is Σ^*, $u \models C_k^r \varphi$ iff the number of indices i with $u_i \cdots u_n \models \varphi$ is congruent to r modulo k. Similarly, let $u \models L_k^r$ iff $|u|$ is congruent to r modulo k. Let K and M be two subsets of the naturals. We denote by FTL(K, M) the extension of FTL by the modalities C_k^r, $0 \leq r < k$ and L_m^r, $0 \leq r < m$ with $k \in K$ and $m \in M$. Moreover, we denote by **FTL**(K, M) the class of all languages definable by the formulas in FTL(K, M).

For each $n \geq 1$, let Z_n denote a cyclic group of order n. Below we will denote by a a cyclic generator of Z_n and by 1 the identity element. The *division ideal* generated by a set M of naturals consists of all divisors of least common multiples of finite sets of naturals in M. When M is empty, the division ideal generated by M is $\{1\}$.

Theorem 6. *A language L belongs to* **FTL**(K, M) *iff* $Synt(L) = (S, A)$ *satisfies the following condition: There is some m in the division ideal generated by M such that every group contained in the submonoid of S generated by A^m is solvable whose order is a multiple of the prime divisors of the integers in K.*

Proof. We have **FTL**(K, M) = **FTL**($\mathbf{K}$), where $\mathbf{K}$ consists of U^r, E and the mg-pairs $(Z_k, \{1, a\})$ and $(Z_m, \{a\})$, for all $k \in K$ and $m \in M$. Thus, by Theorem 3, **FTL**($\mathbf{K}$) consists of all languages whose syntactic mg-pair is in $\widehat{\mathbf{K}}$. It is shown in Ésik, Ito [6] that the syntactic mg-pair of a language belongs to this variety iff the condition described in the Theorem holds. □

See also Straubing, Therien, Thomas [12], Straubing [11].

For a class $\mathbf{K}$ of finite mg-pairs, let FTL + $\mathbf{K}$ denote the extension of FTL by the modal operators corresponding to the languages recognizable by the mg-pairs in $\mathbf{K}$. Moreover, let **FTL** + $\mathbf{K}$ denote the class of languages definable by the formulas in FTL + $\mathbf{K}$.

Theorem 7. FTL + K *is the class of all regular languages iff the following two conditions hold:*

1. *For each m, $(Z_m, \{a\}) \in \widehat{\mathbf{K}}$.*
2. *For each finite (nonabelian simple) group G there is an mg-pair (M, A) in* **K** *such that G divides M.*

Proof. First, by Theorem 5 and Theorem 2, it follows that a language belongs to **FTL + K** iff its syntactic mg-pair is contained in the closed variety **V** generated by **K** and the aperiodics. Now, by (the dual of) a result proved in Dömösi, Ésik [4], **V** is the class of all finite mg-pairs iff the above two conditions hold. The result now follows by using Theorem 3, since a language is regular iff it is recognizable by a finite mg-pair.   □

Corollary 8. *If* **FTL+K** *is the class of all regular languages, then* **K** *is infinite.*

Example 3. Let **K** consist of the mg-pairs $(S_n, \{\pi, \rho\})$, $n \geq 3$, where S_n denotes the symmetric group of all permutations of the set $[n]$, and where π is a transposition and ρ is a cyclic permutation of $[n]$. Then **FTL + K** is the class of all regular languages.

When **K** is a class of finite mg-pairs, let **FTL + MOD + K** denote the language class **FTL + K′**, where $\mathbf{K}' = \mathbf{K} \cup \{(Z_n, \{1, a\}) : n \geq 2\}$. The proof of the following result is similar to that of Theorem 7.

Theorem 8. *A language L belongs to* **FTL + MOD + K** *iff L is regular and for every finite nonabelian simple group G, if G divides the syntactic monoid of L, then G divides the monoid component of an mg-pair in* **K**.

Given a formula φ of FTL over the alphabet Σ, let $\Diamond\varphi$ denote the formula $\mathsf{tt}\mathsf{U}\varphi$. Thus, for each word $u = u_1 \cdots u_n$ in Σ^*, $u \models \Diamond\varphi$ iff $u_1 \cdots u_n \models \varphi$, for some $i \in [n]$. In [3], Cohen, Perrin and Pin studied the expressive power of the restricted temporal logic RTL whose formulas over an alphabet Σ are constructed from the atomic formulas p_σ, $\sigma \in \Sigma$ by the X and $\Diamond$ modalities. Let **RTL** denote the class of languages definable by the formulas in RTL. Let U_1 denote a two-element semilattice (which may be identified with the mg-pair (U_1, U_1)).

Proposition 10. RTL = FTL$(\{U_1, E\})$.

The proof is based on the observation that U_1 is isomorphic to the syntactic monoid of the two-letter language $L = (a + b)^* b (a + b)^*$, and for any formula φ, $\Diamond\varphi$ is expressible as $L(a \mapsto \neg\varphi, b \mapsto \varphi)$.

Recall that a semigroup S is called **L**-*trivial*, cf. Almeida [1], Eilenberg [5], if Green's **L**-relation on S is the equality relation. Moreover, a semigroup S is *locally* **L**-*trivial* iff for each idempotent e, the monoid eSe is **L**-trivial. Accordingly, we call an mg-pair (M, A) locally **L**-trivial if the subsemigroup of M generated by A is locally **L**-trivial.

It follows from well-known facts (cf. [5,1]) that a finite mg-pair is locally **L**-trivial iff it belongs to the least closed variety containing U_1 and E_1, or U_1 and E. Thus, from Proposition 10 and Theorem 2 we may derive:

Theorem 9. Cohen, Perrin, Pin [3] *A language $L \subseteq \Sigma^*$ belongs to* **RTL** *iff L is regular and $Synt(L)$ is locally* **L**-*trivial.*

Following the definitions of the language classes $\mathbf{FTL}(K, M)$, we may define the classes $\mathbf{RTL}(K, M)$. For lack of space we omit the proof of the following result.

Theorem 10. *A language L belongs to* $\mathbf{RTL}(K, M)$ *iff $Synt(L) = (S, A)$ satisfies the following condition: There is an integer m in the division ideal generated by M such that for each idempotent e of the subsemigroup T of S generated by A^m, it holds that eTe is a solvable group whose order is a multiple of the primes that divide the integers in K.*

References

1. J. Almeida, *Finite Semigroups and Universal Algebra*, World Scientific, 1994.
2. A. Baziramwabo, P. McKenzie and D. Therien, Modular temporal logic, in: *Proc. 1999 IEEE Conference LICS, Trento, Italy.*
3. J. Cohen, J.-E. Pin and D. Perrin, On the expressive power of temporal logic, *J. Computer and System Sciences*, 46(1993), 271–294.
4. P. Dömösi and Z. Ésik, Critical classes for the α_0-product, *Theoretical Computer Science*, 61(1988), 17–24.
5. S. Eilenberg, *Automata, Languages, and Machines*, vol. A and B, Academic Press, 1974 and 1976.
6. Z. Ésik and M. Ito, Temporal logic with cyclic counting and the degree of aperiodicity of finite automata, *Acta Cybernetica*, to appear.
7. Z. Ésik and K. G. Larsen, Regular languages definable by Lindström quantifiers, *BRICS Reports*, RS-02-20, April, 2002.
8. J. A. Kamp, Tense logic and the theory of linear order, Ph. D. Thesis, UCLA, 1968.
9. R. McNaughton and S. Papert, *Counter-Free Automata*, MIT Press, 1971.
10. A. Pnueli, The temporal logic of programs, in: *Proc. 18th FOCS, Providence, RI*, IEEE Press, 1977, 46–57.
11. H. Straubing, *Finite Automata, Formal Logic, and Circuit Complexity*, Birkhauser, 1994.
12. H. Straubing, D. Therien and W. Thomas, Regular languages defined with generalized quantifiers, *Information and Computation*, 118(1995), 289–301.
13. D. Therien and Th. Wilke, Temporal logic and semidirect products: An effective characterization of the until hierarchy, *DIMACS TR-96-28*, 1996.
14. P. Wolper, Temporal logic can be more expressive, *Information and Control*, 56(1983), 72–99.

A Remark about Quadratic Trace Equations

Volker Diekert and Manfred Kufleitner

Universität Stuttgart
Institut für Formale Methoden der Informatik
Breitwiesenstr. 20–22, D–70565 Stuttgart, Germany

Abstract. We present a simple PSPACE-algorithm for computing the most general solution of a system of quadratic trace equations with involution. We extend the known linear time algorithm for quadratic word equations with length constraints to cope with involutions. Finally, we show that the same linear-time result cannot be expected for trace equations. We obtain an NP-hardness result.

1 Introduction

A string graph is the intersection graph of a finite set of curves on a surface. Recently the string graph recognition problem in the plane has been shown to be NP-complete via a surprising polynomial time reduction to word equations with involution by Schaefer, Sedgwick and Štefankovič [9]. The NP-completeness result came quite unexpected because for many years, even decidability was not known. After the reduction in [9] one finds a system of word equations and the NP-algorithm guesses in a first step the lengths in the solution. Starting from string graphs these lengths are bounded by a single exponential function, so in binary notation the NP-algorithm can indeed guess them. Then it is possible to apply the techniques developed by Plandowski and Rytter in [7] for solving word equations in polynomial time when lengths are specified. However, inspecting the reduction from string graph recognition to word equations, one sees that only quadratic word equations appear and this simplifies the situation, so that as an alternative one can also apply the method for solving quadratic equations of [8] in linear time when lengths are specified. The only problem is that the authors of [8] did not consider the presence of an involution. But this is not essential as it is shown in Sect. 3.

For surfaces of higher genus the reduction of [9] leads to quadratic systems of trace equations, again with involution. Trace monoids are free monoids with a partial commutation between letters. (The partial commutation corresponds to the crossings of the curves.) It turns out that these quadratic trace equations can be solved rather straightforwardly, too: We present a simple PSPACE-algorithm for computing the most general solution when the input is a partial commutation relation with a system of quadratic trace equations with involution. By this

M. Ito and M. Toyama (Eds.): DLT 2002, LNCS 2450, pp. 59–66, 2003.

result[1] the string graph recognition problem for surfaces of higher genus is in PSPACE.

In contrast to the situation in the plane, there is no a priori bound known on the lengths in a solution. Nevertheless, it is natural to ask what happens, if we encounter as input a system of quadratic trace equations where the lengths in a solution are specified in binary. We show in Sect. 4 that still we cannot expect any polynomial time algorithm in the trace case. The satisfiability problem for single quadratic matching equation remains NP-hard, even when the length information is given. For the NP-hardness result the trace monoid can be fixed.

2   Quadratic Trace Equations with Involution

Free partially commutative monoids are commonly denoted as *trace monoids* and well-studied in mathematics and computer science [3]. A trace monoid is a free monoid modulo a partial commutation between generators. Motivated by extending results to free partially commutative groups (*graph groups*) one has been led to consider trace monoids with involution. The involution corresponds to taking inverse elements. In the group case the involution is usually without fixed points on generators since in the free case there is no non-trivial element of order two. But when starting the discussion by string graphs then the involution is reading crossing points from right-to-left, so all letters are fixed points. The best is therefore to make no special assumption how the underlying involution behaves on letters. The central object here are trace monoids with involution and quadratic equations for them.

We let Γ be a finite alphabet which is equipped with an involution $\bar{} : \Gamma \to \Gamma$. An *involution* is a mapping such that $\bar{\bar{a}} = a$ of all $a \in \Gamma$. By Ω we denote the set of variables and by $\overline{\Omega}$ its disjoint copy, so $\Omega \cup \overline{\Omega}$ is a set with an involution without fixed points. The involution is extended to $(\Gamma \cup \Omega \cup \overline{\Omega})^*$ by letting $\overline{x_1 \cdots x_n} = \overline{x_n} \cdots \overline{x_1}$. Note that, in contrast to variables, we allow fixed points on letters (constants). As a special case we have $\bar{a} = a$ for all $a \in \Gamma$.

An *equation* $U = V$ is a pair (U, V) with $U, V \in (\Gamma \cup \Omega \cup \overline{\Omega})^*$, and a *system of equations* is a finite collections $\{U_1 = V_1, \ldots, U_k = V_k\}$ of equations. It is a *quadratic* system if every variable $x \in \Omega$ appears at most twice in the system. A variable may appear either as x or in the form $\overline{x}$, but both forms are counted for x.

By $I \subseteq \Gamma \times \Gamma$ we denote an *independence relation* which is an irreflexive and symmetric relation. Moreover, we demand that I is compatible with the involution, i.e., we have $(a, b) \in I$ if and only if $(a, \overline{b}) \in I$. Since I is irreflexive, this implies $(a, \overline{a}) \notin I$ for all $a \in \Gamma$.

By $M(\Gamma, I)$ we denote the *trace monoid* which is defined by $M(\Gamma, I) = \Gamma^* / \{ab = ba \mid (a, b) \in I\}$. Thus each trace $t \in M(\Gamma, I)$ can be represented by a word $a_1 \cdots a_n$ with $a_i \in \Gamma$. Moreover, we can define its *length* $|t|$ by n, its *alphabet* alph(t) by $\{a_1, \ldots, a_n\} \subseteq \Gamma$, and we define $\bar{t}$ by the trace $\overline{a_n} \cdots \overline{a_1}$.

---

[1]   Compare also [9] with its journal version to appear in the Journal of Computer and System Sciences and which will contain this PSPACE-statement, too.

Note that all three notions are well-defined. Hence, the trace monoid $M(\Gamma, I)$ is also equipped with an involution.

We shall use a basic tool from trace theory which is called Levi's Lemma. For a proof see e.g. [5, Thm 1.3.4].

Lemma 1. *Let* $x, y, u, v \in M(\Gamma, I)$ *be traces. Then the following assertions are equivalent:*

(1) We have $xy = yv$.
(2) There are $p, q, r, s \in M(\Gamma, I)$ *such that* $\mathrm{alph}(r) \times \mathrm{alph}(s) \subseteq I$ *and*

$$x = pr, \quad y = sq, \quad u = ps, \quad v = rq.$$

By σ we mean a mapping $\sigma : \Omega \to M(\Gamma, I)$. The mapping is extended to a homomorphism $\sigma : (\Gamma \cup \Omega \cup \overline{\Omega})^* \to M(\Gamma, I)$ by defining $\sigma(a) = a$ for $a \in \Gamma$ and $\sigma(\overline{x}) = \overline{\sigma(x)}$ for $x \in \Omega$.

We say that σ is a *solution* of a system $\{U_1 = V_1, \ldots, U_k = V_k\}$, if $\sigma(U_1) = \sigma(V_1), \ldots, \sigma(U_k) = \sigma(V_k)$ are true in $M(\Gamma, I)$. Since solvability is asked in $M(\Gamma, I)$ we speak henceforth about a system of trace equations with involution.

By [2] we know how to solve such a system in PSPACE, if (Γ, I) is not part of the input. Nevertheless the algorithm in [2] is very complex: It is a quite involved reduction to word equations with involution. It relies on Plandowski's method [6] and on the result in [1]. Here we consider the uniform problem where (Γ, I) is part of the input (and no PSPACE bound is known, in general). But we restrict ourselves to quadratic systems (which are much easier to handle): We obtain a very simple PSPACE-algorithm. The algorithm is an analogue to the classical PSPACE-algorithm for quadratic word equations in [4].

Theorem 1. *The following problem can be solved in* PSPACE.

Input: *A structure* $(\Gamma, {}^-, I)$ *and a quadratic system of trace equations with involution over this structure.*
Question: *Does the system have a solution?*

Proof. After a preprocessing step we may assume that all equations are triangulated, i.e., they have the form $z = xy$ with $x, y, z \in \Gamma \cup \Omega \cup \overline{\Omega}$. Note that if e.g. there were an equation $x = \overline{x}$ for $x \in \Omega$, then this equation was simply removed, since $x = \overline{x}$ is satisfiable and neither x nor $\overline{x}$ appeared elsewhere in the system. Next we guess for each variable $x \in \Omega$ an alphabet $\alpha(x) \subseteq \Gamma$. We let $\alpha(a) = \{a\}$ for $a \in \Gamma$ and $\alpha(\overline{x}) = \overline{\alpha(x)}$ where $\overline{\{a_1, \ldots, a_n\}}$ is the set $\{\overline{a_1}, \ldots, \overline{a_n}\}$. The idea behind $\alpha(x)$ is that for a solution σ we have $\mathrm{alph}(\sigma(x)) = \alpha(x)$ for all $x \in \Gamma \cup \Omega \cup \overline{\Omega}$.

For all equations $z = xy$ we must have $\alpha(z) = \alpha(x) \cup \alpha(y)$. If we write $(r, s) \in I$ we must satisfy $\alpha(r) \times \alpha(s) \subseteq I$. If alphabetic constraints as above cannot be satisfied, we stop the following procedure without success.

In another preprocessing step we remove all variables x with $\alpha(x) = \emptyset$. As a consequence some more variables and equations may disappear. For each equation $z = xy$ we define its weight by the product $|\alpha(x)| \cdot |\alpha(y)|$. The total weight

of the system is the sum of the weights of all equations. We fix this weight as
the space bound for the algorithm. The algorithm performs Levi's Lemma and
removes equations of weight zero as soon as they are created. This procedure
is described by four rules. The first three remove equations of weight zero. We
must apply them as long as possible before we are allowed to apply the fourth
rule which is Levi's Lemma.

Here are four the rules:

Rule 1: If there is an equation $z = xy$ with $z \in \Gamma$, then we must have
$|\alpha(x)| \cdot |\alpha(y)| = 0$ (weight zero) and $|\alpha(x)| + |\alpha(y)| = 1$. We remove the equation
and we replace all occurrences of x (or of y resp.) by the corresponding constant.
(The output is here $x := a$ (or $y := a$ resp.), if $z = a \in \Gamma$. We omit mostly the
discussion of the output in the following.)

Rule 2: We may assume (by Rule 1 and by replacing $\overline{z} = xy$ by $z = \overline{y}\overline{x}$) that
all equations $z = xy$ satisfy $z \in \Omega$. If there is an equation $z = xy$ of weight zero,
then we can remove it. More precisely, if, say $|\alpha(y)| = 0$, then we replace z in the
remaining equations by x, so we have also one variable less. Note that exactly
here we may encounter a situation $z = \overline{z}$, which is a satisfiable constraint if and
only if $\alpha(z)$ is closed under involution. The condition $z = \overline{z}$ has to be respected
when computing a most general solution.

Rule 3: By applying Rules 1 and 2 as long as possible we may assume that
$\alpha(x) \neq \emptyset$ for all remaining variables. If there is an equation $z = xy$ and z appears only once, then we simply remove this equation. (It can be viewed as a
definition $z := xy$ moving to the output.)

Rule 4: After Rule 3 there must be a doubly defined variable or the system is
empty. Otherwise the system is not solvable. So assume we have two equations:

$$z = xy,$$
$$z = uv.$$

We apply Levi's Lemma. Thus we replace these two equations by four new
equations with four new variables:

$$x = pr, \quad y = sq, \quad u = ps, \quad v = rq.$$

We then guess the alphabetical constraints for the new variables p, r, s and q
such that we satisfy (or we have to stop):

$$\alpha(x) = \alpha(p) \cup \alpha(r),$$
$$\alpha(y) = \alpha(s) \cup \alpha(q),$$
$$\alpha(u) = \alpha(p) \cup \alpha(s),$$
$$\alpha(v) = \alpha(r) \cup \alpha(q),$$
$$\alpha(r) \times \alpha(s) \subseteq I.$$

The last condition implies $\alpha(r) \cap \alpha(s) = \emptyset$ and this is enough to verify that the total weight did not increase. In fact, no pair $(a, b) \in \Gamma \times \Gamma$ is counted more than twice in the weight for the new equations and if it is counted twice, then it has been counted twice before. If a pair $(a, b) \in \Gamma \times \Gamma$ is counted exactly once in the weight of the new equations, then it was counted once before. There are also some cases, where the weight actually decreases.

Before we allow to apply Rule 4 again, we have to use Rules 1 to 3 as long as possible since some of the four new equations may have weight zero.

It is clear that the procedure can be implemented in PSPACE. If the empty system is calculated, then the original system is satisfiable.

We now consider the converse: Let us assume that the system is satisfiable. Then there is a solution σ and we may define a length by $|x| = 2|\sigma(x)|$ for $x \in \Omega$ and $|a| = 1$ for a constant $a \in \Gamma$. We may define a length of the system by a pair (m, n) where m the maximal length $|x|$ for some variable in the system and n is sum of the lengths of all right-hand sides of the equations. Then we can apply the rules in such a form that at each step at least one component of the length decreases and no component increases. This proves termination and the number of necessary steps is at most linear in the length of the original system.   □

Remark 1. From the algorithm described above, we can deduce a doubly exponential upper bound for the length of a minimal solution.

Remark 2. Alphabetic constraints as used in the algorithm above are special instances of recognizable constraints. Assume that in addition, the input contains a list of recognizable trace languages $L(x) \subseteq M(\Gamma, I)$ for all $x \in \Omega$ (each specified, say, by some NFA for all representing words of $L(x)$ in Γ^*). A solution has now to respect the recognizable constraints, too. It is possible to extend the PSPACE-algorithm in such a way that it copes with the more general problem as well. Details are left to the reader.

3   Quadratic Word Equations with Length Constraints

In this section, we will describe a linear-time algorithm for solving a system of quadratic word equations with involution. This is essentially the very same algorithm and result as in [8]. The only minor modification is due to the involution.

Theorem 2. *There is a linear time algorithm to solve the following problem (on a unit cost RAM).*

Input: *A quadratic system of word equations E over the structure $(\Gamma, ^-)$ and a list of natural numbers $(\ell_x)_{x \in \Omega}$, written in binary.*

Output: *If there exists a solution σ with $|\sigma(x)| = \ell_x$ for every $x \in \Omega$ then return an equivalent system with no doubly defined variables, else if there is no such solution: Stop.*

Proof. (The proof is nearly identical to the one in [8, Thm. 2]. However we explain the entire algorithm for convenience of the reader.) The first step is to split every equation into equations of the form $z = xy$, $z \in \Omega$, $x, y \in \Gamma \cup \Omega \cup \overline{\Omega}$. After each step we eliminate variables x with $\ell_x = 0$ and verify that $\ell_z = \ell_x + \ell_y$, where $\ell_{\overline{x}} = \ell_x$ for $x \in \Omega$ and $\ell_a = 1$ for $a \in \Gamma$. In addition, we can output a definition for the eliminated variables after each step. If all variables in the system E are at most defined once, i.e. occur at most once on a left-hand side, the whole systems consists of definitions. In this case we can output E and we are done. Let in the following be $c = 0.55$.

Consider a doubly defined variable z and its two equations:

$$z = xy, \tag{1}$$

$$z = uv. \tag{2}$$

If $\ell_x = \ell_u$, then we either have an immediate contradiction or we can eliminate at least one variable. Therefore, without restriction, $0 < \ell_u < \ell_x$.

If $\ell_x - \ell_u > c \cdot \ell_z$, we have to distinguish different cases. First if $x = v$, then u and y are conjugates. Since x has no other occurrences we can replace (1) and (2) by

$$u = rs, \quad y = sr,$$

where r and s are new variables with $\ell_r = \ell_x \pmod{\ell_u}$ and $\ell_s = \ell_u - \ell_r$.

The next case (and new case w.r.t. [8]) is $\overline{x} = v$. We first have to check whether there exists a word s with $s = \overline{s}$ and $|s| = \ell_x - \ell_u$. If $\ell_x - \ell_u$ is even we can always find such an s and if it is odd this is possible if and only if the involution $^-$ has a fixed point, i.e. there is an $a \in \Gamma$ with $a = \overline{a}$. If no such s exists, we have an immediate contradiction, else we have $u = \overline{y}$ since $\ell_u < \ell_x$. So we can remove (1) and (2) and the variables z, x and u from E. Note, that in a solution we then have $x = us = \overline{y}\,\overline{s}$ and $z = us\overline{u}$. In particular, $z = \overline{z}$.

We will next handle the case $\ell_x - \ell_u > c \cdot \ell_z$, $x \neq v$, $\overline{x} \neq v$ and x or v has a second definition. By symmetry we may assume that there is an equation

$$x = pq. \tag{3}$$

If $\ell_p \geq c \cdot \ell_z$ we introduce a new variable r with $\ell_r = \ell_z - \ell_p$ and replace (1) and (3) by

$$z = pr, \quad r = qy.$$

If $\ell_p < c \cdot \ell_z$ we have to distinguish the sub-cases $\ell_u = \ell_p$, $\ell_u < \ell_p$ and $\ell_u > \ell_p$. If $\ell_u = \ell_p$, then again, either there is an immediate contradiction or we can eliminate a variable. The next sub-case is $\ell_u < \ell_p$. We now introduce new variables w and r with $\ell_w = \ell_x - \ell_u$ and $\ell_r = \ell_p - \ell_u$ and replace (1), (2) and (3) by

$$p = ur, \quad w = rq, \quad v = wy.$$

The last of our three sub-cases is $\ell_u > \ell_p$. In analogy to the previous sub-case we introduce new variables w and r with $\ell_w = \ell_x - \ell_u$ and $\ell_r = \ell_u - \ell_p$ and replace (1), (2) and (3) by

$$u = pr, \quad q = rw, \quad v = wy.$$

If $\ell_x - \ell_u \leq c \cdot \ell_z$ or if we are in none of the cases above, i.e. $x \neq v$, $\overline{x} \neq v$ and neither x nor y has a second definition, we replace (1) and (2) by

$$x = uw, \quad v = wy,$$

where w is a new variable with $\ell_w = \ell_x - \ell_u$.

Let $\mathrm{dd}(E)$ be the number of doubly defined variables in E and k be some value such that $k \geq -3/\ln(c) \approx 5.01$. We define the weight of E as follows: $W(E) = |\Omega| + \mathrm{dd}(E) + k \sum_{x \in \Omega} \ln \ell_x$. This weight is linear in the size of the input. The algorithm reduces the weight after each transformation by at least one: In the case $\ell_x - \ell_u > c \cdot \ell_z$ and $\overline{x} = v$, the overall effect is that z, v $(= \overline{x})$ and u $(= \overline{y})$ have been removed and the number of doubly defined variables did not increase, since we eliminated (1) and (2) but u may now be doubly defined. Hence, we have reduced $W(E)$ by at least 3. The calculations for the other cases can be found in [8]. Since a single transformation needs only constant time this yields a linear time algorithm. □

4   Quadratic Trace Equations with Length Constraints

In this section we show that one cannot hope for a polynomial time algorithm to solve quadratic trace equations with length constraints. This happens for a simple matching problem and when underlying trace monoid is generated by three letters only; this number of letters is minimal.

Theorem 3. *Let $M(\Gamma, I) = \{a, b\}^* \times \{c\}^*$. The following problem is NP-hard.*

Input: *A single quadratic trace equation where the right-hand side is a trace, and a list of natural numbers $(\ell_x)_{x \in \Omega}$, where Ω denotes the set of variables.*

Question: *Does the system have a solution σ satisfying $|\sigma(x)| = \ell_x$ for every $x \in \Omega$?*

Proof. In [8] it has been shown that (using a reduction the from $3 - \mathrm{SAT}$) the following matching problem is NP-hard.
Input: A quadratic word equation $L = r$ over the alphabet $\{a, b\}$ with a word $r \in \{a, b\}^*$.
Question: Is there a solution $\sigma : \Omega \rightarrow \{a, b\}^*$?

We will use this problem for a reduction to a quadratic equation over the trace monoid $\{a, b\}^* \times \{c\}^*$.

Let $w_0 x_1 w_1 x_2 w_2 \cdots x_k w_k = r$ be a word equation where the w_i, $0 \leq i \leq k$, and r are words over $\{a, b\}$ and the x_i, $1 \leq i \leq k$, are variables. Without restriction, we can assume $k \geq 1$. An upper bound for every solution of the x_i is $\ell = |r| - \sum_{i=0}^{k} |w_i|$. If ℓ is negative, the above equation has no solution. We transform the equation into the following trace equation over $\{a, b\}^* \times \{c\}^*$:

$$w_0 x_1 w_1 x_2 w_2 \cdots x_k w_k = r c^{(k-1) \cdot \ell}. \tag{4}$$

We set $\ell_{x_i} = \ell$ for $1 \leq i \leq k$.

Clearly, the original system has a solution if and only if (4) has a solution satisfying the length constraints. □

Conclusion

We have seen that quadratic trace equations seem to behave as good (or as bad) as quadratic word equations. However, if lengths constraints are considered then the word case is simpler and we have no analogue to Theorem 2 for traces. We conjecture however that for every trace monoid there is a polynomial time algorithm for solving quadratic trace equations, when, as part of the input, for every variable $x \in \Omega$ and for every letter $a \in \Gamma$ it is specified by some number $\ell_{x,a}$ written in binary how many a occur in the solution for x.

References

1. V. Diekert, C. Gutiérrez, and C. Hagenah. The existential theory of equations with rational constraints in free groups is PSPACE-complete. In A. Ferreira and H. Reichel, editors, *Proc. 18th Annual Symposium on Theoretical Aspects of Computer Science (STACS'01), Dresden (Germany), 2001*, number 2010 in Lecture Notes in Computer Science, pages 170–182. Springer, 2001.
2. V. Diekert and A. Muscholl. Solvability of equations in free partially commutative groups is decidable. In F. Orejas, P. G. Spirakis, and J. van Leeuwen, editors, *Proc. 28th International Colloquium on Automata, Languages and Programming (ICALP'01)*, number 2076 in Lecture Notes in Computer Science, pages 543–554, Berlin-Heidelberg-New York, 2001. Springer.
3. V. Diekert and G. Rozenberg, editors. *The Book of Traces*. World Scientific, Singapore, 1995.
4. Yu. Matiyasevich. A connection between systems of word and length equations and Hilbert's Tenth Problem. *Zap. Nauchn. Sem. Leningrad. Otdel. Mat. Inst. Steklov. (LOMI)*, 8:132–144, 1968. In Russian. English translation in: *Sem. Math. V. A. Steklov, 8*, 61–67, 1970.
5. A. Mazurkiewicz. Introduction to trace theory. In V. Diekert and G. Rozenberg, editors, *The Book of Traces*, chapter 1, pages 3–41. World Scientific, Singapore, 1995.
6. W. Plandowski. Satisfiability of word equations with constants is in PSPACE. In *Proc. 40th Ann. Symp. on Foundations of Computer Science, FOCS'99*, pages 495–500. IEEE Computer Society Press, 1999.
7. W. Plandowski and W. Rytter. Application of Lempel-Ziv encodings to the solution of word equations. In K. G. Larsen et al., editors, *Proc. 25th International Colloquium on Automata, Languages and Programming (ICALP'98), Aalborg (Denmark), 1998*, number 1443 in Lecture Notes in Computer Science, pages 731–742, Berlin-Heidelberg-New York, 1998. Springer.
8. J. M. Robson and V. Diekert. On quadratic word equations. In C. Meinel and S. Tison, editors, *Proc. 16th Annual Symposium on Theoretical Aspects of Computer Science (STACS'99), Trier (Germany), 1999*, number 1563 in Lecture Notes in Computer Science, pages 217–226. Springer-Verlag, 1999.
9. M. Schaefer, E. Sedgwick, and D. Štefankovič. Recognizing string graphs in NP. In *Proceedings 34th Annual ACM Symposium on Theory of Computing, STOC'2002*, pages 1–6, New York, 2002. ACM Press.

Infinite Snake Tiling Problems

Jarkko Kari*

Department of Mathematics, University of Turku
FIN-20014 Turku, Finland
jkari@utu.fi

Abstract. Wang tiles are square tiles with colored edges. We investigate the problems of tiling infinite snakes and cycles where neighboring tiles must have identical colors in their adjacent edges. A snake is a non-overlapping sequence of tiles. We show that it is undecidable if such snakes or cycles are possible using copies of tiles from a given finite tile collection.

1   Introduction

Wang tiles are unit size square tiles with colored edges. Tiles are placed on the two-dimensional infinite grid $\mathbb{Z} \times \mathbb{Z}$. They are oriented and may not be rotated. The tiling rule is that two neighboring tiles must have the same color on the adjacent edges. A tiling system consists of a finite number of different tiles. There is an infinite supply of copies of each tile available. The classic tiling problem by Hao Wang [11] is the decision problem that asks whether a given tiling system admits at least one tiling of the entire plane. R.Berger proved that this question is in general undecidable [3].

In this paper the tiles are not required to cover the whole plane. Rather, we ask whether it is possible to form an infinite *snake* using tiles of a given tiling system. A snake is a non-overlapping sequence of neighboring grid positions. There are two basic variants of the question depending on whether we require neighboring tiles to match even if they are not consecutive in the snake. Both variants are undecidable, as proved in [2]. We outline this proof in Section 3. In Section 2 we discuss various equivalent ways of stating the infinite snake tiling questions. In Section 4 we introduce a new variant where we ask whether it is possible to form a non-trivial cycle, i.e. a closed loop. Using a similar technique we are able to show that this question is undecidable as well.

Snake tiling questions were first introduced by D.Myers in [9] and extensively treated in [5,6]. The main emphasis was in connectivity questions: Given a tiling system and two points, does there exist a snake connecting the given points ? Undecidability of the infinite snake tiling question was correctly conjectured in [6]: "We conjecture that Problem 4.1 [infinite snake tiling problem] is undecidable, but it seems that this would be difficult to prove."

---

* Research supported by NSF Grant CCR 97-33101 and the Academy of Finland Grant 54102

The infinite snake tiling question was reintroduced by L.Adleman in the effort to model *self-assembly* using Wang tiles [1]. Self-assembly refers to the process in which simple objects, e.g. molecules, autonomously come together to form complex structures. Adleman proposed to model this process using Wang tiles. The tiles represent the basic objects, and their colors model the local matching rules that determine how the objects stick to each other. This idea gives rise to several new decision and complexity questions concerning Wang tiles. A basic decision question is to determine if a given set of objects allows uncontrollable growth. In tiling systems this is equivalent to the infinite snake tiling question.

2   Variants of the Infinite Snake Tiling Problem

A *tile* is a 4-tuple
$$t = (c_N, c_E, c_S, c_W) \in C^4$$
where C is a finite set of colors, and the components c_N, c_E, c_S and c_W are the colors of the north, east, south and west edge of tile t. A *tiling system* $T \subseteq C^4$ is a finite set of tiles. Tiles are placed at the integer grid positions, addressed by $\mathbb{Z}^2$. Two positions $(x_1, y_1), (x_2, y_2) \in \mathbb{Z}^2$ are *neighbors* iff $|x_1 - x_2| + |y_1 - y_2| = 1$. Hence each position (x, y) has exactly four neighbors $(x, y+1)$, $(x+1, y)$, $(x, y-1)$ and $(x-1, y)$, referred to as the north, east, south and west neighbor, respectively.

A full *tiling* of the plane is a placement $f : \mathbb{Z}^2 \longrightarrow T$ of tiles on the integer grid positions, and tiling f is *valid* if for each pair of neighboring positions the tiles are such that colors in the adjacent edges match. More precisely, tiling f is valid if for every $(x, y) \in \mathbb{Z}^2$

$$\left.\begin{array}{l} f(x, y) = (c_N, c_E, c_S, c_W) \\ f(x + 1, y) = (d_N, d_E, d_S, d_W) \end{array}\right\} \Longrightarrow c_E = d_W,$$

and

$$\left.\begin{array}{l} f(x, y) = (c_N, c_E, c_S, c_W) \\ f(x, y + 1) = (d_N, d_E, d_S, d_W) \end{array}\right\} \Longrightarrow c_N = d_S.$$

If at least one valid tiling exists then we say that the tiling system T admits a tiling. A classic result by R.Berger states that there is no algorithm to determine if a given tiling system admits a tiling [3]:

Theorem 1. *It is undecidable if a given tiling system T admits a tiling.*   □

Next we define snakes. Let $I = \mathbb{Z}$, $I = \mathbb{N}$ or $I = \{1, 2, \ldots, n\}$ for some $n \geq 1$. A *snake* is an injective function $s : I \longrightarrow \mathbb{Z}^2$ such that for every $i, i + 1 \in I$ positions $s(i)$ and $s(i + 1)$ are neighbors. Positions $s(i)$ and $s(i + 1)$ are called consecutive positions of the snake. Injectivity of s means that the snake may not overlap itself. The snake is called *two-way infinite*, *one-way infinite* or *finite* if $I = \mathbb{Z}$, $I = \mathbb{N}$ or $I = \{1, 2, \ldots, n\}$, respectively. In the last case, n is the length of the snake.

Any subset $P \subseteq \mathbb{Z}^2$ of positions is called a *region* of the plane. The region can be finite or infinite. Region P is *connected* if any two positions in P are connected by a snake contained in P. In particular, if $s : I \longrightarrow \mathbb{Z}^2$ is a snake then its range $s(I)$ is a connected region, called a *snake region*.

A tiling of region P is an assignment $f : P \longrightarrow T$ of tiles into all positions in P. Tiling f is valid if the colors of any two neighboring tiles in P match. If $P = s(I)$ is a snake region then a valid tiling of P is called a *strong tiling* of snake s.

Notice that a strong tiling of a snake requires that the tiles in any two neighboring positions of the snake region match, even if the positions are not consecutive in snake s. In other words, if the snake makes a loop and returns back to touch itself then the colors have to match at the touching edges. In contrast, if only the consecutive positions of a snake are required to match then the tiling is a *weak tiling* of the snake. So a weak tiling of snake s is an assignment $f : s(I) \longrightarrow T$ of tiles such that for every $i, i + 1 \in I$ the adjacent edges of tiles in positions $s(i)$ and $s(i + 1)$ have the same color. Notice that a weak tiling of snake s is not necessarily a valid tiling of region $s(I)$, and that a tiling of a snake region P can be either weakly valid or not weakly valid depending on the ordering of the elements of P in the snake.

Theorem 2. *Let T be a tiling system. The following are equivalent:*

(a) T admits a strong tiling of some two-way infinite snake,
(b) T admits a strong tiling of some one-way infinite snake,
(c) T admits strong tilings of finite snakes of all lengths,
(d) T admits a tiling of some infinite connected region P,

Proof. Implications $(a) \Longrightarrow (b) \Longrightarrow (c)$ are trivial as any two-way infinite snake contains a one-way infinite snake as a subsequence, and similarly every one-way infinite snake contains finite snakes of all lengths. Let us prove $(c) \Longrightarrow (a)$. Assume that for every $n \in \mathbb{Z}$ tiling system T admits a strong tiling of a snake s_n of length $2n + 1$. We can translate the positions of the snakes on the plane without changing their shapes so that the middle position of each snake becomes the origin $(0, 0)$. Let us also re-index the snakes in such a way that instead of $I = \{1, 2, \ldots, 2n + 1\}$ we use $I = \{-n, \ldots, n\}$. Hence the snake s_n is a function $s_n : \{-n, \ldots, n\} \longrightarrow \mathbb{Z}^2$ that satisfies $s_n(0) = (0, 0)$. The sequence $s_1, s_2, \ldots$ of snakes of increasing length has an infinite subsequence that converges into a two-way infinite snake $s : \mathbb{Z} \longrightarrow \mathbb{Z}^2$. More precisely, for every $m \geq 1$ there are infinitely many snakes s_n, $n \geq m$, such that $s_n(i) = s(i)$ for all $-m \leq i \leq m$. We can then extract a strong tiling of s from strong tilings of this converging subsequence.

We still have to prove that (d) is equivalent to (a)–(c). It is clear that $(b) \Longrightarrow (d)$ as any infinite snake region is an infinite connected region. Conversely, it is easy to see that any infinite, connected region P contains a one-way infinite snake region as a subset, so $(d) \Longrightarrow (b)$. $\qquad \square$

Analogously to Theorem 2 we have

Theorem 3. *Let T be a tiling system. The following are equivalent:*

(a) T admits a weak tiling of some two-way infinite snake,
(b) T admits a weak tiling of some one-way infinite snake,
(c) T admits weak tilings of finite snakes of all lengths,   □

In terms of self-assembly we have the following interpretations: Let us say that a tiling system admits *unbounded growth* (under strong/weak matching rule) if it is possible to incrementally grow a tiling of an infinite connected region by adding tiles one by one in such a way that the tiled region remains always connected and

- each new tile must match with all previously placed neighboring tiles (strong matching rule), or
- each new tile must match with at least one previously placed neighboring tile (weak matching rule).

It is straightforward to see that unbounded growth under strong (weak) matching is possible if and only if the tiling system admits a strong (weak, respectively) tiling of some infinite snake.

The infinite snake tiling problems now refer to the following two decision problems: Does a given tiling system T admit a strong/weak tiling of some infinite snake? The snake is not part of the input instance. Both variants are undecidable.

3   Undecidability of the Infinite Snake Tiling Problem

In this Section we sketch the proof of the main result in [2]:

Theorem 4. *It is undecidable if a given tiling system T admits a strong (weak) tiling of an infinite snake.*

Berger's tiling problem will be reduced to the infinite snake tiling problems in two stages. First we prove the undecidability of a directed variant, which is then reduced further to the actual snake tiling question. A *directed tile* is a Wang tile that has a direction N, E, S or W attached to it, representing north, east, south and west direction, respectively. The directions will be shown as horizontal and vertical arrows, and the idea is that the direction will point to the neighboring tile that follows the present tile in the snake. More formally, a directed tile d is an element of $C^4 \times \{N, E, S, W\}$. A *directed tiling system* is a finite set of directed tiles.

A (strong) *directed tiling* of a snake $s : I \longrightarrow \mathbb{Z}^2$ with directed tiling system D is an assignment $f : s(I) \longrightarrow D$ of directed tiles to the snake such that the colors match between any two neighboring positions both in $s(I)$, and for every $i, i + 1 \in I$ the direction of tile $f(s(i))$ points to position $s(i + 1)$. We only need the strong snake tiling variant here – weak directed tiling could be defined analogously.

Lemma 1. [2] *There exists a directed tiling system D_0 such that*

(i) D_0 admits a directed tiling of some one-way infinite snake, and
(ii) every one-way infinite snake s with a valid directed tiling is plane filling, that is, for every n there exists an $n \times n$ square

$$P = \{a+1, a+2, \ldots, a+n\} \times \{b+1, b+2, \ldots, b+n\} \subseteq \mathbb{Z}^2$$

covered by the snake: $P \subseteq s(I)$.   □

We say that D_0 of the Lemma satisfies the *strong plane filling property*. The directed tile set D_0 constructed in [2] is based on the plane filling tiles introduced in [7,8] to prove undecidability results concerning two-dimensional cellular automata. Tile set D_0 is such that any correctly tiled infinite snake is forced to have the shape of the self-similar plane filling curve of Figure 1.

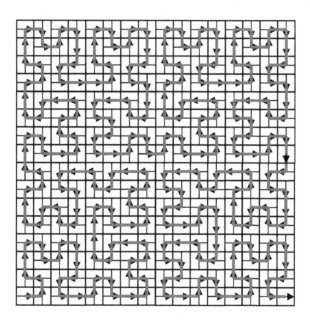

F ig.1. Plane filling snake forced by D_0.

Theorem 5. *It is undecidable if a given directed tiling system admits a directed tiling of some one-way infinite snake.*

Proof. [2] We reduce Berger's undecidable tiling problem to the directed snake tiling question: Let T be an arbitrary (undirected) tiling system, which is the input instance to the tiling problem. Let D_0 be a directed tiling system with the strong plane filling property. Let us construct the directed tiling system $D = T \times D_0$ consisting of "sandwich" tiles: For every $t = (c_1, c_2, c_3, c_4) \in T$ and $d = (e_1, e_2, e_3, e_4, a) \in D_0$ we have the directed tile

$$(t, d) = [(c_1, e_1), (c_2, e_2), (c_3, e_3), (c_4, e_4), a] \in D.$$

This means that two neighboring sandwich tiles match if the colors match in both T and D_0 components, and (t, d) inherits its direction from $d \in D_0$.

Let us show that D admits a directed tiling of a one-way infinite snake if and only if T admits a valid tiling of the full plane. If T correctly tiles the full plane then it also correctly tiles every region, including every snake region. Let us take any snake s that D_0 correctly tiles (property (i) of Lemma 1), and combine the correct tilings according to T and D_0 to form a valid directed tiling of s according to D.

Conversely, assume that D admits a directed tiling of some one-way infinite snake s. Then also the D_0 components of D form a directed tiling of snake s. According to property (ii) of Lemma 1 snake s is plane filling. The T components of D form a valid tiling of $s(I)$ so that T admits a valid tiling of a square of any size n. Therefore T also admits a tiling of the entire plane.

The result now follows from the undecidability of the tiling problem (Theorem 1). □

To prove Theorem 4 we use a *motif construction*. For any given directed tiling system D we construct an undirected tiling system T that admits a tiling of an infinite snake if and only if D admits a directed tiling of an infinite snake. Let n be some large integer. Each directed tile $d \in D$ (called a *macrotile*) will be simulated by a sequence of tiles of T (called *minitiles*) that are forced to form a finite snake (called *motif*) that follows the borders of an $n \times n$ square (see Figure 2).

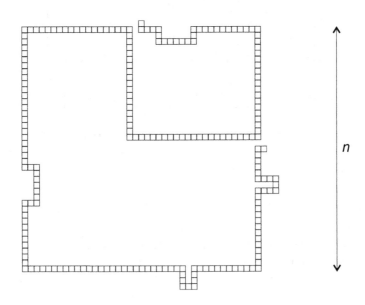

Fig.2. A motif.

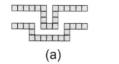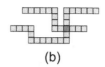

(a)   (b)

Fig.3. A bump and a dent when the colors (a) match, (b) do not match.

On each side of the square the motif forms a bump or a dent. A bump fits inside a dent if they are aligned properly. The south and the east sides of a motif contains a bump and the north and the west sides a dent. The position of the bump and the dent is determined by the color of the corresponding edge in the macrotile. Each color used in D corresponds to a unique location for bumps and dents. As a result, two motifs can be placed side-by-side without overlaps if and only if the colors of the corresponding macrotiles match (see Figure 3).

Each motif has two *free ends* from which it may be linked to another motif. The exit end is in the middle of the side given by the direction of the macrotile, and the entry point is in the middle of any other side of the motif. Hence each macrotile is represented by three different motifs: one for each possible entry direction. The free ends are labeled in such a way that a motif exit may be connected only to the entry of another motif. This can be established by using labels N,E,S and W to indicate direction: label N is used on any exit tile on the north side and any entry tile on the south side of the motifs. The other labels E,S and W are used analogously.

Formation of motifs by minitiles can be forced by using sufficiently many unique colors. Each minitile in every motif is unique, and its colors only fit with the next and the previous minitile in the motif. There is freedom of choice only at the free ends of the motifs where one can choose the next motif to link into the previous motif.

It is easy to see that the minitiles admit a tiling of a one-way infinite snake if and only if D admits a directed tiling of a one-way infinite snake. The minitile snake is obtained from the macrotile snake by replacing each macrotile by its motif. Notice that there is no difference between weak and strong tiling of an infinite snake with the minitiles because the tiles are such that no weakly tiled infinite snake can return to touch itself. This completes the sketch of the proof of Theorem 4.   □

Notice that in the preceeding construction the minitiles were of the specific type that each tile has two sides that do not match with any other tile. Hence each tile uniquely determines the direction where the snake continues. It follows that the infinite snake tiling question is undecidable even when restricted to such special types of tiles.

4 Cycle Tiling Problems

In this section we tile finite snakes that form loops. A *cycle* c of length $n \geq 3$ is a snake of length n whose first and last positions are neighbors. In other words, it is an injective function

$$c : I \longrightarrow \mathbb{Z}^2$$

where $I = \{1, 2, \ldots, n\}$, and for all $i \in \mathbb{Z}$ positions $c(i \bmod n)$ and $c(i+1 \bmod n)$ are neighbors. A *strong tiling* of cycle c is an assignment

$$f : c(I) \longrightarrow T$$

of tiles in the cycle such that tiles in any two neighboring positions $c(i)$ and $c(j)$ match. A *weak tiling* only requires that tiles in consecutive positions $c(i \bmod n)$ and $c(i+1 \bmod n)$ match.

We have the following natural decision question: Is there an algorithm to determine if a given tiling systems admits a valid (weak or strong) tiling of some cycle ? In the following we demonstrate that the problem is undecidable in both its weak and strong variants. Notice that there always exists a semi-algorithm for the positive instances of the cycle tiling problem: There are only a countable number of different cycles and they can be easily enumerated. Each cycle can be tiled in a finite number of different ways so there is a recursive enumeration of all potential tilings of cycles. This is in contrast to the infinite snake tiling problem where the negative instances are semi-decidable: if no infinite snake tiling exists then by Theorem 2 there exists n such that no finite snake of length n can be tiled. For each n there are only a finite number of different snakes of length n to check.

The difference in the direction of recursive enumerability hence seems to indicate that there is no direct reduction between the infinite snake tiling questions and the cycle tiling questions. This difference then suggests that instead of Berger's tiling result we should choose another variant of the tiling problem as the basis of the reductions. This variant is called a *finite tiling problem* and its undecidability is easier to establish than Berger's classic result. In the finite tiling question we are given a tiling system T, and one of the tiles $b \in T$ is selected as a *blank* tile. All four edges of the blank tile must be colored with the same color, that is, $b = (c, c, c, c)$ for some $c \in C$. A tiling $f : \mathbb{Z}^2 \longrightarrow T$ is called finite if there are only a finite number of positions that contain a non-blank tile, that is, $\{(x, y) \mid f(x, y) \neq b\}$ is finite. Tiling $f(x, y) = b$ for all $(x, y) \in \mathbb{Z}^2$ is called the trivial tiling. The following undecidability result concerning finite tilings was introduced in [8] in order to to show that is is undecidable if a given two-dimensional cellular automaton is surjective:

Theorem 6. *It is undecidable if a given tiling system with a blank tile admits a non-trivial finite tiling of the plane.* □

Notice – in contrast to the classic tiling problem – positive instances of the finite tiling problem are recursively enumerable.

As in Section 3 we start with a directed variant of the cycle tiling problem. A (strong) *directed tiling* of cycle c by a directed tiling system D is an assignment

$$f : c(I) \longrightarrow D$$

such that (i) the colors of the tiles in any two neighboring positions $c(i)$ and $c(j)$ match, and (ii) for every $i \in \mathbb{Z}$ the direction of the tile in position $c(i \bmod n)$ points to its neighbor in position $c(i+1 \bmod n)$.

Theorem 7. *It is undecidable if a given directed tiling systems admits a directed tiling of some cycle.*

Proof. We start by constructing a specific directed tiling system D_1. This system plays the same role in the cycle tiling problem as the directed tiling system D_0 of Lemma 1 played in the infinite snake tiling problem. It would be possible to modify D_0 for this purpose (as was done in connection to the cellular automata problems in [8]), but B.Durand has provided a simpler tiling system [4] that can be adapted to our purpose.

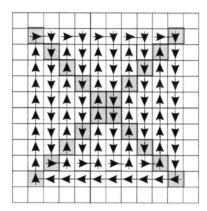

F ig . 4 . Basic cycle of size 10×10.

Consider the square region of tiles with arrows shown in Figure 4. The arrows specify a cycle through a square region of size 10×10. We call this a basic cycle of size 10×10. Analogously we define basic cycles through squares of any even size $2m \times 2m$. Tiles on the diagonals are painted gray. Directed tiles $d \in D_1$ are all different 3×3 blocks of tiles that appear in such basic cycles. The direction of d is given by the arrow in the center position of the 3×3 block. The colors of the edges are the overlap regions between 3×3 blocks centered in neighboring positions (see Figure 5).

With a straightforward case analysis one can verify that tiling system D_1 admits a directed tiling of cycle c if and only if c is a basic cycle.

Let T be any given (undirected) tiling system with a blank tile, which is an input instance to the finite tiling problem. Let us construct sandwich tiles $D \subseteq T \times D_1$, where D contains all elements $(t, d) \in T \times D_1$ with the following restrictions:

(i) if d is a tile on the border of the basic cycle (which is detected by the fact that the 3×3 block it represents contains a tile without an arrow) then t must be the blank tile, and

(ii) if d is the tile in the center of the basic cycle (which is detected by the presence of 2×2 grey tiles in the lower left corner of the 3×3 block) then t must be non-blank.

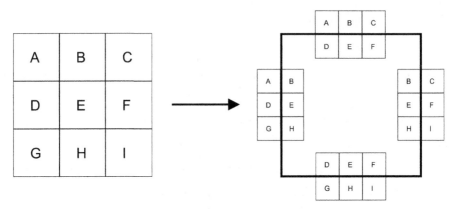

Fig.5. A tile corresponding to a 3 × 3 block.

Then D admits a directed tiling of a cycle if and only if T admits a finite, non-trivial tiling. Namely, if T admits a finite, non-trivial tiling then we can combine such a tiling with a basic cycle formed by D_1 into a directed tiling of the cycle with the sandwich tiles in D. Restrictions (i) and (ii) can be satisfied by choosing the basic cycle sufficiently large so that a non-blank tile can be placed in the center with blank tiles on the border of the square.

Conversely, if D admits a directed tiling of a cycle then the D_1 components force this cycle to be a basic cycle. Then the tiling of this basic cycle with the T components, surrounded by a blank background, is a finite, non-trivial tiling with T. Restrictions (i) and (ii) above guarantee that the tiling is finite and non-trivial, respectively.

The undecidability of directed cycle tiling problem now follows from Theorem 6. ◻

With a motif construction as in Section 3 we can now prove the undirected variant undecidable. Given a directed tiling system D we construct motifs for all $d \in D$. The minitiles are then undirected tiles that admit a tiling of a cycle (in the weak and in the strong sense) if and only if D admits a directed tiling of a cycle. We have

Theorem 8. *It is undecidable if a given tiling system admits a strong/weak tiling of some cycle.* ◻

5 Discussion

We have proved that there are no algorithms to determine if a given finite tiling system can tile some infinite snake or some cycle. The problems are undecidable if only consecutive tiles are required to match, or if all neighboring tiles that belong to the snake or the cycle are required to match. The proofs are based on particular tiling systems that force the snake to be plane filling or, in the case of cycles, to loop through a square. It is interesting to observe a similarity with the undecidability proofs for the injectivity and the surjectivity problems of two-dimensional cellular automata [7,8]. This may suggest the existence of simple reductions between these problems and the snake and cycle tiling problems.

References

1. L. Adleman. Towards a mathematical theory of self-assembly. *Technical Report* 00-722 (2000), Department of Computer Science, University of Southern California.
2. L.Adleman, J.Kari, L.Kari, D.Reishus. On the decidability of self-asssembly of infinite ribbons. *Proceedings of FOCS'2002, 43rd Annual Symposium on Foundations of Computer Science* (2002), 530–537.
3. R.Berger. The undecidability of the domino problem. *Mem.Amer.Math.Soc.* 66 (1966).
4. B.Durand. The surjectivity problem for 2D Cellular Automata. *Journal of Computer and System Sciences* 49 (1994), 149–182.
5. Y.Etzion. On the solvability of domino snake problems. *MSc. Thesis.* Dept. of Applied Math. and Computer Science, The Weizmann Institute of Science, Rehovot, Israel, 1991.
6. Y.Etzion-Petruschka, D.Harel, D.Myers. On the solvability of domino snake problems. *Theoretical Computer Science* 131 (1994), 243–269.
7. J.Kari. Reversibility of 2D cellular automata is undecidable. *Physica D* 45 (1990), 379–385.
8. J.Kari. Reversibility and surjectivity problems of cellular automata. *Journal of Computer and System Sciences* 48 (1994), 149–182.
9. D.Myers. Decidability of the tiling connectivity problem. Abstract 79T-E42, *Notices Amer. Math. Soc.* 195 (26)(1979), A-441.
10. R.M.Robinson. Undecidability and nonperiodicity for tilings of the plane. *Inven.Math.* 12 (1971), 177–209.
11. H.Wang. Proving theorems by pattern recognition. II. *Bell Systems Technical Journal* 40 (1961), 1–42.

Decision Problems
for Linear and Circular Splicing Systems*

Paola Bonizzoni[1], Clelia De Felice[2], Giancarlo Mauri[1], and Rosalba Zizza[2]

[1] Dipartimento di Informatica Sistemistica e Comunicazione
Università degli Studi di Milano - Bicocca
Via Bicocca degli Arcimboldi 8, 20126 Milano - Italy
{bonizzoni,mauri}@disco.unimib.it
[2] Dipartimento di Informatica ed Applicazioni,
Università di Salerno, 84081 Baronissi (SA), Italy
{defelice,zizza}@unisa.it

Abstract. We will consider here the *splicing systems*, generative devices inspired by cut and paste phenomena on DNA molecules under the action of restriction and ligase enzymes. A DNA strand can be viewed as a string over a four letter alphabet (the four deoxyribonucleotides), therefore we can model DNA computation within the framework of formal language theory. In spite of a vast literature on splicing systems, briefly surveyed here, a few problems related to their computational power are still open. We intend to evidence how classical techniques and concepts in automata theory are a legitimate tool for investigating some of these problems.

1 Introduction

Inspired by a recombinant mechanism among DNA molecules and proteins, in 1987 Tom Head introduced the notion of *splicing system*. In nature, splicing occurs in two steps. First, restriction enzymes (proteins) recognize a specific pattern in a (linear or circular) DNA sequence and cut the molecule in a point inside the pattern, specific for each enzyme. Then ligase enzymes paste together properly matched fragments, under specific chemical conditions. When this operation is performed on two linear DNA molecules, two new linear DNA sequences are produced. When this operation is performed on two circular DNA molecules, both molecules are opened inside the pattern and then ligase enzymes paste the opened molecules together, producing a new circular DNA sequence (see [19,24,38] for further details). This phenomenon can be easily translated into a formal language framework. A DNA strand can be considered as a word on a finite alphabet and as DNA occurs in both linear and circular form, two kinds of splicing operations (linear and circular splicing) were defined [19,24]. Different variants of the original definition have been proposed briefly. In general, a linear

---

* Partially supported by MIUR Project *"Linguaggi Formali e Automi: teoria ed applicazioni"* and by the contribution of EU Commission under The Fifth Framework Programme, project MolCoNet IST-2001-32008.

(resp. circular) splicing system or H-system is a triple $S = (A, I, R)$, where A is a finite alphabet, I is the initial (resp. circular) language over A and $R \subseteq (A')^*$, $A \subseteq A'$, is the set of rules for the linear (resp. circular) splicing operation (see Sections 3, 4 for the definitions). The formal language generated by the splicing system is the smallest language containing I and closed under the splicing operation. There are at least three definitions of linear (resp. circular) splicing systems, given by Head, Paun and Pixton respectively and the computational power of these systems is different in the three cases [34,37,38,42]. Furthermore, this computational power also depends on which level in the Chomsky hierarchy the initial set I and the set of the rules R belong to, as shown in [24,38]. According to some hypotheses on I, R, linear splicing systems can reach the same power of the Turing machines [24,34], while when we restrict ourselves to finite linear splicing systems (splicing systems with both R and I finite sets), we get a proper subclass of regular languages [11,15,30,42]. In the same framework, other models have been introduced and proved to be computationally equivalent to Turing machines [14,24,28,35,36].

The idea of using biological processes to compute is the basis of Adleman's experiment, performed some years after the introduction of splicing systems and showing how to solve the (NP-complete) Hamiltonian Path Problem by manipulating DNA sequences in a lab [1]. This research represents part of the general trend towards the proposal of new (biological and quantistic) models of computation. These computational models do not affect the validity of the Church-Turing thesis, but they rather try to reduce time and space required for the solutions of the so-called intractable problems. Nowadays, developments in both directions have been achieved. New theoretical models (such as Watson-Crick Automata, Insertion and Deletion Systems, Sticker systems, P-systems) have been presented [38] and new experiments in the lab have been performed, thus solving other NP-complete problems [22,25,27,38].

Considering that linear splicing has already been investigated thoroughly, we will focus here on its relationships with regular languages. It is known that this class of languages coincides with the class of the languages generated by splicing systems with a regular initial set of words and a finite set of rules, whereas finite linear splicing systems generate a proper subclass of regular languages [42]. The search for a characterization of the subclass of regular languages generated by finite linear splicing systems is one of the open problems proposed by Head. Partial results are known on this problem and these results show that classical notions in automata theory can be useful in solving this problem, investigated in Section 3.3.

We will then consider circular splicing systems limited to the aspect of their relationships with regular languages. Obviously a molecule of circular DNA will be represented as a circular word, i.e., as an equivalence class with respect to the conjugacy relation $\sim$, defined by $xy \sim yx$, for $x, y \in A^*$ [31]. A circular language is a set of circular words and we will notice that looking at regular circular languages is the same as looking at regular languages closed under the conjugacy relation. Except for one special case considered in [3,43], the structure of the latter languages is still unknown and so the search for a characterization of the structure of regular languages closed under the conjugacy relation is one of the interesting problems in this framework (see Section 5). Few results are known on the computational power of circular splicing systems and one important result states that with a regular initial set and finite set of rules, under additional hypotheses (reflexivity, symmetry) on the rules and adding self-splicing, the language generated is regular [42]. Even in the finite case (I, R finite sets), the computational power of circular splicing systems is unknown and both regular and non regular languages may be generated by finite circular splicing systems. Not all regular circular languages can be generated, namely $\sim((A^2)^* \cup (A^3)^*))$ cannot be generated by finite circular splicing systems [6].

A particular class of regular circular languages generated by finite circular splicing systems is provided by the so-called star languages $\sim X^*$ [4], which include free monoids X^* generated by regular group codes X (see Section 5.1 for further details). Languages over a one-letter alphabet generated by finite (Paun) circular splicing systems have also been characterized. They are exactly the regular languages of the form $L_1 \cup \{a^g \mid g \in G\}^+$, where L_1 is a finite subset of a^*, all the words in L_1 are shorter than the words in $\{a^g \mid g \in G\}^+$, G is a set of representatives of the elements in G', G' being a subgroup of Z_n and n being a positive integer. A characterization of these languages in terms of automata also exists, proving that we can decide whether a regular language over a one-letter alphabet can be generated by finite (Paun) circular splicing systems [5,6].

In this paper basics on words and languages are gathered in Section 2; Section 3 is devoted to linear splicing and Section 3.2 presents the general framework and a short description of a few results obtained on the relationships between splicing systems and regular languages.

In Section 3.3 we have limited our study to the case of regular languages generated by finite linear splicing systems. The last two sections concern circular splicing. Definitions are given in Section 4 and the results on the relationships between finite circular splicing systems and regular circular languages are presented in Section 5.

2   Basics

2.1   Words

Let A^* be the free monoid over a finite alphabet A and let $A^+ = A^* \setminus 1$, where 1 is the empty word. For a word (or string) $w \in A^*$, $|w|$ is the length of w and, for a subset X of A^*, we denote $|X|$ the cardinality of X.

In the following $\mathcal{A} = (Q, A, \delta, q_0, F)$ will be a finite state automaton, where Q is a finite set of states, $q_0 \in Q$ is the initial state, $F \subseteq Q$ is the set of final states and δ is the transition function [26,39]. $\mathcal{A}$ is *trim* if each state is accessible and coaccessible, i.e., if for each state $q \in Q$ there exist $x, y \in A^*$ such that $\delta(q_0, x) = q$ and $\delta(q, y) \in F$. Furthermore, $\mathcal{A}$ is *deterministic* if, for each $q \in Q$, $a \in A$, there is at most one state $q' \in Q$ so that $\delta(q, a) = q'$. Given a regular language $L \subseteq A^*$, it is well known that there is a *minimal* finite state automaton $\mathcal{A}$ recognizing it, i.e., $L = L(\mathcal{A})$. The *syntactic monoid* associated with L is the quotient monoid $\mathcal{M}(L)$ with respect to the *syntactic congruence* $\equiv_L$ [39]. Let us recall that two words w, w' are equivalent with respect to the *syntactic congruence* if they have the same set of *contexts* $C(w) = \{(x, y) \in A^* \times A^* \mid xwy \in L\}$, i.e., $w \equiv_L w' \Leftrightarrow [\forall x, y \in A^*, xwy \in L \Leftrightarrow xw'y \in L] \Leftrightarrow C(w) = C(w')$. In the following, $[w]$ will denote the congruence class of w modulo $\equiv_L$. If L is a regular language then the index (i.e., the number of congruence classes) of the syntactic congruence is finite and therefore $\mathcal{M}(L)$ is a finite monoid [32].

Let us recall the definition of a constant. Let $\mathcal{A}$ be the minimal finite state automaton recognizing a regular language L and let $w \in A^*$. We will set $Q_w(\mathcal{A}) = \{q \in Q \mid \delta(q, w) \ is \ defined \}$, simply indicated Q_w when the context makes the meaning evident. Observe that the notation can be extended to $Q_{[x]}$ with $x \in A^*$. The *left* and *right* contexts of a word $w \in A^*$ are therefore defined as follows: $C_L(w) = \{z \in A^* \mid \exists q \in Q_w : \delta(q_0, z) = q\}$, $C_{R,q}(w) = \{y \in A^* \mid \delta(q, wy) \in F\}$ and $C_R(w) = \bigcup_{q \in Q_w} C_{R,q}(w)$. Notice that these definitions are slightly different from the ones given in [39]. A word $w \in A^*$ is a *constant* for a regular language L if $C(w) = C_L(w) \times C_R(w)$ [44]. Obviously, if w is a constant for L and $w' \in A^*$ with $w \equiv_L w'$, then w' is also a constant for L.

We will now give a definition for handling of the labels c (cycles) of some special closed paths in the transition diagram of a finite state automaton $\mathcal{A}$ [5,6]. Indeed, if a rational infinite language L is generated by splicing, we must exhibit a finite number of rules which are able to generate words with an unbounded number of occurrences of cycles as factors. We recall that $x \in A^*$ is a *factor* of a word $w \in A^*$ if there are words $u, u' \in A^*$ such that $w = uxu'$.

Definition 1 (Cycle). *[5] A word $c \in A^+$ is a* cycle *(resp. simple cycle) in $\mathcal{A}$ if $q \in Q$ exists such that $\delta(q, c) = q$ and the internal states crossed by the transition are different from q (resp. from each other and with respect to q). In addition, if c is a cycle in $\mathcal{A}$ and c is not simple, then there are $u_1, \ldots, u_{k+1} \in A^*$, cycles $c_1, \ldots, c_k \in A^+$ and positive integers $p_1, \ldots, p_k$ such that $c = u_1 c_1^{p_1} u_2 \cdots u_k c_k^{p_k} u_{k+1}$, where each u_i is the label of a path in which no state is crossed twice and $u_1, u_{k+1} \neq 1$. Furthermore, if $i, j \in \{1, \ldots, k\}$ exist such that $c_i = c_j$, $j \geq i + 2$ and $c_i \neq c_{i'}$, for $i < i' < j$, then $u_{i+1} \cdots u_j \neq 1$.*

Finally, we will name $FIN, REG, LIN, CF, CS, RE$ the classes of finite, regular, linear, context-free, context sensitive, recursive enumerable languages, respectively.

2.2 Circular Words

Circular words have already been exhaustively examined in formal language theory (see [2,10,31]). For a given word $w \in A^*$ the circular word $\tilde{w}$ is the equivalence class of w with respect to the *conjugacy* relation $\sim$ defined by $xy \sim yx$, for $x, y \in A^*$. The length $|\tilde{w}|$ of a circular word will be defined as the length $|w|$, for any representative w of $\tilde{w}$. When the context does not make it ambiguous, we will use the notation w for a circular word $\tilde{w}$. Furthermore,

$^\sim A^*$ is the set of all circular words over A, i.e., the quotient of A^* with respect to $\sim$. If $L \subseteq A^*$, $^\sim L = \{^\sim w \mid w \in L\}$ is the *circularization* of L, i.e., the set of all circular words corresponding to the elements of L, while every language L such as $^\sim L = C$, for a given circular language $C \subseteq {}^\sim A^*$, is called a *linearization* of C. The set $Lin(^\sim L) = \{w' \in A^* \mid \exists w \in L.\ w' \sim w\}$ of all the strings in A^* corresponding to the elements of $^\sim L$ is the *full linearization* of $^\sim L$.

If FA is a family of languages in the Chomsky hierarchy, $FA^\sim$ is the set of all those circular languages C which have some linearization in FA. In this paper we only deal with $REG^\sim$. It is known that given a regular language $L \subseteq A^*$, its circularization $^\sim L$ is a regular circular language if and only if its full linearization $Lin(^\sim L)$ is a regular language [26]. Consider that circular splicing deals with circular strings and circular languages, i.e., with formal languages which are closed under the conjugacy relation, acting on their circularization. The result is that handling of a regular circular language is the same as handling of a regular language closed under the conjugacy relation (see Section 5).

3 Linear Splicing

3.1 Linear Splicing Systems

Below there are three definitions of linear splicing systems, respectively by Head, Paun and Pixton. The difference between the three definitions depends on the biological phenomena that they want to model, but these biological motivations will not be discussed here.

`Head's definition [19]`. A *Head splicing system* is a 4-uple $S_H = (A, I, B, C)$, where $I \subset A^*$ is a finite set of strings, called *initial language*, B and C are finite sets of triples (α, μ, β), called *patterns*, with $\alpha, \beta, \mu \in A^*$ and μ called the *crossing* of the triple. Given two words $u\alpha\mu\beta v$, $p\alpha'\mu\beta'q \in A^*$ and two patterns (α, μ, β) and (α', μ, β') that have the same crossing and are both in B or both in C, the splicing operation produces $u\alpha\mu\beta'q$ and $p\alpha'\mu\beta v$ (we also say that $\alpha\mu\beta$, $\alpha'\mu\beta'$ are *sites* of splicing).

`Paun's definition [34]`. A *Paun splicing system* is a triple $S_{PA} = (A, I, R)$, where $I \subset A^*$ is a set of strings, called *initial language*, R is a set of *rules* $r = u_1|u_2\$u_3|u_4$, with $u_i \in A^*, i = 1, 2, 3, 4$ and $|, \$ \notin A$. Given two words $x = x_1u_1u_2x_2, y = y_1u_3u_4y_2, x_1, x_2, y_1, y_2 \in A^*$ and the rule $r = u_1|u_2\$u_3|u_4$, the splicing operation produces $w = x_1u_1u_4y_2$ and $w' = y_1u_3u_2x_2$.

`Pixton's definition [42]`. A *Pixton splicing system* is a triple $S_{PI} = (A, I, R)$, where $I \subset A^*$ is a set of strings, called *initial language*, R is a finite collection of rules $r = (\alpha, \alpha'; \beta)$, $\alpha, \alpha', \beta \in A^*$. Given two words $x = \epsilon\alpha\eta$, $y = \epsilon'\alpha'\eta'$ and the rule $r = (\alpha, \alpha'; \beta)$, the splicing operation produces $w = \epsilon\beta\eta'$ and $w' = \epsilon'\beta\eta$.

Given a splicing system S_X, with $X \in \{PA, PI\}$, $(x, y) \vdash_r (w', w'')$ indicates that the two strings w', w'' are produced from (or spliced by) x, y when a rule r is applied. Let $L \subseteq A^*$ and let S_X be a splicing system. We denote

$$\sigma'(L) = \{w, w' \in A^* \mid (x, y) \vdash_r (w, w'),\ x, y \in L, r \in R\}$$

We can analogously define $\sigma'(L)$ for Head's splicing systems S_H by substituting R with the sets B, C of triples and r with two patterns. Thus, we also denote $\sigma^0(L) = L$, $\sigma^{i+1}(L) = \sigma^i(L) \cup \sigma'(\sigma^i(L))$, $i \geq 0$, and

$$\sigma^*(L) = \bigcup_{i \geq 0} \sigma^i(L)$$

Definition 2 (Linear splicing language). *Given a splicing system S_X, with initial language I and $X \in \{H, PA, PI\}$, the language $L(S_X) = \sigma^*(I)$ is the language generated by S_X. A language L is S_X generated (or L is a linear splicing language) if a splicing system S_X exists such that $L = L(S_X)$.*

The problem of comparing the computational power of the three types of splicing operations is given here below.

Problem 1. Given $L = L(S_X)$, with $X \in \{H, PA, PI\}$, does S_Y exist, with $Y \in \{H, PA, PI\} \setminus X$, such that $L = L(S_Y)$?

Problem 1 has been solved for finite linear splicing systems (recall that, in this case, I, R, B, C are finite sets): the computational power increases when we substitute Head's systems with Paun's systems and Paun's systems with Pixton's systems. This result has been demonstrated in [9], together with examples of regular languages separating the above mentioned classes of splicing languages.

3.2 Computational Power of H-Systems

In this section we will briefly describe the well-known results on the computational power of (iterated) linear splicing systems. Our study will consider having an unlimited number of copies of each word in the set, so that a pair of strings (x, y) can generate more than one pair of words with the use of different rules. We will consider words which model single stranded structures of DNA instead of double stranded structures. A model for the latter could be domino languages, as already used in [11]. It would be more realistic to consider the bidimensional structure of DNA, but this is too difficult. Furthermore, as observed in [19,36], there is no loss of information if the double stranded DNA molecule is identified with a single stranded DNA sequence.

Let F_1, $F_2 \in \{FIN, REG, LIN, CF, CS, RE\}$ and denote $H(F_1, F_2) = \{L(S_{PA}) \subseteq A^* \mid S_{PA} = (A, I, R)$ with $I \in F_1, R \in F_2\}$ the class of languages generated by Paun splicing systems, where the initial language I belongs to F_1 and the set of rules R belongs to F_2. In a similar way, we can give analogous definitions with respect to Head's or Pixton's splicing systems, but they will be not used in the next part of this paper. In [24] a table is reported which collects the results, obtained in separate papers, on the level of the Chomsky hierarchy $H(F_1, F_2)$ belongs to. As a consequence, these results allows us to answer some classical decision problems regarding computational models, namely whether or not emptyness, membership and equivalence are decidable or undecidable for a given language generated by a splicing system. For example, it is obvious that the above mentioned questions are decidable for splicing systems which are equivalent to finite state automata. We quote that $H(REG, FIN) = REG$ for Paun's splicing systems. The same identity for Pixton's splicing systems is stated below and was presented for the first time in [42].

Theorem 1. *[42] (Regularity preserving theorem) Given a Pixton splicing system $S_{PI} = (A, I, R)$, where $I \in REG$ and $R \in FIN$, we have $L(S_{PI}) \in REG$.*

Theorem 1 has been proved after partial results on the characterization of the subclass of the splicing systems which are equivalent to finite state automata. These partial results can be found in [11,13,19,30,40]. In some of these papers, many efforts have been made to effectively construct a finite state automaton associated to a given splicing system. In particular, Head and Gatterdam studied a specific type of finite Head splicing systems (persistent in [19]; persistent, reduced and crossing disjoint in [15]), whereas non persistent Head splicing systems have been studied by Culik and Harju [11], via domino languages. For Head's splicing systems S_H with regular initial language and a finite set of triples, Kim showed the regularity of $L(S_H)$ by designing an algorithm which constructs a finite state automaton recognizing the splicing language generated by S_H [30]. An extension of Theorem 1 for both Paun's and Pixton's definitions to full AFL is given by Pixton in Theorem 2. Observe that REG, CF and RE are full AFL.

Theorem 2. *[24,40,41] If FA is a full AFL, then $H(FA, FIN) \subseteq FA$.*

Finally, in [8] the authors show that REG coincides with finite marked splicing systems (systems which have a different definition of the splicing operation and with "marked rules").

3.3   Finite Linear Splicing Systems

The already known results stress that $H(FIN, FIN)$ is strictly intermediate between FIN and REG [24]. Infact, $(aa)^*b$ is a regular language which belongs to $H(FIN, FIN)$ (Example 1). Furthermore, no finite linear splicing system can generate $(aa)^*$. This result has been proved for Head's systems in [15] and it is easy to state it for Pixton's systems. Now two problems arise and, as we have already said, Problem 3 is the special question we will focus on in this paper.

Problem 2. Given $L \in REG$, can we decide whether $L \in H(FIN, FIN)$?

Problem 3. Characterize regular languages which belong to $H(FIN, FIN)$.

Kim has faced Problem 2 for Head's systems designing an algorithm that, for a given regular language L and a given finite set of triples X, allows us to decide whether a language $I \subseteq L$ exists such that L is generated by $S_H = (A, I, X)$ (Kim's algorithm does not need to separate the triples into the two disjoint sets B and C) [29].

Another partial result has been provided by Head in [20]. In that paper, the author considers the generating power of the (Paun) finite splicing systems with a *one-sided context*, i.e., each rule has the form $u|1\$v|1$ (resp. $1|u\$1|v$). It is also supposed that for each rule having the above form, rules $u|1\$u|1$, $v|1\$v|1$ (resp. $1|u\$1|u$, $1|v\$1|v$) are also in R (reflexive hypothesis). Head proved that we can decide whether a regular language is generated by a subclass of one-sided context splicing systems. Very recently, Goode and Pixton have shown that we can decide whether a regular language is generated by a finite Paun linear splicing system [16,23]. Their characterization uses the notions of constant and syntactic monoid (see Section 2.1).

As far as Problem 3 is concerned, efforts have been made to compare the class of languages generated by finite splicing systems with already known subclasses of regular languages (we refer to [32,39] for the definitions of these subclasses). In [19,21] Head has given different characterizations of the family of SLT (strictly locally testable) languages [12], by using also results in [33], generalized in [17].

In [6] the authors introduced a family of languages, called *marker languages*, and showed that they are generated by (Paun) finite linear splicing systems. Let us sketch their definition that uses the above mentioned classical notions of syntactic congruence and constants. Consider a regular language L and the minimal finite state automaton $\mathcal{A}$ recognizing it. Suppose that a word w and a syntactic congruence class $[x]$ exist such that in the transition diagram of $\mathcal{A}$ we find a unique path π, starting from $q \in Q$, with label wx. Also suppose that x is the label of a cycle. In addition, when we consider the transition graph of the classical automaton recognizing the reverse of L, the part of this transition graph starting from q is deterministic. Thus, $w[x]$ is a *marker*. Let q_0, q_f be the initial state and a final state in $\mathcal{A}$, respectively. Then we can generate, through a finite linear splicing system, the language $L(w[x])$ of all the words in L which are labels of paths going from q_0 to q_f and having wx' as a factor, with x' in $[x]$.

Theorem 3. *[6] Let L be a regular language and let $w[x]$ be a marker for L. Then there is a finite splicing system $S_{PA} = (A, I, R)$ such that $L(w[x]) = \{z \in L \mid z = z_1 m z_2, m \in w[x]\} = L(S_{PA})$.*

Example 1. The regular language $L = (aa)^*b$ is a marker language (with marker b). L is generated by $S_{PA} = (A, I, R)$, where $I = \{b, aab\}$ and $R = \{1 \mid aab\$1 \mid b\}$.

The construction of a splicing system generating $L(w[x])$, is indicated in the proof of Theorem 3. This construction is obtained thanks to a relationship between syntactic congruence and linear splicing. A generalization of Theorem 3 will be discussed in a future paper [7].

4   Circular Splicing

4.1   Models

As for linear splicing, there are at least three basic definitions for circular splicing; many variants are also known, some of which are obtained by adding new hypotheses on R.

Head's definition [46]. A *Head circular splicing system* is a 4-uple $SC_H = (A, I, T, P)$, where $I \subseteq {}^\sim A^*$ is the *initial* circular language, $T \subseteq A^* \times A^* \times A^*$ and P is a binary relation on T, such that, if $(p, x, q), (u, y, v) \in T$ and $(p, x, q)P(u, y, v)$ then $x = y$. Thus, given ${}^\sim hpxq, {}^\sim kuxv \in {}^\sim A^*$ with $(p, x, q)P(u, x, v)$, the splicing operation produces ${}^\sim hpxvkuxq$.

Paun's definition [24,40]. A *Paun circular splicing system* is a triple $SC_{PA} = (A, I, R)$, where $I \subseteq {}^\sim A^*$ is the *initial* circular language and $R \subseteq A^*|A^* \$A^*|A^*$, with $|, \$ \notin A$, is the set of rules. Then, given a rule $r = u_1|u_2\$u_3|u_4$ and two circular words ${}^\sim hu_1u_2, {}^\sim ku_3u_4$, the rule cuts and linearizes the two strings obtaining u_2hu_1 and u_4ku_3, and pastes and circularizes them obtaining ${}^\sim u_2hu_1u_4ku_3$.

Pixton's definition [42]. A *Pixton circular splicing system* is a triple $SC_{PI} = (A, I, R)$ where A is a finite alphabet, $I \subseteq {}^\sim A^*$ is the *initial* circular language, $R \subseteq A^* \times A^* \times A^*$ is the set of rules. R is such that if $r = (\alpha, \alpha'; \beta) \in R$ then there exists β' such that $\bar{r} = (\alpha', \alpha; \beta') \in R$. Thus, given two circular words ${}^\sim \alpha\epsilon, {}^\sim \alpha'\epsilon'$, the two rules $r, \bar{r}$ cut and linearize the two strings, obtaining $\epsilon\alpha, \epsilon'\alpha'$ and then paste, substitute and circularize them, producing ${}^\sim \epsilon\beta\epsilon'\beta'$.

As for the linear case, we suppose we have an unlimited number of copies of each word in the set, so that a pair of words (x, y) can generate more than one pair of words by applying different rules. Usually T and R are finite sets in the three definitions of the circular splicing indicated above. This paper is limited to examining finite (resp. regular) circular splicing systems according to Paun's and Pixton's definitions, without additional hypotheses (even without self-splicing, which is usually incorporated in Paun's definition). We recall that a finite (resp. regular) circular splicing system is a circular splicing system with a finite (resp. regular) initial circular language.

Given a circular splicing system SC_X, with $X \in \{PA, PI\}$, $(w', w'') \vdash_r z$ indicates that z is spliced by w', w'', through the use of rule r, with respect to one of the two definitions. It is clear that in Pixton's systems, the splicing operation is the combined action of a pair of rules, r and $\bar{r}$, not necessarily different. Given a circular splicing system SC_X, with $X \in \{PA, PI\}$, and a language $C \subseteq {}^\sim A^*$, we denote $\sigma'(C) = \{z \in {}^\sim A^* \mid \exists w', w'' \in C, \exists r \in R. (w', w'') \vdash_r z\}$. We can analogously define $\sigma'(C)$ for Head's circular splicing systems SC_H by substituting R with the set T of triples and r with two triples. Thus, we define $\sigma^*(C)$ as in the linear case.

Definition 3 (Circular splicing language). *Given a splicing system SC_X, with initial language $I \subseteq {}^\sim A^*$ and $X \in \{H, PA, PI\}$, the circular language $C(SC_X) = \sigma^*(I)$ is the language generated by SC_X. A circular language C is C_X generated (or C is a circular splicing language) if a splicing system SC_X exists such that $C = C(SC_X)$.*

Naturally, we can compare the three definitions of the circular splicing and we have that the computational power of circular splicing systems increases when we substitute Head's systems with Paun's systems and Paun's systems with Pixton's systems [4].

Example 2. We can check that $\{^\sim(aa)^n|n \geq 0\}$ is C_{PA} generated, by choosing $SC_{PA} = (A, I, R)$, with $A = \{a\}, I = \{^\sim aa\} \cup 1, R = \{aa|1\$1|aa\}$ [5] (see [46] for Head's systems). Moreover, $C =^\sim(aa)^*b$ cannot be C_H or C_{PA} generated (see also [5]), but C is C_{PI} generated by $SC_{PI} = (A, I, R)$, with $A = \{a, b\}$, $I = \{^\sim b, ^\sim a^2 b, ^\sim a^4 b\}, R = \{(a^2, a^2 b; a^2), (a^2 b, a^2; 1)\}$.

As a matter of fact, linear and circular splicing operations are not comparable: we have already seen that the regular language $(aa)^*$ is not generated by a finite linear splicing system (see Section 3.3), but the corresponding circular language $^\sim(aa)^*$ is C_{PA} generated (Example 2). Furthermore, the regular language $(aa)^*b$ is generated by a finite linear splicing system (Example 1), but the corresponding circular language $^\sim((aa)^*b)$ cannot be C_{PA} generated [5] . Some ideas on this matter can be found in the proof of Theorem 4.

Notice that any set R of rules in Paun's or Pixton's circular splicing system is implicitly supposed to be symmetric, i.e., for each rule $r = u_1|u_2\$u_3|u_4$ (resp. $r = (\alpha, \alpha'; \beta)$) in the splicing system SC_{PA} (resp. SC_{PI}), rule $\bar{r} = u_3|u_4\$u_1|u_2$ (resp. $\bar{r} = (\alpha', \alpha; \beta')$) belongs to R. Furthermore, in [24,40,41,42] additional hypotheses are considered for Paun's or Pixton's definitions (reflexivity hypothesis and self-splicing operation).

In [18,42,46], the so-called *mixed splicing*, in which both linear and circular words are allowed, is considered. In [46] the authors proposed several splicing operations, either acting on circular strings only (for instance self-splicing) or dealing with mixed splicing. Finally, in [47] a new type of circular extended splicing system (CH systems) is proposed and the universality of this model is demonstrated. Starting from this work, the CH systems have been considered as components of a Communicated Distributed circular *H*-system, i.e., the circular splicing counterpart of the parallel communicating grammar systems [45].

4.2   Computational Power of Circular Splicing Systems

As for the linear case, one main question concerns the investigation of the computational power of circular splicing systems. Once again, we denote $C(F_1, F_2) = \{C(SC_{PA}) \subseteq ^\sim A^* \mid SC_{PA} = (A, I, R)$ with $I \in F_1^\sim, R \in F_2\}$ where F_1, F_2 are families of languages in the Chomsky hierarchy. Clearly, the same definition may be given for the other two definitions of circular splicing systems. Few results are known and the problem can be split into several subproblems, each of them concerning a level of the Chomsky hierarchy containing I and by adding or not adding other hypotheses. As in the linear case, in this paper we will focus on finite circular splicing systems and on their relationship with regular circular languages.

Problem 4. Characterize $REG^\sim \cap C(FIN, FIN)$.

Note that an answer to this problem does not describe the computational power of finite circular splicing systems. Indeed, it is known that in contrast with the linear case, $C(FIN, FIN)$ is not intermediate between two classes of languages in the Chomsky hierarchy. For example, $^\sim a^n b^n$ is a context-free circular language which is not a regular circular language (since its full linearization is not regular). Moreover, it is quite easy to check that $^\sim a^n b^n$ is generated by $SC_H = (A, I, R)$ with $I = 1 \cup ^\sim ab$ and $R = \{(a, 1, b), (b, 1, a)\}$ (also see [46]). On the other hand, $^\sim (aa)^* b$ is a regular circular language that cannot be generated by a finite SC_{PA} splicing system (Example 2), whereas the regular circular language $^\sim ((A^2)^* \cup (A^3)^*)$ cannot be generated by a finite SC_{PI} splicing system (without any additional hypotheses) [6]. Furthermore, there are examples of regular circular languages which cannot be generated by finite circular splicing systems with additional hypotheses [42,46]. It is not yet clear if $C(FIN, FIN)$ contains any context-sensitive or recursive enumerable language which is not context-free.

One of the first results regarding the computational power of Head circular splicing systems states that a finite Head circular splicing system with rules having the form $(1, x, 1)$, with $|x| = 1$, always generates regular circular languages [46]. This result has been reinforced by Pixton as stated in Theorem 4 below, which is a counterpart for circular splicing of Theorem 2.

Theorem 4. *[42] Let $SC_{PI} = (A, I, R)$ be a circular splicing system with I a regular circular language and R reflexive and symmetric. Then $C = C(SC_{PI})$ is regular.*

Proof of Theorem 4 is obtained by indicating the construction of an automaton which accepts the circular splicing language. In the same paper, a link between regular circular languages and automata is pointed out, along with examples of circular splicing languages which are not regular and which are generated by finite circular splicing systems, leading to add some hypotheses to circular splicing operation. A similar closure property of splicing has been demonstrated for Paun's systems with I in a full AFL [24,40,41]. In [41] the author gives a simpler demonstration of Theorem 4 together with a version of this theorem for mixed splicing.

5 Finite Circular Splicing Systems

5.1 Star Languages and Codes

We will now present a class of regular languages whose circularization is C_{PA} generated [4]. As we have already said, we will consider languages L closed under the conjugacy relation and the action of circular splicing over their circularization $^\sim L = C$.

Definition 4. *[4] A star language is a language $L \subseteq A^*$ which is closed under the conjugacy relation and such that $L = X^*$, where X is a regular language.*

As already observed, the crucial point in generating languages through splicing is the existence of a finite set of rules producing an unbounded number of occurrences of labels c of a closed path in $\mathcal{A}$. Even if we limit ourselves to a fixed automaton $\mathcal{A}$, c can be a cycle in $\mathcal{A}$ and also the label of another (not necessarily closed) path in $\mathcal{A}$. In [4] the authors overcome this difficulty, by considering star languages L satisfying the requirement that a special representation of each cycle c (fingerprint of c) belongs to L.

Given $L = L(\mathcal{A})$, for every cycle c in $\mathcal{A}$, the *fingerprint* $F(c)$ of c is inductively defined as follows: if c is a simple cycle, then $F(c) = \{c\}$; if c is not simple (i.e., $c = u_1 c_1^{p_1} u_2 c_2^{p_2} \cdots u_k c_k^{p_k} u_{k+1}$, where u_i are labels of transitions in which no state is crossed twice, $u_1, u_{k+1} \neq 1$, c_i are cycles, $p_i \geq 1$, and if $c_i = c_j$, $j \geq i+2$ and $c_i \neq c_{i'}$, for $i < i' < j$, then $u_{i+1} \cdots u_j \neq 1$), $F(c_1), \cdots, F(c_k)$ are fingerprints of $c_1, \ldots, c_k$, then $F(c) = u_1 F(c_1) u_2 F(c_2) \cdots u_k\ F(c_k) u_{k+1}$.

A language $L = L(\mathcal{A})$ is a *fingerprint closed language* with respect to $\mathcal{A}$ whenever $\mathcal{C}(L) = \bigcup_{c\ cycle} F(c) \subseteq L$, where $\mathcal{C}(L)$ is the union of the fingerprints $F(c)$ for every cycle c in $\mathcal{A}$.

Theorem 5. *[4,5] Let X^* be a fingerprint closed star language with respect to a trim deterministic automaton $\mathcal{A}$ recognizing X^*. Then $\sim X^* \in C(FIN, FIN)$.*

As an example of languages satisfying Theorem 5, in [4,5] the authors show that star languages X^* with X a *regular group code* are fingerprint closed languages (see [2,3,43] for definitions and results on regular group codes).

5.2   The Case of a One-Letter Alphabet

Contrarily to the general case, the structure of unary languages generated by finite (Paun) circular splicing systems can be described exhaustively. Each language $L \subseteq a^*$ is closed under the conjugacy relation since we have $\sim w = \{w\}$, so $L = \sim L$ and we can identify each word (resp. language) with its circularization. Results already obtained describe the structure of every deterministic finite state automaton $\mathcal{A}$ recognizing a regular language $L \subseteq a^*$: infact the transition graph of $\mathcal{A}$ has a (frying-)pan shape.

In the proposition below, $\mathbf{Z}_n$ indicates the *cyclic group* of order $n \in \mathbf{N}$, for $D \subseteq \mathbf{N}$, we can set $a^D = \{a^d \mid d \in D\}$ and $Fact(L)$ is the set of all the factors of the elements of $L \subseteq A^*$.

Proposition 1. *[5] A subset $L = \sim L$ of a^* is C_{PA} generated if and only if there exists a finite subset L_1 of a^*, a positive integer n and a (periodic) subgroup G' of $\mathbf{Z}_n$ such that $L = L_1 \cup (a^G)^+$, where G is a set of representatives of the elements in G' and $a^G \cap Fact(L_1) = \emptyset$. Furthermore, $G = \{m + \lambda_m n \mid m \in G'\}$ with $\lambda_m = min\{\mu \mid a^{m+\mu n} \in (a^G)^+\}$.*

Proposition 1, easily extended in [6] to languages $L = A^J = \bigcup_{j \in J} A^j$, with $J \subseteq \mathbf{N}$, obviously suggests a decidability question, namely, if we can decide whether $L \subseteq a^*$ is a C_{PA} generated language. If no hypothesis is made over L, the answer to this question is no, according to the Rice theorem [26]. When L is supposed to be a regular language, Proposition 2 gives a positive answer to this question.

Proposition 2. *Let $L \subseteq a^*$ be a regular language. Let $\mathcal{A} = (\{a\}, Q, \delta, q_0, F)$ be the minimal finite state automaton recognizing L, where $Q = \{q_0, q_1, \ldots, q_v\}$, for each $i \in \{0, \ldots v - 1\}$ we have $\delta(q_i, a) = q_{i+1}$ and there exists $l \in \{0, \ldots, v\}$ such that $\delta(q_v, a) = q_l$. We have that $L = L(\mathcal{A})$ is C_{PA} generated if and only if either $q_h \notin F$, for all $h \geq l$ or there exist $n', p, s \in \mathbf{N}$ such that $\mathcal{A}$ satisfies the two conditions that follow.*

1. $\{q_h \mid q_h \in Q, \ h \geq l\} \cap F = q_{n'}$.
2. $n' = ps$ and $v - l + 1 = p$.

In [5] it has been proved that unary languages which are C_{PA} generated can be generated by a splicing system with at most one rule. This result takes into account the concept of *minimal splicing system*. This notion was introduced in [34,38], where the minimality was referred to the minimal cardinality of the set of the rules (resp. triples) or to a set of rules (resp. triples) with shorter length. Some initial steps towards this notion and for Head linear splicing systems were implicitly taken in [15].

References

1. Adleman, L. M.: Molecular computation of solutions to combinatorial problems. Science **226** (1994) 1021 − 1024
2. Berstel, J., Perrin, D.: Theory of codes. Academic Press, New York (1985)
3. Berstel, J., Restivo, A.: Codes et sousmonoides fermes par conjugaison. Sem. LITP **81-45** (1981) 10 pages
4. Bonizzoni, P., De Felice, C., Mauri, G., Zizza, R.: DNA and circular splicing. In: Condon, A., Rozenberg, G. (eds.): DNA Computing. Lecture Notes in Computer Science, Vol. 2054. Springer-Verlag, New York (2001) 117 − 129
5. Bonizzoni, P., De Felice, C., Mauri, G., Zizza, R.: Circular splicing and regularity (2002) submitted
6. Bonizzoni, P., De Felice, C., Mauri, G., Zizza, R.: On the power of linear and circular splicing (2002) submitted
7. Bonizzoni, P., De Felice, C., Mauri, G., Zizza, R.: The structure of regular reflexive splicing languages via Schützenberger constants (2002) manuscript
8. Bonizzoni, P., Ferretti, C., Mauri, G.: Splicing systems with marked rules. Romanian Journal of Information Science and Technology **1:4** (1998) 295 − 306
9. Bonizzoni, P., Ferretti, C., Mauri, G., Zizza, R.: Separating some splicing models. Information Processing Letters **76:6** (2001) 255 − 259
10. Choffrut, C., Karhumaki, J.: Combinatorics on Words. In: Rozenberg, G., Salomaa, A., (eds.): Handbook of Formal Languages, Vol. 1. Springer-Verlag, (1996) 329−438
11. Culik, K., Harju, T.: Splicing semigroups of dominoes and DNA. Discrete Applied Mathematics **31** (1991) 261 − 277
12. De Luca, A., Restivo, A.: A characterization of strictly locally testable languages and its application to subsemigroups. Information and Control **44** (1980) 300 − 319
13. Denninghoff, K.L., Gatterdam, R.W.: On the undecidability of splicing systems. International Journal of Computer Math. **27** (1989) 133 − 145
14. Freund, R., Kari, L., Paun, Gh.: DNA Computing based on splicing. The existence of universal computers. Theory of Computing Systems **32** (1999) 69 − 112

15. Gatterdam, R.W.: Algorithms for splicing systems. SIAM Journal of Computing **21:3** (1992) 507 − 520

16. Goode, E., Head, T., Pixton D.: private communication

17. Goode, E., Pixton, D.: Semi-simple splicing systems. In: Martin-Vide, C., Mitrana, V. (eds.): Where Mathematics, Computer Science, Linguistics and Biology Meet. Kluwer Academic Publ., Dordrecht (2001) 343 − 357

18. Head, T.: Splicing schemes and DNA. In: Lindenmayer Systems. Impacts on Theoretical Computer Science and Developmental Biology. Springer-Verlag, Berlin (1992) 371 − 383

19. Head, T.: Formal Language Theory and DNA. An analysis of the generative capacity of specific recombinant behaviours. Bull. Math. Biol. **49** (1987) 737 − 759

20. Head, T.: Splicing languages generated with one sided context. In: Paun, Gh. (ed.): Computing with Bio-molecules. Theory and Experiments, Springer-Verlag Singapore (1998)

21. Head, T.: Splicing representations of strictly locally testable languages. Discrete Applied Math. **87** (1998) 139 − 147

22. Head, T.: Circular suggestions for DNA Computing. In: Carbone, A., Gromov, M., Pruzinkiewicz, P. (eds.): Pattern Formation in Biology, Vision and Dynamics. World Scientific, Singapore and London, to appear

23. Head, T.: Splicing systems. Regular languages and below. In: Pre-Proc. of DNA8 (2002)

24. Head, T., Paun, Gh., Pixton, D.: Language theory and molecular genetics. Generative mechanisms suggested by DNA recombination. In: Rozenberg, G., Salomaa, A. (eds.): Handbook of Formal Languages, Vol. 2. Springer-Verlag (1996) 295 − 360

25. Head, T., Rozenberg, G., Bladergroen, R., Breek, C., Lommerse, P., Spaink, H.: Computing with DNA by operating on plasmids. BioSystems **57** (2000) 87 − 93

26. Hopcroft, J.E., Motwani, R., Ullman, J.D.: Introduction to Automata Theory, Languages, and Computation. 2nd edn. Addison-Wesley, Reading, Mass. (2001)

27. Jonoska, N., Karl, S.: A molecular computation of the road coloring problem. In: Proc. of 2nd Annual Meeting on DNA Based Computers. DIMACS Workshop (1996) 148 − 158

28. Kari, L.: DNA computing. An arrival of biological mathematics. The Mathematical Intelligence **19:2** (1999) 9 − 22

29. Kim, S.M.: An Algorithm for Identifying Spliced Languages. In: Proc. of Cocoon97. Lecture Notes in Computer Science, Vol. 1276. Springer-Verlag (1997) 403 − 411

30. Kim, S.M.: Computational modeling for genetic splicing systems. SIAM Journal of Computing **26** (1997) 1284 − 1309

31. Lothaire, M.: Combinatorics on Words. Encyclopedia of Math. and its Applications. Addison Wesley Publishing Company (1983)

32. McNaughton, R., Papert, S.: Counter-Free Automata. MIT Press, Cambridge, Mass. (1971)

33. Mateescu, A., Paun, Gh., Rozenberg, G., Salomaa, A.: Simple splicing systems. Discrete Applied Mathematics **84** (1998) 145 − 163

34. Paun, Gh.: On the splicing operation. Discrete Applied Mathematics **70** (1996) 57 − 79

35. Paun, Gh.: Regular extended H systems are computationally universal. Journal of Automata, Languages and Combinatorics. **1:1** (1996) 27 − 36

36. Paun, Gh., Salomaa, A.: DNA computing based on the splicing operation. Mathematica Japonica **43:3** (1996) 607 − 632

37. Paun, Gh., Rozenberg, G., Salomaa, A.: Computing by splicing. Theoretical Computer Science **168:2** (1996) 321 − 336

38. Paun, Gh., Rozenberg, G., Salomaa, A.: DNA computing, New Computing Paradigms. Springer-Verlag (1998)
39. Perrin, D.: Finite Automata. In: van Leeuwen, J. (ed.): Handbook of Theoretical Computer Science, Vol. B. Elsevier (1990) 1 − 57
40. Pixton, D.: Splicing in abstract families of languages. Theoretical Computer Science **234** (2000) 135 − 166
41. Pixton, D.: Linear and Circular Splicing Systems. In: Proc. of 1st Int. Symp. on Int. in Neural and Biological Systems (1996) 181 − 188
42. Pixton, D.: Regularity of splicing languages. Discrete Applied Mathematics **69** (1996) 101 − 124
43. Reis, C., Thierren, G.: Reflective star languages and codes. Information and Control **42** (1979) 1 − 9
44. Schützenberger, M.-P.: Sur certaines opérations de fermeture dans le langages rationnels. Symposia Mathematica **15** (1975) 245 − 253
45. Siromoney, R.: Distributed Circular Systems. In: Pre-Proc. of Grammar Systems 2000, Austria
46. Siromoney, R., Subramanian, K.G., Dare, A.: Circular DNA and Splicing Systems. In: Proc. of ICPIA. Lecture Notes in Computer Science, Vol. 654. Springer-Verlag (1992) 260 − 273
47. Yokomori, T., Kobayaski, S., Ferretti, C.: On the power of circular splicing systems and DNA computability. In: Proc. of IEEE Int. Conf. Evol. Comp. ICEC99

Finite Automata Models of Quantized Systems: Conceptual Status and Outlook

Karl Svozil

Institut für Theoretische Physik, University of Technology Vienna
Wiedner Hauptstraße 8-10/136, A-1040 Vienna, Austria
svozil@tuwien.ac.at

Abstract. Since Edward Moore, finite automata theory has been inspired by physics, in particular by quantum complementarity. We review automaton complementarity, reversible automata and the connections to generalized urn models. Recent developments in quantum information theory may have appropriate formalizations in the automaton context.

1 Physical Connections

Physics and computer science share common interests. They may pursue their investigations by different methods and formalisms, but once in a while it is quite obvious that the interrelations are pertinent. Take, for example, the concepts of information and computation. Per definition, any theory of information and computation, in order to be applicable, should refer to physically operationalizable concepts. After all, information and computation is physical [1].

Conversely, concepts of computer science have increasingly influenced physics. Two examples for these developments have been the recent developments in classical continuum theory, well known under the term "deterministic chaos," as well as quantum information and computation theory. Quantum systems nowadays are often perceived as very specific and delicate (due to decoherence; i.e., the irreversible loss of state information in measurements) reversible computations.

Whether or not this correspondence resides in the very foundations of both sciences remains speculative. Nevertheless, one could conjecture a correspondence principle by stating that *every feature of a computational model should be reflected by some physical system. Conversely, every physical feature, in particular of a physical theory, should correspond to a feature of an appropriate computational model.* This is by no means trivial, as for instance the abundant use of nonconstructive continua in physics indicates. No finitely bounded computation could even in principle store, process and retrieve the nonrecursively enumerable and even algorithmically incompressible random reals, of which the continuum "mostly" exists. But also recent attempts to utilize quantum computations for speedups or even to solve problems which are unsolvable within classical recursion theory [2] emphasize the interplay between physics and computer science.

M. Ito and M. Toyama (Eds.): DLT 2002, LNCS 2450, pp. 93–102, 2003.

Already quite early, Edward Moore attempted a formalization of quantum complementarity in terms of finite deterministic automata [3]. Quantum complementarity is the feature of certain microphysical systems not to allow the determination of all of its properties with arbitrary precision at once. Moore was interested in the initial state determination problem: given a particular finite automaton which is in an unknown initial state; find that initial state by the analysis of input-output experiments on a single such automaton. Complementarity manifests itself if different inputs yield different properties of the initial automaton state while at the same time steering the automaton into a state which is independent of its initial one.

Moore's considerations have been extended in many ways. Recently, different complementarity classes have been characterized [4] and their likelihood has been investigated [5,6]. We shall briefly review a calculus of propositions referring to the initial state problem which resembles quantum logic in many ways [7,8]. Automaton theory can be liked to generalized urn models [9]. In developing the analogy to quantum mechanics further, reversible deterministic finite automata have been introduced [10]. New concepts in quantum mechanics [11] suggest yet different finite automaton models.

2   Automaton Partition Logics

Consider a Mealy automaton $\langle S, I, O, \delta, \lambda \rangle$, where S, I, O are the sets of states, input and output symbols, respectively. $\delta(s, i) = s'$ and $\lambda(s, i) = o$, $s, s' \in S$, $i \in I$ and $o \in O$ are the transition and the output functions, respectively.

The initial state determination problem can be formalized as follows. Consider a particular automaton and all sequences of input/output symbols which result from all conceivable experiments on it. These experiments induce a state partition in the following natural way. Every distinct set of input/output symbols is associated with a set of initial automaton states which would reproduce that sequence. This set of states may contain one or more states, depending on the ability of the experiment to separate different initial automaton states. A partitioning of the automaton states associated with an input sequence is obtained if one considers the variety of all possible output sequences. Stated differently: given a set of input symbols, the set of automaton states "decays" into disjoint subsets associated with the possible output sequences. This partition can then be identified with a Boolean algebra, with the elements of the partition interpreted as atoms. By pasting the Boolean algebras of the "finest" partitions together, one obtains a calculus of proposition associated with the particular automaton. This calculus of propositions is referred to as *automaton partition logic*.

The converse is true as well: given any partition logic, it is always possible to (nonuniquely) construct a corresponding automaton with the following specifications: associate with every element of the set of partitions a single input symbol. Then take the partition with the highest number of elements and associate a single output symbol with any one element of this partition. The automaton output function can then be defined by associating a single output symbol per element of the partition (corresponding to a particular input symbol). Finally, choose a transition function which completely looses the state information after only one transition; i.e., a transition function which maps all automaton state into a single one. We just mention that another, independent, way to obtain automata from partition logics is by considering the set of two-valued states.

In that way, a multitude of worlds can be constructed, many of which feature quantum complementarity. For example, consider the Mealy automaton $\langle\{1,2,3\},\{1,2,3\},\{0,1\},\delta = 1,\lambda(s,i) = \delta_{si}\rangle$ (the Kronecker function $\delta_{si} = 1$ if $s = i$, and zero otherwise). Its states are partitioned into $\{\{1\},\{2,3\}\}$, $\{\{2\},\{1,3\}\}$, $\{\{3\},\{1,2\}\}$, for the inputs 1, 2, and 3, respectively. Every partition forms a Boolean algebra 2^2. The partition logic depicted in Fig. 1 is obtained by "pasting" the three algebras together; i.e., by maintaining the order structure and by identifying identical elements; in this case $\emptyset, \{3,1,2\}$. It is a modular, nonboolean lattice MO$_3$ of the "chinese lantern" form. A systematic study [8, pp. 38-39] shows that automata reproduce (but are not limited to) all finite subalgebras of Hilbert lattices of finite-dimensional quantum logic.

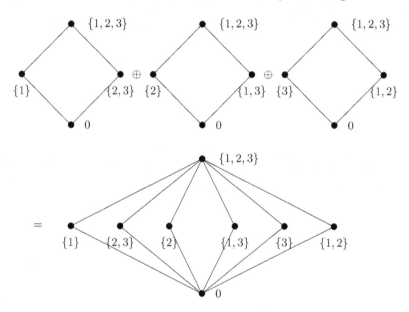

Fig. 1.   Automaton partition logic corresponding to the Mealy automaton $\langle\{1,2,3\},\{1,2,3\},\{0,1\},\delta = 1,\lambda(s,i) = \delta_{si}\rangle$.

Mealy automata are logically equivalent to generalized urn model (GUM) [12,13] $\langle U,C,L,\Lambda\rangle$ which is an ensemble U of ball types with black background color. Printed on these balls are some symbols from a symbolic alphabet L. These symbols are colored. The colors are elements of a set of colors C. A particular ball type is associated with a unique combination of mono-spectrally (no mixture of wavelength) colored symbols printed on the black ball background.

Suppose you have a number of colored eyeglasses built from filters for the $|C|$ different colors. They should absorb every other light than one of a particular single color. When a spectator looks at a particular ball through such an eyeglass, the only recognizable symbol will be the one in the particular color which is transmitted through the eyeglass. All other colors are absorbed, and the symbols printed in them will appear black and therefore cannot be differentiated from the black background. Hence the ball appears to carry a different "message" or symbol, depending on the color at which it is viewed. The above procedure could be formalized by a "lookup" function $\Lambda(u,c) = v$, which depends on the ball type $u \in U$ and on the color $c \in C$, and which returns the symbol $v \in L$ printed in this color. (Note the analogy to the output function λ.)

Again, an empirical logic can be constructed as follows. Consider the set of all ball types. With respect to a particular colored eyeglass, this set gets partitioned into those ball types which can be separated by the particular color of the eyeglass. Every such state partition can then be identified with a Boolean algebra whose atoms are the elements of the partition. A pasting of all of these Boolean algebras yields the calculus of propositions. The corresponding correlation polytope formed by the set of two-valued states characterizes all probability measures.

In order to define an automaton partition logic associated with a Mealy automaton $\langle S, I, O, \delta, \lambda \rangle$ from a GUM $\langle U, C, L, \Lambda \rangle$, let $u \in U$, $c \in C$, $v \in L$, and $s, s' \in S$, $i \in I$, $o \in O$, and assume $|U| = |S|$, $|C| = |I|$, $|L| = |O|$. The following identifications can be made with the help of the bijections t_S, t_I and t_O:

$$
\begin{aligned}
&t_S(u) = s, \; t_I(c) = i, \; t_O(v) = o, \\
&\delta(s,i) = s_i \quad \text{for fixed } s_i \in S \text{ and arbitrary } s \in S, \; i \in I, \\
&\lambda(s,i) = t_O\left(\Lambda(t_S^{-1}(s), t_I^{-1}(i))\right).
\end{aligned} \tag{1}
$$

Conversely, consider an arbitrary Mealy automaton $\langle S, I, O, \delta, \lambda \rangle$. Just as before, associate with every single automaton state $s \in S$ a ball type u, associate with every input symbol $i \in I$ a unique color c, and associate with every output symbol $o \in O$ a unique symbol v; i.e., again $|U| = |S|$, $|C| = |I|$, $|L| = |O|$. The following identifications can be made with the help of the bijections τ_U, τ_C and τ_L:

$$
\tau_U(s) = u, \; \tau_C(i) = c, \; \tau_L(o) = v, \; \Lambda(u,c) = \tau_L(\lambda(\tau_U^{-1}(u), \tau_C^{-1}(c))). \tag{2}
$$

A direct comparison of (1) and (2) yields

$$
\tau_U^{-1} = t_S, \; \tau_C^{-1} = t_I, \; \tau_L^{-1} = t_O. \tag{3}
$$

For example, for the automaton partition logic depicted in Fig. 1, the above construction yields a GUM with three ball types: type 1 has a red "0" printed on it and a green and blue "1;" type 2 has a green "0" printed on it and a red and blue "1;" type 3 has a blue "0" printed on it and a red and green "1." Hence, if an experimenter looks through a red eyeglass, it would be possible to differentiate between ball type 1 and the rest (i.e., ball types 2 and 3).

3   Reversible Finite Automata

The quantum time evolution between two measurements is reversible, but the previously discussed Moore and Mealy automata are not. This has been the motivation for another type of deterministic finite automaton model [10,8] whose combined transition and output function is bijective and thus reversible.

The elements of the Cartesian product $S \times I$ can be arranged as a linear list Ψ of length n, just like a vector. Consider the automaton $\langle S, I, I, \delta, \lambda \rangle \equiv \langle \Psi, U \rangle$ with $I = O$ and $U : (s, i) \rightarrow (\delta(s, i), \lambda(s, i))$, whose time evolution can in be rewritten as

$$\sum_{j=1}^{n} U_{ij} \Psi_j \qquad (4)$$

U is a $n \times n$-matrix which, due to the requirement of determinism, uniqueness and invertability is a *permutation matrix*. The class of all reversible automata corresponds to the group of permutations, and thus reversible automata are characterized by permutations.

Do reversible automata feature complementarity? Due to reversibility, it seems to be always possible to measure a certain aspect of the initial state problem, copy this result to a "save place," revert the evolution and measure an arbitrary other property. In that way, one effectively is able to work with an arbitrary number of automaton copies in the same initial state. Indeed, as already Moore has pointed out, such a setup cannot yield complementarity.

The environment into which the automaton is embedded is of conceptual importance. If the environment allows for copying; i.e., one-to-many evolutions, then the above argument applies and there is no complementarity. But if this environment is reversible as well, then in order to revert the automaton back to its original initial state, all information has to be used which has been acquired so far; i.e., all changes have to be reversed, both in the automaton as well as in the environment. To put it pointedly: there is no "save place" to which the result of any measurement could be copied and stored and afterwards retrieved while the rest of the system returns to its former state. This is an automaton analogue to the no-cloning theorem of quantum information theory.

For example, a reversible automaton corresponding to the permutation whose cycle form is given by $(1,2)(3,4)$ corresponds to an automaton

$$\langle \{1, 2\}, \{0, 1\}, \{0, 1\}, \delta(s, i) = s, \lambda(s, i) = (i + 1) \bmod 2 \rangle \quad ,$$

or, equivalently,

$$\left\langle \begin{pmatrix} (1,0) \\ (1,1) \\ (2,0) \\ (2,1) \end{pmatrix}, \begin{pmatrix} 0\,1\,0\,0 \\ 1\,0\,0\,0 \\ 0\,0\,0\,1 \\ 0\,0\,1\,0 \end{pmatrix} \right\rangle . \qquad (5)$$

A pictorial representation may be given in terms of a flow diagram which represent the permutation of the states in one evolution step.

4   Counterfactual Automata

Automaton partition logic is nondistributive and thus nonclassical (if Boolean logic is considered to be classical). Yet it is not nonclassical as it should be, in particular as compared to quantum mechanics. The partitions have a set theoretic interpretation, and automaton partition logic allows for a full set of two-valued states; i.e., there are a sufficient number of two-valued states for constructing lattice homomorphisms. This is not the case for Hilbert lattices associated with quantum mechanical systems of dimension higher than two. Probably the most striking explicit and constructive demonstration of this fact has been given by Kochen and Specker [14]. They enumerated a finite set of interconnected orthogonal tripods in threedimensional real Hilbert space with the property that the tripods cannot be colored consistently in such a way that two axes are green and one is red. Stated differently, the chromatic number of the threedimensional real unit sphere is four. The entire generated system of associated properties is not finite. Just as for all infinite models, denumerable or not, it does not correspond to any finite automaton model.

Since the colors "green" and "red" can be interpreted as the truth values "false" and "true" of properties of specific quantum mechanical systems, there has been much speculation as to the existence of complementary quantum physical properties. One interpretation supposes that counterfactual properties different from the ones being measured do not exist. While this discussion may appear rather philosophical and not to the actual physical point, it has stimulated the idea that a particle does not carry more (counterfactual) information than it has been prepared for. I.e., *a n-state particle carries exactly one nit of information, and k n-state particles carry exactly k nits of information* [11,15,16]. As a consequence, the information may not be coded into single particles alone (one nit per particle), but in general may be distributed over ensembles of particles, a property called entanglement. Furthermore, when measuring properties different from the ones the particle is prepared for, the result may be irreducible random. The binary case is $n = 2$. Note that, different from classical continuum theory, where the base of the coding is a matter of convention and convenience, the base of quantum information is founded on the n-dimensionality of Hilbert space of the quantized system, a property which is unique and measurable.

A conceivable automaton model is one which contains a set of input symbols I which themselves are sets with n elements corresponding to the n different outputs of the n-ary information. Let $\langle I \times O, I, O, \delta, \lambda \rangle$ with $S = I$ to reflect the issue of state preparation versus measurement. Since the information is in base n, $O = \{1, \ldots, n\}$. Furthermore, require that after every measurement, the automaton is in a state which is characterized by the measurement; and that random results are obtained if different properties are measured than the ones the automaton has been prepared in. I.e.,

$$
\begin{aligned}
\delta((s, o_s), i) &= \begin{cases} (i, \text{RANDOM}(O)) & \text{if } s \neq i, \\ (s, o_s) & \text{if } s = i, \end{cases} \\
\lambda((s, o_s), i) &= \begin{cases} \text{RANDOM}(O) & \text{if } s \neq i, \\ o_s & \text{if } s = i, \end{cases}
\end{aligned}
\tag{6}
$$

The state $(s, o_s) \in I \times O$ is characterized by the mode s the particle has been prepared for, and the corresponding output o_s. The input i determines the "context" (a term borrowed from quantum logic) of measurement, whereas the output value $1 \leq o_s \leq n$ defines the actual outcome. RANDOM(O) is a function whose value is a random element of O.

This straightforward implementation may not be regarded very elegant or even appropriate, since it contains a random function in the definition of a finite automaton. A conceptually more appealing Ansatz might be to get rid of the case $s \neq i$ in which the automaton is prepared with a state information different from the one being retrieved. One may speculate that the randomization of the output effectively originates from the environment which "translates" the "wrong" question into the one which can be answered by the automaton (which is constrained by $s = i$) at the price of randomization. It also remains open if the automaton analogue to entanglement is merely an automaton which cannot be decomposed into parallel single automata.

The formalism developed for quantum information in base n defined by state partitions can be fully applied to finite automata [16]. A k-particle system whose information is in base n is described by k nits which can be characterized by k comeasurable partitions of the product state of the single-particle states with n elements each; every such element has n^{k-1} elements. Every complete set of comeasurable nits has the property that (i) the set theoretic intersection of any k elements of k different partitions is a single particle state, and (ii) the union of all these n^k intersections is just the set of single particle states. The set theoretic union of all elements of a complete set of comeasurable nits form a state partition. The set theoretic union of all of the $n^k!$ partitions (generated by permutations of the product states) form an automaton partition logic corresponding to the quantum system.

We shall demonstrate this construction with the case $k = 2$ and $n = 3$. Suppose the product states are labeled from 1 through 9. A typical element is formed by $\{\{1, 2, 3\}, \{4, 5, 6\}, \{7, 8, 9\}\}$ for the first trit, and $\{\{1, 4, 7\}, \{2, 5, 8\}, \{3, 6, 9\}\}$ for the second trit. Since those trits are comeasurable, their Cartesian product forms the first partition $\{\{\{1, 2, 3\}, \{1, 4, 7\}\}, \ldots, \{\{7, 8, 9\}, \{3, 6, 9\}\}\}$. The complete set of $9!/(2 \cdot 3! \cdot 3!) = 5040$ different two-trit sets can be evaluated numerically; i.e., in lexicographic order,

$$\{\{\{1, 2, 3\}, \{4, 6, 8\}, \{5, 7, 9\}\} \times \{\{1, 4, 5\}, \{2, 6, 7\}, \{3, 8, 9\}\}\}, \qquad (7)$$
$$\{\{\{1, 2, 3\}, \{4, 6, 9\}, \{5, 7, 8\}\} \times \{\{1, 4, 5\}, \{2, 6, 7\}, \{3, 8, 9\}\}\}, \qquad (8)$$

$$\vdots$$

$$\{\{\{1, 2, 3\}, \{4, 5, 6\}, \{7, 8, 9\}\} \times \{\{\{1, 4, 7\}, \{2, 5, 8\}, \{3, 6, 9\}\}\}\}, \qquad (9)$$

$$\vdots$$

$$\{\{\{1, 6, 9\}, \{2, 5, 7\}, \{3, 4, 8\}\} \times \{\{1, 7, 8\}, \{2, 4, 9\}, \{3, 5, 6\}\}\}, \qquad (10)$$
$$\{\{\{1, 6, 9\}, \{2, 5, 8\}, \{3, 4, 7\}\} \times \{\{1, 7, 8\}, \{2, 4, 9\}, \{3, 5, 6\}\}\}\}. \qquad (11)$$

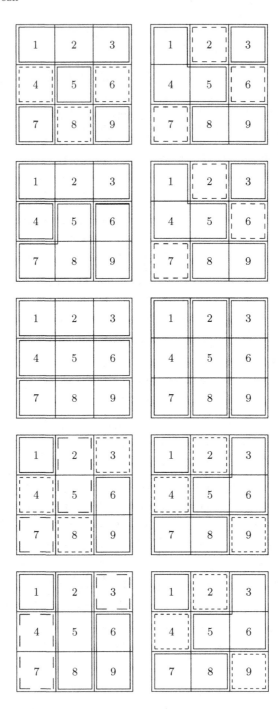

Fig. 2. Two trits yield a unique tessellation of the two-particle product state space; the two trits per line correspond to Eqs. (7)—(11).

The associated partition logic is the horizontal sum of 5040 Boolean algebras with nine atoms (i.e., 2^9) and corresponds to a rather elaborate but structurally simple automaton with 9 states $\{1, \ldots, 9\}$, 2×5040 input symbols $\{1, \ldots, 10080\}$ and 9 output symbols $\{1, \ldots, 9\}$. Every one of the Boolean algebras is the quantum analogue of a particular set of comeasurable propositions associated with the two trits. Together they form a complete set of trits for the two-particle three-state quantized case. A graphical representation of the state single-particle state space tesselation is depicted in Fig 2.

5   Applicability

Despite some hopes, for instance stated by Einstein [17, p. 163], to express finite, discrete physical systems by finite, algebraic theories, a broader acceptance of automata models in physics would require concrete, operationally testable consequences. One prospect would be to search for phenomena which cannot happen according to quantum mechanics but are realizable by finite automata. The simplest case is characterized by a Greechie hyperdiagram of triangle form, with three atoms per edge. Its automaton partition logic is given by

$$\{\{\{1\}, \{2\}, \{3, 4\}\}, \{\{1\}, \{2, 4\}, \{3\}\}, \{\{1, 4\}, \{2\}, \{3\}\}\}. \tag{12}$$

A corresponding Mealy automaton is $\langle\{1, 2, 3, 4\}, \{1, 2, 3\}, \{1, 2, 3\}, \delta = 1, \lambda\rangle$, where $\lambda(1, 1) = \lambda(3, 2) = \lambda(2, 3) = 1$, $\lambda(3, 1) = \lambda(2, 2) = \lambda(1, 3) = 2$, and $\lambda(2, 1) = \lambda(4, 1) = \lambda(1, 2) = \lambda(4, 2) = \lambda(3, 3) = \lambda(4, 3) = 3$.

Another potential application is the investigation of the "intrinsic physical properties" of virtual realities in general, and computer games in particular. Complementarity and the other discussed features are robust and occur in many different computational contexts, in particular if one is interested in the intrinsic "look and feel" of computer animated worlds.

References

1. Landauer, R.: Information is physical. Physics Today **44** (1991) 23–29
2. Calude, C.S., Pavlov, B.: Coins, quantum measurements, and Turing's barrier. Quantum Information Processing **1** (2002) 107–127 `quant-ph/0112087`.
3. Moore, E.F.: Gedanken-experiments on sequential machines. In Shannon, C.E., McCarthy, J., eds.: Automata Studies. Princeton University Press, Princeton (1956)
4. Calude, C., Calude, E., Svozil, K., Yu, S.: Physical versus computational complementarity I. International Journal of Theoretical Physics **36** (1997) 1495–1523
5. Calude, E., Lipponen, M.: Deterministic incomplete automata: Simulation, universality and complementarity. In Calude, C.S., Casti, J., Dinneen, M.J., eds.: Unconventional Models of Computation, Singapore, Springer (1998) 131–149
6. Calude, C.S., Calude, E., Khoussainov, B.: Finite nondeterministic automata: Simulation and minimality. Theoret. Comput. Sci. **242** (2000) 219–235
7. Svozil, K.: Randomness & Undecidability in Physics. World Scientific, Singapore (1993)

8. Svozil, K.: Quantum Logic. Springer, Singapore (1998)
9. Svozil, K.: Logical equivalence between generalized urn models and finite automata. preprint (2002)
10. Svozil, K.: The Church-Turing thesis as a guiding principle for physics. In Calude, C.S., Casti, J., Dinneen, M.J., eds.: Unconventional Models of Computation, Singapore, Springer (1998) 371–385
11. Zeilinger, A.: A foundational principle for quantum mechanics. Foundations of Physics **29** (1999) 631–643
12. Wright, R.: The state of the pentagon. A nonclassical example. In Marlow, A.R., ed.: Mathematical Foundations of Quantum Theory. Academic Press, New York (1978) 255–274
13. Wright, R.: Generalized urn models. Foundations of Physics **20** (1990) 881–903
14. Kochen, S., Specker, E.P.: The problem of hidden variables in quantum mechanics. Journal of Mathematics and Mechanics **17** (1967) 59–87 Reprinted in [18, pp. 235–263].
15. Donath, N., Svozil, K.: Finding a state among a complete set of orthogonal ones. Physical Review A **65** (2002) 044302 `quant-ph/0105046`, co-published by APS in [19].
16. Svozil, K.: Quantum information in base n defined by state partitions. Physical Review A **66** (2002) in print `quant-ph/0205031`.
17. Einstein, A.: Grundzüge der Relativitätstheorie. 1st edition edn. Vieweg, Braunschweig (1956)
18. Specker, E.: Selecta. Birkhäuser Verlag, Basel (1990)
19. Donath, N., Svozil, K.: Finding a state among a complete set of orthogonal ones. Virtual Journal of Quantum Information **2** (2002)

Automata on Linear Orderings

Véronique Bruyère and Olivier Carton

[1] Institut d'Informatique, Université de Mons-Hainaut,
Le Pentagone, 6 avenue du Champ de Mars, B-7000 Mons, Belgium
http://staff.umh.ac.be/Bruyere.Veronique/
[2] LIAFA
Université Paris 7 - case 7014, 2 place Jussieu, F-75251 Paris cedex 05, France
http://www.liafa.jussieu.fr/~carton/

Abstract. We consider words indexed by linear orderings. These extend finite, (bi-)infinite words and words on ordinals. We introduce automata and rational expressions for words on linear orderings.

1   Introduction

The theory of automata finds its origin in the paper of S. C. Kleene of 1956 where the basic theorem, known as Kleene's theorem, is proved for finite words [17]. Since then, automata working on infinite words, trees, traces ... have been proposed and this theory is a branch of theoretical computer science that has developed into many directions [19, 23].

In this paper we focus on automata accepting objects which can be linearly ordered. Examples are finite, infinite, bi-infinite and ordinal words, where the associated linear ordering is a finite ordering, the ordering of the integers, the ordering of the relative integers and the ordering of an ordinal number respectively. Each such class of words has its automata and a Kleene-like theorem exists. Büchi introduced the so-called Büchi automata to show the decidability of the monadic second order theory of $\langle \mathbb{N}, < \rangle$ [10]. He later extended the method to countable ordinals thanks to appropriate automata [11]. Büchi already introduced ω-rational operations in [11]. Rational operations and a Kleene's theorem are proposed in [12] for words indexed by an ordinal less than ω^{ω}, they are extended to any ordinal in [25]. The case of bi-infinite words is solved in [18, 14].

Our goal is a unified approach through the study of words indexed by any linear ordering. We propose a new notion of automaton which is simple, natural and includes previous automata. It is possible to define rational expressions and get a related Kleene's theorem for a large class of linear orderings.

Words indexed by a countable linear ordering were introduced in [13]. They are exactly the frontier of a labeled binary tree read from left to right. Some rational expressions are already studied in [13, 16, 22] with the nice property that they are total functions. For instance, an ω-power can be concatenated with a finite word and the resulting word can be iterated thanks to a reversed ω-power. These operations lead to a characterization of the words which are frontier of regular trees.

M. Ito and M. Toyama (Eds.): DLT 2002, LNCS 2450, pp. 103–115, 2003.

Our automata work as follows. Consider the case of a finite word w of length n. The underlying ordering is $1 < 2 < \cdots < n$. The $n + 1$ states of a path for w are inserted between the letters of w, i.e., at the cuts of the ordering. For a word w indexed by a linear ordering, the associated path has its states indexed by the cuts of the ordering. The automaton has three types of transitions: the usual successor transitions, left limit and right limit transitions. For two states consecutive on the path, there must be a successor transition labeled by the letter in between. For a state q which has no predecessor on the path, the left limit (with respect to q) set P of states is computed and there must be a left limit transition between P and q. Right limit transitions are used when a state has no successor on the path. For a Muller automaton, such a left limit set P is nothing else than the states appearing infinitely often along the path. In our case, the path then ends with an additional left limit transition to a state q which is final.

Recently ordinal words (called Zeno words) were considered as modeling infinite sequences of actions which occur in a finite interval of time [15, 4]. While the intervals of time are finite, infinite sequences of actions can be concatenated. A Kleene's theorem already exists for classical timed automata (where infinite sequences of actions are supposed to generate divergent sequences of times) [2, 1, 5]. In [4], automata of Choueka and Wojciechowski are adapted to Zeno words. A kind of Kleene's theorem is proved, that is, the class of Zeno languages is the closure under an operation called refinement of the class of languages accepted by classical timed automata.

The paper is organized as follows. We recall the basic notions on linear orderings in Section 2. Words indexed by linear orderings are introduced in Section 3. Automata are defined in Section 4. In section 5, we mention some Kleene's theorems.

2   Linear Orderings

In this section, we recall the definitions and fix the terminology. We refer the reader to [20] for a complete introduction to linear orderings.

A *linear ordering* J is a set equipped with an ordering $<$ which is total, that is, for any $j \neq k$ in J, either $j < k$ or $k < j$ holds. The ordering of the integers, of the relative integers and of the rational numbers are linear orderings, respectively denoted by ω, ζ and η. We recall that an *ordinal* is a linear ordering which is well-ordered, that is, any nonempty subset has a least element. Notation $\mathcal{N}$ and $\mathcal{O}$ is used for the class of finite linear orderings and the class of ordinals. In the sequel, we freely say that two orderings are equal if they are actually isomorphic.

Let J and K be two linear orderings. We denote by $-J$ the backwards linear ordering obtained by reversing the ordering relation. This backwards ordering is usually denoted J^* in the literature on linear orderings. Since the star is already used in formal language theory, we prefer the notation $-J$. The linear ordering $J + K$ is the ordering on the disjoint union $J \cup K$ extended with $j < k$ for any

$j \in J$ and any $k \in K$. More generally, let K_j be a linear ordering for any $j \in J$. The linear ordering $\sum_{j \in J} K_j$ is the set of pairs (k, j) such that $k \in K_j$. The relation $(k_1, j_1) < (k_2, j_2)$ holds iff $j_1 < j_2$ or $j_1 = j_2$ and $k_1 < k_2$ in K_{j_1}.

Example 1. The ordering $-\omega + \omega$ is equal to the ordering ζ. The ordering ζ^2 is equal to the sum $\sum_{j \in \zeta} \zeta$. It is made of ζ copies of the ordering ζ.

A *Dedekind cut* or simply a *cut* of a linear ordering J is a pair (K, L) of intervals such that J is the disjoint union $K \cup L$ and for any $k \in K$ and $l \in L$, $k < l$. The set of all cuts of the ordering J is denoted by $\hat{J}$. The set $\hat{J}$ can be linearly ordered as follows. For any cuts $c_1 = (K_1, L_1)$ and $c_2 = (K_2, L_2)$, we define the relation $c_1 < c_2$ iff $K_1 \subsetneq K_2$. It is pure routine to check that this relation endows $\hat{J}$ with a linear ordering. The cuts $(\emptyset, J)$ and $(J, \emptyset)$ are then the first and the last element of $\hat{J}$. It is sometimes convenient to ignore these two cuts. We denote by $\hat{J}^*$ the set $\hat{J} \setminus \{(\emptyset, J), (J, \emptyset)\}$.

Example 2. Let J be the ordinal ω. The set $\hat{J}$ contains the cut $(\{0, \ldots, n - 1\}, \{n, n+1, \ldots\})$ for any integer n and the last cut $(\omega, \emptyset)$. The ordering $\hat{J}$ is thus the ordinal $\omega + 1$. The ordering $\hat{J}$ for $J = \eta$ is not countable since it contains the usual ordering on the set of real numbers.

2.1 The Ordering $J \cup \hat{J}$

The orderings of J and $\hat{J}$ can be extended to an ordering on the disjoint union $J \cup \hat{J}$ as follows. For $j \in J$ and a cut $c = (K, L)$, define the relations $j < c$ and $c < j$ by respectively $j \in K$ and $j \in L$. Note that exactly one of these two relations holds since (K, L) is a partition of J. These relations together with the orderings of J and $\hat{J}$ endows $J \cup \hat{J}$ with a linear ordering.

Example 3. For the ordering $J = \zeta^2$, there is a cut between each consecutive elements in each copy of ζ, but there is also a cut between consecutive copies of ζ. There are also the first and the last cuts. The ordering $J \cup \hat{J}$ is pictured in Figure 1 where each element of J is represented by a bullet and each cut by a vertical bar.

For any element j, there are two cuts c_j^- and c_j^+ obtained by cutting J on the left and on the right of j. These two cuts c_j^- and c_j^+ are defined by $c_j^- = (K, \{j\} \cup L)$ and $c_j^+ = (K \cup \{j\}, L)$ where K and L are the sets of elements smaller and greater than j, that is, $K = \{k \mid k < j\}$ and $L = \{k \mid j < k\}$. The cuts c_j^- and c_j^+ are consecutive in the ordering $\hat{J}$. Conversely, for any consecutive

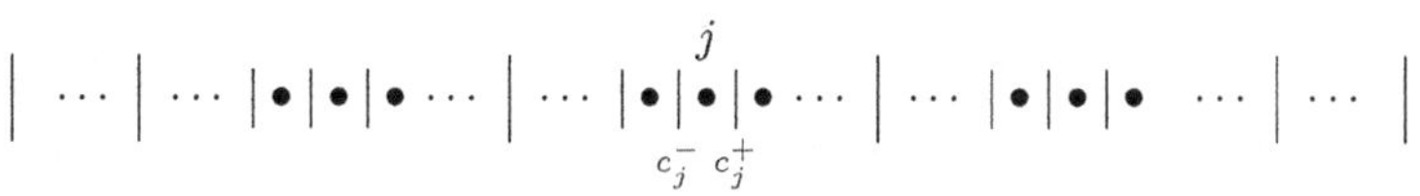

Fig. 1. Ordering $J \cup \hat{J}$ for $J = \zeta^2$

cuts c and c' in $\hat{J}$, there is a unique element j in J such that $c = c_j^-$ and $c' = c_j^+$. In the ordering $J \cup \hat{J}$, one has $c_j^- < j < c_j^+$. Note that if j and j' are consecutive elements in J, then the equality $c_j^+ = c_{j'}^-$ holds.

We denote by $J \cup \hat{J}^*$ the set $J \cup \hat{J} \setminus \{(\emptyset, J), (J, \emptyset)\}$.

3   Words on Linear Orderings

Let A be a finite alphabet whose elements are called letters. For a linear ordering J, a *word* of *length* J over A is a function from J to A which maps any element j of J to a letter a_j of A. A word $(a_j)_{j \in J}$ on a linear ordering J can be seen as a labeled ordering where each point of J has been decorated by a letter. The word whose length is the empty set is called the *empty word* and it is denoted by ε. The set of all words is denoted by $A^\diamond$.

The notion of word we have introduced generalizes the notions of word already considered in the literature. If the ordering J is finite with n elements, a words of length J is a finite sequence $a_1 \ldots a_n$ of letters [19]. A word of length ω is an ω-sequence $a_0 a_1 a_2 \ldots$ of letters which is usually called an ω-word or an infinite word [23]. A word of length ζ is a sequence $\ldots a_{-2} a_{-1} a_0 a_1 a_2 \ldots$ of letters which is usually called a bi-infinite word. An ordinal word is a word indexed by a ordinal.

Example 4. The word $x = b^{-\omega} a b^\omega$ is the word of length $J = \zeta$ defined by $x_j = a$ if $j = 0$ and by $x_j = b$ otherwise.

Let $x = (a_j)_{j \in J}$ and $y = (b_k)_{k \in K}$ be two words of length J and K. The *product* xy (or the *concatenation*) of x and y is the word $z = (c_i)_{i \in J + K}$ of length $J + K$ such that $c_i = a_i$ if $i \in J$ and $c_i = b_i$ if $i \in K$. More generally, let J be a linear ordering and for each $j \in J$, let x_j be a word of length K_j. The *product* $\prod_{j \in J} x_j$ is the word z of length $K = \sum_{j \in J} K_j$ defined as follows. Suppose that each word x_j is equal to $(a_{k,j})_{k \in K_j}$ and recall that K is the set of all pairs (k, j) such that $k \in K_j$. The product z is then equal to $(a_{k,j})_{(k,j) \in K}$.

Example 5. Let J be the ordering ζ and for $j \in J$ define the word x_j by $x_j = b^{-\omega}$ if j is even and by $x_j = ab^\omega$ if j is odd. The product $\prod_{j \in J} x_j$ is the word $(b^{-\omega} a b^\omega)^\zeta$ of length ζ^2.

Two words $x = (a_j)_{j \in J}$ and $y = (b_k)_{k \in K}$ of length J and K are *isomorphic* if there is an ordering isomorphism f from J into K such that $a_j = b_{f(j)}$ for any j in J. This obviously defines an equivalence relation on words. In this paper, we identify isomorphic words and a word is actually a class of isomorphic words. It makes sense to identify isomorphic words since automata and rational expressions that we introduce do not distinguish isomorphic words.

Note that some orderings like ζ have non trivial internal isomorphisms. For instance, let x be the word of Example 4 and y be the word defined by $y_j = a$ if $j = 1$ and by $y_j = b$ otherwise. The two words x and y are isomorphic since the function f given by $f(x) = x + 1$ is an isomorphism from ζ to ζ.

4   Automata

In this section, automata on words on linear orderings are defined. They are a natural generalization of Muller automata on ω-words and of automata introduced by Büchi [11] on ordinal words. The latter automata are usual (Kleene) automata with additional limit transitions of the form $P \to p$ used for limit ordinals. The automata that we introduce have limit transitions of the form $P \to p$ as well as of the form $p \to P$. We point out that these automata make sense for all linear orderings.

Definition 6. *Let A be a finite alphabet. An* automaton *$\mathcal{A}$ over A is a 4-tuple (Q, E, I, F) where Q is a finite set of states, $E \subseteq (Q \times A \times Q) \cup (\mathcal{P}(Q) \times Q) \cup (Q \times \mathcal{P}(Q))$ is the set of transitions, $I \subseteq Q$ is the set of initial states and $F \subseteq Q$ is the set of final states.*

Since the alphabet and the set of states are finite, the set of transitions is also finite. Transitions are either *successor* transitions of the form $p \xrightarrow{a} q$, or *left limit* transitions of the form $P \to q$, or *right limit* transitions of the form $q \to P$, where P is a subset of Q.

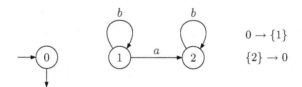

$$0 \to \{1\}$$
$$\{2\} \to 0$$

Fig. 2. Automaton of Example 7

Example 7. The automaton pictured in Figure 2 has 3 successor transitions which are pictured like a labeled graph. It has also a left limit transition $\{2\} \to 0$ and a right limit transition $0 \to \{1\}$. The state 0 is the only initial state and the only final state.

In order to define the notion of path in such an automaton, the following notion of limits is needed. We define it for an arbitrary linear ordering J but we use it when the considered ordering is actually the ordering $\hat{J}$ of cuts of a given ordering J. Let Q be a finite set, let J a linear ordering and let $\gamma = (q_j)_{j \in J}$ be a word over Q. Let j be a fixed element of J. The *left* and *right* limit set of γ at j are the two subsets $\lim_{j^-} \gamma$ and $\lim_{j^+} \gamma$ of Q defined as follows.

$$\lim_{j^-} \gamma = \{q \in Q \mid \forall k < j \; \exists i \quad k < i < j \text{ and } q = q_i\},$$

$$\lim_{j^+} \gamma = \{q \in Q \mid \forall j < k \; \exists i \quad j < i < k \text{ and } q = q_i\}.$$

Note that if the element j has a predecessor, the limit set $\lim_{j-} \gamma$ is empty. Conversely, if the element j is not the least element of J and if it has no predecessor, the limit set $\lim_{j-} \gamma$ is nonempty since the set Q is finite. Similar results hold for right limit sets.

We now come to the definition of a path in an automaton on linear orderings. Let x be a word of length J. Roughly speaking, a path associated with x is a labeling of each cut of J by a state of the automaton such that local properties are satisfied. If two cuts are consecutive there must be a successor transition labeled by the letter of x in between the two cuts. If a cut does not have a predecessor or a successor, there must be a limit transition between the left or right limit set of states and the state of the cut.

Definition 8. *Let $\mathcal{A}$ be an automaton and let $x = (a_j)_{j \in J}$ be a word of length J. A path γ labeled by x is a sequence of states $\gamma = (q_c)_{c \in \hat{J}}$ of length $\hat{J}$ such that*

- *For any consecutive cuts c_j^- and c_j^+, $q_{c_j^-} \xrightarrow{a_j} q_{c_j^+}$ is a successor transition.*
- *For any cut c which is not the first cut and which has no predecessor, $\lim_{c-} \gamma \to q_c$ is a left limit transition.*
- *For any cut c which is not the last cut and which has no successor, $q_c \to \lim_{c+} \gamma$ is a right limit transition.*

By the previous definition, there is a transition entering the state q_c for any cut c which is not the first cut. This transition is a successor transition if the cut c has a predecessor in $\hat{J}$ and it is a left limit transition otherwise. Similarly, there is a transition leaving q_c for any cut c which is not the last cut.

Since the ordering $\hat{J}$ has a first and a last element, a path always has a first and a last state which are indexed by the first and the last cut. A path is *successful* iff its first state is initial and its last state is final. A word is *accepted* by the automaton iff it is the label of a successful path. A set of words is *recognizable* if it is the set of words accepted by some automaton.

Example 9. Consider the automaton $\mathcal{A}$ of Figure 2 and let x be the word $(b^{-\omega}ab^\omega)^2$ of length $\zeta + \zeta$. A successful path γ labeled by x is pictured in Figure 3. This path is made of two copies of the path $01^{-\omega}2^\omega0$. A path for the word $(b^{-\omega}ab^\omega)^\zeta$ cannot be made by ζ copies of $01^{-\omega}2^\omega0$ because the right limit set of the first state would be $\{0, 1, 2\}$ and the automaton has no transition of the form $q \to \{0, 1, 2\}$. The automaton $\mathcal{A}$ thus recognizes the set $(b^{-\omega}ab^\omega)^*$.

4.1   Comparison with Usual Automata

The aim of this section is to show that these automata are a very natural generalization of automata on finite and infinite words.

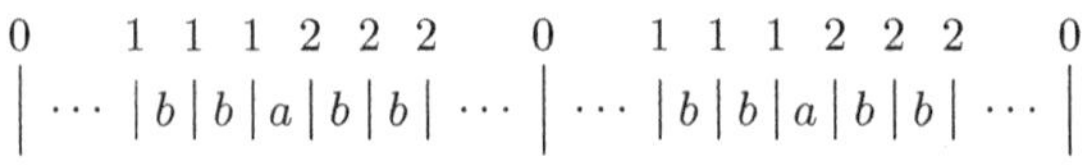

$$0 \quad 1\ 1\ 1\ 2\ 2\ 2 \quad 0 \quad 1\ 1\ 1\ 2\ 2\ 2 \quad 0$$
$$\Big| \cdots \big| b \big| b \big| a \big| b \big| b \big| \cdots \Big| \cdots \big| b \big| b \big| a \big| b \big| b \big| \cdots \Big|$$

Fig. 3. A path labeled by $(b^{-\omega}ab^\omega)^2$

Note first that our automata are usual automata on finite word with additional limit transitions. We claim that the notion of path we have introduced for words on orderings coincide with the usual notion of paths considered in the literature for finite words, ω-words and ordinal words.

Consider first a finite word $x = a_1 \ldots a_n$. The set of cuts of the finite ordering $\{1, \ldots, n\}$ can be identified with $\{1, \ldots, n+1\}$. In our setting, a path labeled by x is then a finite sequence $q_1, \ldots, q_{n+1}$ of states such that $q_j \xrightarrow{a_j} q_{j+1}$ is a successor transition for any j in $\{1, \ldots, n\}$. This matches the usual definition of a path in an automaton [19, p. 5]. The limit transitions of the automaton are not used in this finite path. Indeed, if J is the finite ordering $\{1, \ldots, n\}$, any cut but the first has a predecessor and any cut but the last has a successor in the ordering $\hat{J}$. Furthermore, this property characterizes finite orderings. This implies that a path labeled by a nonfinite word must use a limit transition. This shows that usual automata on finite words are exactly our automata without any limit transition.

Let $x = a_0 a_1 a_2 \ldots$ be an ω-word. The set of cuts of the ordering $J = \omega$ is the ordinal $\omega + 1 = \{0, 1, 2, \ldots, \omega\}$ (see Example 2). The pairs of consecutive cuts are the pairs $(j, j+1)$ for $j < \omega$ whereas the cut $c = \omega$ has no predecessor. In our setting, a path γ labeled by x is a sequence $q_0, q_1, q_2, \ldots, q_\omega$ of states such that $q_j \xrightarrow{a_j} q_{j+1}$ is a successor transition for any $j < \omega$ and such that $\lim_{\omega-} \gamma \to q_\omega$ is a left limit transition. Note that $\lim_{\omega-} \gamma$ is the set of states which occur infinitely many times in γ. This path is successful iff q_0 is initial and q_ω is final. Define the family $\mathcal{T}$ of subsets of states by

$$\mathcal{T} = \{P \mid \exists q \in F \text{ such that } P \to q \in E\}.$$

The path γ is then successful iff q_0 is initial and if the set $\lim_{\omega-} \gamma$ of states belongs to the family $\mathcal{T}$. This matches the definition of a successful path in a Muller automaton [23, p. 148]. A Muller automaton with a acceptance condition given by a table $\mathcal{T}$ can be viewed as an automaton in our setting. It suffices to add to this automaton two new states q_- and q_+ (q_+ being the only final state) and the left limit transitions $P \to q_+$ for $P \in \mathcal{T}$ and $P \to q_-$ for $P \notin \mathcal{T}$.

The set of cuts of an ordinal α is the ordinal $\alpha + 1$. Therefore, the notion of path we have introduced coincide for ordinal words with the notion of path considered in [3].

4.2   Examples of Automata

In this section, we provide some examples of automata.

Example 10. Consider the automaton $\mathcal{A}$ pictured in Figure 4. This automaton has no right limit transition. It recognizes the words whose length is an ordinal since a linear ordering J is an ordinal iff any of its cuts (except the last one) has a successor in $\hat{J}$.

A *gap* in a linear ordering is a cut (K, L) such that K has no greatest element and L has no least element. An ordering is *complete* if it has no gap. For instance,

Fig. 4. Automaton of Example 10

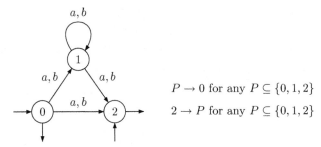

Fig. 5. Automaton of Example 11

Fig. 6. Automaton of Example 12

the ordering $\mathbb{Q}$ of the rational numbers is not complete whereas the ordering $\mathbb{R}$ of the real numbers is complete.

Example 11. Consider the automaton $\mathcal{A}$ pictured in Figure 5. This automaton accepts words whose length is a complete ordering. In this automaton, there is no limit transitions $P \to q$ and $q' \to P'$ with $q = q'$. Therefore, the length of an accepted word cannot have a gap and it is a complete ordering. Conversely, let x whose length J is complete. Since J is complete, any cut c of $\hat{J}$ is of the form c_j^+ or/and of the form $c_{j'}^-$ for some element j of J. Let $\gamma = (q_c)_{c \in \hat{J}}$ be the path defined by $q_c = 0$ if c is of the form $c_{j'}^-$ but not of the form c_j^+, $q_c = 2$ if c is of the form c_j^+ but not of the form $c_{j'}^-$ and $q_c = 1$ if c is both of the form c_j^+ and $c_{j'}^-$. This path is successful and its label is the word x.

Example 12. Consider the automaton $\mathcal{A}$ pictured in Figure 6. This automaton accepts words whose length is a dense and complete ordering without any first and last element. For the same reason as for the automaton of Example 11, the length of an accepted word must be complete. Furthermore, this automaton has no consecutive successor transitions. Therefore any words accepted by this automaton cannot have consecutive letters and its length must be a dense ordering. Conversely, any word with an dense and complete length is accepted by this automaton.

5 Kleene's Theorems

In this section, the rational operations used to define rational sets of words on linear orderings are introduced. These operations include of course the usual Kleene operations for finite words which are the union $+$, the concatenation $\cdot$ and the star operation $*$. They also include the omega iteration $^\omega$ usually used to construct ω-words and the ordinal iteration $^\sharp$ introduced by Wojciechowski [25] for ordinal words. Three new operations are also needed: the backwards omega iteration $^{-\omega}$, the backwards ordinal iteration $^{-\sharp}$ and a last binary operation denoted $\diamond$ which is a kind of iteration for all linear orderings.

We first define the iterations in a unified framework. The general iteration $X^{\mathcal{J}}$ with respect to a class $\mathcal{J}$ of linear orderings can be defined by

$$X^{\mathcal{J}} = \{\prod_{j \in J} x_j \mid J \in \mathcal{J} \text{ and } x_j \in X\}.$$

The sets X^*, X^ω, $X^{-\omega}$, $X^\sharp$ and $X^{-\sharp}$ are then respectively equal to $X^{\mathcal{J}}$ for $\mathcal{J}$ equal to the class $\mathcal{N}$ of all finite linear orderings, the class $\{\omega\}$ which only contains the ordering ω, the class $\{-\omega\}$, the class $\mathcal{O}$ of all ordinals and the class $-\mathcal{O} = \{-\alpha \mid \alpha \in \mathcal{O}\}$.

We now define the binary operations. Let X and Y be two sets of words. The sets $X + Y$, $X \cdot Y$ and $X \diamond Y$ are defined by

$$X + Y = X \cup Y \quad \text{and} \quad X \cdot Y = \{xy \mid x \in X \text{ and } y \in Y\},$$

$$X \diamond Y = \{ \prod_{j \in J \cup \hat{J}^*} z_j \mid J \in \mathcal{S} \text{ and } z_j \in X \text{ if } j \in J \text{ and } z_j \in Y \text{ if } j \in \hat{J}^*\}.$$

A word x belongs to $X \diamond Y$ iff there is a linear ordering J such that x is the product of a sequence of length $J \cup \hat{J}^*$ of words where each word indexed by an element of J belongs to X and each word indexed by a cut in $\hat{J}^*$ belongs to Y.

An abstract *rational expression* is a well-formed term of the free algebra over $A \cup \{\varepsilon\}$ with the symbols denoting the rational operations as function symbols. Each rational expression denotes a set of words which is inductively defined by the above definitions of the rational operations. A set of words is *rational* if it can be denoted by a rational expression. We use the abbreviation X^ς for $X^{-\omega}X^\omega$ and $X^\diamond$ for $X \diamond \varepsilon$.

Example 13. The expressions $A^\diamond$, $A^\diamond a A^\diamond$ respectively denote the set of all words and the set of words having an occurrence of the letter a. The expression $(A^\omega)^{-\omega}$ denote the set of words which are sequences of length $-\omega$ of ω-words. A word x belongs to the set denoted by $a^\varsigma \diamond b$ iff, for some linear ordering J, x is a sequence of length $J \cup \hat{J}^*$ of words x_j such that $x_j = a^\varsigma$ for any $j \in J$ and $x_j = b$ for any $j \in \hat{J}^*$.

Example 14. The expression AA^* denotes the set of nonempty finite words and the expression $(A^\diamond)^\omega A^\diamond + A^\diamond (A^\diamond)^{-\omega}$ denotes its complement. Indeed, a linear ordering $J \neq \emptyset$ is not finite if it has at least a cut (K, L) such that either K does not have a greatest element or L does not have a least element. The rational expression $(A^\diamond)^\omega$ denotes the set of words whose length does not have a last element. Therefore, the expression $(A^\diamond)^\omega A^\diamond$ denotes the set whose length has a cut (K, L) such that K does not have a greatest element. Symmetrically, the expression $A^\diamond (A^\diamond)^{-\omega}$ denotes the set whose length has a cut (K, L) such that L does not have a least element.

Recall that a linear ordering J is said to be *dense* if for any $i < k$ in J, there is $j \in J$ such that $i < j < k$. It is *scattered* if it contains no dense subordering. In [6, 7], we have been able to prove the following extension of Kleene's theorem.

Theorem 15. *A set of words on countable scattered linear orderings is rational iff it is recognizable.*

In [8], we have proposed a hierarchy among rational sets of words on linear orderings. As for star-free sets, this hierarchy is obtained by restricting the rational operations that can be used. Each class contains the rational sets that can be described by a given subset of the rational operations.

We give a characterization of each class of the hierarchy by a corresponding class of automata. A set of words belongs to the given class if and only if it is recognized by an automaton of the corresponding class. Each of these characterizations is thus a Kleene's theorem which holds for that class. For well-known classes, these Kleene's theorems were already proved by Wojciechowski [25] for words on ordinals or by Choueka [12] for words on ordinals smaller than ω^ω. In each case, the corresponding class of automata is obtained naturally by restricting the kind of transitions that can be used. For instance, the automata for words on ordinals do have left limit transitions but no right limit transitions as there were defined by Büchi [11].

A natural generalization of the result would be to remove the restrictions on the orderings and to first consider words on countable linear orderings and then words on all linear orderings. Automata that we have introduced are suitable for all linear orderings. It seems however that new rational operations are then needed. An operation like the η-shuffle introduced in [16] is necessary.

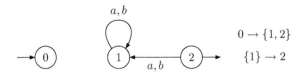

Fig. 7. Automaton recognizing $(A^\omega)^{-\omega}$

Example 16. The automaton pictured in Figure 7 recognizes the set denoted by the rational expression $(A^\omega)^{-\omega}$. The part of the automaton given by state 1 and the left limit transition $\{1\} \to 2$ accepts the set A^ω. The successor transition from state 2 to state 1 allows to concatenate two words of A^ω. The right limit transition $0 \to \{1, 2\}$ leads to a sequence of length $-\omega$ of words of A^ω.

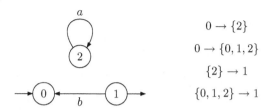

$$0 \to \{2\}$$
$$0 \to \{0, 1, 2\}$$
$$\{2\} \to 1$$
$$\{0, 1, 2\} \to 1$$

Fig. 8. Automaton recognizing $a^\varsigma \diamond b$

Example 17. The automaton pictured in Figure 8 recognizes the set $a^\varsigma \diamond b$. The part of the automaton given by state 2 and the two limit transitions $0 \to \{2\}$ and $\{2\} \to 1$ accepts the word a^ς whereas the part given by the successor transition from state 1 to state 0 accepts the word b. Any occurrence of a^ς is preceded and followed by an occurrence of b in the automaton. More generally, thanks to the limit transitions $0 \to \{0, 1, 2\}$ and $\{0, 1, 2\} \to 1$, the occurrences of a^ς are indexed by a linear ordering J, the occurrences of b are indexed by the ordering $\hat{J}^*$ and they are interleaved according to the ordering $J \cup \hat{J}^*$.

Conclusion

As a conclusion, we mention some open problems. Automata on infinite words have been introduced by Büchi to prove the decidability of the monadic second-order theory of the integers [9]. Since then, automata and logics have been shown to have strong connections [23]. It is known that the monadic second-order theory of all linear orderings is decidable [21] but the proof is based on a model theoretic approach, called the composition method, avoiding the use of automata. We refer to [24] for a review of the composition method and the Shelah's proof of Büchi's decidability result. The next step is to investigate the connections between logics and the automata that we have introduced. Such a study has to begin with the closure of the class of recognizable sets under the boolean operations. We do not know whether this class is closed under the complementation.

References

1. E. Asarin. Equations on timed languages. In T. Henzinger and S. Sastry, editors, *Hybrid Systems : Computation and Control*, number 1386 in Lect. Notes in Comput. Sci., pages 1–12, 1998.

2. E. Asarin, P. Caspi, and O. Maler. A Kleene theorem for timed automata. In *Proceedings, Twelth Annual IEEE Symposium on Logic in Computer Science*, pages 160–171, 1997.
3. N. Bedon and O. Carton. An Eilenberg theorem for words on countable ordinals. In Cláudio L. Lucchesi and Arnaldo V. Moura, editors, *Latin'98: Theoretical Informatics*, volume 1380 of *Lect. Notes in Comput. Sci.*, pages 53–64. Springer-Verlag, 1998.
4. B. Bérard and C. Picaronny. Accepting Zeno words without making time stand still. In *Mathematical Foundations of Computer Science 1997*, volume 1295 of *Lect. Notes in Comput. Sci.*, pages 149–158, 1997.
5. P. Bouyer and A. Petit. A Kleene/Büchi-like theorem for clock languages. *J. of Automata, Languages and Combinatorics*, 2001. To appear.
6. V. Bruyère and O. Carton. Automata on linear orderings. Technical Report 2000–12, Institut Gaspard Monge, 2000. Submitted.
7. V. Bruyère and O. Carton. Automata on linear orderings. In Jiří Sgall, Aleš Pultr, and Petr Kolman, editors, *MFCS'2001*, volume 2136 of *Lect. Notes in Comput. Sci.*, pages 236–247, 2001.
8. V. Bruyère and O. Carton. Hierarchy among automata on linear orderings. In R. Baeza-Yates, U. Montanari, and N. Santoro, editors, *TCS'2002/IFIP'2002*, pages 107–118. Kluwer Academic Publishers, 2002.
9. J. R. Büchi. Weak second-order arithmetic and finite automata. *Z. Math. Logik und grundl. Math.*, 6:66–92, 1960.
10. J. R. Büchi. On a decision method in the restricted second-order arithmetic. In *Proc. Int. Congress Logic, Methodology and Philosophy of science, Berkeley 1960*, pages 1–11. Stanford University Press, 1962.
11. J. R. Büchi. Transfinite automata recursions and weak second order theory of ordinals. In *Proc. Int. Congress Logic, Methodology, and Philosophy of Science, Jerusalem 1964*, pages 2–23. North Holland, 1965.
12. Y. Choueka. Finite automata, definable sets, and regular expressions over ω^n-tapes. *J. Comput. System Sci.*, 17(1):81–97, 1978.
13. B. Courcelle. Frontiers of infinite trees. *RAIRO Theoretical Informatics*, 12(4):319–337, 1978.
14. D. Girault-Beauquier. Bilimites de langages reconnaissables. *Theoret. Comput. Sci.*, 33(2–3):335–342, 1984.
15. M. R. Hansen, P. K. Pandya, and Z. Chaochen. Finite divergence. *Theoret. Comput. Sci.*, 138(1):113–139, 1995.
16. S. Heilbrunner. An algorithm for the solution of fixed-point equations for infinite words. *RAIRO Theoretical Informatics*, 14(2):131–141, 1980.
17. S. C. Kleene. Representation of events in nerve nets and finite automata. In C.E. Shannon, editor, *Automata studies*, pages 3–41. Princeton University Press, Princeton, 1956.
18. M. Nivat and D. Perrin. Ensembles reconnaissables de mots bi-infinis. In *Proceedings of the Fourteenth Annual ACM Symposium on Theory of Computing*, pages 47–59, 1982.
19. D. Perrin. Finite automata. In J. van Leeuwen, editor, *Handbook of Theoretical Computer Science*, volume B, chapter 1, pages 1–57. Elsevier, 1990.
20. J. G. Rosenstein. *Linear Orderings*. Academic Press, New York, 1982.
21. S. Shelah. The monadic theory of order. *Annals of Mathematics*, 102:379–419, 1975.
22. W. Thomas. On frontiers of regular sets. *RAIRO Theoretical Informatics*, 20:371–381, 1986.

23. W. Thomas. Automata on infinite objects. In J. van Leeuwen, editor, *Handbook of Theoretical Computer Science*, volume B, chapter 4, pages 133–191. Elsevier, 1990.

24. W. Thomas. Ehrenfeucht games, the composition method, and the monadic theory of ordinal words. In *Structures in Logic and Computer Science, A Selection of Essays in Honor of A. Ehrenfeucht*, number 1261 in Lect. Notes in Comput. Sci., pages 118–143. Springer-Verlag, 1997.

25. J. Wojciechowski. Finite automata on transfinite sequences and regular expressions. *Fundamenta Informaticæ*, 8(3-4):379–396, 1985.

Some Properties of Ciliate Bio-operations*

Mark Daley and Lila Kari

Department of Computer Science,
University of Western Ontario,
London, ON, N6A 5B7 Canada,
{lila,daley}@csd.uwo.ca

Abstract. The process of gene unscrambling in ciliates (a type of unicellular protozoa), which accomplishes the difficult task of re-arranging gene segments in the correct order and deleting non-coding sequences from an "encrypted" version of a DNA strand, has been modeled and studied so far from the point of view of the computational power of the DNA bio-operations involved. Here we concentrate on a different aspect of the process, by considering only the linear version of the bio-operations, that do not involve thus any circular strands, and by studying the resulting formal operations from a purely language theoretic point of view. We namely investigate closure operations of language families under the mentioned bio-operations and study language equations involving them. Among the problems addressed, we study the decidability of existence of solutions to equations of the form $L \diamond Y = R$, $X \diamond L = R$ where L and R are given languages, X and Y are unknowns, and $\diamond$ signifies one of the defined bio-operations.

1 Introduction

The stichotrichous ciliates are a group of primitive single-celled organisms which have generated a great deal of interest due to their unique genetic mechanisms. Most organisms store their genomic DNA in a linear sequence consisting of coding regions interspersed with non-coding regions. Several ciliate genes, however, are stored in a scrambled form. For example, if a functional copy of a gene consists of the coding regions arranged in the order 1-2-3-4-5, it may appear in the order 3-5-4-1-2 in the genome. This presents an interesting problem for the organism, who must somehow descramble these genes in order to generate functional proteins required for its continued existence.

The details of the biological mechanism underlying the unscrambling process are still unknown. For further information on the biology of the descrambling process in ciliates the reader is referred to [12,13,14]. The two existing formal models for gene unscrambling, by Kari and Landweber [5,10], respectively Ehrenfeucht, Harju, Petre, Prescott and Rozenberg [11,3,15] are consistent with the existing

---

* Research supported by Natural Sciences and Engineering Council of Canada Grants. All correspondence to L.K.

biological data. Each proposes a set of two, respectively three, atomic operations the combination of which can lead to the unscrambling of an arbitrarily scrambled gene. The bio-operations proposed by the first model are circular insertions and deletions, i.e. insertions and deletions of circular strands into/from linear strands, guided by the presence of certain pointers [5,10]. The second model focuses more on the properties of pointers and proposes three operations: hi(hairpin loop with inverted pointers) which reverses a substring between two inverted pointer sequences, ld(loop with direct pointers)-excision which deletes a substring between two pointers and dlad(double loop with alternating direct pointers)-excision/reinsertion which swaps two substrings marked by pointer-pairs. In both cases, the operations presented are based on real biological events that can occur and change a DNA molecule.

This paper does not address the biological aspects and implications of the proposed operations. Instead, we continue in the style of Dassow et. al's work on properties of operations inspired by general DNA recombination events [2] and focus on some of their properties as word/language operations. We namely consider, in Section 2, closure properties of languages in the Chomsky hierarchy under the defined operations. Moreover, in Section 3 we consider language equations of the type $L \diamond Y = R$, $X \diamond L = R$ where L and R are given languages and X and Y are the unknowns. We study the decidability of the existence of solutions to these equations as well as the existence of singleton solutions.

The notations used in the paper are summarized as follows. An alphabet Σ is a finite non-empty set. A word w over Σ is an element of the free semigroup (denoted Σ^+) generated by the letters of Σ and the catenation operation. The length of a word, written $|w|$ is equal to the number of letters in the word. In the free monoid Σ^* we also allow the empty word λ where $|\lambda| = 0$. A language L is a, possibly infinite, set of words over a given alphabet. The complement of a language L is written L^c and is defined as $L^c = \Sigma^* \setminus L$.

We will consider here the classic families of the Chomsky Hierarchy, that is, the families of: regular languages (REG), context-free languages (CF), context-sensitive languages (CS) and recursively enumerable languages (RE). For further details on basic formal language theory, the reader is referred to [16].

2   Closure Properties

This section will define the two bio-operations of the [5,10] model and a generalization of an operation in the [11,3,15] model and investigate closure properties of families in the Chomsky hierarchy under them.

2.1   Synchronized Insertion and Deletion

Two basic operations have been defined in [5,10] that model the processes of intramolecular respectively intermolecular DNA recombinations that are thought to accomplish gene unscrambling.

The operation modeling the intermolecular recombination accomplishes the insertion of a circular sequence vx in a linear string uxw, resulting in a string

$uxvxw$. If we ignore the fact that the inserted string is circular, the operation, called *synchronized insertion* (the word "synchronized" points out that insertion can only happen if the sequence x is present in both input strands), is formally defined as follows.

Definition 1. *Let α, β be two nonempty words in Σ^*. The synchronized insertion of β into α is defined as: $\alpha \oplus \beta = \{uxvxw | \alpha = uxw, \beta = vx, x \in \Sigma^+, u, v, w \in \Sigma^*\}$.*

The operation modeling intramolecular recombination accomplishes the deletion of a sequence vx from the original strand $uxvxw$, in the form of a circular strand. Ignoring the differences between the linear and circular strands, the resulting operation, that of *synchronized deletion*, is defined as follows.

Definition 2. *Let α, β be two nonempty words in Σ^*. The synchronized deletion of β from α is defined as: $\alpha \ominus \beta = \{uxw | \alpha = uxvxw, \beta = vx, x \in \Sigma^+, u, v, w \in \Sigma^*\}$.*

The operations differ from the original ones defined in [10] and [5] in that no circular strands are present here. The above two definitions can be extended to languages in the natural way. This section examines the closure properties of the families of regular, context-free, context-sensitive and recursively enumerable languages under the synchronized insertion and deletion.

We begin by recognizing that we can consider, without loss of generality, only one-symbol contexts instead of arbitrarily-sized contexts.

Lemma 1. *For any $\alpha, \beta \in \Sigma^+$, $\alpha \oplus \beta = \{u'av'aw' | \alpha = u'aw', \beta = v'a, a \in \Sigma, u', v', w' \in \Sigma^*\}$.*

Using the preceding result we can now show that the synchronized insertion can be expressed in terms of the controlled sequential insertion, defined in [9] as follows. Let $L \subseteq \Sigma^+$ and, for each $a \in \Sigma$, let $\Delta(a) \subseteq \Sigma^*$. The controlled sequential insertion into L, according to Δ, is defined as $L \longleftarrow \Delta = \cup_{u \in L}(u \longleftarrow \Delta)$, where $u \longleftarrow \Delta = \{u_1 a v_a u_2 | u = u_1 a u_2, v_a \in \Delta(a)\}$.

Proposition 1. *REG, CF, CS and RE are closed under synchronized insertion.*

Proof. We claim that $L_1 \oplus L_2 = L_1 \longleftarrow \Delta$ where $\Delta(a) = (L_2\{a\}^{-1}) \{a\}$, and the right quotient of two languages in Σ^*, denoted by $L_1 L_2^{-1}$, is defined as

$$L_1 L_2^{-1} = \{u | uv \in L_1, v \in L_2\}.$$

"$\subseteq$" Given $\gamma \in L_1 \oplus L_2$, by Lemma 1 γ is of the form $uavaw$ where $uaw \in L_1$, $va \in L_2$. As $va \in L_2$, $v \in L_2\{a\}^{-1}$ and therefore $va \in (L_2\{a\}^{-1})\{a\}$. Consequently, $\gamma = uavaw \in L_1 \longleftarrow \Delta$.

"$\supseteq$" Suppose $\gamma \in L_1 \longleftarrow \Delta$. Then γ is of the form $uav_a w$ where $uaw \in L_1$ and $v_a \in \Delta(a)$. Since $\Delta(a) = (L_2\{a\}^{-1})\{a\}$, $v_a = va$, $v \in L_2/\{a\}$ and therefore $va \in L_2$. Consequently, $\gamma \in uaw \oplus va$, $uaw \in L_1, va \in L_2, a \in \Sigma^+, u, v, w \in \Sigma^*$.

Since REG, CF, CS, RE are closed under controlled sequential insertion [9], they are closed under synchronized insertion as well.

The closure of the family of regular languages under synchronized deletion can be similarly ascertained by expressing synchronized deletion in terms of the controlled sequential deletion, [9]. The controlled sequential deletion is defined similarly to the controlled sequential insertion, with the difference that

$$u \longrightarrow \Delta = \{u_1 a u_2 | \ u = u_1 a v_a u_2, u_1, u_2 \in \Sigma^*, a \in \Sigma, v_a \in \Delta(a)\}.$$

By [9], REG are closed under controlled sequential deletion, while CF is not (but it is closed under controlled sequential deletion with regular languages) and CS is not even closed under controlled sequential deletion with regular languages (but it is closed under controlled sequential deletion with singleton languages). The first of these closure properties leads to the next closure result, preceded by a lemma that simplifies synchronized deletion similarly to the synchronized insertion.

Lemma 2. *For any* $\alpha, \beta \in \Sigma^+$ $\alpha \ominus \beta = \{u'aw | \alpha = u'av'aw, \beta = v'a, a \in \Sigma, u', v', w \in \Sigma^*\}.$

We are now ready to address the closure properties of the families in the Chomsky hierarchy under synchronized deletion.

Proposition 2. *REG and RE are closed under synchronized deletion.*

Proof. We claim that $L_1 \ominus L_2 = L_1 \longrightarrow \Delta$ where $\Delta(a) = (L_2\{a\}^{-1})\{a\}$.
   "$\subseteq$" Given $\gamma \in L_1 \ominus L_2$, by Lemma 2, we have $\gamma = uaw$ where $uavaw \in L_1$ and $va \in L_2$. Given that $\Delta(a) = (L_2\{a\}^{-1})\{a\}$, we have $va \in \Delta(a)$ and thus $\gamma = uaw \in L_1 \longrightarrow \Delta$.
   "$\supseteq$" Suppose $\gamma \in L_1 \longrightarrow \Delta$ where $\Delta(a) = (L_2\{a\}^{-1})\{a\}$. Then γ is of the form uaw where $uavaw \in L_1$ and $va \in (L_2\{a\}^{-1})\{a\}$, thus $\gamma \in L_1 \ominus L_2$. The result follows as REG and RE are closed under controlled sequential deletion [9].

Proposition 3. *CF is not closed under synchronized deletion.*

Proof. Let L_1, L_2 be the context free languages:

$$L_1 = \#(\{a^i b^{2i} | i > 0\}^* \amalg \{\#\}),$$

$$L_2 = a\{b^i a^i | i > 0\}^* \#,$$

where the shuffle operation is defined as

$$u \amalg v = \{u_1 v_1 \dots u_n v_n | \ u = u_1 \dots u_n, v = v_1 \dots v_n, u_i, v_i \in \Sigma^*, 1 \le i \le n\}.$$

(Similar languages were used in [4] to show that CF is not closed under left quotient).
   The language
$$(L_1 \ominus L_2) \cap \#b^* = \{b^{2^n} | n > 0\}$$
is not context-free and the result follows.

Proposition 4. *CS is not closed under synchronized deletion with regular languages.*

Proof. Let $L \subseteq \Sigma^*$, with $a, b \notin \Sigma$, be a recursively enumerable language which is not context-sensitive.

There exists a context-sensitive language L_1 such that L_1 consists of words of the form $a^i b\alpha$ where $i \geq 0$ and $\alpha \in L$. Furthermore, for all $\alpha \in L$ there exists some $i \geq 0$ such that $a^i b\alpha \in L_1$ [16].

Suppose # is a symbol not in $\Sigma \cup \{a, b\}$. Consider the language

$$\#(L_1 \amalg \{\#\}) \ominus a^* b\#.$$

Clearly, $a^* b\#$ is regular and moreover, $\#(L_1 \amalg \{\#\}) \ominus a^* b\# = \#L$ which, by the definition of L, is not context-sensitive.

Even though CS is not closed under synchronized deletion with regular languages, it is closed under synchronized deletion with singleton languages. Indeed, this follows directly from Proposition 2 and the closure of context-sensitive languages under controlled sequential deletion with singleton languages [9].

2.2   Hairpin Inversion

We next consider a generalization of the hairpin inversion operation hi defined in [11]. The name of the operation reflects the fact that it models the process of a DNA strand forming a hairpin, having the end of the hairpin cleaved and then re-attached with the sticky ends switched. This results in the sequence of the cleaved and re-attached region now being the mirror image of what it was prior to the operation.

If $w = a_1 a_2 \ldots a_n$, $a_i \in \Sigma$, $1 \leq i \leq n$ is a word in Σ^+ the *reverse* or *mirror image* of w is denoted by $\tilde{w}$ and defined as $\tilde{w} = a_n \ldots a_2 a_1$.

Definition 3. *Let α be a word in Σ^+. The hairpin inverse of α, denoted by $hi(\alpha)$ is defined as $hi(\alpha) = \{x p \tilde{y} \tilde{p} z | \alpha = x p y \tilde{p} z \text{ and } x, y, z \in \Sigma^*, p \in \Sigma^+\}$.*

This definition can be extended to languages in Σ^+ in the natural way. Similarly to Lemma 1, we first show that it is enough to consider pointers of length one only.

Lemma 3. *If $\alpha \in \Sigma^+$, $hi(\alpha) = \{x a \tilde{y} a z | \alpha = x a y a z, a \in \Sigma, x, y, z \in \Sigma^*\}$.*

The mirror image operation has the property that $\tilde{\tilde{L}} = L$. The hairpin inversion operation is a variation of the mirror image operation in that it inverts subwords inside words of a language. The following lemma answers the question of whether or not applying hairpin inversion twice to a language yields the original language. As it turns out, while the hairpin invertible words of a language L are included in $hi(hi(L))$, the reverse does not hold.

Lemma 4. *If $L \subseteq \Sigma^+$, then for all $a \in \Sigma$, $L \cap \Sigma^* a \Sigma^* a \Sigma^* \subseteq hi(hi(L))$ while $hi(hi(L))$ is not included in $L \cap \Sigma^* a \Sigma^* a \Sigma^*$.*

The following propositions address the closure properties of families in the Chomsky hierarchy under the operation of hairpin inversion.

Proposition 5. *REG is closed under hairpin inversion.*

Proof. Let L be the regular language accepted by an automaton $A = (\Sigma, S, s_o, s_f, P)$, where Σ is the alphabet, S is the set of states, s_0 is the initial state, s_f is the final state, and P is the set of productions of the form $sa \longrightarrow s'$, $s, s' \in S$, $a \in \Sigma$. For every two states $s_i, s_j \in S$, define $L_{s_i, s_j} = \{w \in \Sigma^* | s_i w \Rightarrow^* s_j\}$. In other words, L_{s_i, s_j} consists of those words which will cause the automaton A to move from state s_i to state s_j when read. We claim that

$$hi(L) = \bigcup_{s_i, s_j, s_k, s_l \in S} \quad \bigcup_{a \in L_{s_i, s_j} \cap L_{s_k, s_l} \cap \Sigma} L_{s_0, s_i} \{a\} \widetilde{L_{s_j, s_k}} \{a\} L_{s_l, s_f}.$$

"$\subseteq$" Let $\alpha \in hi(L)$. Then there exists $xayaz \in L, a \in \Sigma, x, y, z \in \Sigma^*$ such that $\alpha = xa\tilde{y}az$. As $xayaz \in L(A) = L$, there exists a derivation $s_0 xayaz \Rightarrow^* s_f$, i.e. there exist $s_i, s_j, s_k, s_l \in S$ such that $s_0 xayaz \Rightarrow^* s_i ayaz \Rightarrow^* s_j yaz \Rightarrow^* s_k az \Rightarrow^* s_l z \Rightarrow^* s_f$.

This implies $x \in L_{s_0, s_i}, a \in L_{s_i, s_j}, y \in L_{s_j, s_k}, a \in L_{s_k, s_l}, z \in L_{s_e}, s_f$.

We then have $\tilde{y} \in \widetilde{L_{s_j, s_k}}$ and $\alpha \in L_{s_0, s_i} a \widetilde{L_{s_j, s_k}} a L_{s_l, s_f} \subseteq$ RHS.

"$\supseteq$" Consider $\alpha \in$ RHS. Take $x \in L_{s_0, s_i}, \tilde{y} \in \widetilde{L_{s_j, s_k}}, z \in L_{s_l, s_f}$ and $a = a$. This means $xayaz \in L_{s_0, s_i} L_{s_i, s_j} L_{s_j, s_k} L_{s_k, s_e} L_{s_l, s_f}$ which means there exists a derivation $s_0 xayaz \Rightarrow^* s_f$ which implies $xayaz \in L$ and $xa\tilde{y}az \in hi(L)$. The proposition now follows as REG is closed under finite union, catenation and mirror image.

Proposition 6. *CF is not closed under hairpin inversion.*

Proof. Consider the language $L = \{wpc\tilde{w}dp | w \in \{a, b\}^*, p, c, d \in \Sigma\}$ where $\Sigma = \{a, b, p, c, d\}$ and $\tilde{w}$ denotes the mirror image of w.
Clearly L is context-free. However,

$$hi(L) \cap \Sigma^* pd\Sigma^* cp = \{wpdwcp | w \in \{a, b\}^*, p, c, d \in \Sigma\}$$

which is a classic example of a non context-free language.

Proposition 7. *CS is closed under hairpin inversion.*

Proof. Let $L = L(G)$ where $G = (N, T, S, P)$ is generated by a context-sensitive grammar in a normal form where every production in P containing letters of T is of the form $X \to a$ where $X \in N^*, a \in T$, [16].

We construct a context-sensitive grammar $G' = (N', T', S', P')$ as follows: $N' = N \cup \{X_a, Y_a, Z_a, B_a, a'', a' | a \in T\} \cup \{S', \$, C, C', C'', B, B'\}$, $P' = \{A \to X_a | A \to a \in P\} \cup \{A \to Y_a | A \to a \in P\} \cup (P \setminus \{A \to a | A \to a \in P\}) \cup \{X_a \to a, Y_a \to a, a' \to a\}$, $T' = T \cup \{\#\}$.

In addition, for all $a, d \in \Sigma$, P' contains the following rules:

1. $S' \to \#BS\#$
2. $Ba' \to a'B$ (B will have to check if the sentential form is in the correct form. Here B traverses x' in a sentential form $\#Bx'X_ay'Y_az'\#$ corresponding to a word $xayaz \in L(G)$).
3. $BX_a \to X_aB_a$ (B meets X_a, this change is registered by its transformation in B_a which stores knowledge about a in its subscript.)
4. $B_ac' \to c'B_a$ (B_a reads y')
5. $B_aY_a \to Y_aB'$ (B_a reads Y_a if Y_a has same index with X_a)
6. $B'a' \to a'B'$ (B' reads z')
7. $B'\# \to C\#$ (B' reaches the end of the sentential form and changes to C)
8. $\alpha C \to C\alpha$, $\alpha \in \Sigma' \cup \{Y_a\}$ (C moves left until it reaches X_a, when it will start reversing y')
9. $X_aC \to X_aC'$ (C' starts to reverse the word y')
10. $C'd' \to Z_dC'$ (Store d' in Z_d, where d' is the first letter of y')
11. $Z_dC'b' \to b'Z_dC'$ (Z_dC' moves right until it reaches Y_a)
12. $Z_dC'Y_a \to d''C''Y_a$ (the move of the first letter d of y' to the end of the word has been completed)
13. $\alpha C'' \to C''\alpha$, $\alpha \in \{a'', a'|a \in \Sigma\}$ (C'' moves left until it reaches X_a)
14. $X_aC'' \to X_aC'$ (start again)
15. $Z_dC'b'' \to b''Z_dC'$ (Z_dC' should be able to move also over b'')
16. $X_aC'd'' \to X_a\$d''$ (When there are no letters in y' to invert anymore, transform C' into $\$$).
17. $\$a \to a\$$, $\alpha \in \{a', a'', Y_a|a \in \Sigma\}$, $\$\# \to \#\#$ (the dollar sign moves to the left until it reaches $\#$ when it changes into $\#$).
18. $X_a \to a, Y_a \to a, a' \to a, a'' \to a.$

We claim that $L(G') = \#hi(L)\#\#$. Indeed, a derivation according to G' can only proceed as follows.

$$S' \overset{1}{\Rightarrow} \#BS\# \overset{P}{\Rightarrow}^* \#Bx'X_ay'Y_az'\# \overset{2}{\Rightarrow}^* \#x'BX_ay'Y_az'\# \overset{3}{\Rightarrow}$$

$$\#x'X_aB_ay'Y_az'\# \overset{4}{\Rightarrow}^* \#x'X_ay'B_aY_az'\# \overset{5}{\Rightarrow} \#x'X_ay'Y_aB'z'\# \overset{6}{\Rightarrow}^*$$

$$\#x'X_ay'Y_az'B'\# \overset{7}{\Rightarrow} \#x'X_ay'Y_az'C\# \overset{8}{\Rightarrow}^* \#x'X_aCy'Y_az'\# \overset{9}{\Rightarrow}$$

$$\#x'X_aC'y'Y_az'\# = \#x'X_aC'd'y_1'Y_az'\# \overset{10}{\Rightarrow} \#x'X_aZ_dC'y_1'Y_az'\# \overset{11}{\Rightarrow}^*$$

$$\#x'X_ay_1'Z_dC'Y_az'\# \overset{12}{\Rightarrow} \#x'X_ay_1'd''C''Yaz'\# \overset{13}{\Rightarrow}^* \#x'X_aC''y_1'd''Y_az'\# \overset{14}{\Rightarrow}$$

$$\#x'X_aC'y_1'd''Y_aZ'\# \overset{8-14}{\Rightarrow}^* \#x'X_aC'y''Y_az'\# \overset{16}{\Rightarrow} \#x'X_a\$y''Y_az'\# \overset{17}{\Rightarrow} \#x'y''Y_az'\$$$

$$\# \overset{17}{\Rightarrow} \#x'X_ay''Y_az''\#\# \overset{18}{\Rightarrow}^* \#xayaz\#\#.$$ The proposition now follows as $\#hi(L)\#\#$ is a context-sensitive language and therefore $hi(L)$ is a context-sensitive language as it is the image of $\#hi(L)\#\#$ through a homomorphism that erases $\#$ and leaves all other letters unchanged.

3   Language Equations

We begin this section by investigating equations of the type $hi(X) = R$, where R is a given language and X is the unknown.

Proposition 8. *Let $R \subseteq \Sigma^*$ be regular language. If there exists a language $L \subseteq \Sigma^*$ such that $hi(L) = R$ then there exists a regular language $R', L \subseteq R' \subseteq \Sigma^*$, with the same property.*

Proof. Construct the language $R' = [hi(R^c)]^c$.

(i) We show that $hi(R') \subseteq R$ by way of contradiction. Assume there exists some $u \in hi(R')$ such that $u \notin R$. Since $u \notin R$, it must be the case that $u \in R^c$. As $u \in hi(R')$ it must be of the form $u = x a \tilde{y} a z$ where $x a y a z \in R'$. However, $u = x a \tilde{y} a z$ implies $x a y a z \in hi(u) \subseteq hi(R^c)$ a contradiction since $R' = [hi(R^c)]^c$.

(ii) We show now that every language $L \subseteq \Sigma^*$ such that $hi(L) \subseteq R$ is included in R'. Indeed, assume there exists $L \subseteq \Sigma^*$ with $hi(L) \subseteq R$ but $L \not\subseteq R'$. Then there must exist a word $u \in L \setminus R'$. As $u \notin R', u \in hi(R^c)$ implies that $u = x a \tilde{y} a z$ with $x a y a z \in R^c$. However, as $u \in L$, by the definition of hairpin inversion, $x a y a z \in hi(L) \subseteq R$ a contradiction.

Return now to the proof of the proposition. If there exists L with $hi(L) = R$ then, by (ii), $L \subseteq R'$. By (i) we have that $R = hi(L) \subseteq hi(R') \subseteq R$ which means $hi(R') = R$. By Proposition 5, and closure of REG under complement, R' is regular. Moreover, it follows from the proof that R' can be effectively constructed.

The preceding proposition aids us in deciding whether an equation $hi(X) = R$ has a solution X in case R is a regular language.

Proposition 9. *If $R \subseteq \Sigma^*$ is a regular language, the problem of whether or not the equation $hi(X) = R$ has a solution $X \subseteq \Sigma^*$ is decidable.*

Proof. Construct $R' = [hi(R^c)]^c$. If $hi(X) = R$ has a solution then R' is also, by Proposition 8, a solution. An algorithm for deciding our problem will consist in effectively constructing R' and then checking whether or not $hi(R') = R$. The problem is thus decidable as the equality of regular languages is decidable.

We now investigate equations of the form $X \diamond L = R$, $L \diamond Y = R$ where L and R are given languages, X and Y unknowns, and $\diamond$ signifies the synchronized insertion or deletion operation. To find their solutions, we proceed similarly to solving algebraic equations $x + a = b$. Namely, we must employ an operation "inverse" to addition (in this case subtraction) to determine the solution $x = b - a$. As, unlike addition, the operations of synchronized insertion and deletion are not commutative, we will need to define two separate notions: the notion of a left inverse for solving equations $X \diamond L = R$, and of right inverse for solving equations of the form $L \diamond Y = R$. In the interest of space, we will omit proofs in the sequel.

Definition 4. *Let $\diamond$, $*$ be two binary word operations. The operation $*$ is said to be the left-inverse of the operation $\diamond$ if, for all words u, v, w over the alphabet Σ, the following relation holds:*

$$w \in (u \diamond v) \text{ iff } u \in (w * v).$$

In other words, the operation $*$ is the left-inverse of the operation $\diamond$ if, given a word w in $u \diamond v$, the left operand u belongs to the set obtained from w and the other operand v, by using the operation $*$. The relation "is the left-inverse of" is symmetric.

Proposition 10. *The left-inverse of the operation $\oplus$ of synchronized insertion is the operation $\ominus$ of synchronized deletion.*

We can now use Proposition 10 and the following theorem, [8], to investigate solutions of language equations of the type $X \oplus L = R$ where L and R are given languages in Σ^* and X is the unknown.

Theorem 1. *Let L, R be languages over an alphabet Σ and $\diamond, *$ be two binary word (language) operations, left-inverses to each other. If the equation $X \diamond L = R$ has a solution $X \subseteq \Sigma^*$, then also the language $R' = (R^c * L)^c$ is a solution. Moreover, R' includes all the other solutions of the equation (set inclusion).*

Corollary 1. *If the equation $X \oplus L = R$ (respectively $X \ominus L = R$) has a solution, then $R' = (R^c \ominus L)^c$ (respectively $R' = (R^c \oplus L)^c$) is a maximal solution to the equation.*

We shall use the above results to investigate the decidability of the following problems: Given languages L and R over Σ, R regular, *Does there exist a solution X to the equation $X \oplus L = R$?*, and *Does there exist a singleton solution $X = \{w\}$ to the equation $X \oplus L = R$?*

Proposition 11. *The problem "Does there exist a solution X to the equation $X \oplus L = R$?", is decidable for regular languages L and R.*

Proposition 12. *The problem "Does there exist a singleton solution $X = \{w\}$ to the equation $X \oplus L = R$?" is decidable for regular languages L and R.*

The study of the existence of solutions to the equation $X \oplus L = R$, when R is regular, is completed by the following undecidability results.

Proposition 13. *The problem "Does there exist a solution X to the equation $X \oplus L = R$?" is undecidable for context-free languages L and regular languages R.*

If L is a language over an alphabet Σ, the word $x \in \Sigma^+$ is called *left-useful* with respect to $\ominus$ and L (shortly, left-useful) if there exists a $y \in L$ such that $x \ominus y \neq \emptyset$. A language X is called left-useful with respect to $\ominus$ and L (shortly, left-useful), if it consists only of left-useful words. From the above definitions it follows that the problem "Does there exist a solution X to the equation $X \ominus L = R$?" and its singleton version are equivalent to the corresponding problems where the existence of a left-useful language or word are investigated. Therefore, in the sequel, we will mean a left-useful language when referring to a language or word whose existence is sought.

An argument similar to Proposition 11, and based on the effectiveness of the proofs of closure of REG under $\oplus$ and $\ominus$, shows that the problem "Does there exist a solution X to the equation $X \ominus L = R$?" is decidable for regular languages L and R. For the context-free case the following result holds.

Proposition 14. *The problem "Does there exist a language X such that $X \ominus L = R$" is undecidable for context-free languages L and regular languages R.*

The following decidability result is basically a consequence of the fact that the result of a synchronized deletion from a word is a finite set.

Proposition 15. *The problem "Does there exist a word w such that $w \ominus L = R$?" is decidable for regular languages L and R.*

To investigate symmetric equations of the type $L \oplus Y = R$ and $L \ominus Y = R$ where L and R are given languages and Y is an unknown language, we shall make use of the following result from [8], keeping in mind that, in the case of synchronized deletion, we are actually investigating the existence of right-useful solutions (the notion is defined similarly to that of left-useful solutions).

Theorem 2. *Let L, R be languages over Σ and $\diamond, *$ be two binary word (language) operations right-inverses to each other. If the equation $L \diamond Y = R$ has a solution Y, the language $R' = (L * R^c)^c$ is a maximal solution.*

The notion of right-inverse in the preceding theorem, similar to the notion of left-inverse, is formally defined in [8] as follows.

Definition 5. *Let $\diamond, *$ be two binary word operations. The operation $*$ is said to be right-inverse of the operation $\diamond$ if, for all words u, v, w in Σ^* the following relation holds: $w \in (u \diamond v)$ iff $v \in (u * w)$.*

By using Theorem 2 we could find solutions to equations of the form $L \oplus Y = R, L \ominus Y = R$ if we found the right inverses of $\oplus$ and $\ominus$.

Definition 6. *Let $u, v \in \Sigma^+$. The synchronized bi-deletion of v from u is defined as*

$$u \boxminus v = \{w \mid u = xayaz, v = xaz, w = ya, a \in \Sigma, \; x, y, z \in \Sigma^*\}.$$

Definition 7. *Let $\diamond$ be a binary operation. The word operation $\diamond^r$ defined by $u \diamond^r v = v \diamond u$ is called <u>reversed $\diamond$</u>.*

We can now find out the right inverses of synchronized insertion and deletion and thus solutions to our language equations.

Proposition 16. *The right-inverse of synchronized deletion $\ominus$ is synchronized bi-deletion. The right-inverse of synchronized insertion is reversed synchronized bi-deletion.*

Corollary 2. *If the equation $L \oplus Y = R$ (respectively $(L \ominus Y = R)$) has a solution, then $R' = (R^c \boxminus L)^c$ (respectively $(L \boxminus R^c)^c$) is a maximal solution.*

Before solving our decidability problems, we need to first determine the closure properties of the families in the Chomsky hierarchy under synchronized bi-deletion.

Proposition 17. *The family of regular languages is closed under synchronized bi-deletion while the family of context-free language is not closed under synchronized bi-deletion and the family of context-sensitive languages is not closed under synchronized bi-deletion with regular languages.*

The preceding results on synchronized bi-deletion lead to the following proposition.

Proposition 18. *The problem of whether or not there exists a solution Y to the equations $L \oplus Y = R$, $L \ominus Y = R$ is decidable for regular languages L and R.*

Proposition 19. *The existence of a solution Y to the equations $L \oplus Y = R$ and $L \ominus Y = R$ is undecidable for regular languages R and context-free languages L.*

4   Conclusion

We have considered the properties of three operations used in the modeling of the ciliate gene descrambling process: synchronized insertion, synchronized deletion and hairpin inversion. We found that all the families of the Chomsky hierarchy are closed under synchronized insertion while only the families of regular and recursively enumerable languages are closed under synchronized deletion. Additionally we showed that only the family of context-free languages was not closed under hairpin inversion. In order to consider language equations involving each of the three operations we have also defined the operation of synchronized bi-deletion (the right-inverse of synchronized deletion) and showed only the families of regular and recursively enumerable languages to be closed under this operation.

We demonstrated that the existence of a solution X to the equation $hi(X) = R$, where R is a regular language is decidable. Additionally, the existence of a solution was shown to be decidable for equations of the form $L \diamond Y = R$ and $X \diamond L = R$ where $\diamond$ is one of synchronized insertion or synchronized deletion operations and L, R are regular languages. The same problems are undecidable in the case that L is a context-free language.

By investigating the properties of these formal operations, we have provided some insight into the nature of the bio-operations that must be present in the ciliate gene descrambling mechanism. Continued theoretical study of the gene descrambling problem combined with improved biological results will hopefully lead to a better understanding of this fascinating process.

References

1. J.M. Autebert, J. Berstel, L. Boasson 1997. Context-free languages and Pushdown Automata. *Handbook of formal languages*, 1: 111–174, Springer, Berlin.

2. J. Dassow, V. Mitrana, A. Salomaa. 2002. Operations and language generating devices suggested by the genome evolution. *Theoretical Computer Science*, 270: 701-738.

3. A. Ehrenfeucht, D.M. Prescott, G. Rozenberg. 2001. Computational aspects of gene (un)scrambling is ciliates. In *Evolution as Computation* (L.F. Landweber, E. Winfree eds.) Springer-Verlag, Berlin, Heidelberg, 45-86.

4. S. Ginsburg. 1975. *Algebraic and Automata-Theoretic Properties of Formal Languages*. North-Holland, Amsterdam.

5. L. Kari, L.F. Landweber. 2000. Computational power of gene rearrangment In *DNA5, DIMACS series in Discrete Mathematics and Theoretical Computer Science*)(E. Winfree, D. Gifford eds.), American Mathematical Society, 54: 207-216.

6. L. Kari, G. Thierrin. 1996. Contextual insertions/deletions and computability. *Information and Computation*, 131: 47–61.

7. L. Kari. 1992. Insertion and deletion of words: determinism and reversibility. *Lecture Notes in Computer Science*, 629:315-327.

8. L. Kari. 1994. On language equations with invertible operations. *Theoretical Computer Science*, 132: 129-150.

9. L. Kari. 1991. *On insertions and deletions in formal languages*. PhD thesis, University of Turku, Finland.

10. L.F. Landweber, L. Kari. 1999. The evolution of cellular computing: nature's solutions to a computational problem. *DNA4 Biosystems* (L. Kari, H. Rubin, D.H. Wood eds.), Elsevier, 52(1-3):3-13.

11. I. Petre, A. Ehrenfeucht, T. Harju and G. Rozenberg. 2002. Patterns of micronuclear genes in cilliates. In *DNA7, Lecture Notes in Computer Science* (N. Jonoska, N. Seeman eds.), Springer-Verlag, 2340: 279-289.

12. D.M. Prescott. 1992. Cutting, splicing, reordering, and elimination of DNA sequences in hypotrichous ciliates. *BioEssays*, 14(5): 317-324.

13. D.M. Prescott. 1992. The unusual organization and processing of genomic DNA in hypotrichous ciliates. *Trends in Genet.*, 8:439-445.

14. D.M. Prescott. 2000. Genome gymnastics: Unique modes of DNA evolution and processing in ciliates. *Nature Reviews Genetics*, 1:191-198.

15. D.M. Prescott, A. Ehrenfeucht, G. Rozenberg. 2001. Molecular operations for DNA processing in hypotrichous ciliates. To appear in *European Journal of Protistology*.

16. A. Salomaa. 1973. *Formal languages*, Academic Press, New York.

On the Descriptional Complexity
of Some Variants of Lindenmayer Systems

Jürgen Dassow[1], Taishin Nishida[2], and Bernd Reichel[1]

[1] Fakultät für Informatik, Otto-von-Guericke-Universität Magdeburg, PSF 41 20,
D-39016 Magdeburg, Germany
{dassow,reichel}@iws.cs.uni-magdeburg.de
[2] Faculty of Engineering, Toyoma Prefectural University,
Kosugi-machi, 939-0398 Toyoma, Japan
nishida@pu-toyama.ac.jp

Abstract. We define the number of productions and the number of symbols as complexity measures for interactionless Lindenmayer systems with a completely parallel derivation process and for some variants of limited Lindenmayer systems with a partially parallel derivation process and for the associated languages. We prove that up to an initial part any natural number can occur as complexity of some language. Moreover, we show the existence of languages with small descriptional complexity with respect to one mechanism and large complexity with respect to the other device.

1 Introduction

Lindenmayer systems (L systems for short) have been introduced in 1968 as a grammatical model to describe the development of (lower) organisms. Most results concern the effect of determinism, tables and extensions by "nonterminals", relations to other classes of formal languages, closure properties, decision problems, growth functions and combinatorial properties. For summaries we refer to [4,9,10].

There are only few papers on the syntactical complexity of L languages only considering the following topics: the study of parameters which are of interest for tabled systems as the number of active symbols, the number of tables and the degree of nondeterminism (see [6,13,4,1]) and the descriptional complexity of (variants of) adult languages (which coincide with context-free languages, and both mechanisms are compared by means of syntactic measures; see [5,7]).

Surprisingly, there is no investigation of the basic type of L systems, the interactionsless 0L systems, with respect to measures of descriptional complexity which are well known from the sequential grammars of the Chomsky hierarchy. In this paper we start the study of 0L systems and 0L languages with respect to the number of productions and the total number of symbols.

First, we show that up to an initial part any natural number can occur as the descriptional complexity of some 0L language.

M. Ito and M. Toyama (Eds.): DLT 2002, LNCS 2450, pp. 128–140, 2003.

Furthermore, we are interested in a comparison of the complexities of 0L systems and context-free grammars generating the same language. Since context-free grammars have a purely sequential derivation process we contribute to the relationship between parallelism and sequentiality.

To refine the comparison we also take into consideration k-limited and uniformly k-limited 0L systems, which have a partially parallel derivation process and were introduced in 1988/90 by WÄTJEN (see [11,12]).

Essentially we prove that there are sequences of languages such that the descriptional complexity with respect to one type of grammars/systems grows to infinity whereas it is bounded if the other type is used.

Analogous results for Indian parallel grammars which are a further type of grammars with partial parallelism in the derivation are given in [8].

For lack of space we omit some proofs, especially in Section 4. A complete version of this paper with additional results is given in [2].

2   Definitions

For an alphabet V, we denote the set of all (non-empty) words over V by V^* (and V^+, respectively). The length of a word $w \in V^*$ is designated by $|w|$. For a letter $a \in V$ and a word $w \in V^*$, $\#_a(w)$ denotes the number of occurrences of a in w.

We now recall some concepts on context-free grammars and 0L systems. Mostly, we only present notations.

A *context-free grammar* is specified as a quadruple $G = (N, T, P, S)$ where N and T are disjoint alphabets of nonterminals and terminals, respectively, P is a finite subset of $N \times (N \cup T)^*$ and S is an element of N. The elements of P are called productions or rules. As usual, the pairs (A, v) of P are written as $A \rightarrow v$.

We say that $x \in V^+$ directly derives $y \in V^*$, written as $x \Longrightarrow y$, if $x = x_1 A x_2$ for some $x_1, x_2 \in (N \cup T)^*$ and $A \in N$, $y = x_1 v x_2$ and $A \rightarrow v \in P$, i.e., we replace one occurrence of a nonterminal according to the rules of P. Thus we have a sequential derivation process. By $\overset{*}{\Longrightarrow}$ we denote the reflexive and transitive closure of $\Longrightarrow$. The language $L(G)$ generated by G is defined as

$$L(G) = \{z \mid z \in T^*, \ S \overset{*}{\Longrightarrow} z\}.$$

It consists of all terminal words which can be generated from the axiom S.

A *0L system* is specified as a triple $G = (V, P, w)$ where V is an alphabet, w is a non-empty word over V and P is a finite subset of $V \times V^*$ such that, for any $a \in V$, there is at least one element $a \rightarrow v$ in P. We have no distinction between terminals and nonterminals, but a completeness condition which ensures that any letter can be rewritten.

We say that $x \in V^+$ directly derives $y \in V^*$, written as $x \Longrightarrow y$, if $x = x_1 x_2 \ldots x_n$ for some $n \geq 0$, $x_i \in V$, $1 \leq i \leq n$, $y = y_1 y_2 \ldots y_n$ and $x_i \rightarrow y_i \in P$ for $1 \leq i \leq n$, i.e., any letter of x is replaced according to the rules of P. Thus the derivation process in a 0L system is purely parallel. By $\overset{*}{\Longrightarrow}$ we denote the reflexive and transitive closure of $\Longrightarrow$. The language $L(G)$ generated by G is defined as $L(G) = \{z \mid w \overset{*}{\Longrightarrow} z\}$.

A *k-limited 0L system* (abbreviated as kl0L system) is specified as a quadruple $G = (V, P, w, k)$ where (V, P, w) is a 0L system and k is a positive integer. A derivation step of a k-limited 0L system differs from that of a 0L systems in such a way that instead of the fully parallel rewriting, now exactly $\min\{\#_a(x), k\}$ occurrences of each letter a in the word x are replaced according to rules of P. The relation $\overset{*}{\Longrightarrow}$ and the language $L(G)$ are defined as in the case of 0L systems.

A *uniformly k-limited 0L system* (abbreviated as ukl0L system) is specified as a k-limited 0L systems, but each derivation step consists in a rewriting of $\min\{k, |x|\}$ letters of the current sentential form x. The relation $\overset{*}{\Longrightarrow}$ and the language $L(G)$ are again defined as in the case of 0L systems.

All types of L systems as given above do not make a distinction between terminals and nonterminals. To get such an aspect one has introduced extended versions of the systems. An *extended* 0L system (E0L system for short) is specified as a quadruple $G = (V, T, P, w)$ where $G' = (V, P, w)$ is a 0L system and T is a subset of V. The language $L(G)$ is defined as $L(G) = L(G') \cap T^*$, i. e., the extended 0L language consists only of those words which can be generated from the axiom and contain only symbols of the terminal set T. Analogously, we define the extended versions of kl0L and ukl0L systems.

By CF, $0L$, $kl0L$ and $ukl0L$ we denote the families of all context-free grammars, 0L systems, k-limited 0L systems and uniformly k-limited 0L systems, respectively. We set

$$\mathcal{G} = \{CF, 0L\} \cup \bigcup_{k \geq 1} \{kl0L, ukl0L\} \quad \text{and} \quad \mathcal{G}' = \{0L\} \cup \bigcup_{k \geq 1} \{kl0L, ukl0L\}.$$

For $X \in \mathcal{G}'$, we denote the families of extended X systems and of languages generated by extended X systems by EX and $\mathcal{L}(EX)$, respectively. We set $E\mathcal{G}' = \bigcup_{X \in \mathcal{G}'} EX$.

Let T be an arbitrary alphabet. By CF_T we designate the family of context-free grammars whose terminal set is T. Analogously, for $X \in \mathcal{G}'$, by X_T and EX_T we denote the families of X systems of the form $G = (T, P, w)$ and $G = (V, T, P, w)$ or $G = (T, P, w, k)$ and $G = (V, T, P, w, k)$, respectively, with arbitrary alphabets V, arbitrary sets P of productions and arbitrary axioms w. We set $\mathcal{G}'_T = \bigcup_{X \in \mathcal{G}' \cup E\mathcal{G}'} X_T$ and $\mathcal{G}_T = \{CF_T\} \cup \mathcal{G}'_T$.

For $X \in \mathcal{G} \cup E\mathcal{G}' \cup \bigcup_T \mathcal{G}_T$, the set of languages generated by elements of X is denoted by $\mathcal{L}(X)$.

The measures of descriptional complexity for context-free grammars are the number of nonterminals, the number of productions and the number of symbols in a "complete" description of the grammar. Since we are interested in L systems and their complexity we do not consider the number of nonterminals because – also in extended systems – there are rules for any letters of the alphabet V. We now formally define the other measures.

For a context-free grammar $G' = (N, T, P', S)$ and a 0L (or kl0L or ukl0L) system $G = (V, P, w)$ (or $G = (V, P, w, k)$) and an extended 0L (or kl0L or ukl0L) system $G = (V, T, P, w)$ (or $G = (V, T, P, w, k)$) we set

$$Prod(G') = \#(P') \quad \text{and} \quad Symb(G') = 1 + \sum_{A \to v \in P'} (|v| + 2),$$

$$Prod(G) = \#(P) \quad \text{and} \quad Symb(G) = |w| + \sum_{a \to v \in P} (|v| + 2).$$

Thus *Prod* gives the number of productions, and *Symb* gives the length of the word consisting of the axiom and all productions. Hence *Symb* is the length of a complete description of the grammar/system (if we do not use superfluous symbols and the difference between terminals and nonterminals in a grammar can be seen from notation).

Let $X \in \mathcal{G} \cup E\mathcal{G}' \cup \bigcup_T \mathcal{G}_T$. Then, for a language $L \in \mathcal{L}(X)$, we set

$$Prod_X(L) = \min\{Prod(G) \mid L(G) = L, \ G \in X\},$$
$$Symb_X(L) = \min\{Symb(G) \mid L(G) = L, \ G \in X\}.$$

Thus the complexity of a language $L \in \mathcal{L}(X)$ is given by the minimal X system/grammar which generates L.

3   On the Range of Descriptional Complexity of Languages

First we show that some initial numbers cannot occur as the descriptional complexity of languages of a certain type.

Lemma 1. *For $X \in \mathcal{G}'$, any alphabet V and any language $L \in \mathcal{L}(X_V)$,*

$$Prod_X(L) \geq \#(V) \quad \text{and} \quad Symb_X(L) \geq 2 \#(V) + 1.$$

Proof. By the completeness condition for Lindenmayer systems, for any $a \in V$, there is a production with left side a. Thus we need at least $\#(V)$ productions. Moreover, any productions requires at least two symbols (if the right side is empty) and the axiom has at least the length 1. This implies the second relation.
□

We now give a definition ensuring that – up to a finite initial segment – all natural numbers can occur as the descriptional complexity of some language.

Definition 1. *Let $X \in \mathcal{G}' \cup E\mathcal{G}' \cup \bigcup_T \mathcal{G}_T'$ be a family of grammars, and let $K \in \{Prod, Symb\}$ be a measure of descriptional complexity. We say that K is complete with respect to X if there is a natural number n_0 such that, for any $n \geq n_0$, there exists a language $L_n \in \mathcal{L}(X)$ with $K_X(L_n) = n$.*

The following lemma immediately follows from the definitions.

Lemma 2. *Let $X \in \mathcal{G}' \cup E\mathcal{G}'$, $K \in \{Prod, Symb\}$. If K is complete with respect to X_T for some alphabet T, then K is complete for X, too.*
□

Lemma 3. i) *For $n \geq 1$, $k \geq 2$, let $L_{n,k} = \{a, b^k, b^{2k}, \ldots, b^{nk}, \lambda\}$. Then*

$$Prod_{0L}(L_{n,k}) = n + 1 \quad and \quad Symb_{0L}(L_{n,k}) = \frac{k \cdot n^2 + (k+4) \cdot n + 6}{2}.$$

ii) *For $n \geq 1$ and $k \geq 1$, let $L'_{n,k} = \{a, b^{2k}, b^{4k}, \ldots, b^{2nk}\}$. Then*

$$Prod_{kl0L}(L'_{n,k}) = Prod_{ukl0L}(L'_{n,k}) = n + 1 \quad and$$
$$Symb_{kl0L}(L'_{n,k}) = Symb_{ukl0L}(L'_{n,k}) = k \cdot n^2 + (k+2) \cdot n + 4.$$

Proof. i) Let $H_{n,k}$ be a 0L system generating $L_{n,k}$. Let $b \to w$ be a rule. Applying this rule to any occurrence of b in b^{nk} we get w^{nk}. This implies $w = b$ or $w = \lambda$. If both rules are present, then we can generate $b^{nk-1} \notin L_{n,k}$ by applying $b \to b$ to $nk - 1$ occurrences of b and $b \to \lambda$ to one occurrence of b in b^{nk}. If only one of the rules $b \to b$ and $b \to \lambda$ is in the set of rules, then we have $b^{ik} \Longrightarrow b^{ik}$ or $b^{ik} \Longrightarrow \lambda$ for $1 \leq i \leq n$. Thus all words b^{ik} have to be generated from the axiom a in one step, i. e., we have the rules $a \to b^{ik}$ for $1 \leq i \leq n$. Thus

$$H_{n,k} = (\{a, b\}, \{a \to b^{ik} \mid 1 \leq i \leq n\} \cup \{b \to \lambda\}, a)$$

is the unique 0L system with $L(H_{n,k}) = L_{n,k}$, $Prod(H_{n,k}) = Prod(L_{n,k})$ and $Symb(H_{n,k}) = Symb(L_{n,k})$. Now the statement follows from $Prod(H_{n,k}) = n + 1$ and $Symb(H_{n,k}) = \left(\sum_{i=1}^{n}(2 + ik)\right) + 2 + 1 = \frac{k \cdot n^2 + (k+4) \cdot n + 6}{2}$.

ii) The proofs of the statements for k-limited 0L systems and uniformly k-limited 0L systems can be given in an analogous way.   □

Theorem 1. *For any alphabet T with at least two symbols and any $X \in \mathcal{G}'$, the measure Prod is complete with respect to X_T.*

Proof. Let $X = 0L$. W. l. o. g. we assume $T = \{a, b, a_1, a_2, \ldots, a_{m-2}\}$. For $n \geq n_0 = m$ and $m \geq 2$, let $R_n = L_{n-m+1,2} \cup \{a_1, a_2, \ldots, a_{m-2}\}$ where $L_{n-m+1,2}$ is defined in Lemma 3. As in the proof of Lemma 3 i) one can show that $Prod_{0L}(R_n) \geq n$. We need at least $n - m + 1$ rules to generate b^{2i}, $1 \leq i \leq n - m + 1$, the rule $b \to \lambda$ and $m - 2$ rules for the generation of $m - 2$ symbols of $\{a\} \cup \{a_j \mid 1 \leq j \leq m - 2\}$ different from the axiom of length 1. On the other hand (V, P, a_1), where

$$P = \{a_i \to a_{i+1} \mid 1 \leq i \leq m - 3\} \cup \{a_{m-2} \to a, b \to \lambda\}$$
$$\cup \{a \to b^{2j} \mid 1 \leq j \leq n - m + 1\},$$

generates R_n and has n productions.

For $k \geq 1$ and $X = kl0L$ and $X = ukl0L$, we consider the language

$$R_{n,k} = L'_{n-m+1,k} \cup \{a_1, a_2, \ldots, a_{m-2}\}$$

and give an analogous proof using Lemma 3 ii).   □

Note that the proof of Theorem 1 is optimal in that sense that the bound n_0 is minimal as can be seen from Lemma 1.

Theorem 2. *For any alphabet T and any $X \in \mathcal{G}'$, the measure Symb is complete with respect to X_T.*

Proof. Let $X = 0L$, $T = \{a_1, a_2, \ldots, a_m\}$, $n \geq n_0 = 2m + 1$. The language $U_n = \{a_1^{n-2m}, \lambda\}$ over T is generated by the 0L system

$$J_n = (T, \{a_i \to \lambda \mid 1 \leq i \leq n\}, a_1^{n-2m})$$

with $Symb(J_n) = 2m + n - 2m = n$. This implies $Symb_{0L}(U_n) \leq n$.

Moreover, since any 0L system for U_n has at least m rules, any rule requires at least 2 symbols and a_1^{n-2m} has to be the axiom, we get $Symb_{0L}(U_n) \geq 2m + n - 2m = n$.

Analogously, we prove the statement for (uniformly) k-limited 0L systems.

$\square$

Note again that the proof of Theorem 2 is optimal in that sense that the used bound n_0 is minimal by Lemma 1.

In the case of extended 0L (kl0L or ukl0L) systems we have no completeness result for the measure *Symb*, and for *Prod*, we can only prove the completeness with respect to the families without a restriction of the underlying alphabet.

Lemma 4. *For $X \in \mathcal{G}'$, Prod is complete with respect to EX.*

Proof. We only give the proof for $X = 0L$. The proof in the other case can be given analogously.

We consider, for $n \geq 1$, the language $W_n = \{a_1 a_2 \ldots a_n\}$, where $a_1, a_2, \ldots, a_n$ are pairwise different letters. Assume that the E0L system $G_n = (V, T, P, w)$ generates W_n. Then $T = \{a_1, a_2, \ldots, a_n\}$. By $T \subseteq V$ and the completeness condition P has to contain a rule $a_i \to v_i$ for any i, $1 \leq i \leq n$. Thus $Prod_{E0L}(W_n) \geq n$. On the other hand, n productions are sufficient since the E0L system

$$(\{a_1, \ldots, a_n\}, \{a_1, \ldots, a_n\}, \{a_1 \to a_1, \ldots, a_n \to a_n\}, a_1 \ldots a_n)$$

generates W_n.

$\square$

4 Complexity of Some Special Languages

In this section, we give some results (without proofs with the exception of two) about the complexity of some special languages which are used later.

Lemma 5. *For $n \geq 1$, let*

$$L_n = \{a^{2^i} \mid 1 \leq i \leq n\}$$
$$\cup \{x_1 x_2 \ldots x_m \mid m \geq 1, \ x_i = b \ or \ x_i = a^{2^{n+1}}, \ 1 \leq i \leq m\} \cup \{c\}.$$

Then $Prod_{0L}(L_n) \leq 6$ and $Prod_{CF}(L_n) \geq n$.

Proof. Because the 0L system

$$G_n = (\{a, b, c\}, \{a \to a^2, b \to b, b \to b^2, b \to a^{2^{n+1}}, c \to a, c \to b\}, c)$$

generates L_n we have $Prod_{0L}(L_n) \leq 6$.

Let $G'_n = (N, \{a, b, c\}, P, S)$ be a context-free grammar with $L(G'_n) = L_n$ and $Prod(G'_n) = Prod_{CF}(L_n)$. We note that, for every nonterminal $A \in N$, there are at least two words w_A and w'_A over the terminal set $\{a, b, c\}$ with $A \overset{*}{\Longrightarrow} w_A$ and $A \overset{*}{\Longrightarrow} w'_A$. (If there is only one such word w_A, then we replace any occurrence of A in the right-hand side of every production which has occurrences of A by w_A and cancel all rules with left-hand side A. This procedure does not change the language, but it leads to a decrease of the number of productions.)

We first show that, for any integer i with $0 \leq i \leq n$, there is no derivation

$$S \overset{*}{\Longrightarrow} xAyBz \overset{*}{\Longrightarrow} xuyvz = a^{2^i}. \tag{1}$$

Let us assume the contrary. Then we can assume that $w_A = u = a^s$ and $w_B = v = a^t$ for some integers s and t. We discuss the following four situations:

Case 1: w'_A and w'_B are contained in a^*, too. Then we can derive a contradiction as in the proof of Lemma 4.3 in [8].

Case 2: $w'_A = a^{s'}$ for some integer s' and $w'_B = a^p w' a^q$ where w' starts and ends with the letter b. Then we have the derivations

$$S \overset{*}{\Longrightarrow} xAyBz \overset{*}{\Longrightarrow} xa^s ya^p w' a^q z \in L_n \text{ and}$$
$$S \overset{*}{\Longrightarrow} xAyBz \overset{*}{\Longrightarrow} xa^{s'} ya^p w' a^q z \in L_n.$$

Hence $|xy| + s + p = j_1 2^{n+1}$ and $|xy| + s' + p = j_2 2^{n+1}$ and therefore $|s - s'| = j2^{n+1}$. Since $s \leq 2^i < 2^{n+1}$, we get $s' > s$ and $s' - s = j2^{n+1}$ Moreover, there are the derivations

$$S \overset{*}{\Longrightarrow} xAyBz \overset{*}{\Longrightarrow} xa^s ya^t z = a^{2^i} \text{ and}$$
$$S \overset{*}{\Longrightarrow} xAyBz \overset{*}{\Longrightarrow} xa^{s'} ya^t z = a^{2^i + j2^{n+1}}.$$

Since $2^i + j2^{n+1}$ is not a multiple of 2^{n+1}, $a^{2^i + j2^{n+1}} \notin L_n$ contradicts the derivation of this word in G'_n.

Case 3: $w'_B = a^{t'}$ for some integer t' and $w'_A = a^p w' a^q$ where w' starts and ends with the letter b. Then we can obtain a contradiction in analogy to Case 2.

Case 4: $w'_A = a^k w' a^l$ and $w'_B = a^p w'' a^q$ where w' and w'' start and end with the letter b. Then we have the derivations

$$S \overset{*}{\Longrightarrow} xAyBz \overset{*}{\Longrightarrow} xa^s ya^p w'' a^q z \in L_n,$$
$$S \overset{*}{\Longrightarrow} xAyBz \overset{*}{\Longrightarrow} xa^k w' a^l ya^t z \in L_n,$$
$$S \overset{*}{\Longrightarrow} xAyBz \overset{*}{\Longrightarrow} xa^k w' a^l ya^p w'' a^q z \in L_n.$$

Thus $|xy| + s + p = j_1 2^{n+1}$, $|yz| + l + t = j_2 2^{n+1}$ and $|y| + l + p = j_3 2^{n+1}$. This equations imply

$$p = j_1 2^{n+1} - s - |xy|,$$
$$l = j_2 2^{n+1} - t - |yz|,$$
$$j_3 2^{n+1} = j_1 2^{n+1} - s - |xy| + j_2 2^{n+1} - t - |yz| + |y|$$
$$= (j_1 + j_2) 2^{n+2} - |xa^s y a^t z| = (j_1 + j_2) 2^{n+1} - 2^i.$$

Hence $(j_1 + j_2 - j_3) 2^{n+1} = 2^i$ which is impossible by $0 < 2^i < 2^{n+1}$ and $(j_1 + j_2 - j_3) 2^{n+1} \leq 0$ or $(j_1 + j_2 - j_3) 2^{n+1} \geq 2^{n+1}$.

Since there are no derivation of the form (1) for the words a^{2^i}, derivations of these words have to be linear. We now prove that there is no nonterminal A with

$$S \overset{*}{\Longrightarrow} xAy \overset{*}{\Longrightarrow} a^{2^i} \quad \text{and} \quad S \overset{*}{\Longrightarrow} x'Ay' \overset{*}{\Longrightarrow} z \in L_n$$

with x, y, x', y' and $xy \neq x'y'$.

Assume the contrary. Let $w_A = a^s$ and $a^{2^i} = xa^s y$. By assumption, we have the derivation $S \overset{*}{\Longrightarrow} x'Ay' \overset{*}{\Longrightarrow} x'a^s y' \in L_n$, where $x', y' \in a^*$ or at least one of x' and y' has the form $a^k w' a^l$ where w' starts and ends with b. Now we discuss four cases, again, depending on these possibilities for x' and y' and whether $w'_A \in a^*$ or $w'_A = a^p w'' a^q$ for some w'' starting and ending with b. By considerations similar to those given above, we show a contradiction in any case.

Therefore, by combining the above considerations and the note that every nonterminal A has at least two words $w_A, w_{A'} \in \{a, b, c\}^*$ such that $A \overset{*}{\Longrightarrow} w_A$ and $A \overset{*}{\Longrightarrow} w_{A'}$, we need $n + 1$ rules $S \to a$, $S \to a^2$, ..., $S \to a^{2^n}$ to generate a, a^2, a^4, ..., a^{2^n}. □

Lemma 6. For $k \geq 1$ and $n \geq 1$, let

$$R_{n,k} = \{a^{2^i} \mid r \leq i \leq r + n\}$$
$$\cup \{x_1 x_2 \ldots x_m \mid m \geq 1, \ x_i = b \text{ or } x_i = a^{2^{r+n+1}}, \ 1 \leq i \leq m\} \cup \{c\}$$

where r is an integer such that $2^r \geq k + 2$. Then $\text{Prod}_{0L}(R_{n,k}) \leq 6$, $\text{Prod}_{kl0L}(R_{n,k}) \geq n + 1$ and $\text{Prod}_{ukl0L}(R_{n,k}) \geq n + 1$. □

Lemma 7. For $k \geq 1$ and $n \geq 1$, let $T_{n,k} = \{a^{ik} b^{ik} c^{ik} \mid 0 \leq i \leq n\} \cup \{d\}$. Then

i) $\text{Prod}_{kl0L}(T_{n,k}) \leq 4$ and $\text{Symb}_{kl0L}(T_{n,k}) \leq 9 + 3nk$,

ii) $\text{Prod}_{CF}(T_{n,k}) \geq n$ and $\text{Symb}_{CF}(T_{n,k}) \geq \frac{kn^2 + (k+4)n}{2}$,

iii) for $n \geq k' \geq 2$, we have $\text{Prod}_{uk'l0L}(T_{n,k}) = n$ and $\text{Symb}_{uk'l0L}(T_{n,k}) = \frac{3kn^2 + (3k+4)n + 10}{2}$.

Proof. i) follows from the $kl0L$ system

$$(\{a, b, c, d\}, \{a \to \lambda, b \to \lambda, c \to \lambda, d \to a^{nk} b^{nk} c^{nk}\}, d, k)$$

which generates $T_{n,k}$ and has four productions and $9 + 3nk$ symbols.

ii) If $U_{n,k} = (N, \{a, b, c, d\}, P, S)$ is a context-free grammar with $L(U_{n,k}) = T_{n,k}$ and $Prod(U_{n,k}) = Prod_{CF}(T_{n,k})$, then we can easily prove that any derivation $S \overset{*}{\Longrightarrow} uAv \Longrightarrow a^{ik}b^{ik}c^{ik}$ (where $1 \leq i \leq k$) applies in the last step a rule $A \rightarrow u'b^{ik}v'$. Therefore we need at least n productions. Moreover, $Symb(U_{n,k}) \geq \sum_{i=1}^{n}(2 + ik) = \frac{kn^2+(k+4)n}{2}$ holds.

iii) It is easy to show that, for any uk'l0L system $U'_{n,k}$ with $L(U'_{n,k}) = T_{n,k}$, $x \rightarrow x$ is the only rule for $x \in \{a, b, c\}$. Therefore any word $a^{ik}b^{ik}c^{ik}$ has to be generated from d. Thus

$$U'_{n,k} = (\{a, b, c, d\}, P_{n,k}, d, k'),$$

where

$$P_{n,k} = \{a \rightarrow a, b \rightarrow b, c \rightarrow c\} \cup \{d \rightarrow a^{ik}b^{ik}c^{ik} \mid 1 \leq i \leq n\},$$

is the minimal uk'l0L system with respect to $Prod$ and $Symb$. The statement follows from $Prod(U'_{n,k}) = n$ and $Symb(U'_{n,k}) = \frac{3kn^2+(3k+4)n+10}{2}$.   □

Lemma 8. Let $k \geq 2$, $k' \geq 2$, $k \neq k'$, and $n \geq 2$ and

$$K_{n,k} = \{e\} \cup \{w \mid w = x_1 x_2 \ldots x_{n \cdot k} y_1 y_2 \ldots y_{n \cdot k}, \; x_i \in \{a, c\}, \; y_i \in \{b, d\},$$
$$1 \leq i \leq nk, \; \#_c(w) + \#_d(w) = rk \; for \; some \; r\}.$$

Then

i) $Prod_{ukl0L}(K_{n,k}) \leq 4$ and $Symb_{ukl0L}(K_{n,k}) \leq 15 + 2nk$, and
ii) $Prod_{k'l0L}(K_{n,k}) \geq n$ and $Symb_{k'l0L}(T_{n,k}) \geq n^2$.   □

Lemma 9. There is a constant $c > 0$ such that, for any integer $n \geq 2$, $Symb_{CF}(\{a^n\}) \geq c \cdot \log(n)$.   □

Lemma 10. For $n \geq 8$, let $S_n = \{a_i^{2^i} \mid 1 \leq i \leq n\}$. Then

$$Symb_{0L}(S_n) \leq 4n + 1 \quad and \quad Symb_{CF}(S_n) \geq d \cdot n^2$$

for some constant $d > 0$.   □

5   Comparison of Families of Grammars

In order to compare families of grammars/systems with respect to measures of descriptional complexity we define the following relations.

Definition 2. Let X and Y be two families of $\mathcal{G} \cup E\mathcal{G}' \cup \bigcup_T \mathcal{G}_T$, and let K be a measure of descriptional complexity. We say that

i) $K(X,Y) \geq 1$ *if there are a sequence of languages* $L_n \in \mathcal{L}(X) \cap \mathcal{L}(Y)$, $n \geq 1$, *and constants* $c > 0$ *and* $n_0 \in \mathbb{N}$ *such that, for* $n \geq n_0$,

$$K_Y(L_n) - K_X(L_n) \geq c \cdot n,$$

ii) $K(X,Y) \geq 2$ *if there is a sequence of languages* $L_n \in \mathcal{L}(X) \cap \mathcal{L}(Y)$, $n \geq 1$, *such that*

$$\lim_{n \to \infty} \frac{K_X(L_n)}{K_Y(L_n)} = 0,$$

iii) $K(X,Y) = 3$ *if there are a sequence of languages* $L_n \in \mathcal{L}(X) \cap \mathcal{L}(Y)$, $n \geq 1$, *and constants* $c \in \mathbb{N}$ *and* $n_0 \in \mathbb{N}$ *such that, for* $n \geq n_0$,

$$K_X(L_n) \leq c \quad and \quad K_Y(L_n) \geq n.$$

In Definition 2, i) states that the difference of the language can grow to infinity, ii) tells that the growths have different asymptotic behaviour, and iii) describes the situation where the difference cannot be bounded by a function.

Obviously, $K(X,Y) = 3$ implies $K(X,Y) \geq 2$ and $K(X,Y) \geq 2$ implies $K(X,Y) \geq 1$.

Thus we can give the following notion.

Definition 3. *Let* X, Y *and* K *be as in Definition 2. We set*

i) $K(X,Y) = 2$ *if* $K(X,Y) \geq 2$ *holds and* $K(X,Y) = 3$ *is not valid.*
ii) $K(X,Y) = 1$ *if* $K(X,Y) \geq 1$ *holds and* $K(X,Y) \geq 2$ *is not valid.*

Remark. We note that with respect to the measure *Symb* the relation $K(X,Y) = 3$ cannot occur.

This can be seen as follows. Given a constant c, there is only a finite number of elements G from Y with $Symb(G) \leq c$. Hence there is no infinite sequence of languages as required if $K(X,Y) = 3$.

Definition 4. *Let* X *and* Y *be two families of* $\mathcal{G} \cup E\mathcal{G}' \cup \bigcup_T \mathcal{G}_T$ *such that* $\mathcal{L}(Y) \subseteq \mathcal{L}(X)$, K *a measure of descriptional complexity and* $i \in \{1, 2, 3\}$. *We say that*

i) $X \leq_K^b Y$ *if there is a constant* c *such that, for all* $L \in \mathcal{L}(Y)$,

$$K_X(L) \leq K_Y(L) + c,$$

ii) $X \leq_K^i Y$ *if* $X \leq_K^b Y$ *and* $K(X,Y) = i$

The following lemma follows immediately from the definitions.

Lemma 11. *Let* X *and* Y *be two families of* $\mathcal{G} \cup E\mathcal{G}'$, K *a measure of descriptional complexity and* $i \in \{1, 2, 3\}$.

i) *If there is an alphabet* T *with* $K(X_T, Y_T) \geq i$, *then* $K(X,Y) \geq i$ *also holds. Especially,* $K(X_T, Y_T) = 3$ *for some* T *implies* $K(X,Y) = 3$.
ii) *Let* $\mathcal{L}(Y) \subseteq \mathcal{L}(X)$. *If there is an alphabet* T *with* $X_T \leq_K^i Y_T$, *then* $X \leq_K^j Y$ *for some* $j \geq i$. *Especially,* $X_T \leq_K^3 Y_T$ *for some* T *implies* $X \leq_K^3 Y$. $\quad \square$

We start with an comparison of each of the families $0L$, $kl0L$ and $ukl0L$, $k \geq 1$, with CF.

Theorem 3. *For any alphabet T with at least three letters,*

i) $Prod(0L_T, CF_T) = 3$ and $Prod(CF_T, 0L_T) = 3$,
ii) $Symb(0L_T, CF_T) = 2$ and $Symb(CF_T, 0L_T) = 2$.   □

Theorem 4. *For any $k \geq 2$ and any alphabet T with at least three letters,*

i) $Prod(CF_T, kl0L_T) = 3$ and $Prod(kl0L_T, CF_T) = 3$,
ii) $Symb(CF_T, kl0L_T) = 2$.   □

Theorem 5. *For any $k \geq 2$ and any alphabet T with at least two letters, we have $Prod(CF_T, ukl0L_T) = 3$ and $Symb(CF_T, ukl0L_T) = 2$.*   □

We note that all relations given in the Theorems 3, 4 and 5 are optimal with respect to the value of $K(X, Y)$, however some values also hold for alphabets with two letters only.

We mention as an *open problem* to determine the values for $Symb(kl0L, CF)$, $Prod(ukl0L, CF)$ and $Symb(ukl0L, CF)$.

We now compare each of the families $E0L$, $Ekl0L$ and $Eukl0L$, $k \geq 1$, with CF. First we recall that, for $X \in \mathcal{G}'$, $\mathcal{L}(CF) \subset \mathcal{L}(EX)$. However, with respect to the descriptional complexity measure $Prod$ we obtain incomparability.

Theorem 6.   i) *For any $X \in \mathcal{G}'$, $Prod(CF, EX) = 3$.*
ii) *For any $X \in \{0L\} \cup \bigcup_{k \geq 1}\{kl0L\}$, $Prod(EX, CF) = 3$.*   □

The determination of $Prod(Eukl0L, CF)$ and the values for the measure $Symb$ remains as an *open problem*.

The situation changes completely if we restrict to a fixed alphabet.

Lemma 12. *For $X \in \mathcal{G}'$ and any context-free language L over T, $Prod_{EX}(L) \leq Prod_{CF}(L) + \#(T)$ and $Symb_{EX}(L) \leq Symb_{CF}(L) + 3\#(T)$.*

Proof. If $G = (N, T, P, S)$ is a context-free grammar, then $H = (N \cup T, P \cup \{a \rightarrow a \mid a \in T\}, S, T)$ is an EX system with $L(H) = L(G)$.   □

Theorem 7.   i) *For any alphabet T, $E0L_T \leq^3_{Prod} CF_T$ and $E0L_T \leq^2_{Symb} CF_T$.*
ii) *For any alphabet T and any $k \geq 1$, $Ekl0L_T \leq^3_{Prod} CF_T$.*

Proof. The statements are direct consequences of Lemma 12 and Theorems 3 and 4.   □

Now we compare the different types of L systems with each other.

Theorem 8. *For $k \geq 1$ and any alphabet T with at least two letters,*

i) $Prod(0L_T, kl0L_T) = 3$ and $Prod(kl0L_T, 0L_T) = 3$,
ii) $Prod(0L_T, ukl0L_T) = 3$ and $Prod(ukl0L_T, 0L_T) = 3$,
iii) $Symb(kl0L_T, 0L_T) = Symb(ukl0L_T, 0L_T) = 2$.   □

Theorem 9. *For any $k \geq 2$ and $k' \geq 2$ and any alphabet T consisting of at least 5 letters,*

i) $Prod(kl0L_T, uk'l0L_T) = 3$ and $Prod(uk'l0L_T, kl0L_T) = 3$,
ii) $Symb(kl0L_T, uk'l0L_T) = 2$ and $Symb(uk'l0L_T, kl0L_T) = 2$.

Proof. The statements are reformulations of parts of Lemmas 7 and 8.   □

We finish with the statement that the use of "nonterminals" in the case of L system can essentially decrease the descriptional complexity.

Theorem 10. *For $X \in \mathcal{G}'$ and any alphabet T with at least two letters, $EX_T \leq_{Prod}^3 X_T$ and $EX_T \leq_{Symb}^2 X_T$.*

Proof. We give the proof for *Prod* only, the modification for *Symb* are left to the reader.

Obviously, since any X system is an EX system, too, $EX \leq_{Prod}^b X$.

By Lemma 3 and the proofs of Theorems 3 and 4 we know that, for any X, there are languages K_n with $Prod_{CF}(K_n) \leq c$ for some constant c and $Prod_X(K_n) \geq n$. By Lemma 12, $Prod_{EX}(K_n) \leq c + \#(T)$.   □

References

1. H. BORDIHN and J. DASSOW, A note on the degree of nondeterminism. In: G. ROZENBERG and A. SALOMAA (Eds.), *Developments in Language Theory*. World Scientific, Singapore, 1994, 70–79.

2. J. DASSOW, T. Y. NISHIDA, and B. REICHEL, Two Papers on the Descriptional Complexity of Lindenmayer Systems. Technical Report, Department of Computer Science, Otto-von-Guericke-University Magdeburg, to appear.

3. J. GRUSKA, Some classifications of context-free languages. *Inform Control* **14** (1969), 152–179.

4. G. T. HERMAN and G. ROZENBERG, *Developmental Systems and Languages*. North-Holland, Amsterdam, 1974

5. A. KELEMENOVA and M. REMOVCIKOVA, A0L and CFG-size of languages. In [10], 177–182.

6. J. KLEIJN and G. ROZENBERG, A study in parallel rewriting systems. *Inform. Control* **44** (1980), 134–163.

7. T. Y. NISHIDA, Comparisons of sizes of context-free languages among context-free grammars and 0L systems with stable and recurrent termination. *J. Automata, Languages and Combinatorics* **6** (2001), 507–518.

8. B. REICHEL, Some classifications of Indian parallel languages. *J. Inf. Process. Cybern.* (EIK) **26** (1990), 85–99.

9. G. ROZENBERG and A. SALOMAA, *The Mathematical Theory of L Systems*. Academic Press, New York, 1980.

10. G. ROZENBERG and A. SALOMAA (Eds.), *Lindenmayer Systems*. Springer-Verlag, Berlin, 1992.

11. D. WÄTJEN, k-limited 0L systems and languages. *J. Inform. Process. Cybern.* EIK **24** (1988), 267–285.

12. D. WÄTJEN, On k-uniformly-limited T0L systems and languages. *J. Inform. Process. Cybern.* EIK **26** (1990), 229–238.

13. T. YOKOMORI, D. WOOD and K.-J. LANGE, A three-restricted normal form for ET0L languages. *Inform. Proc. Letters* **14** (1982), 97–100.

Carriers and Counters*
P Systems with Carriers vs. (Blind) Counter Automata

Hendrik Jan Hoogeboom

Institute for Advanced Computer Science
Universiteit Leiden, The Netherlands
http://www.liacs.nl/~hoogeboo/

Abstract. P systems with carriers are shown to be computationally complete with only a single membrane and two carriers. We argue that their power is due to the maximal parallelism required in the application of the rules. In fact, with a single carrier, these systems are equivalent to blind counter automata. Finally, with a single passenger restriction the systems are again computationally complete.

1   Introduction

Membrane systems (or P systems) are a computational framework inspired by cells and their biochemical processes. The system is composed of a set of hierarchically nested membranes, and within each membrane objects evolve according to specific rules, and these objects may travel from one membrane into another under certain conditions. Starting from [Pă00], the field has been rapidly growing, as witnessed by the bibliography [BibP].

In general, the objects can be strings that are manipulated by some form of rewriting, either as in classic formal language theory, or other forms of rewriting based on natural processes like splicing, recombination, or mutation – please consult [MP01], or the recent monograph [Pă02]. Here we consider *membrane systems with carriers*, the variant introduced in [MPR02]. Objects in the system are unstructured abstract symbols that cannot be deleted, copied, or changed in any way. Instead, they can only be moved from one membrane into the other by so-called carriers. Each system has a finite number of these carriers and they are used to move specific combinations of objects as passengers from one membrane into the other. It has been proved that this model is computationally complete (characterizing recursively enumerable sets of natural numbers) even with only two membranes, three carriers, and at most three passengers per carrier.

In Section 3 we reconsider this result, and show this can be improved to a single membrane, two carriers and (again) three passengers per carrier, Theorem 1. We view the single membrane as a counter automaton (or equivalently, as a register machine) where the contents of the membrane code state and counter values. A single carrier shuttles in and out of the membrane to change state and to update counter values. Another carrier is instrumental in performing the zero

---

* Work partially Supported by contribution of EU Commission under the Fifth Framework Programme, project "MolCoNet" IST-2001-32008

M. Ito and M. Toyama (Eds.): DLT 2002, LNCS 2450, pp. 140–151, 2003.

tests. The real value of our construction is that it suggests that the key feature of P systems responsible for universality is the requirement of maximal parallelism, that is, a computational step is illegal if it can be extended by applying another rule to objects not yet involved in the step. To support this conjecture, we show in Section 4 that carrier P systems that do not require maximal parallelism in their computational steps can be simulated by a *blind* counter automaton, i.e., a counter automaton unable to perform zero tests. This makes the situation similar to Petri nets, where adding maximal parallelism also gives universality [Bu81]. In the single carrier case there is no room for parallelism and we characterize the languages of single carrier systems as the blind counter languages, Corollary 4.

Finally, in the last section, we demonstrate that the third parameter, the number of passengers per carrier, can be dropped to one without loosing computational completeness from the unrestricted model, Theorem 5. (Unfortunately we cannot bound the number of membranes and carriers.) This is quite surprising as it disables types of synchronization used to implement counter machines in the single membrane case.

2   Preliminaries

Counter Automata. The family of *recursively enumerable* sets is denoted by $\mathbb{N} \cdot \mathsf{RE}$, where we use the prefix $\mathbb{N}$ to recall ourselves that, in this paper, we are considering subsets of the natural numbers. The natural device to define recursively enumerable sets is the Turing machine. In [HU79, Theorem 7.9] it is shown (following Minsky and Fischer) how a Turing machine can be simulated by a machine equipped with (at least) two counters, each storing a natural number. In [FP01] the importance of that simulation for regulated rewriting in general, and its applications to P systems in particular, has been explained. Indeed, in our paper we deal directly with counter automata, rather than with matrix grammars which seem to be the usual formalism in the area.

Transitions of the counter automaton have the syntax $\rho = (p \xrightarrow{\alpha} q, \iota)$, with intended meaning to change state from p to q, read α from the input (where α may be either a letter or the empty word λ), and perform the instruction ι to the counters. This instruction is of one of four types: ε, do not change any of the counters; $+A$, increment counter A by one; $-A$, decrement counter A by one, defined only if the counter has a positive value; $A = 0$, test whether counter A is empty, defined only if the counter has value zero.

The automaton accepts $n \in \mathbb{N}$ if it has a computation from initial to final state reading input a^n, starting and ending with empty counters.

If we disallow the zero test to the counters (i.e., the instruction $A = 0$) then we obtain the *blind* counter automaton [Gr78], BCA for short. These are strictly less powerful than the counter automaton (or Turing machine). The family of $\mathbb{N}$ subsets accepted by these automata (with arbitrary number of counters) is denoted by $\mathbb{N} \cdot \mathsf{BC}$. Note that BCA have no (explicit) zero tests, but a single (implicit) zero test is performed as we only consider computations ending with empty counters.

Membrane Systems. The structure of a *membrane system,* or P system [Pǎ00], is given by a set of hierarchically nested *membranes.* Each membrane contains items that evolve and move to neighbouring membranes according to rules specified for the particular membrane. Outside the out-most membrane we consider the *environment,* subject to its own set of rules. A catalogue of system types is given in the recent overview by Martín-Vide and Pǎun [MP01]; we refer the reader to that paper for further intuition and motivation.

For P systems with carriers, see [MPR02] for a more formal definition, one distinguishes two kind of items within the system, and we have to specify an alphabet O of *objects,* and an alphabet V of *carriers,* or *vehicles.*

The system evolves as the carriers attach to objects and transport these objects into neighbouring membranes (or from the outer membrane into the environment). This process is governed by a finite set of *rules* specified for the environment and each of the membranes. These rules can be of one of four types. Let $v \in V$, $k \geq 0$, $a_1, \ldots, a_k \in O$; let a finite multiset $\sigma : O \to \mathbb{N}$ be represented by a word over the alphabet O in an obvious way (i.e., by counting multiplicities of the symbols). We distinguish the following rules:

attaching rules, with syntax $v a_1 \ldots a_k \to [v a_1 \ldots a_k]$,
meaning that the multiset $a_1 \ldots a_k$ of objects attaches to the carrier v,
detaching rules, with syntax $[v a_1 \ldots a_k] \to v a_1 \ldots a_k$,
meaning that the multiset $a_1 \ldots a_k$ of objects detaches from the carrier v,
carry-in rules, with syntax $[v a_1 \ldots a_k] \to \text{in}$,
meaning that carrier v moves its multiset of passengers to a membrane (non-deterministically chosen) within the present region (which may be either membrane or environment),
carry-out rules, with syntax $[v a_1 \ldots a_k] \to \text{out}$,
meaning that carrier v moves its multiset of passengers to the region (either membrane or environment) in which the present membrane is contained.

The *initial configuration* of the system is given by a multiset over $O \cup V$ for each of the membranes, specifying the amounts of objects and carriers there. No such amounts are given for the environment; initially it is supposed to contain no carriers, and unlimited amounts of each object.

In each step rules are applied to the configuration in parallel, assuming maximal parallelism. This means that no carriers are left untouched that can be involved in the application of a rule (taking into account the multiplicities of the available objects).

The system reaches a *halting configuration* if in that configuration no rules can be applied. The number computed in that computation is given by the number of objects (elements of O) in a designated membrane of the system. The language defined by the system consists of all numbers computed in halting computations starting in the initial configuration. We use $\mathbb{N}\cdot\mathsf{CP}_m(p, k)$ to denote that family of languages computed by P systems with carriers, having at most m membranes, p carriers, each carrying at most k objects. We replace a parameter by $*$ if we do not impose a bound on it.

3   Single Membrane

In [MPR02, Theorem 1] the equality $\mathbb{N} \cdot \mathsf{RE} = \mathbb{N} \cdot \mathsf{CP}_2(3,3)$ is obtained by simulating matrix grammars. During that simulation carriers make non-deterministic guesses on the future of the computation. The trick is then to force conflicting guesses into infinite (non-halting) computations utilizing the required maximal parallelism. We basically use the same technique in our single membrane result.

Theorem 1. $\mathbb{N} \cdot \mathsf{RE} = \mathbb{N} \cdot \mathsf{CP}_1(2,3)$.

Proof. In order to show that $\mathbb{N} \cdot \mathsf{RE} \subseteq \mathbb{N} \cdot \mathsf{CP}_1(2,3)$, we simulate an multicounter automaton by a P system with carriers.

The single membrane of the system contains symbols that represent the values of the counters, and a symbol that represents the state of the counter automaton. The P system has two carriers. The first carrier v performs the simulation. It shuttles back and forth between membrane R and environment E, in each cycle replacing the state symbol inside the membrane by a new state symbol, and adding or removing counter symbols according to the simulated transition. Additionally, v carries a symbol that represents that transition.

The second carrier ∂ (for 'divergence') remains inside the membrane during successful computations. It is used under the rule of maximal parallelism to perform the zero tests to the counters. If, at the time of such a test, a relevant counter symbol is present, the carrier ∂ will pick it up, starting an infinite computation carrying its passenger in and out forever.

As discussed, the counter automaton has transitions of the form $(p \overset{\alpha}{\to} q, \iota)$ with α either a single letter or the empty word, and ι one of four types of counter instructions. Without loss of generality we make the following four assumptions. Final states have no outgoing transitions; for each transition $(p \overset{\alpha}{\to} q, \iota)$ we assume that either $\alpha = \lambda$ or $\iota = \varepsilon$ (or both), and we assume $p \neq q$; for each transition $(p \overset{\alpha}{\to} q, A = 0)$ we assume that all transitions that lead to q have the test $A = 0$. These assumptions can be satisfied by standard construction, for instance in case of the latter two, by leading transitions that are not in compliance to a new state, and adding a transition from the new state back to the original target.

Let a be the tape symbol of the counter automaton. For the P system we introduce a new 'output counter' a to represent the number of symbols read by the counter automaton (and generated by the P system). Accordingly, we replace every transition $(p \overset{a}{\to} q, \varepsilon)$ by the transition $(p \to q, +a)$, and we deal with the a counter as with the other counters (except that we do not require it to be empty at the end of the simulation).

The P system contains the following symbols: a state symbol q for each state of the counter automaton, a transition symbol a_ρ for each transition ρ, an initial symbol a_{in}, a counter symbol A for each counter, and a halting symbol $\hbar$. The system has two carriers v and ∂.

Initially v and ∂ are inside the membrane. The only objects inside the membrane are single copies of p_{in} and a_{in} representing the initial state of the counter automaton.

We introduce 'initial' rules, and 'default' rules for each transition symbol a_ρ and each state r.

	in E :	in R :
initial	$[va_{in}] \to va_{in}$	$va_{in} \to [va_{in}], [va_{in}] \to$ out
default	$[va_\rho r] \to$ in	$va_\rho r \to [va_\rho r], [va_\rho r] \to$ out

For each transition $\rho = (p \to q, \iota)$ we add rules that allow v to take into R the target state q and remove the source state p from R. Further passenger may be a counter symbol A depending on instruction ι:

ι	in E :	in R :
ε	$va_\rho q \to [va_\rho q], [va_\rho q] \to$ in $[va_\rho p] \to va_\rho p$	$[va_\rho q] \to va_\rho q$
$+A$	$va_\rho Aq \to [va_\rho Aq], [va_\rho Aq] \to$ in $[va_\rho p] \to va_\rho p$	$[va_\rho Aq] \to va_\rho Aq$
$-A$	$va_\rho q \to [va_\rho q], [va_\rho q] \to$ in $[va_\rho Ap] \to va_\rho Ap$	$[va_\rho q] \to va_\rho q$ $va_\rho Ap \to [va_\rho Ap],$ $[va_\rho Ap] \to$ out
$A = 0$	$va_\rho q \to [va_\rho q], [va_\rho q] \to$ in $[va_\rho p] \to va_\rho p$ $[\partial qA] \to$ in	$[va_\rho q] \to va_\rho q$ $\partial qA \to [\partial qA], [\partial qA] \to$ out

We discuss the simulation: how the halting computations of the P system correspond to successful computations of the counter automaton.

States. Carrier v never moves without taking exactly one state symbol (except when choosing the final state test $\hbar$, discussed below). This implies there is always a unique state symbol inside the membrane (when v is in the environment). Note that v can take any state symbol from the membrane into the environment. Only the proper symbol, corresponding to the source state p of the transition, can be detached from the carrier in the environment. Note in particular that the computation will not halt when the transition does not match the state, but rather the mismatch will be carried forever in and out (using the default rules) making the computation void. In this way the successive state symbols inside the membrane are consistent with consecutive transitions.

Counters. For each instruction the proper number of counter symbols is taken into or out from the membrane. In case of the instruction $-A$ the computation cannot halt when the corresponding counter is empty; instead v will take out one of the states without A symbol using the default rules, starting an infinite computation. For the test $A = 0$ the state q is brought into the membrane (state q has only incoming transitions with that test). The presence of both q and A inside the membrane leads to 'divergence' as these symbols will attach to ∂. (This divergence is not avoided if v immediately shuttles q back to the environment by the default rule as v cannot detach the improper state q.)

Final state. For each final state p and each non-final state r we add the rules

to E :	to R :
$v\hbar \to [v\hbar], [v\hbar] \to \mathrm{in}$	$[v\hbar] \to v\hbar$
	$v\hbar p \to [v\hbar p], [v\hbar p] \to \mathrm{out}$
$[v\hbar r] \to \mathrm{in}$	$v\hbar r \to [v\hbar r], [v\hbar r] \to \mathrm{out}$

The system is tested for final state if the carrier v takes $\hbar$ inside. There it attaches to the final state, takes it outside, and the computation halts (lacking possibilities to apply rules). If a non-final state is present inside at the moment v arrives with $\hbar$, the system will go into an infinite computation by 'default' rules. Observe that emptiness of the counters can be (and has to be) tested explicitly by the system before switching to a final state. Once the counter automaton has emptied all its counters when accepting, the only remaining objects in the membrane are those on the designated output counter a. $\qquad\Box$

As in [MPR02], the carrying index can be lowered to two, provided we allow an unbounded number of carriers. Indeed, in the above proof the carrier v always carries a transition symbol a_ρ (or $\hbar$). This pair va_ρ can be replaced by a new carrier v_ρ, making the number of carriers dependent on the size of the counter automaton that is simulated. (See Section 5 for single passengers rather than two; but there we loose the bound for membranes.)

Corollary 2. $\mathbb{N}\cdot\mathrm{RE} = \mathbb{N}\cdot\mathrm{CP}_1(*, 2)$.

Remark. The construction in the proof of Theorem 1 above reminds us of the classical wolf, cabbage, and goat problem. attributed to [AY99, Propositio XVIII][1]. The carrier v crossing the membrane echoes the little boat crossing the river, whereas the carrier ∂ models the conflicting presence of goat with either wolf or cabbage on the banks of the river (here the membrane).

4   Single Carrier

If we do not require maximal parallelism in the computational steps, P systems with carriers loose their computational completeness. We can simulate their work by blind counter automata (or equivalently, by Petri nets or matrix grammars without appearance checking).

Theorem 3. *Without maximal parallelism the behaviour of any P system with carriers can be simulated by a BCA.*

---

[1] Propositio de Homine et Capra et Lupo. Homo quidam debebat ultra flavium transferre lupum, capram, et fasciculum cauli. Et non potuit aliam navem invenire, nisi quae duos tantum ex ipsis ferre valebat. Praeceptum itaque ei fuerat ut omnia haec ultra illaesa omnino transferret. Dicat, qui potest, quomodo eis illaesis transire potuit.

Proof. For each object of the P system, and each of its membrane regions, the BCA uses a counter to record the number of that object in that particular region. The finite state of the BCA stores the positions and passengers of the carriers. Rules of the P system can thus be simulated in a straightforward manner.

The main technical problem to solve is that of halting. The BCA must verify that the P system is halted before moving to a final state. In other words, it must check that no rule is applicable.

Rules that carry-in or carry-out, detaching rules in general, and attaching rules in the environment can be checked in the finite state, independent of the further configuration of the system (the contents of the membranes as stored in the counters). So we only consider those carriers that are in one of the membranes not carrying any symbols. In order to accept it must be verified that for *each* attaching rule $vA_1^{i_1} \ldots A_n^{i_n} \to [vA_1^{i_1} \ldots A_n^{i_n}]$ in the membrane at least one of the objects (say A_k) has less than i_k copies, so the rule can not be applied (here $A_1, \ldots, A_n$ lists the objects in the membrane).

As the blind counter automaton is not able to test the value of its counters this halting test can only be done indirectly: the automaton guesses the contents of its counters (from a bounded number of possibilities as we explain shortly) decreases its counters by that amount, and goes into final state. Now the computation is only successful if the guess was right, which means that the counters are empty in the end, a necessary condition for accepting blind counter computations.

To be more precise, let M be the maximal number of occurrences of a single counter symbol in the carrier rules in a particular membrane. This means $i_k \leq M$ for $k = 1, \ldots, n$ in each rule $vA_1^{i_1} \ldots A_n^{i_n} \to [vA_1^{i_1} \ldots A_n^{i_n}]$. The BCA now chooses a vector $(\sigma_1, \ldots, \sigma_n)$ over $\{0, \ldots, M-1, \infty\}$, such that for each rule there is a k with $\sigma_k < i_k$. Note that the number of such vectors is finite, and can be hard-wired in the BCA; each halting configuration of the P system is represented by such a vector.

Before moving to an accepting state, the BCA decreases its counters according to the guess, that is, if $\sigma_k \neq \infty$, then the counter for A_k is decreased by exactly σ_k, and otherwise ($\sigma_k = \infty$) the A_k counter is decreased non-deterministically by an arbitrary number. The important observation is that in each halting configuration of the P system, and only in halting configurations, a vector as above can be chosen according to which the counters can be decreased to zero, leading to an accepting computation of the BCA.

Remember that the P system generates the computed output as the number of objects present in a designated membrane. While the BCA decreases its counters non-deterministically to zero in the accepting phase described above, it reads input symbols simultaneously to decreasing the counters corresponding to objects in the designated membrane. In this way, the input read corresponds to the number generated.   □

With a single carrier there is no parallelism in the system. Hence the previous result shows (the first) right-to-left inclusion of the following characterization.

Corollary 4. $\mathbb{N} \cdot \text{BC} = \mathbb{N} \cdot \text{CP}_*(1, *) = \mathbb{N} \cdot \text{CP}_1(1, 3)$.

Proof. The inclusion $\mathbb{N}\cdot\mathsf{BC} \subseteq \mathbb{N}\cdot\mathsf{CP}_1(1,3)$ is implicit in the proof of Theorem 1: after dropping the zero test from the available instructions, there is no need to include the divergence carrier ∂ in the simulation. However, we have to redesign the zero tests upon acceptance.

For convenience we assume first that the empty word λ (representing 0) is not in the language of the simulated BCA. This allows us to code a^n as $n-1$ copies of the output counter a and a final state symbol p.

For each counter A (other than the output counter a) and each state r which is not final here we have the rules:

in E :	in R :
$v \rightarrow [v], [v] \rightarrow \text{in}$	$[v] \rightarrow v$
$[vr] \rightarrow \text{in}$	$vr \rightarrow [vr], [vr] \rightarrow \text{out}$
$[vA] \rightarrow \text{in}$	$vA \rightarrow [vA], [vA] \rightarrow \text{out}$

Now the end of the computation is signalled by v moving in without passenger. The presence of a non-final state r or a counter symbol A leads to an infinite computation. Otherwise, in a final state with all counters empty, we reach an halting configuration. The membrane then contains copies of a, a final state symbol p and the carrier v. The carrier is not counted for the output generated, but we should subtract one from the a's before going into the final state to compensate for the presence of p.

If the BCA accepts λ then we make sure the P system is able to remove the contents of the membrane at its first step while reaching a halting configuration. Add to R another symbol Z and add the rules $va_{in}Z \rightarrow [va_{in}Z], [va_{in}Z] \rightarrow \text{out}$. After that, Z is considered as counter which has to be emptied by adding the proper instructions. $\qquad\square$

By [HJ94] the family $\mathbb{N}\cdot\mathsf{BC}$ consists of semilinear languages, and hence is equal to $\mathbb{N}\cdot\mathsf{REG}$, the one-letter regular languages. Like the other results of this paper, Corollary 4 can be generalized to Parikh sets using a set $\{a_1, \ldots, a_k\}$ of output symbols to obtain subsets of $\mathbb{N}^k$, $k \geq 1$. Extending our notation in the obvious way, we have $\mathbb{N}^k\cdot\mathsf{BC} = \mathbb{N}^k\cdot\mathsf{CP}_*(1,*)$. Recall that $\mathbb{N}^2\cdot\mathsf{BC}$ contains non-semilinear sets, such as $\{\binom{i}{j} \mid 1 \leq i, 1 \leq j \leq 2^i\}$.

5   Single Passenger

Having discussed the single membrane and single carrier cases, it seems natural to examine the power of systems that are allowed at most a single passenger. This seems to prohibit any synchronization in moving objects in and out, or any tests on the simultaneous presence of two symbols in a membrane (as used for ∂ in the proof of Theorem 1).

Nevertheless, single passenger systems are computationally complete, a rather surprising result. Unfortunately, both the number of membranes and the number of carriers depend on the simulated automaton. As before, maximal parallelism plays a major role in handling the zero tests.

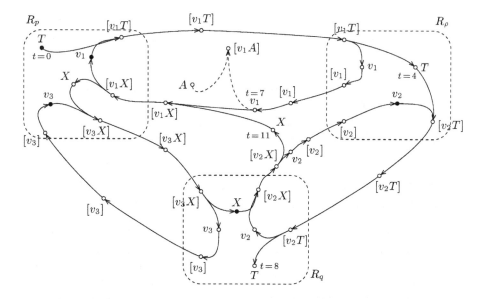

Fig. 1. Zero test as performed by a P system with carrier index 1 (Theorem 5). Shown is the simulation of the transition $\rho : (p \rightarrow q, A = 0)$. Upon arrival (at $t = 0$) of T' in R_p it is brought to R_q via R_ρ. The carrier v_1 arrives in the outer membrane at $t = 7$. It takes X arriving at $t = 11$, unless A's are present which force a non-halting computation.

Theorem 5. $\mathbb{N} \cdot \mathsf{RE} = \mathbb{N} \cdot \mathsf{CP}_*(*, 1)$.

Proof. We prove $\mathbb{N} \cdot \mathsf{RE} \subseteq \mathbb{N} \cdot \mathsf{CP}_*(*, 1)$ by simulating a counter automaton. We make assumptions on the counter automaton, as in the proof on Theorem 1.

For each state p the outer membrane contains a membrane R_p representing p. The change of state of the counter automaton is modelled by moving an object, the state token T, from membrane to membrane. This is done by carriers associated to the transitions of the counter automaton. So, somewhat simplified, the carrier v associated to $(p \rightarrow q, \iota)$ resides in R_p and, upon arrival of T carries it via the outer membrane to R_q. Afterwards it returns to R_p, and on its way carries a counter symbol A to or from the outer membrane when $\iota = \pm A$.

Direct addressing. The carry-in rules in this proof were written with a designated target membrane in mind, whereas the syntax of those rules does not allow us to specify a specific destination; it is chosen non-deterministically. When a carrier enters the wrong membrane by mistake, we should avoid a situation where the computation is blocked as this will lead to a halting configuration, accepting rather than rejecting the computation. There are two ways to simulate direct addressing.

First, it is possible to nest all membranes inside one another, in a fixed order. Traveling from R_p to R_q is forced by placing proper carry-in or carry-out instructions at the intermediate membranes (depending on the order of nesting).

Second, we may place within each membrane a new membrane, used to trap misplaced carriers into an infinite loop. Whenever a carrier enters the wrong membrane by mistake it is sent in and out of that new membrane, so that the computation does not reach a halting state.

We did not implement those rules for direct addressing below, but they should be provided for a properly working system.

Zero test. There are three designated carriers associated to the single transition $\rho = (p \rightarrow q, A = 0)$; we will denote them here by v_1, v_2 and v_3. First, v_1 resides in the state membrane R_p. Upon arrival of the token T it brings it to the transition membrane R_ρ. Carrier v_1 returns to R_p via the outer membrane R but only if it attaches to a return token X, which will arrive there later in the process. If instead a counter symbol A is present, then v_1 will attach to that first (thanks to maximal parallelism) and start an infinite computation. Carrier v_2 resides in the transition membrane R_ρ. When token T arrives v_2 brings it to the state membrane R_q. Then, v_2 returns to R_ρ carrying the token X to the outer membrane. Finally, carrier v_3 resides in R_p. When the return token X arrives, v_3 travels back and forth to R_q to return X for future use. (See Figure 1.)

We list the necessary rules here.
The v_1-cycle:

in R_p :	in R :	in R_ρ :	in R :	in R_p :
$v_1 T \rightarrow [v_1 T]$	$[v_1 T] \rightarrow$ in	$[v_1 T] \rightarrow v_1 T$	$[v_1] \rightarrow v_1$	$[v_1 X] \rightarrow v_1 X$
$[v_1 T] \rightarrow$ out		$v_1 \rightarrow [v_1]$	$v_1 X \rightarrow [v_1 X]$	
		$[v_1] \rightarrow$ out	$[v_1 X] \rightarrow$ in	

Zero test: in R the rules $v_1 A \rightarrow [v_1 A]$, $[v_1 A] \rightarrow$ out, in E the rule $[v_1 A] \rightarrow$ in.
The v_2-cycle:

in R_ρ :	in R :	in R_q :	in R :	in R_ρ :
$v_2 T \rightarrow [v_2 T]$	$[v_2 T] \rightarrow$ in	$[v_2 T] \rightarrow v_2 T$	$[v_2 X] \rightarrow v_2 X$	$[v_2] \rightarrow v_2$
$[v_2 T] \rightarrow$ out		$v_2 X \rightarrow [v_2 X]$	$v_2 \rightarrow [v_2]$	
		$[v_2 X] \rightarrow$ out	$[v_2] \rightarrow$ in	

The v_3-cycle:

in R_p :	in R :	in R_q :	in R :	in R_p :
$v_3 X \rightarrow [v_3 X]$	$[v_3 X] \rightarrow$ in	$[v_3 X] \rightarrow v_3 X$	$[v_3] \rightarrow$ in	$[v_3] \rightarrow v_3$
$[v_3 X] \rightarrow$ out		$v_3 \rightarrow [v_3]$		
		$[v_3] \rightarrow$ out		

Decrement counter. There are two designated carriers associated to the transition $\rho = (p \rightarrow q, -A)$, we will denote them here by v_1 and v_2. First, v_1 resides in the state membrane R_p. Upon arrival of the token T it brings it to the transition membrane R_ρ. Carrier v_1 returns to R_p via the outer membrane R where it exchanges a return token X for a counter symbol A. This symbol is brought to R_p and thus removed from the counter. When no A's are present, the carrier is forced into an infinite loop. Carrier v_2 resides in the transition membrane R_ρ. When token T arrives v_2 brings it to the state membrane R_q. Then, v_2 returns to R_ρ picking up the token X in the outer membrane. We have introduced the return token X to enable the infinite loop of v_1 without passenger. We list the necessary rules here.

The v_1-cycle:

in R_p :	in R :	in R_ρ :	in R :	in R_ρ :
$v_1 T \to [v_1 T]$	$[v_1 T] \to$ in	$[v_1 T] \to v_1 T$	$[v_1 X] \to v_1 X$	$[v_1 A] \to v_1 A$
$[v_1 T] \to$ out		$v_1 X \to [v_1 X]$	$v_1 A \to [v_1 A]$	
		$[v_1 X] \to$ out	$[v_1 A] \to$ in	

No A's: in R the rules $v_1 \to [v_1]$, $[v_1] \to$ out, in E the rule $[v_1] \to$ in.

The v_2-cycle:

in R_ρ :	in R :	in R_q :	in R :	in R_ρ :
$v_2 T \to [v_2 T]$	$[v_2 T] \to$ in	$[v_2 T] \to v_2 T$	$[v_2] \to v_2$	$[v_2 X] \to v_2 X$
$[v_2 T] \to$ out		$v_2 \to [v_2]$	$v_2 X \to [v_2 X]$	
		$[v_2] \to$ out	$[v_2 X] \to$ in	

Increment counter. There are two designated carriers associated to the transition $\rho = (p \to q, +A)$, we will denote them here by v_1 and v_2. First, v_1 resides in the state membrane R_p. Upon arrival of the token T it brings it to the transition membrane R_ρ. Carrier v_1 returns to R_p via the outer membrane R where it drops a passenger, a counter symbol A that was picked up in R_ρ. Carrier v_2 resides in the transition membrane R_ρ. When token T arrives v_2 brings it to the state membrane R_q. Then, v_2 returns to R_ρ picking up a new counter token A in the environment bringing it to R_ρ for future use. We list the necessary rules here.

The v_1-cycle:

in R_p :	in R :	in R_ρ :	in R :	in R_p :
$v_1 T \to [v_1 T]$	$[v_1 T] \to$ in	$[v_1 T] \to v_1 T$	$[v_1 A] \to v_1 A$	$[v_1] \to v_1$
$[v_1 T] \to$ out		$v_1 A \to [v_1 A]$	$v_1 \to [v_1]$	
		$[v_1 A] \to$ out	$[v_1] \to$ in	

The v_2-cycle:

in R_ρ :	in R :	in R_q :	in R :	in E :
$v_2 T \to [v_2 T]$	$[v_2 T] \to$ in	$[v_2 T] \to v_2 T$	$[v_2] \to$ out	$[v_2] \to v_2$
$[v_2 T] \to$ out		$v_2 \to [v_2]$		$v_2 A \to [v_2 A]$
		$[v_2] \to$ out		$[v_2 A] \to$ in

in R :	in R_ρ :
$[v_2 A] \to$ in	$[v_2 A] \to v_2 A$

Final state. As before, we assume that all counters have been tested for zero before entering the final state. In the corresponding membrane we have no rules for token T, thus the system enters a halting configuration (assuming that no previous actions lead to infinite loops).

□

Conclusion. We have separately studied the power of the features of carrier P systems, characterizing the power of single membrane, single carrier, and single passenger systems. Somewhat surprisingly, only the single carrier case was not able to reach computational completeness. Additionally, we have argued that P systems with carriers are similar to Petri nets: in the absence of synchronization they can be simulated by blind counter automata, whereas under the rule of maximal parallelism they are computationally complete. We suggest this is true for other classes of P systems, at least if no other powerful features are present, like priorities on rules, or rewriting on string objects.

In [FH02] we continue this line of research, and apply our techniques to the related P systems with symport/antiport rules.

References

AY99.   Alcuin of York (735-804). *Propositiones ad Acuendos Iuvenes*, around 799.

BibP.   Bibliography of P Systems, at: *The P Systems Web Page*, C. Zandron (ed.), `http://psystems.disco.unimib.it`

Bu81.   H.-D. Burkhard. Ordered Firing in Petri Nets, *Elektr. Informationsverarbeitung und Kybernetik EIK* 17 (1981) 71–86.

FP01.   R. Freund, Gh. Păun, On the Number of Non-terminals in Graph-controlled, Programmed, and Matrix Grammars, *in* [MR01], pages 214–225.

FH02.   P. Frisco, H.J. Hoogeboom, Simulating Counter Automata by P Systems with Symport/Antiport. In: Gh Păun, C. Zandron (eds.) *Pre-proceedings of the Second Workshop on Membrane Computing*, Curtea de Argeş Romania, Molconet Publication 1, 2002, pages 237–248. Updated papers to appear in LNCS.

Gr78.   S.A. Greibach, Remarks on Blind and Partially Blind One-Way Multicounter Machines, *Theoretical Computer Science* 7 (1978) 311–324.

HJ94.   D. Hauschildt, M. Jantzen, Petri Net Algorithms in the Theory of Matrix Grammars, *Acta Informatica* 8 (1994) 719–728.

HU79.   J.E. Hopcroft, J.D. Ullman. *Introduction to Automata Theory, Languages, and Computation*, Addison-Wesley, 1979.

MP01.   C. Martín-Vide, Gh. Păun, Computing with Membranes (P Systems): Universality Results, *in* [MR01], pages 214–225.

MPR02. C. Martín-Vide, Gh. Păun, G. Rozenberg, Membrane Systems with Carriers, *Theoretical Computer Science* 270 (2002) 779–796.

MR01.   M. Margenstern, Y. Rogozhin (eds.) *Procedings Universal Computational Models*, Chişinău, *Lecture Notes in Computer Science*, Volume 2055, Springer Verlag, 2001.

Pă00.   Gh. Păun, Computing with Membranes, *Journal of Computer and System Sciences*, 61 (2000) 108–143.

Pă02.   Gh. Păun. *Membrane Computing, an Introduction*. Natural Computing Series, Springer Verlag, 2002.

On the Separation between k-Party and $(k − 1)$-Party Nondeterministic Message Complexities*

Henry N. Adorna[1,2]

[1] Lehrstuhl für Informatik I, RWTH-Aachen,
Ahornstraße 55, D-52074, Aachen, Germany
henri@i1.informatik.rwth-aachen.de
[2] Department of Mathematics, Univ. of the Philippines,
Diliman 1101, Q.C., Philippines
henri@math.upd.edu.ph

Abstract. We introduce (reasonable) generalizations of the one-way uniform two-party protocols introduced in [6,7], which provide a closer relationship between communication complexity and finite automata as regular language recognizers. A superpolynomial separation between k-party and $(k − 1)$-party message complexities of the nondeterministic model is established by exhibiting a sequence of concrete languages to witness it, thus a strong hierarchy result. As a consequence, the new model provides an essentially better lower bound method for estimating $ns(L)$, for some regular languages L. We remark that in the deterministic case hierarchy is not realized.

1 Introduction

Communication protocols accepting languages have been considered in [6,12,14]. The first, however, rather simple application of communication complexity in this respect was given in [5]. That for every regular language L, there is a positive integer m s.t. a standard 2-party protocol can recognized it in at most m communication.

Half a decade ago, *one-way uniformity* was introduced in [6,7] to communication complexity of 2-party protocols, which was started in [16] (and also in [1]). This effort was motivated by giving a closer relationship between communication complexity and finite automata.

Uniformity leads to an equality relation between the minimal size of a deterministic finite automata for L, $s(L)$ and its so-called **deterministic message complexity,** $dmc_2(L)$, while the **nondeterministic message complexity** of L, $nmc_2(L)$ remains to provide a *lower bound* for the minimal size of a nondeterministic finite automaton for L, $ns(L)$. Until today most lower bound techniques for $ns(L)$ are based on communication technique.

---

* Supported by the Deutscher Akademischer Austauschdienst (DAAD) and DFG Project HR 14/16-1

M. Ito and M. Toyama (Eds.): DLT 2002, LNCS 2450, pp. 152–161, 2003.

In several ocassions, it has been shown that there are languages whose $\mathrm{nmc}_2(L)$ is much smaller than $\mathrm{ns}(L)$ [9,10,11], though. In particular, a superpolynomial gap was demonstrated between them in [9,10].

In [3], a reasonable extension of the uniform model in [7] has been introduced. It has been observed, however, that, for some regular languages L, an additional processor helps in providing a better estimate of the lower bound for $\mathrm{ns}(L)$.

This is corollary to the fact that the language $PART_n$ from [9] witnesses a similar gap between $\mathrm{nmc}_2(PART_n)$ and $\mathrm{nmc}_3(PART_n)$ [3]. The increase in complexity is due mainly to the fact that the input can be awkwardly partitioned between the participating computers.

Intuitively, for $k \geq 3$, our k-party model P_k computes on any $w \in \Sigma^*$ always with respect to (w.r.t.) a k-partition of w by passing messages from the left to right, beginning from the first computer C_1 consecutively until the last computer C_k. The message/decision to be transmitted by every C_i, $1 \leq i \leq k-1$, depends on α_i and the message obtained from C_{i-1}. Only C_k will have to decide on the acceptance of the input, once a message from C_{k-1} is received.

We represent a computation of our model as:

$$C \colon \Sigma^* \to 2^{\{0,1\}^*} \cup \{\texttt{accept}, \texttt{reject}\} \quad \text{s.t.} \quad \forall w \in \Sigma^*,$$

$$m_1 \$ m_2 \$ \cdots \$ m_{k-1} \$ m_f \in C(w).$$

A set of words or language is accepted by our model iff at least an accepting computation for each word w in it ends with $m_f = \texttt{accept}$. A computation for a word which terminates with $m_f = \texttt{reject}$, connotes nonacceptance of such a word by our model. For $1 \leq i \leq k-1$, $m_i \in 2^{\{0,1\}^*}$.

We denoted by $M(P_k)$ the set of messages spent by P_k in recognizing an L and $\min |M(P_k)|$ will be our message complexity for L.

An almost similar k-party model was presented by Tiwari [15], wherein for $w = x\,y$, the mode of communication is two-way aside from the first computer, (with x) who serves messages one-way to the right and the last one (with y) to its left. The middle ones have nothing at the start. However, our model is strictly one-way and every computer can have part of the input at the outset.

This paper aims to establish a superpolynomial gap between message complexities for all $k \geq 2$. Specifically, we will show that for all $k \geq 2$, there is a sequence of regular languages, $L_{(k,n)}$ such that all k-party protocols will require a large amount of messages to recognized it, however, it has an efficient $(k-1)$-party protocol. This possibility was mentioned in [2,3].

The remainder of the paper will be as follows: We will first introduce our model and complexity measure, then we show that a language is regular iff it is accepted by some k-party protocol with finite message complexity (Section 2). Afterwards, we give a prelude (Section 3) on the possibility of a strong hierarchy on the nondeterministic message complexity. Finally, we prove the main result (Section 4), by showing

$$\mathrm{nmc}_k(L_{(k,n)}) = 2^{\Omega(\sqrt{\mathrm{nmc}_{(k-1)}(L_{(k,n)})})}.$$

A remark on the deterministic case is provided at the end (Section 5).

In this paper, we will deal primarily with nondeterministic models of computations. Thus, models of computation and complexity measures are deemed nondeterministic, unless otherwise specified. All protocols will be 1-way uniform k-party nondeterministic protocols and will be denoted by P_k.

2   Preliminaries

Let $L \subseteq \{0,1\}^*$ be a regular language which is to be recognized by k linearly connected independent computers. Every computers C_i ($1 \leq i \leq k$), however, obtains part α_i ($0 \leq |\alpha_i| \leq |w|$) of the whole input $w \in \Sigma^*$ s.t. $w = \alpha_1\alpha_2\ldots\alpha_k$. We denote a k-**partition** of w by a k-tuple $(\alpha_1, \alpha_2, \ldots, \alpha_k)$.

The Model

A 1-**way uniform k-party nondeterministic protocol over** Σ is a triple $P_k = \langle \Phi, \{\phi_i\}_{i=2}^{k-1}, \varphi \rangle$ where
- $\Phi \colon \Sigma^* \longrightarrow 2^{\{0,1\}^*}$ is a function.
- $\phi_i \colon \Sigma^* \times \{0,1\}^* \longrightarrow 2^{\{0,1\}^*}$ are functions.
- $\varphi \colon \Sigma^* \times \{0,1\}^* \longrightarrow \{\texttt{accept}, \texttt{reject}\}$ is a function.

Intuitively, P_k computes on any $w \in \Sigma^*$ always w.r.t. a k-partition of w by passing messages from the left to right, beginning from the first computer C_1 consecutively until the last computer C_k, which will be the only one to decide on the acceptance of the input, once a message from C_{k-1} is received. The message to be transmitted by every C_i, $1 \leq i \leq k-1$, depends on α_i and the message obtained from C_{i-1}.

A *computation* C of P_k on an input $w \in \Sigma^*$ w.r.t. a k-partition $(\alpha_1, \alpha_2, \ldots, \alpha_k)$ is denoted by a function

$$C \colon \Sigma^* \to 2^{\{0,1\}^*} \cup \{\texttt{accept}, \texttt{reject}\} \quad \text{s.t.} \quad \forall w \in \Sigma^*,$$

$$m_1 \$ m_2 \$ \cdots \$ m_{k-1} \$ m_f \in C(w), \quad \text{where}$$

- $m_1 \in \Phi(\alpha_1)$.
- $m_i \in \phi_i(\alpha_i, m_{i-1})$, $2 \leq i \leq k-1$.
- $m_f = \varphi(\alpha_k, m_{k-1}) \in \{\texttt{accept}, \texttt{reject}\}$.

We denote the set of all messages that a P_k uses in the computation by

$$M(P_k) = \Phi(\Sigma^*) \cup \bigcup_{i=2}^{k-1} \phi_i(\Sigma^* \times \{0,1\}^*) \quad \text{s.t.}$$

- $|M(P_k)|$ is finite.
- $M(P_k) \cap M(P_k) \cdot \Sigma^+ = \emptyset$ or all messages must be *prefix-free*.

If C yields to $m_f = \texttt{accept}$, then C is an **accepting computation** for w. Else, it is **rejecting**. A language L is **accepted** by P_k, denoted $L = L(P_k)$ iff for all $w \in L$ there is a computation of P_k which yields to $\texttt{accept}$ w.r.t. all

k-partitions of w and for all $w \notin L$, no accepting computation may exists. We define

$$\mathbf{nmc}_k(P_k) = |M(P_k)|$$

as the **1-way uniform k-party message complexity** of P_k. The minimum of all these message complexities for a fixed language is the **1-way uniform k-party nondeterministic message complexity** of L, i.e.

$$\mathbf{nmc}_k(L) = \min\{|M(P_k)| \,|\, L = L(P_k)\}.$$

For any positive integer m, the set of all languages recognized by a P_k within communication message m is denoted by $\mathrm{NMCOMM}_k^1(m)$.

Theorem 1. *For all $k \geq 2$, L is regular iff $\exists\, m \in Z^+$, $L \in \mathrm{NMCOMM}_k^1(m)$.*

Proof. ($\Longrightarrow$) Let N be an NFA accepting L with finite set of states Q. Let $m = |Q|$. By a simulation technique, a protocol P_k can be constructed s.t. every computer in it simulates N on their respective local part of the input beginning from the message sent by the preceding computer, which is in binary code representation of length $l = \lceil \log_2 |Q| \rceil$, of every state in Q of N. Then it is obvious that the number of messages will be exactly the size of N.

($\Longleftarrow$) Suppose for $k \geq 2$ and some $m \in Z^+$, $L \in \mathrm{NMCOMM}_k^1(m)$. We define a relation R_{P_k} on Σ^* s.t. for any $x, y \in \Sigma^*$, $x\, R_{P_k}\, y$ iff $C(x) = C(y)$. The fact that equality is an equivalence relation, then so is R_{P_k}. The finiteness of the set of messages implies the finiteness of the number of equivalence classes of R_{P_k}, too. Suppose for any $x, y \in \Sigma^*$ we have $x\, R_{P_k}\, y$. Then for all $z \in \Sigma^*$, the protocol P_k accepting L will not distinguish between xz and yz in computing them w.r.t. all their k-partitions, i.e. $C(x) = C(y)$ implies $C(xz) = C(yz)$. Moreover, if $C(x) = C(y)$ then for all $z \in \Sigma^*$ $xz \in L$ iff $yz \in L$. Certainly, R_{P_k} is a refinement of the natural right invariant equivalence relation R_L on L. Eventually, by Myhill-Nerode Theorem, the result follows.

From the above result, it is an easy observation that, for all regular languages L, there is a positive integer m, s.t. for all $k \geq 2$,

$$L \in \mathrm{NMCOMM}_k^1(m).$$

If we choose m to be the size of the minimal NFA for L, then we have the following Corollary.

Corollary 1. *For all $k \geq 2$ and every regular language L,*

$$L \in \mathrm{NMCOMM}_k^1(\mathrm{ns}(L)).$$

It is worth noting that, by a simple simulation technique, we have for the complement L^C of L that

$$\text{if } L \in \mathrm{NMCOMM}_k^1(\mathrm{ns}(L)), \text{ then } L^C \in \mathrm{NMCOMM}_k^1(2^{\mathrm{ns}(L)}).$$

Theorem 2. *For all $m \in Z^+$, there is an L s.t. $L \notin \mathrm{NMCOMM}_k^1(m)$.*

Proof. Let $L = \{0^m\}$. It is not hard to see that a protocol P_k will require $m + 1$ messages for L, since every computer in it will have to send the amount of 0's they have, however, in binary code of length $l = \lceil \log_2(m + 1) \rceil$. The number of bits in every computer in P_k extends from 0 to m.

Below we consider message complexities of regular languages under regular operations.

Lemma 1. *For all* $k \geq 2$, *for integers* m *and* m', *let* $L_1 \in \text{NMCOMM}_k^1(m)$ *and* $L_2 \in \text{NMCOMM}_k^1(m')$, *then the following holds,*
1. $L_1 \cup L_2 \in \text{NMCOMM}_k^1(m + m')$.
2. $L_1 L_2 \in \text{NMCOMM}_k^1(m + m')$.
3. $L_1^* \in \text{NMCOMM}_k^1(m)$.

Proof. To show that $L_1 \cup L_2 \in \text{NMCOMM}_k^1(m+m')$, we construct a protocol P_k s.t. $\text{nmc}_k(P_k) \leq m+m'$. For an input $w \in \Sigma^*$, the first computer of the protocol P_k guesses whether it is in L_1 or in L_2. A bit 1 will be sent if it guessed that it is in L_1, else 0. Then the subsequent computers will be working on recognizing the input either in L_1 or in L_2 depending on the bit affixed by C_1. Noticed that the amount of messages that P_k will be spending is equal to $m + m'$. Thus, $\text{nmc}_k(L_1 \cup L_2) \leq m + m'$.

Next we show that $L_1 L_2 \in \text{NMCOMM}_k^1(m + m')$, too. Let P_k be a protocol that accepts this concatenation. Let $w = xy \in \Sigma^*$. Noticed that if x is a subword of α_1 in C_1, then C_1 can immediately decide on it as partially accepted. Then it (C_1) sends a message 1 to signal that the other partners must now work on the recognition of $y \in L_2$. Else, everybody will have to reject w and just let C_k announce it. Thus, P_k needs at the most m' messages to perform a favorable computation on the input. If y is a subword of α_k in C_k, then a similar computation can be performed which, however, needs m messages. In the general case, for any input $w = xy$, the first, say $j \leq k$, computers of P_k verify, if the prefix of w is in L_1. If anyone of them proves that it is, then it prompts the next one by sending, say a "1" in order for them to verify $y \in L_2$. Obviously, they didn't spend more than $m + m'$ of messages.

Similarly, if $\text{nmc}_k(L_1) \leq m$, then for all $w \in L_1^*$, w will just be a multiple copies or concatenations of words in L_1. Then the protocol constructed for the concatenation works well for the star of L_1, too. However, it uses only at the most m messages.

3   Gaps between Message Complexities

Lemma 2. [3] *For all* $k \geq 2$ *and any regular language* L,

$$\text{nmc}_{(k-1)}(L) \leq \text{nmc}_k(L).$$

Proof. Let $L \subseteq \Sigma^*$ be a regular language accepted by a P_k. We can construct a protocol P_{k-1} s.t. the first $k - 2$ computers in P_{k-1} do the job of the first $k - 2$ computers in P_k, and the work of the last two computers in P_k will be done only by the C_{k-1} in P_{k-1}. Then the amount of messages that the $k-2$ computers will send until it reaches C_{k-1} in P_{k-1} will be no more than that which will be transmitted by the $k-1$ processors of P_k so that C_k can decide for acceptance.

The preceding Lemma can be restated in terms of $\mathrm{NMCOMM}_k^1(m)$, as follows:

Theorem 3. *For all* $k \geq 3$, $\forall m \in Z^+$

$$\mathrm{NMCOMM}_k^1(m) \subseteq \mathrm{NMCOMM}_{k-1}^1(m).$$

Proof. By the previous Lemma, for all L, $\mathrm{nmc}_{(k-1)}(L) \leq \mathrm{nmc}_k(L) \leq m$. Then the inclusion follows.

Theorem 4. *For all* $k \geq 3$, $\forall m_1 \in Z^+$ $\exists m_2 \in Z^+$ *s.t.* $m_1 \leq m_2$

$$\mathrm{NMCOMM}_{k-1}^1(m_1) \subseteq \mathrm{NMCOMM}_k^1(m_2).$$

Proof. Let $L \in \mathrm{NMCOMM}_{k-1}^1(m_1)$ and let $m_2 = \mathrm{nmc}_k(L)$.

Example 1. *Let* $L_{(3,n)} = \{xyz \mid x, y, z \in \{0,1\}^n, x = y \vee x \neq z\}$. *A remarkably message efficient running protocol was issued in* [9] *which certified* $\mathrm{nmc}_2(L_{(3,n)}) \in O(n^2)$. *However,* $\mathrm{nmc}_3(L_{(3,n)}) \in 2^{\Omega(n)}$ [3].

If we restrict ourselves to finite languages, then an invariance on message complexity for some k can be reached.

Theorem 5. *For finite languages,* $\exists k \in Z^+$ *and* $m \in Z^+$ $\forall j \geq 0$, *s.t.*

$$\mathrm{NMCOMM}_k^1(m) = \mathrm{NMCOMM}_{k+j}^1(m).$$

Proof. Let $w \in L \subseteq 2^{\Sigma^{\leq n}}$, $n \in Z^+$. Without loss of generality, let $k = |w| + 1$. Then the computers in P_k share parts of the bits of w as required by the model. Obviously, if everyone must have at least a bit, then there will be a processor without any bit from w, i.e. its local part is λ (the empty string). In effect, this machine will not contribute essentially in the sending of messages to accept L. Similar event is expected if there are r of them, $1 \leq r \leq k - 1$.

Let $m = \mathrm{nmc}_k(P_k)$, s.t. $L \in \mathrm{NMCOMM}_k^1(m)$. An addition of at least $j \geq 0$ processors in the protocol will only increase the number of possible machines obtaining no bit from w, hence will play only as a reliable relay processors in the procedure. This leads us to conclude that $\mathrm{nmc}_k(P_k) = \mathrm{nmc}_{(k+j)}(P_{k+j}) = m$ and therefore, $L \in \mathrm{NMCOMM}_{k+j}^1(m)$. By Theorem 3, equality follows.

4   The Main Result

In this part, we will prove the following result;

Theorem 6. *For all non-constant $k \geq 3$, there exist regular languages L_j, $j \in Z^+$, s.t.*

$$\mathrm{nmc}_{(k-1)}(L_j) \in O(j^2) \quad and \quad \mathrm{nmc}_k(L_j) \in 2^{\Omega(j)}.$$

This will follow from the series of the subsequent lemmas.

Lemma 3. Main Lemma
For all non-constant $k \geq 3$, $\exists L$ s.t.

$$\mathrm{nmc}_k(L) = 2^{\Omega(\sqrt{\mathrm{nmc}_{(k-1)}(L)})}.$$

Proof. Let $n \in Z^+$. Define $L = L_{(k,n)}$, for non-fixed $k \geq 3$ s.t.
$$L_{(k,n)} = \{x_1 x_2 \cdots x_k \mid x_i \in \{0,1\}^n, \forall i \in \{1,2,\ldots,k\}$$
$$\text{and } \exists i \in \{1,2,\ldots,k-2\}, \text{ s.t. } x_i = x_{i+1} \vee x_i \neq x_{i+2} \}.$$
We need to show an existence of a protocol P_{k-1} s.t. $\mathrm{nmc}_{(k-1)}(P_{k-1}) = O(n^2)$ and for all protocols P_k, $\mathrm{nmc}_k(P_k) \geq 2^{\frac{n}{k-1}}$.

In the manner that we have constructed $L_{(k,n)}$, if there is some index $i \in \{1,2,\ldots,k-2\}$ s.t. $x_i \neq x_{i+2}$, then P_{k-1} will not have a hard time catching this inequality nondeterministically. Similarly, if at the most $k-1$ of the x_i's coincide, then we can also be sure of meeting an inequality, soon. The worst case is when all of the x_i's coincide and no computer has the ability to recognize any of these equalities.

We will continue the proof in the series of Lemmas below. For the succeeding statements, k is non-constant.

Lemma 4. *For all $k \geq 3$, $n \in N$, there is a P_{k-1} s.t. $L(P_{k-1}) = L_{(k,n)}$ and $\mathrm{nmc}_{(k-1)}(P_{k-1}) = O(n^2)$*

Proof. The existence of P_{k-1} is assured by the fact that $L_{(k,n)}$ is regular. Due to space constraint, we simply describe how P_{k-1} works.

Firstly, every computer in P_{k-1} compute the length l_i of their local substring α_i of the input $w \in \Sigma^*$. Computers in P_{k-1} will use the code message $(T, j, b_j, h, f(b_j) g(b_j))$, where C_1 picks the jth bit b_j from the prefix of α_1 of length at most n. This bit b_j together with the integer j will be sent from C_1 all the way to C_{k-1}. T stands for the total amount of bits so far consumed from $|w|$ by the transmitting computer. h is a binary value which will take 1 iff the prefixes of some x_i and x_{i+1} coincide and 0, otherwise. (Note that this is possible for all C_i's in P_{k-1} whenever $|\alpha_i| \geq 2n$.)

The functions $f(b_j)$ and $g(b_j)$ indicate the result of comparing the bit b_j to its corresponding bit found in the local parts of the input found in the subsequent computers. Specifically, $f(b_j) = b_{j+rn-T}$ and $g(b_j) = b_{j+(r+1)n-T}$, for $r = 1, 2, \ldots, k-1$. If the functions yields to identities, they will have 1 as value, else 0. Now, let us focus on the last three pieces of information found in the code message, namely $h, f(b_j) g(b_j)$ for every transmission made by every computer C_i.

As was noted earlier, the existence of inequalities among the x_i's in w is an easy task for P_{k-1}. Suppose our input is of the form $aba \cdots abba \cdots ab$, where $a, b \in \{0, 1\}^n$. That is, whenever this equality (bb) exists, the code message will sure to track an inequality nondeterministically. Same event could be expected, if there will be at the most $k-1$ x_i's that coincide.

If the input string is such that all the x_i's are equal, the inequality condition is impossible, this input must have an accepting computation, though. The case will be trivial if there is at least a computer which possesses at least the whole of the two consecutive x_i's.

Let us consider the situation when no one can ever identify any two of these x_i's. A situation could be when $n < l_1, l_{k-1} \leq 2n$ and $0 < l_i \leq n$, for $i = 2, 3, \ldots, k-2$. The code message will have a suffix $1, 11$ from C_1 all the way to C_{k-1}. At C_{k-1} with α_{k-1}, if the suffixes of x_{k-1} and that of x_k coincide and the bit comparison still holds -i.e., equal, it accepts, else it rejects.

However, if $w = abab \cdots abab \in \{0, 1\}^{kn}$, $a, b \in \{0, 1\}^n$, then the last three bits of the code message will either be $0, 01$ from C_1 until it reaches C_{k-1} or $0, 10$. The smooth sequence of transmission of code messages suffixed by any of these pieces of information up to C_{k-1}, suggests only that an inequality relation was not met. In which case, C_{k-1} for that particular nondeterministic computation, rejects the input.

Note that b_j is chosen only by C_1 from its substring of the input, whose length ranges from 0 through kn. And for every $0 < l_1 \leq kn$, it needs to pick a bit from a prefix of α_1 of lenght at most n that corresponds to the chosen j at the outset. This bit together with the others will be sent to C_2 consecutively until it reaches C_{k-1}, which will make the decision for the protocol P_{k-1}. The number of messages is $O(n^2)$.

Lemma 5. *For all $k \geq 3$ and $n \in N$,*

$$\mathrm{nmc}_k(L_{(k,n)}) \in 2^{\Omega(n)}.$$

Proof. Without loss of generality, let k be an even integer. Let P_k be a 1-way uniform k-party nondeterministic protocol accepting $L_{(k,n)}$. Let $Set_{(k,n)} = \{ u^k \mid u \in \{0, 1\}^n \} \subset L_{(k,n)}$. For all $w \in Set_{(k,n)}$, we fix an equal-length partition

$$(u_1, u_2, \ldots u_{k-1}, u_k), \ u_i \in \{0, 1\}^n, \ i = 1, 2, \ldots, k.$$

Our first aim is to show that for every two different words x and y the set of accepting computations of P_k on x and y w.r.t. the equal-length partition of x and y are disjoint.

Let $uu \cdots uu, vv \cdots vv \in Set_{(k,n)}$ be two words s.t. $u \neq v$. Let

$$m_1 \$ m_2 \$ \cdots \$ m_{k-1} \$\texttt{accept} \in C(uu \cdots uu) \cap C(vv \cdots vv)$$

be an accepting computation of P_k for both inputs w.r.t. the equal-length partition. This means, for all $x \in \{u, v\}$, C_1 sends to C_2 a message $m_1 \in \Phi(x)$ and for all $y \in \{u.v\}$ the message $m_2 = \phi_2(y, m_1)$ will be sent by C_2 to C_3

afterwards, which will send $m_3 = \phi_3(w, m_2)$ where $w \in \{u, v\}$, too. Notice that for all $2 \leq i \leq k - 1$, C_i sends a message $m_i \in \phi_i(s, m_{i-1})$, for all $s \in \{u, v\}$. Finally, C_k decides as follows: $\varphi(z, m_{k-1}) = \mathtt{accept}$, for each $z \in \{u, v\}$.

Now, let us construct an accepting computation $C(uv \cdots uv)$ of P_k on $uv \cdots uv$. Since $m_1 \in \Phi(u)$, $C(uv \cdots uv)$ can start with m_1. We know that $m_2 \in \phi_2(v, m_1)$, then $m_1 \$ m_2 \$$ is a possible prefix of the computation. Up until we have the message $m_{k-1} \in \phi_{k-1}(v, m_{k-2})$, $m_1 \$ m_2 \$ \cdots \$ m_{k-1}$ is a possible prefix of $C(uv \cdots uv)$. Eventually, because $\varphi(u, m_{k-1}) = \mathtt{accept}$, we have

$$m_1 \$ m_2 \$ \cdots \$ m_{k-1} \$ \mathtt{accept},$$

which is an accepting computation for $uv \cdots uv$ by P_k w.r.t. the equal-length partition $(u, v \ldots, u, v)$. But this input is not in $L_{(k,n)}$. This is a contradiction to that fact that P_k accepts $L_{(k,n)}$. Hence, these sets of accepting computation must be disjoint. Thus, for each of the 2^n words in $Set_{(k,n)}$, our protocol P_k must have distinct computations w.r.t. the equal-length partitions on $uu \cdots uu$ and $vv \cdots vv$. That is, there is at least 2^n different accepting computations of P_k on them. On the other hand, there are at the most $(\mathrm{nmc}_k(P_k))^{k-1}$ accepting computation of P_k. So,

$$(\mathrm{nmc}_k(P_k))^{k-1} \geq 2^n \implies \mathrm{nmc}_k(P_k) \geq 2^{\frac{n}{k-1}}.$$

By definition of the message complexity of L, we obtain the result.

These results prove our Main Lemma, therefore the Main Results follows, immediately.

By similar arguments as above, the following Observation is easy.

Observation 1. *For all $k \in Z^+$, $k \geq 3$ and every $n \in N$,*

$$\mathrm{ns}(L_{(k,n)}) \geq 2^{\frac{n}{k-1}}.$$

Noticed that this does not make communication complexity a much fitted approximation method for $\mathrm{ns}(L)$ in general, though. However, (for no constant k) our lower bound method is tight.

5   Remark on the Deterministic Case

If we restrict Φ and ϕ_i for all $i = 2, 3, \ldots, k - 1$ to be s.t. the set of images has cardinality 1, then we have a 1-**way uniform** k-**party deterministic protocol** over Σ. The message complexity can be defined analogously to the nondeterministic model. The set of all languages recognized within m message communications will be $\mathrm{MCOMM}_k^1(m)$. The result in [3] implies that the deterministic case shows no hierarchy at all, since for all $k \geq 2$, $\mathrm{dmc}_2(L) = \mathrm{dmc}_k(L) = s(L)$. However, the innate exponential gap [13] between NFA and DFA can also be observed between our models, i.e. for all $k \geq 2$, there is an L s.t.

$$\mathrm{NMCOMM}_k^1(\mathrm{ns}(L)) \subseteq \mathrm{MCOMM}_k^1(2^{\mathrm{ns}(L)}).$$

Acknowledgement

The author would like to express his gratitude to Joachim Kupke for fruitful discussions and comments on this work. Similarly to Juraj Hromkovič for giving ideas to start with, for his comments on the initial version of this paper and generously facilitating this research at his Department.

References

1. H. ABELSON, Lower Bounds on Information Transfer in Distributed Computation. *Proc. 19th IEEE FOCS* (1978), 151-158.
2. H.N. ADORNA, Some Properties of k-Party Message Complexity. Unpublished Manuscript, RWTH-Aachen (2001).
3. H.N. ADORNA, 3-Party Message Complexity is Better than 2-Party Ones for Proving Lower Bounds on the Size of Minimal Nondeterministic Finite Automaton. In: J. DASSOW, D. WOTSCHKE (eds.) *Pre-Proc. 3rd Internatonal Workshop on Descriptional Complexity of Automata, Grammars and Related Structures* Preprint No. 16, Uni. Magdeburg. (2001), 23-34.
4. J.E. HOPCROFT, J.D. ULLMAN, *Introduction Automata Theory, Languages and Computation.* Addison-Wesley, 1979.
5. J. HROMKOVIČ, Relation Between Chomsky Hierarchy and Communication Complexity Hierarchy *Acta Math. Univ. Com.,* Vol **48-49** (1986), 311-317.
6. J. HROMKOVIČ, *Communication Complexity and Parallel Computating.* Springer, 1997.
7. J. HROMKOVIČ, G. SCHNITGER, Communication Complexity and Sequential Computation. In: PRÍARA, P. RUŽIČKA (eds.) *Proc. of Mathematical Foundation of Computer Science,* LNCS **1295,** Springer-Verlag (1997) 71-84.
8. J. HROMKOVIČ, G. SCHNITGER, On the Power of Las Vegas for One-Way Communication Complexity, OBDD's and Finite Automata. *Information and Computation,* **169** (2001), 284-296.
9. J. HROMKOVIČ, S. SEIBERT, J, KARHUMÄKI, H. KLAUCK, G. SCHNITGER, Communication Complexity Method for Measuring Nondeterminism in Finite Autmata. *Information and Computation* **172** (2002), 202-217.
10. G. JIRÁSKOVA, Finite Automata and Communication Protocols. In: C. MARTIN-VIDE, V. MITRANA (eds.) *Words, Sequences, Grammars, Languages: Where Biology, Computer Science, Linguistics and Mathematics Meet II,* to appear.
11. H. KLAUCK, G. SCHNITGER, Nondeterministic Finite Automata versus Nondeterministic Uniform Communication Complexity. Unpublished Manuscript, University of Frankfurt, (1996).
12. E. KUSHILEVITZ, N. NISAN, *Communication Complexity.* Cambridge University Press, 1997.
13. F. MOORE, On the Bounds for State-Set Size in the Proofs of Equivalence Between Deterministic, Nondeterministic and Two-Way Finite Automata. *IEEE Trans. Comput.* **20** (1971), 1211-1214.
14. C. PAPADIMITRIOU, M. SIPSER, Communication Complexity. *Proc. 14th ACM STOC,* (1982), 196-200.
15. P. TIWARI, Lower Bound on Communication Complexity in Distributed Computer Networks. *Journal of the ACM,* Vol.**42**, No. 4 (1987) 921-938.
16. A.C. YAO, Some Complexity Questions Related to Distributed Computing. *Proc. 11th ACM STOC,* (1979), 209-213.

Unary Language Operations
and Their Nondeterministic State Complexity

Markus Holzer[1] and Martin Kutrib[2]

[1] Institut für Informatik, Technische Universität München
Boltzmannstraße 3, D-85748 Garching bei München, Germany
holzer@informatik.tu-muenchen.de
[2] Institut für Informatik, Universität Giessen
Arndtstraße 2, D-35392 Giessen, Germany
kutrib@informatik.uni-giessen.de

Abstract. We investigate the costs, in terms of states, of operations on infinite and finite unary regular languages where the languages are represented by nondeterministic finite automata. In particular, we consider Boolean operations, concatenation, iteration, and λ-free iteration. Most of the bounds are tight in the exact number of states, i.e. the number is sufficient and necessary in the worst case. For the complementation of infinite languages a tight bound in the order of magnitude is shown.

1 Introduction

Finite automata are used in several applications and implementations in software engineering, programming languages and other practical areas in computer science. They are one of the first and most intensely investigated computational models. Nevertheless, some challenging problems of finite automata are still open. An important example is the question how many states are sufficient and necessary to simulate two-way nondeterministic finite automata with two-way deterministic finite automata. The problem has been raised in [14] and partially solved in [17]. A lower bound and an interesting connection with the open problem whether DLOGSPACE equals NLOGSPACE or not is given in [1].

Since regular languages have many representations in the world of finite automata it is natural to investigate the succinctness of their representation by different types of automata in order to optimize the space requirements. It is well known that nondeterministic finite automata (NFA) can offer exponential saving in space compared with deterministic finite automata (DFA), but the problem to convert a given DFA to an equivalent minimal NFA is PSPACE-complete [7]. Since minimization of NFAs is also PSPACE-complete, conversions from nondeterministic to deterministic variants are of particular interest. Concerning the number of states asymptotically tight bounds are $O(n^n)$ for the two-way DFA to one-way DFA conversion, $O(2^{n^2})$ for the two-way NFA to one-way DFA conversion, and 2^n for the one-way NFA to one-way DFA conversion, for example. For finite languages over a k-letter alphabet the NFA to DFA conversion has been

M. Ito and M. Toyama (Eds.): DLT 2002, LNCS 2450, pp. 162–172, 2003.

solved in [15] with a tight bound of $O(k^{\frac{n}{\log_2 k+1}})$. A valuable source for further results and references is [2].

Related to these questions are the costs (in terms of states) of operations on regular languages with regard to their representing devices. For example, converting a given NFA to an equivalent DFA gives an upper bound for the NFA state complexity of complementation. In recent years results for many operations have been obtained. For DFAs state-of-the-art surveys can be found in [20, 21]. The general nondeterministic case has been studied in [4].

When certain problems are computationally hard in general, a natural question concerns simpler versions. To this regard promising research has been done for unary languages. It turned out that this particular case is essentially different from the general case. For example, the minimization of NFAs becomes NP-complete instead of PSPACE-complete [6, 18]. The problem of evaluating the costs of unary automata simulations has been raised in [17]. In [3] it has been shown that the unary NFA to DFA conversion takes $e^{\Theta(\sqrt{n \cdot \ln(n)})}$ states, the NFA to two-way DFA conversion has been solved with a bound of $O(n^2)$ states, and the costs of the unary two-way to one-way DFA conversion reduces to $e^{\Theta(\sqrt{n \cdot \ln(n)})}$. Several more results can be found in [10, 11].

State complexity results concerning operations on unary regular languages represented by DFAs are covered by the surveys [20, 21]. Estimations of the average state complexity are shown in [12].

Here we investigate the costs (in terms of states) of operations on infinite and finite unary regular languages represented by NFAs. In particular, we consider Boolean operations and catenation operations. The reversal of a unary language is trivial. Most of the bounds are tight in the exact number of states, i.e. the number is sufficient and necessary in the worst case. For the complementation of infinite languages a tight bound in the order of magnitude is shown. The technical depth of our results varies from immediate to more subtle extensions to previous work. Indeed the technique to prove minimality for DFAs is not directly applicable to the case of NFAs. Therefore, we mostly have to use counting arguments to prove our results on NFA minimality with respect to the number of states.

In the next section we define the basic notions and present preliminary results. Section 3 is devoted to the study of infinite languages. The NFAs for finite unary languages essentially have a simple structure. By this observation the results for finite languages are derived in Section 4.

2   Preliminaries

We denote the positive integers $\{1, 2, \ldots\}$ by $\mathbb{N}$, the set $\mathbb{N} \cup \{0\}$ by $\mathbb{N}_0$, and the *powerset* of a set S by 2^S. The *empty word* is denoted by λ. For the *length of w* we write $|w|$. We use $\subseteq$ for inclusions and $\subset$ if the inclusion is strict. By $\gcd(x_1, \ldots, x_k)$ we denote the *greatest common divisor* of the integers $x_1, \ldots, x_k$, and by $\operatorname{lcm}(x_1, \ldots, x_n)$ their *least common multiple*. If two numbers x and y are relatively prime (i.e. $\gcd(x, y) = 1$) we write $x \perp y$.

Definition 1. *A* nondeterministic finite automaton (NFA) *is a system* $\mathcal{A} = \langle S, A, \delta, s_0, F \rangle$, *where*

1. S *is the finite set of* internal states,
2. A *is the finite set of* input symbols,
3. $s_0 \in S$ *is the* initial state,
4. $F \subseteq S$ *is the set of* accepting states, *and*
5. $\delta : S \times A \to 2^S$ *is the* transition function.

The set of rejecting states is implicitly given by the partitioning, i.e. $S \setminus F$.

An NFA is called *unary* if its set of input symbols is a singleton. Throughout the paper we use $A = \{a\}$. In some sense the transition function is complete. W.l.o.g. we may require δ to be a total function, since whenever the operation of an NFA is supposed not to be defined, then δ can map to the empty set which, trivially, belongs to 2^S. Thus, in some sense the NFA need not be complete. However, throughout the paper we assume that the NFAs are always *reduced*. This means that there are no unreachable states and that from any state an accepting state can be reached. An NFA is said to be *minimal* if its number of states is minimal with respect to the accepted language. Since every n-state NFA with λ-transitions can be transformed to an equivalent n-state NFA without λ-transitions [5] for state complexity issues there is no difference between the absence and presence of λ-transitions. For convenience, we consider NFAs without λ-transitions only.

As usual the transition function δ is extended to a function $\Delta : S \times A^* \to 2^S$ reflecting sequences of inputs as follows: $\Delta(s, wa) = \bigcup_{s' \in \Delta(s,w)} \delta(s', a)$, where $\Delta(s, \lambda) = \{s\}$ for $s \in S$, $a \in A$, and $w \in A^*$.

Let $\mathcal{A} = \langle S, A, \delta, s_0, F \rangle$ be an NFA, then a word $w \in A^*$ is *accepted* by $\mathcal{A}$ if $\Delta(s_0, w) \cap F \neq \emptyset$. The *language accepted* by $\mathcal{A}$ is:

$$L(\mathcal{A}) = \{w \in A^* \mid w \text{ is accepted by } \mathcal{A}\}$$

The next preliminary result involves NFAs directly. It is a key tool in the following sections, and can be proved by a simple pumping argument.

Lemma 2. *Let $n \geq 1$ be an arbitrary integer. Then $n+1$ resp. n states are sufficient and necessary in the worst case for an* NFA *to accept the language $\{a^n\}^+$ resp. $\{a^n\}^*$.*

3   Operations on Regular Languages

3.1   Boolean Operations

We start our investigations with Boolean operations on NFAs that accept unary regular languages. At first we consider the union. Due to the lack of suited tools and methods such as unique minimization etc. the proof of the lower bound refers specifically to the structures of the witness automata.

Theorem 3. *For any integers* $m, n \geq 1$ *let* $\mathcal{A}$ *be a unary* m-*state and* $\mathcal{B}$ *be a unary* n-*state NFA. Then* $m + n + 1$ *states are sufficient for an NFA* $\mathcal{C}$ *to accept the language* $L(\mathcal{A}) \cup L(\mathcal{B})$.

Proof. To construct an $(m+n+1)$-state NFA for the language $L(\mathcal{A}) \cup L(\mathcal{B})$, we simply use a new initial state and connect it to the states of $\mathcal{A}$ and $\mathcal{B}$ that are reached after the first state transition. During the first transition, $\mathcal{C}$ nondeterministically guesses whether the input may belong to $L(\mathcal{A})$ or $L(\mathcal{B})$. Subsequently, $\mathcal{A}$ or $\mathcal{B}$ is simulated. Obviously, $L(\mathcal{C}) = L(\mathcal{A}) \cup L(\mathcal{B})$ and $|S| = |S_A| + |S_B| + 1 = m + n + 1$. $\hfill\square$

Theorem 4. *Let* $m, n > 1$ *be two integers such that neither* m *is a multiple of* n *nor* n *is a multiple of* m. *Then there exist a unary* m-*state NFA* $\mathcal{A}$ *and a unary* n-*state NFA* $\mathcal{B}$ *such that any NFA* $\mathcal{C}$ *accepting* $L(\mathcal{A}) \cup L(\mathcal{B})$ *needs at least* $m + n + 1$ *states.*

Proof. W.l.o.g. we may assume $m > n$. Since m is not a multiple of n we obtain $n > \gcd(m, n)$.

Let $\mathcal{A}$ be an m-state NFA that accepts the language $\{a^m\}^*$ and $\mathcal{B}$ be an n-state NFA that accepts $\{a^n\}^*$. Let $\mathcal{C}$ be an NFA with initial state s_0 for the language $L(\mathcal{A}) \cup L(\mathcal{B})$. In order to prove that $\mathcal{C}$ has at least $m + n + 1$ states assume that it has at most $m + n$ states.

At first we consider the input a^{m+n} and show that it does not belong to $L(\mathcal{C})$. Since $2m > m + n > m$ the input does not belong to $L(\mathcal{A})$. If it would belong to $L(\mathcal{B})$, then there were a constant $c > 2$ such that $m + n = c \cdot n$. Therefore, $m = (c - 1) \cdot n$ and, thus, m would be a multiple of n what contradicts the assumption of the theorem.

The next step is to show that each of the states $s_0 \vdash s_1 \vdash \cdots \vdash s_{m-1} \vdash s_m$ which are passed through when accepting the input a^m either is not in a cycle or is in a cycle of length m. To this end assume contrarily some state s_i is in a cycle of length $x \neq m$, i.e. s_i is reachable from s_i by processing x input symbols. Due to the number of states we may assume $x \leq m + n$.

By running several times through the cycle a^{m+x}, a^{m+2x} and a^{m+3x} are accepted.

We observe $m + x$ is not a multiple of m. If it is not a multiple of n we are done. Otherwise, x cannot be a multiple of n since m is not a multiple of n. Moreover, now we observe $m + 2x$ is not a multiple of n since $m + x$ is and x is not. If $m + 2x$ is not a multiple of of m we are done.

Otherwise we consider $m + 3x$ which now cannot be a multiple of m since $m + 2x$ is and x is not. If $m + 3x$ is not a multiple of n we are done.

Otherwise we summarize the situation: $m + 3x$ and $m + x$ are multiples of n which implies $2x$ is a multiple of n. Since $m + 2x$ is a multiple of m we conclude that $2x$ is a multiple of m, too. Moreover, x is neither a multiple of m nor of n.

From $x \leq m + n < 2m \implies 2x < 4m$ we derive $2x \in \{m, 3m\}$. If $2x = m$, then m is a multiple of n, a contradiction. So let $2x = 3m$. Since $2x$ is a multiple of n there exists a constant $c \in \mathbb{N}$ such that $3m = cn$. From $x \leq m + n$ follows $\frac{3}{2}m \leq m + n$. Therefore $\frac{1}{2}m \leq n$ which together with $n < m$ implies $c \in \{4, 5, 6\}$.

It holds $\frac{3}{c}m = n$. On the other hand, $m + x = m + \frac{3}{2}m = \frac{5}{2}m$ is a multiple of n. Therefore $\frac{5}{2}/\frac{3}{c} = \frac{5 \cdot c}{6}$ must belong to N for $c \in \{4, 5, 6\}$. Thus $c = 6$, but in this case $\frac{1}{2}m = n \Longrightarrow m = 2n$ and m is a multiple of n what contradicts the assumption of the theorem.

In order to complete the proof of the theorem now we come back to the sequence of states $s_0 \vdash s_1 \vdash \cdots \vdash s_{m-1} \vdash s_m$ passed through when accepting the input a^m. Correspondingly let $s_0 \vdash s_1' \vdash \cdots \vdash s_{n-1}' \vdash s_n'$ be an accepting sequence for a^n.

Since $L(\mathcal{C})$ is an infinite language $\mathcal{C}$ must contain some cycle. If a s_i from $s_0, \ldots, s_m$ is in a cycle, then its length is m states. State s_i may not be in the sequence $s_0 \vdash s_1' \vdash \cdots \vdash s_n'$ because in this case a^{m+n} would be accepted. In order to accept a^n but to reject $a^1, \ldots, a^{n-1}$ in any case, the states $s_0, s_1', \ldots, s_n'$ must be pairwise different. Altogether this results in at least $n + 1 + m$ states.

If on the other hand a s_j from $s_0, s_1', \ldots, s_n'$ is in a cycle, then it may not be in $s_0, s_1, \ldots, s_m$. Otherwise the cycle length must be m and a^{m+n} would be accepted. Concluding as before this case results in at least $m + n + 1$ states, too.

Finally, if neither a state from $s_0, s_1, \ldots, s_m$ nor a state from $s_1', \ldots, s_n'$ is in a cycle, then obviously $s_0, \ldots, s_m$ must be pairwise different, but all states s_j' may or may not appear in $s_0, \ldots, s_m$. This takes at least $m + 1$ states. So there remain at most $n - 1$ states for a cycle. Therefore, the cycle length x is at most $n - 1$. Now we consider an accepting computation for the input a^{mn}. It must have run through the cycle. Running once more through the cycle leads to acceptance of a^{mn+x} which does not belong to $L(\mathcal{C})$.

So in any case we obtain a contradiction to the assumption that $\mathcal{C}$ has at most $m + n$ states. $\square$

Next we are going to prove a tight bound for the intersection. The upper bound is obtained by the simple cross-product construction. The lower bound requires $m \perp n$. In [13] unary languages are studied whose *deterministic* state complexities are not relatively prime.

Theorem 5. *For any integers $m, n \geq 1$ let $\mathcal{A}$ be a unary m-state and $\mathcal{B}$ be a unary n-state NFA. Then $m \cdot n$ states are sufficient for an NFA $\mathcal{C}$ to accept the language $L(\mathcal{A}) \cap L(\mathcal{B})$. If $m \perp n$, then there exist a unary m-state NFA $\mathcal{A}$ and a unary n-state NFA $\mathcal{B}$ such that any NFA $\mathcal{C}$ accepting $L(\mathcal{A}) \cap L(\mathcal{B})$ needs at least $m \cdot n$ states.*

Proof. As witness languages consider $L(\mathcal{A}) = \{a^m\}^*$ and $L(\mathcal{B}) = \{a^n\}^*$. Since $m \perp n$ the intersection $L(\mathcal{A}) \cap L(\mathcal{B})$ is $\{a^{m \cdot n}\}^*$. Due to Lemma 2 any NFA accepting the intersection needs at least $m \cdot n$ states. $\square$

For complementation of unary NFAs a crucial role is played by the function $F(n) = \max\{\mathrm{lcm}(x_1, \ldots, x_k) \mid x_1, \ldots, x_k \in \mathrm{N} \wedge x_1 + \cdots + x_k = n\}$ which gives the maximal order of the cyclic subgroups of the symmetric group of n symbols. For example, the first seven values of F are $F(1) = 1$, $F(2) = 2$, $F(3) = 3$, $F(4) = 4$, $F(5) = 6$, $F(6) = 6$, $F(7) = 12$ due to the sums $1 = 1$, $2 = 2$, $3 = 3$, $4 = 4$, $5 = 2 + 3$, $6 = 1 + 2 + 3$ (or $6 = 6$) and $7 = 3 + 4$.

Since F depends on the irregular distribution of the prime numbers we cannot expect to express $F(n)$ explicitly by n. The function itself has been investigated by Landau [8, 9] who has proved the asymptotic growth rate

$$\lim_{n \to \infty} \frac{\ln(F(n))}{\sqrt{n \cdot \ln(n)}} = 1$$

A bound immediately derived from Landau's result is:

$$\ln(F(n)) \in \Theta(\sqrt{n \cdot \ln(n)})$$

For our purposes the implied rough estimation $F(n) \in e^{\Theta(\sqrt{n \cdot \ln(n)})}$ suffices. Shallit and Ellul [16] pointed out that the bound $F(n) \in O(e^{\sqrt{n \cdot \ln(n)}})$ which is claimed in [3] is not correct. They deduced finer bounds from a result in [19] where the currently best known approximation for F has been proved. Nevertheless, in [3] it has been shown that for any unary n-state NFA there exists an equivalent $O(F(n))$-state deterministic finite automaton.

This upper bound for the transformation to a deterministic finite automaton is also an upper bound for the costs of the unary NFA complementation, since the unary deterministic complementation neither increases nor decreases the number of states (simply interchange accepting and rejecting states). The next theorem is an immediate consequence.

Theorem 6. *For any integer $n \geq 1$ the complement of a unary n-state NFA language is accepted by an $O(F(n))$-state NFA.*

This expensive upper bound is tight in the order of magnitude.

Theorem 7. *For any integer $n > 1$ there exists a unary n-state NFA A such that any NFA accepting the complement of $L(A)$ needs at least $\Omega(F(n))$ states.*

Proof. Let $x_1, \ldots, x_k \in \mathbb{N}$ be integers such that $x_1 + \cdots + x_k = n - 1$ and $\mathrm{lcm}(x_1, \ldots, x_k) = F(n-1)$. If one of the integers x_i may be factorized as $x_i = y \cdot z$ such that $y, z > 1$ and $y \perp z$, then x_i can be replaced by $y, z, 1, 1, \ldots, 1$ where the number of ones is $y \cdot z - (y + z)$. Obviously neither the sum nor the least common multiple is affected. So we may assume that all x_i are powers of prime numbers. If two of the integers are powers of the same prime number, then the smaller one is replaced by a corresponding number of ones. Again neither the sum nor the least common multiple is affected. But now we may assume without loss of generality that all x_i are relatively prime, i.e. $x_i \perp x_j$ for $i \neq j$.

Now define for $1 \leq i \leq k$ the languages $L_i = \{a^{x_i}\}^*$ and consider the union of their complements: $L = \overline{L_1} \cup \overline{L_2} \cup \cdots \cup \overline{L_k}$.

Since $\overline{L_i}$ is acceptable by a x_i-state NFA A_i the language L is acceptable by an NFA A with at most $1 + x_1 + \cdots + x_k = n$ states. To this end we introduce a new initial state and connect it nondeterministically to the states of the A_i that are reached after their first state transition. Since all x_i are relatively prime the complement of L is $\overline{L} = \{a^{\mathrm{lcm}(x_1, \ldots, x_k)}\}^*$. Therefore, the complement of the n-state language L needs $\mathrm{lcm}(x_1, \ldots, x_k) = F(n - 1)$ states. Since $F(n)$ is of order $e^{\Theta(\sqrt{n \cdot \ln(n)})}$ it follows $F(n - 1)$ is of order $\Omega(F(n))$. $\qquad\square$

3.2   Catenation Operations

The lower bound of the concatenation misses the upper bound by one state. It is an open question how to close the gap by more sophisticated constructions or witness languages.

Theorem 8. *For any integers $m, n > 1$ let $\mathcal{A}$ be a unary m-state NFA and $\mathcal{B}$ be a unary n-state NFA. Then $m + n$ states are sufficient for an NFA $\mathcal{C}$ to accept the language $L(\mathcal{A})L(\mathcal{B})$. Moreover, there exist a unary m-state NFA $\mathcal{A}$ and a unary n-state NFA $\mathcal{B}$ such that any NFA $\mathcal{C}$ accepting $L(\mathcal{A})L(\mathcal{B})$ needs at least $m + n - 1$ states.*

Proof. The upper bound is due to the observation that in $\mathcal{C}$ one has simply to connect the accepting states in $\mathcal{A}$ with the states in $\mathcal{B}$ that follow the initial state.

Let $\mathcal{A} = \langle S_A, \{a\}, \delta_A, s_{0,A}, F_A \rangle$ and $\mathcal{B} = \langle S_B, \{a\}, \delta_B, s_{0,B}, F_B \rangle$ such that $S_A \cap S_B = \emptyset$, then $\mathcal{C} = \langle S, \{a\}, \delta, s_0, F \rangle$ is defined according to $S = S_A \cup S_B$, $s_0 = s_{0,A}$, $F = F_A \cup F_B$ if $\lambda \in L(\mathcal{B})$, $F = F_B$ otherwise, and for all $s \in S$:

$$\delta(s, a) = \begin{cases} \delta_A(s, a) & \text{if } s \in S_A \setminus F_A \\ \delta_A(s, a) \cup \delta_B(s_{0,B}, a) & \text{if } s \in F_A \\ \delta_B(s, a) & \text{if } s \in S_B \end{cases}$$

Let $L(\mathcal{A})$ be the m-state language $\{a^k \mid k \equiv m - 1 \pmod{m}\}$ and $L(\mathcal{B})$ be the n-state language $\{a^k \mid k \equiv n - 1 \pmod{n}\}$.

The shortest word in $L(\mathcal{A})$ respectively $L(\mathcal{B})$ is a^{m-1} respectively a^{n-1}. Therefore the shortest word in $L(\mathcal{C})$ is a^{m+n-2}. Assume contrarily to the assertion the NFA $\mathcal{C}$ has at most $m + n - 2$ states. Let $\mathcal{C}$ accept the input a^{m+n-2} by running through the state sequence $s_0 \vdash s_1 \vdash \cdots \vdash s_{m+n-2}$ where all states except s_{m+n-2} are non-accepting. Due to the assumption at least one of the non-accepting states s_i must appear at least twice in the sequence. This implies that there exists an accepting computation that does not run through the cycle $s_i \vdash \cdots \vdash s_i$. So an input whose length is at most $m + n - 3$ would be accepted, a contradiction. □

The constructions yielding the upper bounds for the iteration and λ-free iteration are similar. The trivial difference between both operations concerns the empty word only. Moreover, the difference does not appear for languages containing the empty word. Nevertheless, in the worst case the difference costs one state.

Theorem 9. *For any integer $n > 2$ let $\mathcal{A}$ be a unary n-state NFA. Then $n + 1$ resp. n states are sufficient and necessary in the worst case for an NFA to accept the language $L(\mathcal{A})^*$ resp. $L(\mathcal{A})^+$.*

Proof. Let $\mathcal{A} = \langle S_A, \{a\}, \delta_A, s_{0,A}, F_A \rangle$ be an n-state NFA. Then the transition function of an n-state NFA $\mathcal{C} = \langle S, \{a\}, \delta, s_0, F \rangle$ that accepts the language $L(\mathcal{A})^+$ is for $s \in S$ defined as follows:

$$\delta(s,a) = \begin{cases} \delta_A(s,a) & \text{if } s \notin F_A \\ \delta_A(s,a) \cup \delta_A(s_{0,A},a) & \text{if } s \in F_A \end{cases}$$

The other components remain unchanged, i.e., $S = S_A$, $s_0 = s_{0,A}$, and $F = F_A$.

If the empty word belongs to $L(\mathcal{A})$ then the construction works fine for $L(\mathcal{A})^*$ also. Otherwise an additional state has to be added: Let $s_0' \notin S_A$ and define $S = S_A \cup \{s_0'\}$, $s_0 = s_0'$, $F = F_A \cup \{s_0'\}$ and for $s \in S$

$$\delta(s,a) = \begin{cases} \delta_A(s,a) & \text{if } s \notin F_A \cup \{s_0'\} \\ \delta_A(s,a) \cup \delta_A(s_{0,A},a) & \text{if } s \in F_A \\ \delta_A(s_{0,A},a) & \text{if } s = s_0' \end{cases}$$

In order to prove the tightness of the bounds for any $n > 2$ consider the n-state language $L = \{a^k \mid k \equiv n - 1 \pmod{n}\}$.

At first we show that $n + 1$ states are necessary for $\mathcal{C} = \langle S, \{a\}, \delta, s_0, F \rangle$ to accept $L(\mathcal{A})^*$.

Contrarily, assume $\mathcal{C}$ has at most n states. We consider words of the form a^i with $0 \leq i$. The shortest four words belonging to $L(\mathcal{A})^*$ are λ, a^{n-1}, a^{2n-2}, and a^{2n-1}. It follows $s_0 \in F$. Moreover, for a^{n-1} there must exist a path $s_0 \vdash s_1 \vdash \cdots \vdash s_{n-2} \vdash s_n$ where $s_n \in F$ and $s_1, \ldots, s_{n-2}$ are different non-accepting states. Thus, $\mathcal{C}$ has at least $n - 2$ non-accepting states.

Assume for a moment F to be a singleton. Then $s_0 = s_n$ and for $1 \leq i \leq n-3$ the state s_0 must not belong to $\delta(s_i, a)$. Processing the input a^{2n-1} the NFA cannot enter s_0 after $2n - 2$ time steps. Since $a^1 \notin L(\mathcal{A})^*$ the state s_0 must not belong to $\delta(s_0, a)$.

On the other hand, $\mathcal{C}$ cannot enter one of the states $s_1, \ldots, s_{n-3}$ since there is no transition to s_0. We conclude that $\mathcal{C}$ is either in state s_{n-2} or in an additional non-accepting state s_{n-1}. Since there is no transition such that $s_{n-2} \in \delta(s_{n-2}, a)$ in both cases there exists a path of length n from s_0 to s_0. But a^n does not belong to $L(\mathcal{A})^*$ and we have a contradiction to the assumption $|F| = 1$.

Due to our assumption $|S| \leq n$ we now have $|F| = 2$ and $|S| - |F| = n - 2$. Let us recall the accepting sequence of states for the input a^{n-1}: $s_0 \vdash s_1 \vdash \cdots \vdash s_{n-2} \vdash s_n$. Both s_0 and s_n must be accepting states. Assume $s_n \neq s_0$. Since a^{2n-2} belongs to $L(\mathcal{A})^*$ there must be a possible transition $s_0 \vdash s_1$ or $s_n \vdash s_1$. Thus, a^{2n-2} is accepted by s_n. In order to accept a^{2n-1} there must be a corresponding transition from s_n to s_n or from s_n to s_0. In both cases the input a^n would be accepted. Therefore $s_n = s_0$.

By the same argumentation the necessity of a transition for the input symbol a from s_0 to s_0 or from s_0 to s_n follows. This implies that a^1 is accepted. From the contradiction follows $|S| > n$.

As an immediate consequence we obtain the tightness of the bound for $L(\mathcal{A})^+$. In this case $s_0 \in F$ is not required. Thus, just one accepting state is necessary. $\qquad\square$

4   Operations on Finite Languages

The situation for finite unary languages is easier since essentially the structure of the corresponding NFAs is simple.

In [15] it is stated that every finite unary n-state NFA language is acceptable by some complete deterministic finite automaton with at most $n + 1$ states. A little bit more sophisticated, one observes that the minimum NFA for a finite unary language L has $n+1$ states if the longest word in L is of length n. Otherwise the NFA would run through a cycle when accepting a^n and, thus, L would be infinite. Now we can always construct a minimum NFA $\mathcal{A} = \langle S, \{a\}, \delta, s_0, F \rangle$ for L as follows: $S = \{s_0, s_1, \ldots, s_n\}$, $\delta(s_i, a) = \{s_{i+1}\}$ for $0 \leq i \leq n - 1$, and $F = \{s_i \mid a^i \in L\}$.

From the construction it follows conversely that a minimum $(n + 1)$-state NFA for a non-empty finite unary language accepts the input a^n. An immediate consequence is that we have only to consider the longest words in the languages in order to obtain the state complexity of operations that preserve the finiteness.

Theorem 10. *For any integers $m, n \geq 1$ let $\mathcal{A}$ be a unary m-state NFA and $\mathcal{B}$ be a unary n-state NFA. If $L(\mathcal{A})$ and $L(\mathcal{B})$ are finite, then $\max(m, n)$, $\min(m, n)$ respectively $m + n - 1$ states are sufficient and necessary in the worst case for an NFA $\mathcal{C}$ to accept the language $L(\mathcal{A}) \cup L(\mathcal{B})$, $L(\mathcal{A}) \cap L(\mathcal{B})$ respectively $L(\mathcal{A})L(\mathcal{B})$.*

The complementation and iteration applied to finite languages yield to infinite languages. So in general for the lower bounds we cannot argue with the simple chain structure as before.

Theorem 11. *For any integer $n \geq 1$ let $\mathcal{A}$ be an n-state NFA. If $L(\mathcal{A})$ is finite, then $n + 1$ states are sufficient and necessary in the worst case for an NFA $\mathcal{C}$ to accept the complement of $L(\mathcal{A})$.*

Proof. W.l.o.g. we may assume that $\mathcal{A}$ has the simple chain structure with states from s_0 to s_{n-1} as mentioned before. By interchanging accepting and non-accepting states we obtain an NFA that processes all inputs up to a length $n - 1$ as required. But all longer words a^k, $k \geq n$, are belonging to the complement of $L(\mathcal{A})$. So it suffices to add a new accepting state s_n and two transitions from s_{n-1} to s_n and from s_n to s_n, in order to complete the construction of $\mathcal{C}$.

The tightness of the bound can be seen for the n-state NFA language $L = \{a^k \mid 0 \leq k \leq n - 1\}$. Since a^n is the shortest word belonging to the complement of L from the proof of Theorem 8 follows that $\mathcal{C}$ has at least $n + 1$ states.   □

The state complexity for the iterations in the finite language case is as for infinite languages if the iteration is λ-free. If not the costs are reduced by two states.

Lemma 12. *For any integer $n > 1$ let $\mathcal{A}$ be a unary n-state NFA. If $L(\mathcal{A})$ is finite, then $n - 1$ resp. n states are sufficient and necessary in the worst case for an NFA to accept the language $L(\mathcal{A})^*$ resp. $L(\mathcal{A})^+$.*

Proof. For the upper bounds we can adapt the construction of Theorem 9. The accepting states are connected to the states following the initial state. That is all for λ-free iterations.

For iterations we have to provide acceptance of the empty word. The following two observations let us save two states compared with infinite languages. First, the initial state is never reached again after initial time. Second, since the underlying language is finite and the accepting automaton is reduced there must exist an accepting state s_f for which the state transition is not defined. We can take s_f as new initial state and delete the old initial state what altogether leads to an $(n-1)$-state NFA for the iteration.

The bound for the λ-free iteration is reached for the language $L = \{a^{n-1}\}$ which requires n states. For the accepting $L^+ = \{a^{n-1}\}^+$ at least n states are necessary.

The bound for the iteration is reached for the language $L = \{a^n\}$ that requires $n+1$ states. Clearly, in order to accept $\{a^n\}^*$ at least n states are necessary. $\square$

For the sake of completeness the trivial bounds for the reversal are adduced in Table 1. The results concerning DFAs are covered by [20, 21].

Table 1. Comparison of unary NFA and unary DFA state complexities.

	NFA		DFA	
	finite	infinite	finite	infinite
$\cup$	$\max\{m,n\}$	$m+n+1$	$\max\{m,n\}$	mn
$\sim$	$n+1$	$e^{\Theta(\sqrt{n \cdot \ln(n)})}$	n	n
$\cap$	$\min\{m,n\}$	mn	$\min\{m,n\}$	mn
R	n	n	n	n
$\cdot$	$m+n-1$	$O(m+n)$	$m+n-2$	mn
$*$	$n-1$	$n+1$	$n^2-7n+13$	$(n-1)^2+1$
$+$	n	n		

References

1. Berman, P. and Lingas, A. *On the complexity of regular languages in terms of finite automata.* Technical Report 304, Polish Academy of Sciences, 1977.
2. Birget, J.-C. *State-complexity of finite-state devices, state compressibility and incompressibility.* Mathematical Systems Theory 26 (1993), 237–269.
3. Chrobak, M. *Finite automata and unary languages.* Theoretical Computer Science 47 (1986), 149–158.
4. Holzer, M. and Kutrib, M. *State complexity of basic operations on nondeterministic finite automata.* Implementation and Application of Automata (CIAA '02), 2002, pp. 151–160.

5. Hopcroft, J. E. and Ullman, J. D. *Introduction to Automata Theory, Language, and Computation*. Addison-Wesley, Reading, Massachusetts, 1979.
6. Hunt, H. B., Rosenkrantz, D. J., and Szymanski, T. G. *On the equivalence, containment, and covering problems for the regular and context-free languages*. Journal of Computer and System Sciences 12 (1976), 222–268.
7. Jiang, T. and Ravikumar, B. *Minimal NFA problems are hard*. SIAM Journal on Computing 22 (1993), 1117–1141.
8. Landau, E. *Über die Maximalordnung der Permutationen gegebenen Grades*. Archiv der Math. und Phys. 3 (1903), 92–103.
9. Landau, E. *Handbuch der Lehre von der Verteilung der Primzahlen*. Teubner, Leipzig, 1909.
10. Mereghetti, C. and Pighizzini, G. *Two-way automata simulations and unary languages*. Journal of Automata, Languages and Combinatorics 5 (2000), 287–300.
11. Mereghetti, C. and Pighizzini, G. *Optimal simulations between unary automata*. SIAM Journal on Computing 30 (2001), 1976–1992.
12. Nicaud, C. *Average state complexity of operations on unary automata*. Mathematical Foundations of Computer Science (MFCS '99) LNCS 1672, 1999, pp. 231–240.
13. Pighizzini, G. and Shallit, J. O. *Unary language operations, state complexity and Jacobsthal's function*. International Journal of Foundations of Computer Science 13 (2002), 145–159.
14. Sakoda, W. J. and Sipser, M. *Nondeterminism and the size of two way finite automata*. ACM Symposium on Theory of Computing (STOC '78), 1978, pp. 275–286.
15. Salomaa, K. and Yu, S. *NFA to DFA transformation for finite languages over arbitrary alphabets*. Journal of Automata, Languages and Combinatorics 2 (1997), 177–186.
16. Shallit, J. O. and Ellul, K. Personal communication, October 2002.
17. Sipser, M. *Lower bounds on the size of sweeping automata*. Journal of Computer and System Sciences 21 (1980), 195–202.
18. Stockmeyer, L. J. and Meyer, A. R. *Word problems requiring exponential time*. ACM Symposium on Theory of Computing (STOC '73), 1973, pp. 1–9.
19. Szalay, M. *On the maximal order in S_n and S_n^**. Acta Arith. 37 (1980), 321–331.
20. Yu, S. *State complexity of regular languages*. Journal of Automata, Languages and Combinatorics 6 (2001), 221–234.
21. Yu, S. *State complexity of finite and infinite regular languages*. Bulletin of the EATCS 76 (2002), 142–152.

Constructing Infinite Words
of Intermediate Complexity

Julien Cassaigne

Institut de Mathématiques de Luminy
CNRS UPR 9016 / FRUMAM
Case 907, 163 avenue de Luminy
F-13288 Marseille cedex 9

Abstract. We present two constructions of infinite words with a complexity function that grows faster than any polynomial, but slower than any exponential. The first one is rather simple but produces a word which is not uniformly recurrent. The second construction, more involved, produces uniformly recurrent words and allows to choose the growth of the complexity function in a large family.

1 Introduction

Let A be a finite alphabet, and $\mathbf{u} \in A^{\mathbb{N}}$ an infinite word on A with indices in $\mathbb{N}$. A *factor* of $\mathbf{u}$ is a finite word that occurs as a block of consecutive letters in $\mathbf{u}$. The structure of the language $F(\mathbf{u})$ of all factors of $\mathbf{u}$ plays an essential role in the combinatorial and dynamical properties of $\mathbf{u}$, and in particular the rate of growth of the associated *complexity function* $p_{\mathbf{u}} \colon \mathbb{N} \to \mathbb{N}$, where $p_{\mathbf{u}}(n)$ is the number of factors of $\mathbf{u}$ of length n (this function was introduced in [6], but already appears under the name *block growth* in [10]; see also [1]).

The complexity function of an infinite word is a nondecreasing function which obviously satisfies the inequality $1 \leq p_{\mathbf{u}}(n) \leq \#A^n$. Moreover, its logarithm is sub-additive, i.e., $p_{\mathbf{u}}(m + n) \leq p_{\mathbf{u}}(m)p_{\mathbf{u}}(n)$ for all m and n in $\mathbb{N}$. Not all such functions can be obtained as complexity functions, however, and characterizing the set of possible complexity functions is a challenging and probably very difficult problem. Only partial results in this direction are known currently, either explicit constructions of sequences having certain particular complexity functions [1,4], or classes of functions that cannot be complexity functions [3].

The complexity function of numerous infinite words has been computed, many of which have *low complexity*, i.e., $p_{\mathbf{u}}(n)$ grows at most as a polynomial function of n. This is in particular the case of Sturmian words [9, Chapter 2], of fixed points of substitutions [6,11], of codings of dynamical systems like interval exchanges and billiards [8], etc., see [1,9] for more examples. Other infinite words have *high complexity*, i.e., $p_{\mathbf{u}}(n)$ grows exponentially; for instance, words of topological entropy h for any $h > 0$ can be constructed [7] (the topological entropy is the limit $h_{\text{top}}(\mathbf{u}) = \lim_{n \to \infty} \ln p_{\mathbf{u}}(n)/n$). Another construction [2] yields a word of complexity $n(\sqrt{2})^n$ up to a bounded factor.

M. Ito and M. Toyama (Eds.): DLT 2002, LNCS 2450, pp. 173–184, 2003.

However, not many examples are known which have *intermediate complexity*, i.e., for which $h_{\text{top}}(\mathbf{u}) = 0$ but $\ln p_{\mathbf{u}}(n)/\ln n$ is unbounded. Our purpose is to construct such examples. A typical intermediate complexity function would for instance grow as $2^{\sqrt{n}}$. We were not able to achieve $p_{\mathbf{u}}(n) \sim 2^{\sqrt{n}}$, so instead we will take a weaker notion of growth rate and require only that $\log p_{\mathbf{u}}(n) \sim \sqrt{n}$.

In the sequel, we shall denote by $\ln(t)$ the natural logarithm of t, and by $\log(t)$ its logarithm in base 2. Most words considered here are on the alphabet $A_2 = \{0, 1\}$.

2   First Construction

It is easy to construct a language that contains exactly 2^k words of length k^2. For instance, let Q be the set of all words of the form

$$x_1 x_2 00 x_3 0000 x_4 000000 \ldots x_k 0^{2k-2}$$

where $x_1 x_2 \ldots x_k$ is any word in A_2^k. However, this language is not the language of factors of an infinite word. By concatenating all these words in a certain order, we can obtain an infinite word with a complexity at least equal to $2^{\lceil \sqrt{n} \rceil}$, but we have no guarantee that the complexity is not much higher due to new factors that appear when concatenating elements of Q. It is nevertheless possible to control these new factors.

Let $\mathbf{c} = c_0 c_1 c_2 \ldots$ be a binary infinite word on A_2 of maximal complexity, i.e., $p_{\mathbf{c}}(n) = 2^n$ for all n, for instance the binary Champernowne word [5]:

$$\mathbf{c} = 1.10.11.100.101.110.111.1000.1001.1010.1011.1100.1101.1110.1111\ldots .$$

Consider the family of finite words (q_k) defined as follows:

$$q_k = c_{k-1} c_{k-2} 00 c_{k-3} 0000 \ldots c_0 0^{2k-2}$$

so that $|q_k| = k^2$, and the infinite word $\mathbf{u} = q_1 q_2 q_3 \ldots$ obtained by concatenating them in order,

$$\mathbf{u} = 1.1100.010010000.1000100001000000.11000000010000001000000000\ldots .$$

Theorem 1. *The complexity function of the infinite word $\mathbf{u}$ defined above satisfies*

$$2^{\lceil \sqrt{n} \rceil} \leq p_{\mathbf{u}}(n) \leq n^2 2^{\lceil \sqrt{n} \rceil}$$

for all $n \geq 1$, and consequently $\log p_{\mathbf{u}}(n) \sim \sqrt{n}$.

Proof. First, let us check that $Q \subseteq F(\mathbf{u})$, which implies the lower bound on $p_{\mathbf{u}}(n)$. Let $x = x_1 x_2 \ldots x_k$ be any word in A_2^k. Let $\tilde{x} = x_k x_{k-1} \ldots x_2 x_1$ be the mirror image (also called retrograde) of x. Then $\tilde{x}$ occurs at some position j in $\mathbf{c}$, since $\mathbf{c}$ has maximal complexity, i.e., $c_j = x_k, c_{j+1} = x_{k-1}, \ldots, c_{j+k-1} = x_1$. Then

$x_1 x_2 00 x_3 0000 x_4 000000 \ldots x_k 0^{2k-2}$ is a prefix of q_{j+k}: every element of Q is a prefix of some q_i, therefore a factor of $\mathbf{u}$.

We now have to establish the upper bound. First observe that when $1 \le n \le 10$,

$$p_{\mathbf{u}}(n) \le 2^n \le n^2 2^{\lceil \sqrt{n} \rceil}$$

so that we can assume that $n \ge 11$. Let us denote by a_k the length of the prefix $q_1 q_2 \ldots q_k$, namely $a_k = |q_1 q_2 \ldots q_k| = k(k+1)(2k+1)/6$. Let w be any factor of length n of $\mathbf{u}$, and consider the position j of its first occurrence in $\mathbf{u}$. We discuss on how the first occurrence of w intersects the factorisation of $\mathbf{u}$ by the q_i's: three cases are possible.

Case 1. First, if for some $k \ge 1$ one has $j \le a_{k-1} < a_k \le j+n$, i.e., if some q_k occurs entirely inside w, then $n \ge |q_k| = k^2$ and $j \le a_{k-1} \le k^3/3 \le n^{3/2}/3$. Since w is entirely determined by n and j, the number of such factors for fixed n is at most $n^{3/2}/3 + 1$.

Case 2. Assume now that for some $k \ge 1$ one has $a_{k-1} < j < a_k < j+n < a_{k+1}$, i.e., that the first occurrence of w is across the boundary between q_k and q_{k+1}. Then we can write $w = w_1 w_2$, where w_1 is a non-empty suffix of q_k and w_2 is a non-empty prefix of q_{k+1}. There are two subcases.

Case 2a. If $k \le (n+1)/2$, then $j < a_k \le (n+1)(n+2)(n+3)/24$. As in case 1, since w is entirely determined by n and j, the number of such factors is at most $(n+3)^3/24$.

Case 2b. If $k > (n+1)/2$, then q_k ends in 0^{2k-2}, and $2k - 2 \ge n$, therefore $w_1 = 0^h$ for some h. The word w_2 is a prefix of length $n - h$ of some element of Q, and for fixed $n - h$ there are exactly $2^{\lceil \sqrt{n-h} \rceil}$ possibilities for such a prefix. As $0 < h < n$, there are altogether less than $n 2^{\lceil \sqrt{n} \rceil}$ possibilities for w.

Case 3. Finally, if one has $a_k \le j < j+n \le a_{k+1}$ for some $k \ge 1$, i.e., if w occurs inside some q_k, there are again two subcases.

Case 3a. If w contains at most one 1, then $w = 0^n$ or $w = 0^h 10^{n-h-1}$, which makes $n + 1$ possibilities.

Case 3b. If w contains at least two 1's, then it can be written as

$$w = 0^h c_{k-i-1} 0^{2i} c_{k-i-2} 0^{2i+2} \ldots c_{k-i-l} 0^{h'}$$

with $1 \le i + 1 < i + l \le k$, and

$$
\begin{aligned}
n &= h + (2i+1) + (2i+3) + \cdots + (2i+2l-3) + 1 + h' \\
&= h + 2i(l-1) + (l-1)^2 + 1 + h' > (l-1)^2 \;,
\end{aligned}
$$

hence $l \le \lceil \sqrt{n} \rceil$, and also $0 \le h < n$ and $0 \le i \le (n-2)/2$. As w is completely determined by the choice of h, i, and by the word $c_{k-i-1} c_{k-i-2} \ldots c_{k-i-l}$, there are at most $(n^2/2) 2^{\lceil \sqrt{n} \rceil}$ possibilities for w.

Summing all these bounds, we get

$$p_{\mathbf{u}}(n) \le n^{3/2}/3 + 1 + (n+3)^3/24 + n2^{\lceil\sqrt{n}\rceil} + n + 1 + (n^2/2)2^{\lceil\sqrt{n}\rceil}$$
$$\le n^2 2^{\lceil\sqrt{n}\rceil} \text{ for } n \ge 11$$

since $n^{3/2}/3 + 1 + n + 1 \le (n^2/8)2^{\lceil\sqrt{n}\rceil}$, $(n+3)^3/24 \le (n^2/8)2^{\lceil\sqrt{n}\rceil}$, and $n2^{\lceil\sqrt{n}\rceil} \le (n^2/4)2^{\lceil\sqrt{n}\rceil}$ when $n \ge 11$. ∎

The construction in Theorem 1 has one major drawback: it is often desired, especially for applications in ergodic theory, to have *recurrent* words (where every factor occurs infinitely often), and preferably *uniformly recurrent* words (where every factor occurs infinitely often with bounded gaps; the symbolic dynamical systems generated by such words are also called *minimal* [9, Chapter 1]). Unfortunately, the infinite word **u** is not uniformly recurrent, and not even recurrent.

Proposition 1. *The infinite word* **u** *contains infinitely many factors of the form* 10^h11, *and each of them occurs only once.*

Proof. Let j be the smallest index such that $c_j = 1$. Then $q_1 q_2 \ldots q_j = 0^{a_j}$, and for any $k > j$, q_k ends in $10^{2(j+1)k-j^2-2j-2}$. By construction, 11 can only occur at the beginning of some q_k, and it does occur at the beginning of infinitely many of them. Thus at each of these q_k's there is an occurrence of $10^{2(j+1)k-j^2-4j-2}11$, and this word occurs nowhere else. ∎

It is however possible, with a little modification of the construction, to ensure recurrence. Uniform recurrence is much more difficult to achieve, since it forbids the use of spacer blocks like 0^h: we shall see how to overcome this problem in the next section.

Let us first define a recurrent infinite word **v** on the infinite alphabet $\mathbb{N}$. This word, sometimes called the *Infinibonacci word*, or *dyadic valuation word*, is a very useful tool to construct infinite words with various properties. Define first a sequence of finite words z_j on $\mathbb{N}$ (known as *Zimin words* or *sesquipowers* [9, Chapters 3 and 4]) by $z_1 = 0$ and $z_{j+1} = z_j j z_j$. Then **v** is the limit of the sequence of words (z_j). Alternatively, $\mathbf{v} = v_1 v_2 v_3 \ldots$ where v_n is the dyadic valuation of n, i.e., the largest integer h such that 2^h divides n. A prefix of **v** is as follows:

$$\mathbf{v} = 01020103010201040102010301020105010201030102010401020103\ldots\ .$$

Let now $\mathbf{u}' = q_1 q_2 q_1 q_3 q_1 q_2 q_1 q_4 q_1 q_2 q_1 \ldots$ be the image of **v** under the substitution ψ that maps j to q_{j+1}, where q_{j+1} is defined as before:

$$\mathbf{u}' = 1.1100.1.010010000.1.1100.1.1000100001000000.1.1100.1\ldots\ .$$

Theorem 2. *The infinite word* $\mathbf{u}'$ *defined above is recurrent, and its complexity function satisfies*

$$2^{\lceil\sqrt{n}\rceil} \le p_{\mathbf{u}'}(n) \le n^2 2^{\lceil\sqrt{n}\rceil}$$

for all $n \ge 1$, *and consequently* $\log p_{\mathbf{u}'}(n) \sim \sqrt{n}$.

Proof. The proof is very similar to that of Theorem 1. The lower bound is proved by exactly the same argument. For the upper bound, we can again assume that $n \geq 11$.

Let w be any factor of length n of $\mathbf{u}'$. We discuss again on how w intersects the factorisation of $\mathbf{u}'$ by the q_i's. This time, instead of the position of the first occurrence of w in $\mathbf{u}'$, consider k, the smallest integer such that w is a factor of $\psi(z_k)$. By hypothesis, w is not a factor of $\psi(z_{k-1})$ so it intersects q_k (recall that $\psi(z_k) = \psi(z_{k-1})\psi(k-1)\psi(z_{k-1}) = \psi(z_{k-1})q_k\psi(z_{k-1})$). There are now four cases.

Case 1. Assume first that $w = w_1 q_k w_2$, where w_1 is a suffix of $\psi(z_{k-1})$ and w_2 is a prefix of $\psi(z_{k-1})$. Then w_1 and w_2 are entirely defined by their length, since $\psi(z_{k-1})$ is both a prefix and a suffix of $\psi(z_j)$ for all $j \geq k-1$ (actually, w_2 is a prefix of $\mathbf{u}'$, and w_1 a suffix of the leftward infinite word $\psi(\tilde{\mathbf{v}})$). So w itself is entirely defined by $h = |w_1|$ and k (from which $|w_2|$ can be deduced). As in case 1 of the proof of Theorem 1, we have $n \geq |q_k| = k^2$, so $1 \leq k \leq \sqrt{n}$, and $0 \leq h < n$, therefore the number of possibilities for w is at most $n^{3/2}$.

Case 2. Assume now that w occurs as a factor of $q_k\psi(z_{k-1})$, without being a factor of q_k. Then $w = w_1 w_2$, where w_1 is a suffix of q_k and w_2 is a prefix of $\psi(z_{k-1})$. The factor w is entirely determined by k and $h = |w_2|$, with $0 < h < n$. There are two subcases.

Case 2a. If $k \leq (n+1)/2$, then the number of possibilities is at most $n(n+1)/2$.

Case 2b. If $k > (n+1)/2$, then q_k ends in 0^{2k-2}, and $2k - 2 \geq n$, therefore $w_1 = 0^{n-h}$. Then w is entirely determined by h and there are at most n such factors.

Case 3. Assume now that w occurs as a factor of $\psi(z_{k-1})q_k$, without being a factor of q_k. Then $w = w_1 w_2$, where w_1 is a suffix of $\psi(z_{k-1})$ and w_2 is a prefix of q_k. The factor w is entirely determined by w_2 and $h = |w_1|$, with $0 < h < n$. As in case 2b of the proof of Theorem 1, we have less than $n2^{\lceil\sqrt{n}\rceil}$ possibilities for w.

Case 4. Finally, if w is a factor of q_k, then as in case 3 of the proof of Theorem 1, the number of possibilities is bounded by $n + 1 + (n^2/2)2^{\lceil\sqrt{n}\rceil}$.

Summing all these bounds, we get

$$p_{\mathbf{u}'}(n) \leq n^{3/2} + n(n+1)/2 + n + n2^{\lceil\sqrt{n}\rceil} + n + 1 + (n^2/2)2^{\lceil\sqrt{n}\rceil}$$
$$\leq n^2 2^{\lceil\sqrt{n}\rceil} \text{ for } n \geq 11$$

since $n^{3/2} + n(n+1)/2 + n + n + 1 \leq (n^2/4)2^{\lceil\sqrt{n}\rceil}$ and $n2^{\lceil\sqrt{n}\rceil} \leq (n^2/4)2^{\lceil\sqrt{n}\rceil}$ when $n \geq 11$.   ∎

3   Second Construction

To get uniformly recurrent words, we need a new construction without arbitrarily long spacer blocks 0^h. Instead, we will use prefixes of the infinite word itself, which will be constructed in a doubly recursive way, again based on the Infinibonacci word.

Also, we will choose a more general setting and construct not just one complexity function of intermediate growth, but a whole family of such functions.

For simplicity, we shall use the notation $f(t) \gg g(t)$, or $g(t) \ll f(t)$, meaning that $g(t)/f(t)$ tends to 0 when t tends to $+\infty$.

Theorem 3. *Let $\varphi \colon \mathbb{R}^+ \to \mathbb{R}^+$ be a function such that*

(i) $\varphi(t) \gg \log t$;
(ii) φ is differentiable, except possibly at 0;
(iii) $\varphi'(t) \ll t^{-\beta}$ for some positive constant β;
(iv) φ' is decreasing.

Then there exists a uniformly recurrent infinite word $\mathbf{u} \in A_2^{\mathbb{N}}$ such that

$$\log p_{\mathbf{u}}(n) \sim \varphi(n) \ .$$

In particular, $\varphi(n)$ can be taken equal to n^α for any α with $0 < \alpha < 1$, which includes the case $\varphi(n) = \sqrt{n}$. This construction can also be used with slower growing functions like $\varphi(n) = (\log(n+3))^2$. More exotic functions are also possible, for instance $\varphi(n) = (n+10)^{1/2 + \cos(\ln(\ln(n+10)))/4}$, which oscillates very slowly between $n^{1/4}$ and $n^{3/4}$.

Proof. Without loss of generality, we can assume that $\varphi(0) = 0$, $\varphi(1) = 1$, and $\varphi(t) > \log t$ for all $t > 0$. Note that φ is increasing and continuous, so that its inverse function φ^{-1} is well-defined.

We define inductively the substitution $\psi \colon \mathbb{N}^* \to A_2^*$ and the family $(x_k)_{k \in \mathbb{N}}$ of prefixes of the dyadic valuation word $\mathbf{v}$ (see Section 2) as follows:

  – $\psi(0) = 0$, $\psi(1) = 1$;
  – x_k is the longest prefix of $\mathbf{v}$ such that

$$|\psi(x_k)| \leq \max(\varphi^{-1}(k+1) - \varphi^{-1}(k) - 1, 0) \ ;$$

  – for all $j \geq 1$, $\psi(2j) = \psi(x_{\lfloor \log j \rfloor}) 0 \psi(j)$ and $\psi(2j+1) = \psi(x_{\lfloor \log j \rfloor}) 1 \psi(j)$.

Let then $\mathbf{u} = \psi(\mathbf{v})$. We have to prove that this indeed defines an infinite word $\mathbf{u}$ (see Lemma 1 below), and that $\mathbf{u}$ has the desired properties (see Lemmas 4 and 7 below).

Lemma 1. *The infinite word $\mathbf{u}$ is well-defined.*

Proof. We have to check that the inductive definition is correct. To determine x_k, we need to be able to compute the length of the image under ψ of long enough prefixes of $\mathbf{v}$. Obviously, the constructed substitution ψ will be non-erasing (the image of a letter is never the empty word), so that $|\psi(z_i)| \geq |z_i| = 2^i - 1$. Since we assumed that $\varphi(t) > \log t$, we have $\varphi^{-1}(k+1) - \varphi^{-1}(k) - 1 < 2^{k+1} - 1 \leq |\psi(z_{k+1})|$. Therefore x_k can only be a proper prefix of z_{k+1}, and to determine it exactly we need only the lengths of the images of letters occurring in z_{k+1}, that is up to k. In particular, x_0 and x_1 are readily defined.

Now, assume that, for some $j \geq 1$, $\psi(i)$ is known for $i < 2j$. We can then compute $\psi(x_{\lfloor \log j \rfloor})$ (since $\lfloor \log j \rfloor < 2j$, and we know $\psi(j)$, therefore we can compute $\psi(2j)$ and $\psi(2j+1)$. ∎

Lemma 2. *The sequence* $(|x_k|)$ *is nondecreasing and tends to infinity.*

Proof. Let $b_k = \varphi^{-1}(k+1) - \varphi^{-1}(k) - 1$. It can also be written as

$$b_k = -1 + \int_k^{k+1} \frac{1}{\varphi'(\varphi^{-1}(t))} dt \ .$$

As φ' is decreasing and tends to 0, φ is increasing and so is $\frac{1}{\varphi'} \circ \varphi^{-1}$. Therefore the sequence (b_k) is increasing, so that by construction x_k is a prefix of x_{k+1} and thus $(|x_k|)$ is nondecreasing. Note that as $b_0 = 0$, b_k is never negative and $\max(b_k, 0) = b_k$.

The assumption $\varphi'(t) \ll t^{-\beta}$ implies that $\varphi(t) \ll t^{1-\beta} \ll t$. If (b_k) were bounded by a constant C, then for all k we would have

$$\varphi^{-1}(k) = \sum_{i=0}^{k-1} (b_i + 1) \leq k(C+1) \ ,$$

and consequently $\varphi(k(C+1)) \geq k$, in contradiction with $\varphi(t) \ll t$. Therefore (b_k) tends to infinity, and so does $(|x_k|)$. ∎

Lemma 3. *The sequence* $(|\psi(j)|)$ *is nondecreasing. Moreover, for* $j \geq 1$, $|\psi(j)|$ *depends only on* $\lfloor \log j \rfloor$ *and*

$$1 + \varphi^{-1}(\lfloor \log j \rfloor) - \lfloor \log j \rfloor (1 + \varphi^{-1}(\lfloor \log \lfloor \log j \rfloor \rfloor)) \leq |\psi(j)| \leq 1 + \varphi^{-1}(\lfloor \log j \rfloor) \ .$$

Proof. By construction, for all $j \geq 1$ we have $|\psi(2j)| = |\psi(2j+1)| > |\psi(j)|$. By induction on $k \in \mathbb{N}$, we deduce that $|\psi(j)|$ is constant for j between 2^k and $2^{k+1} - 1$, and that the sequence $(|\psi(2^k)|)$ is strictly increasing. More precisely, $|\psi(2^{k+1})| - |\psi(2^k)| = |\psi(x_k)| + 1 \leq \varphi^{-1}(k+1) - \varphi^{-1}(k)$, hence $|\psi(j)| \leq 1 + \varphi^{-1}(\lfloor \log j \rfloor)$.

Let a be the letter that follows x_k in $\mathbf{v}$. By definition of x_k, $|\psi(x_k a)| > \varphi^{-1}(k+1) - \varphi^{-1}(k) - 1$. As x_k is a proper prefix of z_{k+1} (see the proof of Lemma 1), a is at most k so $|\psi(a)| \leq 1 + \varphi^{-1}(\lfloor \log k \rfloor)$. Then

$$\begin{aligned} |\psi(x_k)| &= |\psi(x_k a)| - |\psi(a)| \\ &\geq \varphi^{-1}(k+1) - \varphi^{-1}(k) - 1 - (1 + \varphi^{-1}(\lfloor \log k \rfloor)) \ . \end{aligned}$$

Summing, we get

$$\begin{aligned} |\psi(2^k)| &= 1 + \sum_{h=0}^{k-1} (|\psi(x_h)| + 1) \\ &\geq 1 + \varphi^{-1}(k) - k(1 + \varphi^{-1}(\lfloor \log k \rfloor)) \ , \end{aligned}$$

hence the lower bound on $|\psi(j)|$. ∎

Lemma 4. *The infinite word* **u** *is uniformly recurrent.*

Proof. Let w be a factor of **u**. We want to prove that w occurs in **u** with bounded gaps.

By Lemma 2, there exists an index k_0 such that the prefix $\psi(x_{k_0})$ of **u** is long enough to contain the first occurrence of w. Let $j_0 = 2^{k_0+1}$. Then, for all $j \geq j_0$, $\psi(x_{k_0})$ is a prefix of $\psi(x_{\lfloor \log j \rfloor - 1})$, which in turn, by construction of ψ, is a prefix of $\psi(j)$. So w is a factor of $\psi(j)$ for all $j \geq j_0$.

Let now $L = |\psi(j_0 z_{j_0} j_0)|$. We claim that every factor of **u** of length L contains an occurrence of w. Let y be such a factor, and consider the first occurrence of y in **u**. Two cases are possible:

Case 1. The first occurrence of y intersects at least two images of letters. Then $y = s\psi(r)p$, where s is a suffix of $\psi(j_1)$, p is a prefix of $\psi(j_2)$, j_1 and j_2 are two letters, and $j_1 r j_2$ is a factor of **v**. Since $|y| = L$, at least one of the following three inequalities holds:

Case 1a. $|s| \geq |\psi(j_0)|$. Then, by Lemma 3, $j_1 \geq j_0$. Let h be the largest integer such that $j_1 \geq 2^h j_0$, and $j = \lfloor 2^{-h} j_1 \rfloor$, so that $|\psi(j)| = |\psi(j_0)|$. Both s and $\psi(j)$ are suffixes of $\psi(j_1)$, and s is longer, so $\psi(j)$ is a suffix of s. As $\psi(j)$ contains w, so does y.

Case 1b. $|p| \geq |\psi(j_0)|$. Then, by Lemma 3, $j_2 \geq j_0$. Both p and $\psi(x_{k_0})$ are prefixes of $\psi(j_2)$, and p is longer, so $\psi(x_{k_0})$ is a prefix of p. As $\psi(x_{k_0})$ contains w, so does y.

Case 1c. $|\psi(r)| > |\psi(z_{j_0})|$. Then r cannot be a factor of z_{j_0}, so it must contain some letter $j \geq j_0$ (indeed, all factors of **v** that do not contain any letter larger than or equal to j_0 are factors of z_{j_0}). Then y contains $\psi(r)$ which contains $\psi(j)$ which in turn contains w.

Case 2. The first occurrence of y is inside some $\psi(j)$. Recall that

$$\psi(j) = \psi(x_{\lfloor \log j \rfloor - 1}) e \psi(\lfloor j/2 \rfloor)$$

where $e = j \bmod 2 \in A_2$. Since this is the first occurrence of y, it is neither inside $\psi(x_{\lfloor \log j \rfloor - 1})$, nor inside $\psi(\lfloor j/2 \rfloor)$, as both have already occurred before. So y can be written as $y = s\psi(r)ep$, where s is a suffix of $\psi(j_1)$ for some letter j_1, p is a prefix of $\psi(\lfloor j/2 \rfloor)$, and $j_1 r$ is a suffix of $x_{\lfloor \log j \rfloor - 1}$. Since $|y| = L$, at least one of the following three inequalities hold:

Case 2a. $|s| \geq |\psi(j_0)|$. Then, as in case 1a above, w occurs in y.

Case 2b. $|p| \geq |\psi(j_0)|$. Then, as in case 1b above (with $j_2 = \lfloor j/2 \rfloor$), w occurs in y.

Case 2c. $|\psi(r)| > |\psi(z_{j_0})|$. Then, as in case 1c above, w occurs in y.

In all cases, we have found an occurrence of w in y. ∎

For all $n \in \mathbb{N}$, define the quantities

$$k(n) = \min\{k : |\psi(2^k)| \geq n\}$$

and

$$m(n) = \min\{k : |\psi(x_{k-1})| \geq n\} .$$

Lemma 5. *The functions $k(n)$ and $m(n)$ satisfy the following relations: $k(n) \sim \varphi(n)$ and $m(n) \ll n^\alpha$ for some $\alpha > 0$.*

Proof. Observe first that $|\psi(2^k)| \geq k + 1$, and thus $k(n) \leq n - 1$. Applying Lemma 3 successively to $j = 2^{k(n)}$ and $j = 2^{k(n)-1}$, we get

$$n \leq |\psi(2^{k(n)})| \leq 1 + \varphi^{-1}(k(n))$$

and

$$1 + \varphi^{-1}(k(n) - 1) - (k(n) - 1)(1 + \varphi^{-1}(\lfloor \log(k(n) - 1) \rfloor)) \leq |\psi(2^{k(n)-1})| \leq n - 1 .$$

The first line yields $k(n) \geq \varphi(n - 1)$, which is equivalent to $\varphi(n)$ since $\varphi'(t) \ll \varphi(t)$. The second one yields

$$k(n) - 1 \leq \varphi\left(n - 2 + (k(n) - 1)(1 + \varphi^{-1}(\lfloor \log(k(n) - 1) \rfloor))\right)$$
$$\leq \varphi\left(n + k(n)(1 + \varphi^{-1}(\log n))\right) .$$

Condition (iv), i.e., the concavity of φ, allows to replace φ with its order one expansion at n:

$$k(n) - 1 \leq \varphi(n) + \varphi'(n)k(n)(1 + \varphi^{-1}(\log n)) .$$

By condition (iii), $\varphi'(n) \ll n^{-\beta}$, and by condition (i), $\varphi^{-1}(\log n) \ll n^\eta$ for any $\eta > 0$, so the order one term is neglectable with respect to $k(n)$, and we conclude that $k(n) \sim \varphi(n)$.

In the proof of Lemma 3, we saw that

$$|\psi(x_k)| \geq \varphi^{-1}(k + 1) - \varphi^{-1}(k) - 1 - (1 + \varphi^{-1}(\lfloor \log k \rfloor)) .$$

A small computation using condition (iii) and the integral formula in the proof of Lemma 2 shows that $\varphi^{-1}(k + 1) - \varphi^{-1}(k) \gg (\varphi^{-1}(k))^\beta \gg k^\beta$. Then

$$k^\beta \ll |\psi(x_k)| + 2 + \varphi^{-1}(\lfloor \log k \rfloor) .$$

Let now $k = m(n) - 2$, so that $|\psi(x_k)| \leq n - 1$. We get

$$(m(n) - 2)^\beta \ll n + 1 + \varphi^{-1}(\lfloor \log(m(n) - 2) \rfloor)$$

where the term $\varphi^{-1}(\lfloor \log(m(n) - 2) \rfloor)$ is neglectable in front of $m(n)$ by condition (i), so $m(n) \ll n^{1/\beta}$. ∎

For $n \in \mathbb{N}$, let $\pi(n)$ be the total number of words of length n which are prefixes of $\psi(j)$ for some j, and similarly $\sigma(n)$ be the total number of words of length n which are suffixes of $\psi(j)$ for some j.

Lemma 6. *The functions $\pi(n)$ and $\sigma(n)$ satisfy the following relations:* $\sigma(n) = 2^{k(n)}$ *and* $\pi(n) \leq 1 + m(n)2^{k(n+2)}$.

Proof. If w is a suffix of length n of $\psi(j)$, then $j \geq 2^{k(n)}$ (otherwise $\psi(j)$ would be too short). Let $j' = \lfloor 2^{k(n)-\lfloor \log j \rfloor}j \rfloor$, so that $2^{k(n)} \leq j' < 2^{k(n)+1}$ and $\psi(j')$ is a suffix of $\psi(j)$. Then w is the suffix of length n of $\psi(j')$, and all these suffixes are different, so there are exactly $2^{k(n)}$ possibilities for w.

If w is a prefix of length n of $\psi(j)$, then $j \geq 2^{k(n)}$ as above. On the other hand, if $j \geq 2^{m(n)}$, then $|\psi(x_{\lfloor \log j \rfloor - 1})| \geq n$ and w is a prefix of $\psi(x_{\lfloor \log j \rfloor - 1})$, hence of $\mathbf{u}$, so it is uniquely defined. If $2^{k(n)} \leq j < 2^{m(n)}$, then w is a prefix of $\psi(x_k)e_1\psi(x_{k-1})e_2 \ldots \psi(x_{k-h+1})e_h\psi(x_{k-h})$, where $k = \lfloor \log j \rfloor - 1 < m(n) - 1$, h is the largest integer such that $|\psi(x_k)e_1\psi(x_{k-1})e_2 \ldots \psi(x_{k-h+1})e_h| \leq n$, and $e_1e_2 \ldots e_h \in A_2^h$. The integer h is at most $k+1$, with in that case the convention $x_{-1} = 1$. By Lemma 2,

$$|\psi(x_k)e_1\psi(x_{k-1})e_2 \ldots \psi(x_{k-h+1})e_h| \geq |\psi(x_{h-1})e_1\psi(x_{h-2})e_2 \ldots \psi(x_0)e_h|$$
$$= |\psi(2^h)| - 1 \ ,$$

so $|\psi(2^h)| \leq n + 1$, which implies that $h < k(n+2)$. Since w is entirely defined by k and $e_1e_2 \ldots e_h$, there are at most $m(n)2^{k(n+2)}$ possibilities for such a w. ∎

Lemma 7. *The complexity function of* $\mathbf{u}$ *satisfies* $\log p_{\mathbf{u}}(n) \sim \varphi(n)$.

Proof. A lower bound is given by Lemmas 5 and 6: obviously $p(n) \geq \sigma(n)$ and $\log \sigma(n) = k(n) \sim \varphi(n)$.

Let w be a factor of length $n \geq 2$ of $\mathbf{u}$. As in the proof of Theorem 2, we consider j, the smallest integer such that w is a factor of $\psi(z_{j+1})$. By hypothesis, w is not a factor of $\psi(z_j)$ so it intersects $\psi(j)$. There are four cases, partly similar to the four cases of Theorem 2:

Case 1. Assume first that $w = w_1\psi(j)w_2$, where w_1 is a suffix of $\psi(z_j)$ and w_2 is a prefix of $\psi(z_j)$. Then w_1 and w_2 are entirely defined by their length, and w itself is entirely defined by $h = |w_1|$ and j. As $|\psi(j)| \leq n$, we have $j < 2^{k(n+1)}$, so the number of possibilities for w is at most $n2^{k(n+1)}$.

Case 2. Assume now that w occurs as a factor of $\psi(jz_j)$, without being a factor of $\psi(j)$. Then $w = w_1w_2$, where w_1 is a suffix of $\psi(j)$ and w_2 is a prefix of $\psi(z_j)$. The factor w is entirely determined by w_1, so there are $\sum_{h=1}^{n} \sigma(h)$ possibilities. By Lemma 6, this is bounded by $n2^{k(n)}$.

Case 3. Assume now that w occurs as a factor of $\psi(z_j j)$, without being a factor of $\psi(j)$. Then $w = w_1w_2$, where w_1 is a suffix of $\psi(z_j)$ and w_2 is a prefix of $\psi(j)$. The factor w is entirely determined by w_2, so there are $\sum_{h=1}^{n} \pi(h)$ possibilities. By Lemma 6, this is bounded by $n(1 + m(n)2^{k(n+2)})$.

Case 4. Finally, if w is a factor of $\psi(j)$, necessarily (see case 2 in the proof of Lemma 4) w can be written as $w = w_1ew_2$, where w_1 is a suffix of $\psi(x_{\lfloor \log j \rfloor - 1})$, $e = j \bmod 2 \in A_2$, and w_2 is a prefix of $\psi(\lfloor j/2 \rfloor)$. We now discuss on j.

Case 4a. If $j \geq 2^{m(n)+1}$, then $n \leq |\psi(x_{\lfloor \log j \rfloor - 2})|$, and w_2 is a prefix of $\psi(x_{\lfloor \log j \rfloor - 2})$, hence also of $\mathbf{u}$, so it is entirely determined by its length. Let j' be the largest letter such that $\psi(j')$ intersects w_1 (viewed as a suffix of $\psi(x_{\lfloor \log j \rfloor - 1})$): then there is a prefix r of $z_{j'}$ such that $z_{j'}j'r$ is a suffix of $x_{\lfloor \log j \rfloor - 1}$, w_1 is a suffix of $\psi(z_{j'}j'r)$, and w_1 is not a suffix of $\psi(r)$. There are again two subcases.

Case 4a1. $\psi(j'r)$ is a proper suffix of w_1. Then, as in case 1 above, w_1 is entirely defined by its length, j', and $|r|$. As $|\psi(j')| < n$, we have $j' < 2^{k(n)}$, so the number of possibilities for w is at most $2n^2 2^{k(n)}$.

Case 4a2. w_1 is a suffix of $\psi(j'r)$. Then $w_1 = w_3\psi(r)$, where w_3 is a suffix of $\psi(j')$, and w is entirely defined by w_3, $|r|$, and e (from which $|w_2|$ can be deduced). The number of possibilities is therefore at most $2n^2 2^{k(n)}$.

Case 4b. If $j < 2^{m(n)+1}$, then $1 \leq \lfloor \log j \rfloor \leq m(n)$. Given its length, the word w_1 is entirely determined by $\lfloor \log j \rfloor$, so there are at most $m(n)$ possibilities for it. There are $\pi(|w_2|)$ possibilities for w_2 given its length, and by Lemma 6 this is bounded by $1 + m(n)2^{k(n+2)}$. Consequently the number of possibilities for w is at most $2nm(n)(1 + m(n)2^{k(n+2)})$.

Summing all these bounds, we get

$$p_{\mathbf{u}}(n) \leq n2^{k(n+1)} + n2^{k(n)} + n(1 + m(n)2^{k(n+2)}) +$$
$$2n^2 2^{k(n)} + 2n^2 2^{k(n)} + 2nm(n)(1 + m(n)2^{k(n+2)})$$
$$\leq 12n^2 m(n)^2 2^{k(n+2)} \ .$$

By Lemma 5, $m(n)^2 \ll n^{2\alpha}$ so $\log(12n^2 m(n)^2)$ grows logarithmically in n, which is neglectable with respect to $\varphi(n)$. It remains

$$\log p_{\mathbf{u}}(n) \sim k(n+2) \sim \varphi(n) \ .$$

∎

Lemmas 1, 4, and 7 conclude the proof of Theorem 3. ∎

Example. Let $\varphi(n) = \sqrt{n}$. Then x_k is the longest prefix of $\mathbf{v}$ such that $|\psi(x_k)| \leq 2k$. The first values of $\psi(j)$ and x_k are computed below:

- by definition, $\psi(0) = 0$ and $\psi(1) = 1$;
- $x_0 = \varepsilon$, $\psi(x_0) = \varepsilon$,
  $x_1 = 01$, $\psi(x_1) = 01$;
- $\psi(2) = \psi(x_0)0\psi(1) = 01$,
  $\psi(3) = \psi(x_0)1\psi(1) = 11$,
  $\psi(4) = \psi(x_1)0\psi(2) = 01001$,
  $\psi(5) = \psi(x_1)1\psi(2) = 01101$,
  $\psi(6) = \psi(x_1)0\psi(3) = 01011$,
  $\psi(7) = \psi(x_1)1\psi(3) = 01111$;
- $x_2 = 010$, $\psi(x_2) = 010$ (as $\psi(0102) = 01001$ is too long),
  $x_3 = 01020$, $\psi(x_3) = 010010$,

$x_4 = 0102010,\ \psi(x_4) = 01001010,$
$x_5 = 01020103,\ \psi(x_5) = 0100101011,$
$x_6 = 0102010301,\ \psi(x_6) = 010010101101,$
$x_7 = 01020103010,\ \psi(x_7) = 0100101011010,$
$x_8 = 0102010301020,\ \psi(x_8) = 0100101011010010,$
$x_9 = 010201030102010,\ \psi(x_9) = 010010101101001010,$
$x_{10} = x_{11} = x_9$ (as $|\psi(x_94)| = 23$ is too long),
$x_{12} = 01020103010201040,\ \psi(x_{12}) = 01001010110100101001 0010,$
etc. (the knowledge of $\psi(0)$ to $\psi(7)$ is enough to compute up to x_{181});
$-\ \psi(8) = \psi(x_2)0\psi(4) = 010001001$, etc.

Then the infinite word $\mathbf{u} = \psi(\mathbf{v})$, which starts as follows:

$$\mathbf{u} = 0.1.0.01.0.1.0.11.0.1.0.01.0.1.0.01001.0.1.0.01.0.1.0.11.0.1.0.01 \ldots$$

is uniformly recurrent and $\log p_{\mathbf{u}}(n) \sim \sqrt{n}$.

References

1. JEAN-PAUL ALLOUCHE, Sur la complexité des suites infinies, *Bull. Belg. Math. Soc.* **1** (1994), 133–143.
2. GEORGES BLANC AND NOËLLE BLEUZEN-GUERNALEC, Production en temps réel et complexité de structure de suites infinies, *RAIRO Inf. Théor. Appl.* **23** (1989), 195–216.
3. JULIEN CASSAIGNE, Special factors of sequences with linear subword complexity, in *Developments in Language Theory II (DLT'95)*, pages 25–34, World Scientific, 1996.
4. JULIEN CASSAIGNE, Complexité et facteurs spéciaux, *Bull. Belg. Math. Soc.* **4** (1997), 67–88.
5. DAVID G. CHAMPERNOWNE, The construction of decimals normal in the scale of ten, *J. London Math. Soc.* **8** (1933), 254–260.
6. ANDRZEJ EHRENFEUCHT, K. P. LEE, AND GRZEGORZ ROZENBERG, Subword complexities of various classes of deterministic developmental languages without interaction, *Theoret. Comput. Sci.* **1** (1975), 59–75.
7. CHRISTIAN GRILLENBERGER, Constructions of strictly ergodic systems — I. Given entropy, *Z. Wahr. verw. Geb.* **25** (1973), 323–334.
8. PASCAL HUBERT, Complexité des suites définies par des billards rationnels, *Bull. Soc. Math. France* **123** (1995), 257–270.
9. M. LOTHAIRE, *Algebraic Combinatorics on Words*, Cambridge University Press, 2002.
10. MARSTON MORSE AND GUSTAV A. HEDLUND, Symbolic dynamics, *Amer. J. Math.* **60** (1938), 815–866.
11. JEAN-JACQUES PANSIOT, Complexité des facteurs des mots infinis engendrés par morphismes itérés, in *ICALP '84*, pages 380–389, *Lect. Notes Comp. Sci.* **172**, Springer-Verlag, 1984.

A Space Lower Bound of Two-Dimensional Probabilistic Turing Machines

Yuji Sasaki[1], Katsushi Inoue[2], Akira Ito[2], and Yue Wang[3]

[1] Chitose Factory, Hitachi Kokusai Electronocs, Chitose, 066-8566 Japan
sasaki.yuji@h-kokusai.com
[2] Department of Computer Science and Systems Engineering,
Faculty of Engineering, Yamaguchi University, Ube, 755-8611 Japan
{inoue,ito}@csse.yamaguchi-u.ac.jp
[3] Media and Information Technology Center, Yamaguchi University,
Ube, 755-8611 Japan
wangyue@yamaguchi-u.ac.jp

Abstract. This paper shows a sublogarithmic space lower bound for two-dimensional probabilistic Turing machines (2-ptm's) over square tapes with bounded error, and shows, using this space lower bound theorem, that a specific set is not recognized by any $o(\log n)$ space-bounded 2-ptm. Furthermore, the paper investigates a relationship between 2-ptm's and two-dimensional Turing machines with both nondeterministic and probabilistic states, which we call "two-dimensional stochastic Turing machines (2-stm's)", and shows that for any $\log\log n \leq L(n) = o(\log n)$, $L(n)$ space-bounded 2-ptm's with bounded error are less powerful than $L(n)$ space-bounded 2-stm's with bounded error which start in nondeterministic mode, and make only one alternation between nondeterministic and probabilistic modes.

1 Introduction

The classes of languages recognized by probabilistic finite automata and probabilistic Turing machines have been studied extensively [2-5]. On the other hand, recently [9,10], two-dimensional probabilistic finite automata (2-pfa's) and two-dimensional probabilistic Turing machines (2-ptm's) were introduced, and several properties of them have been investigated. There seem to be many problems to be solved in the future for 2-pfa's and 2-ptm's.

This paper is concerned with investigating properties of 2-ptm's with bounded errors whose input tapes are restricted to square ones.

Section 2 of the paper presents some definitions and notations necessary for this paper. Dwork and Stockmayer [2] proved an impossibility result for probabilistic finite automata with bounded error. By using an idea similar to the proof of this result, Freivalds and Karpinski [4] proved, for the first time, a sublogarithmic space lower bound for probabilistic Turing machines with bounded error. In Section 3 of this paper, by extending their proof techniques, we first prove a sublogarithmic space lower bound for 2-ptm's with bounded error for recognizing sets of square tapes. By using this space lower bound theorem, we then show

M. Ito and M. Toyama (Eds.): DLT 2002, LNCS 2450, pp. 185–196, 2003.

that a specific set is not recognized by any sublogarithmically space-bounded 2-ptm with bounded error.

In [1,7,8], several properties of machines with both nondeterministic and probabilistic states were investigated. Liskiewicz [7] showed that for sublogarithmic space bounds, there exists an infinite hierarchy (of recognizing power) based on the number of alternations between nondeterministic and probabilistic configurations. Section 4 investigates a relationship between 2-ptm's and two-dimensional Turing machines with both nondeterministic and probabilistic states, which we call 'two-dimensional stochastic Turing machines (2-stm's)', and shows that for any $loglog\ n \leq L(n) = o(log\ n)$, $L(n)$ space-bounded 2-ptm's with bounded error are less powerful than $L(n)$ space-bounded 2-stm's with bounded error which start in nondeterministic mode, and make only one alternation between nondeterministic and probabilistic modes.

Section 5 concludes this paper by posing several open problems.

2   Definitions and Notations

Let Σ be a finite set of symbols. A *two-dimensional tape* over Σ is a two-dimensional rectangular array of elements of Σ. The set of all two-dimensional tapes over Σ is denoted by $\Sigma^{(2)}$. Given a tape $x \in \Sigma^{(2)}$, we let $l_1(x)$ be the number of rows and $l_2(x)$ be the number of columns of x. If $1 \leq i \leq l_1(x)$ and $1 \leq j \leq l_2(x)$, we let $x(i,j)$ denote the symbol in x with coordinates (i,j). For each $m, n(m, n \geq 1)$, let $\Sigma^{(m,n)} = \{x \in \Sigma^{(2)} | l_1(x) = m$ and $l_2(x) = n\}$.

Furthermore, we define $x[(i_1, i_2), (i'_1, i'_2)]$, only when $1 \leq i_1 \leq i'_1 \leq l_1(x)$ and $1 \leq i_2 \leq i'_2 \leq l_2(x)$, as the two-dimensional tape z satisfying the following (i) and (ii):

(i) $l_1(z) = i'_1 - i_1 + 1$ and $l_2(z) = i'_2 - i_2 + 1$,
(ii) for each $i, j(1 \leq i \leq l_1(z), 1 \leq j \leq l_2(z))$, $z(i,j) = x(i_1 + i - 1, i_2 + j - 1)$.

We call $x[(i,j), (i',j')]$ the '$[(i,j), (i',j')]$-segment of x'.

We next recall a two-dimensional probabilistic Turing machine [9] which is a natural extension of a two-way probabilistic Turing machine [5] to two dimension.

Let S be a finite set. A *coin-tossing distribution on S* is a mapping ψ from S to $\{0, \frac{1}{2}, 1\}$ such that $\Sigma_{a \in S} \psi(a) = 1$. The mapping means 'choose a with probability $\psi(a)$'. A *two-dimensional probabilistic Turing machine* (denoted by 2-ptm) is a 7-tuple $M = (Q, \Sigma, \Gamma, \delta, q_0, q_a, q_r)$, where Q is a finite set of *states*, Σ is a finite *input alphabet* ($\# \notin \Sigma$ is the *boundary symbol*), Γ is a finite *storage tape alphabet* ($B \in \Gamma$ is the *blank symbol*), δ is a *transition function*, $q_0 \in Q$ is the *initial state*, $q_a \in Q$ is the *accepting state*, and $q_r \in Q$ is the *rejecting state*.

The machine M has a read-only (rectangular) input tape with boundary symbols $\#$ and one semi-infinite storage tape, initially blank. A position is assigned to each cell of the read-only input tape and the storage tape, as shown in Fig. 1. The transition function δ is defined on $(Q - \{q_a, q_r\}) \times (\Sigma \cup \{\#\}) \times \Gamma$ such that for each $q \in Q - \{q_a, q_r\}$, each $\sigma \in \Sigma \cup \{\#\}$ and each $\gamma \in \Gamma, \delta[q, \sigma, \gamma]$

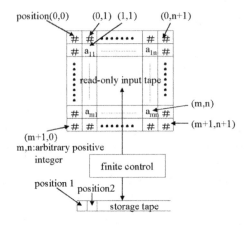

Fig. 1. 2-ptm M

is a coin tossing distribution on $Q \times (\Gamma - \{\#\}) \times D_1 \times D_2$, where $D_1 = \{\text{Left}, \text{Right}, \text{Up}, \text{Down}, \text{Stay}\}$, $D_2 = \{\text{Left}, \text{Right}, \text{Stay}\}$. This meaning is that if M is in state q with the input head scanning the symbol σ and the storage tape head scanning the symbol γ, then with probability $\delta[q, \sigma, \gamma](q', \gamma', d_1, d_2)$ the machine enters state q', rewrites the symbol γ by the symbol γ', moves the input head one symbol in direction d_1, and moves the storage tape head one symbol in direction d_2.

Given an input tape $x \in \Sigma^{(2)}$, M starts in the initial state q_0 with the input head on the upper left-hand corner of x, with all the cells of the storage tape blank and with the storage tape head on the left end of the storage tape. The computation of M on x is then governed (probabilistically) by the transition function δ until M either accepts by entering the accepting state q_a or rejects by entering the rejecting state q_r. We assume that δ is defined so that the input head never falls off an input tape out of the boundary symbols $\#$, the storage tape head cannot write the blank symbol, and fall off the storage tape by moving left. M halts when it enters state q_a or q_r.

The computation of a 2-ptm can be viewed as a tree, where each path from the root to a leaf, which represents a particular computation path, is associated with the probability that the corresponding computation path is chosen. The acceptance (or rejection) probability of a 2-ptm on an input tape is the sum of the probability associated with accepting (or rejecting) computation paths.

We say that a 2-ptm *accepts (or rejects)* an input tape with probability p if the acceptance (or rejection) probability is p. Let $T \subseteq \Sigma^{(2)}$ and $0 \leq \varepsilon < \frac{1}{2}$. A 2-ptm M *recognizes* T with error probability ε if for all $x \in T$, M accepts x with probability at least $1 - \varepsilon$, and for all $x \notin T$, M rejects x with probability at least $1 - \varepsilon$.

Let M be a 2-ptm. A *configuration* of M is a combination of a position of the input head and a storage state of M, where a *storage state* of M is a combination of the state of the finite control, the non-blank contents of the storage tape, and the storage tape head position.

We next introduce a two-dimensional stochastic Turing machine.

A *two-dimensional stochastic Turing machine* (2-stm) is a 2-ptm provided with nondeterministic states. A *strategy* of a 2-stm on an input tape is a function that maps each nondeterministic configuration to one of its successors, where 'nondeterministic (or probabilistic) configuration' is a configuration whose state component is a nondeterministic (or probabilistic) state. We assume that a 2-stm chooses one of its strategies. Therefore, the computation of a 2-stm can be viewed as a collection of trees, each representing the computation as a 2-ptm for a fixed strategy. We say that a 2-stm *accept* (*or reject*) an input tape with probability p if there exists a tree whose acceptance (or rejectance) probability (as a computation for a 2-ptm) is p.

Let $T \subseteq \Sigma^{(2)}$ and $0 \le \varepsilon < \frac{1}{2}$. A 2-stm M *recognizes* T with error probability ε if for all $x \in T$, there exists a strategy of M on x on which M accepts x with probability at least $1 - \varepsilon$, and for all $x \notin T$, on every strategy of M on x, M rejects x with probability at least $1 - \varepsilon$.

In this paper, we are concerned with 2-ptm's and 2-stm's whose input tapes are restricted to square ones. We denote these machines by 2-ptm^s's and 2-stm^s's, respectively. Let $L : N \to N$ be a function, where N denotes the set of all the positive integers. We say that a 2-ptm^s (or 2-stm^s) M is $L(n)$ *space-bounded* if for each $n \ge 1$, and for each input tape x with $l_1(x) = l_2(x) = n$, M uses at most $L(n)$ cells of the storage tape. By 2-$PTM^s(L(n))$, we denote the class of sets of square tapes recognized by $L(n)$ space-bounded 2-ptm^s's with error probability less than $\frac{1}{2}$.

For each positive integer $k \ge 1$, by $MA_k2\text{-}STM^s(L(n))$ (or $AM_k2\text{-}STM^s$ $(L(n))$), we denote the class of sets of square tapes recognized by $L(n)$ space-bounded 2-stm^s's with error probability less than $\frac{1}{2}$ which make at most $k - 1$ alternations between nondeterministic and probabilistic configurations, and start in a nondeterministic (or probabilistic) state. From definitions, it is obvious that for any $L(n)$,

2-$PTM^s(L(n)) = AM_12\text{-}STM^s(L(n))$.

For any set S, $|S|$ denotes the number of elements in S.

3   Space Lower Bound Theorem

In this section, we prove a sublogarithmic space lower bound theorem for 2-ptm^s's with bounded error, and by using this theorem, we show that a specific set is not recognized by any $o(log\ n)$ space-bounded 2-ptm^s with bounded error.

We first give some preliminaries necessary for getting our desired result.

Let M be a 2-ptm and Σ be the input alphabet of M. For each $m \ge 2$ and each $1 \le n \le m - 1$, an (m, n)-*chunk* over Σ is a pattern as shown in Fig. 2, where $v_1 \in \Sigma^{(m-1,n)}$ and $v_2 \in \Sigma^{(m,m-n)}$.

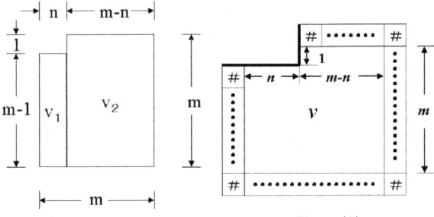

Fig. 2. (m, n)-chunk

Fig. 3. $v(\#)$

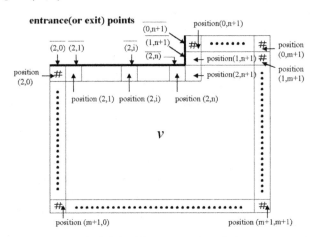

Fig. 4. An illustration for $v(\#)$

By $ch_{(m,n)}(v_1, v_2)$, we denote the (m, n)-chunk as shown in Fig. 2. For any (m, n)-chunk v, we denote by $v(\#)$ the pattern obtained from v by attaching the boundary symbols $\#$ to v as shown in Fig. 3.

Below, we assume without loss of generality that M enters or exits the pattern $v(\#)$ only at the face designated by the bold line in Fig. 3. Thus, the number of the entrance points to $v(\#)$ (or the exit points from $v(\#)$) for M is $n + 3$. We suppose that these entrance points (or exit points) are named $\overline{(2, 0)}, \overline{(2, 1)}, ..., \overline{(2, n)}, \overline{(1, n + 1)}, \overline{(0, n+1)}$ as shown in Fig. 4. Let $PT(v(\#))$ be the set of these entrance points (or exit points). To each cell of $v(\#)$, we assign a position as shown in Fig. 4. Let $PS(v(\#))$ be the set of all the positions of $v(\#)$. For each $n \geq 1$, an n-chunk over Σ is a pattern in $\Sigma^{(1,n)}$. For any n-chunk u, we denote by $u(\#)$ the pattern obtained from u by attaching the boundary

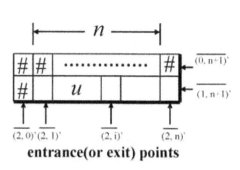

Fig. 5. An illustration for $u(\#)$

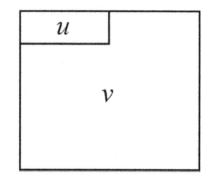

Fig. 6. $v[u]$

symbols $\#$ to u as shown in Fig. 5. We again assume without loss of generality that M enters or exits the pattern $u(\#)$ only at the face designated by the bold line in Fig. 5. The number of the entrance points to $u(\#)$ (or the exit points from $u(\#)$) for M is again $n+3$, and these entrance points (or exit points) are named $\overline{(2,0)'}, \overline{(2,1)'}, ..., \overline{(2,n)'}, \overline{(1,n+1)'}, \overline{(0,n+1)'}$ as shown in Fig. 5. Let $PT(u(\#))$ be the set of these entrance points (or exit points). (Note that the entrance points of an n-chunk are distinguished from the entrance points of an (m,n)-chunk only by 'dash'.) For any (m,n)-chunk v over Σ and any n-chunk u over Σ, let $v[u]$ be the tape in $\Sigma^{(m,m)}$ consisting of v and u as shown in Fig.6.

Let q_a and q_r be the accepting and rejecting states of M, respectively and x be an (m,n)-chunk (or an n-chunk) over the input alphabet of M ($m \geq 2, n \geq 1$). We define the *chunk probabilities* of M on x as follows. A *starting condition* for the chunk probability is a pair (s,l), where s is a storage state of M and $l \in PT(x(\#))$: its intuitive meaning is 'M has just entered $x(\#)$ in storage state s from entrance point l of $x(\#)$'. A *stopping condition* for the chunk probability is either:

(i) a pair (s,l) as above, meaning that M exits from $x(\#)$ in storage state s at exit point l,
(ii) 'Loop' meaning that the computation of M loops forever within $x(\#)$,
(iii) 'Accept' meaning that M halts in the accepting state q_a before exiting from $x(\#)$ at exit points of $x(\#)$, or
(iv) 'Reject' meaning that M halts in the rejecting state q_r before exiting from $x(\#)$ at exit points of $x(\#)$.

For each starting condition σ and each stopping condition τ, let $p(x,\sigma,\tau)$ be the probability that stopping condition τ occurs given that M is started in starting condition σ on an (m,n)-chunk (or n-chunk) x.

Computations of a 2-ptm are modeled by Markov chains [11] with finite state space, say $\{1,2,...,s\}$ for some s. A paticular Markov chain is completely specified by its matrix $R = \{r_{ij}\}_{1 \leq i,j \leq s}$ of transition probabilities. If the Markov chain is in state i, then it next moves to state j with probability r_{ij}. The chains we consider have the designated starting state, say, state 1, and some set T_r of trapping states, so $r_{tt} = 1$ for all $t \in T_r$. For $t \in T_r$, let $p^*[t, R]$ denote the probability that Markov chain R is trapped in state t when started in state 1.

The following lemma which bounds the effect of small changes in the transition probabilities of a Markov chain is used below. Let $\beta \geq 1$. Say that two numbers r and r' are β-close if either (1) $r = r' = 0$ or (2) $r > 0, r' > 0$ and $\beta^{-1} \leq r/r' \leq \beta$. Two Markov chains $R = \{r_{ij}\}_{i,j=1}^s$ and $R' = \{r'_{ij}\}_{i,j=1}^s$ are β-close if r_{ij} and r'_{ij} are β-close for all pairs i, j.

Lemma 3.1 [2] *Let R and R' be two s-state Markov chains which are β-close, and let t be a trapping state of both R and R'. Then $p^*[t, R]$ and $p^*[t, R']$ are B^{2s}-close.*

We are now ready to prove the following space lower bound theorem.

Theorem 3.1 *Let $A, B \subseteq \Sigma^{(2)}$ with $A \cap B = \emptyset$. Suppose that there is an infinite set I of positive integers and a function $H(n)$ such that $H(n)$ is a fixed fuction bounded by some exponential in n, and for each $n \in I$ there is a set $V(n)$ of $(H(n), n)$-chunks over Σ such that:*

(1) there are constants $c > 1$ and $r > 0$ such that $|V(n)| \geq 2^{c^{n^r}}$ for all $n \in I$, and

(2) for every $n \in I$ and every $v, v' \in V(n)$ with $v \neq v'$, there is an n-chunk $u \in \Sigma^{(1,n)}$ such that:

$$either \begin{cases} v[u] \in A \\ v'[u] \in B \end{cases} or \begin{cases} v[u] \in B \\ v'[u] \in A \end{cases}.$$

Then, if an $L(n)$ space-bounded 2-ptm^s with error probability $\varepsilon < \frac{1}{2}$ separates A and B, then $L(H(n))$ can not be $o(n^r)$.

Proof. Suppose that there is an $L(n)$ space-bounded 2-ptm^s M separating A and B with error probability $\varepsilon < \frac{1}{2}$. By $C(n)$ we denote the set of possible storage states of M on input tapes of side-length n ($n \geq 1$), and let $c(n) = |C(n)|$. It is obvious that $c(n) \leq O(exp(L(n)))$. Let I, $H(n)$, and $V(n)$ be as described in the theorem.

Suppose to the contrary that $L(H(n)) = o(n^r)$ and $c(H(n)) = 2^{o(n^r)}$. We shall below consider the computations of M on input tapes of side-length $H(n)$.

Consider the chunk probabilities $p(v, \sigma, \tau)$ defined above. For each $(H(n), n)$-chunk v in $V(n)$, there are a total of

$$d(n) = c(H(n)) \times |PT(v(\#))| \times (c(H(n)) \times |PT(v(\#))| + 3) = O(n^2\{c(H(n))\}^2)$$

chunk probabilities. Fix some ordering of the pairs (σ, τ) of starting and stopping conditions and let $P(v)$ be the vector of these $d(n)$ probabilities according to this ordering.

We first show that if $v \in V(n)$ and if p is a nonzero element of $P(v)$, then $p \geq 2^{-c(H(n))a(n)}$, where $a(n) = |PS(v(\#))| = O(\{H(n)\}^2)$. Form a Markov chain $K(v)$ with states of the form (s, l), where s is a storage state of M and $l \in PS(v(\#)) \cup PT(v(\#))$. The chain state (s, l) with $l \in PS(v(\#))$ corresponds to M being in storage state s scanning the symbol at position l of $v(\#)$. Transition probabilities from such states are obtained from the transition probabilities of M in the obvious way. For example, if the symbol at position (i, j) of $v(\#)$ is e, and if M in storage state s reading the symbol e can move its input head left and enter storage state s' with probability $\frac{1}{2}$, then the transition probability from state $(s, (i, j))$ to state $(s', (i, j-1))$ is $\frac{1}{2}$. Chain states of the form $(s, \overline{(i, j)})$ with $\overline{(i, j)} \in PT(v(\#))$ are trap states of $K(v)$ and correspond to M just having exited from $v(\#)$ at exit point $\overline{(i, j)}$ of $v(\#)$. Now consider, for example, $p = p(v, \sigma, \tau)$, where $\sigma = (s, \overline{(i, j)})$ and $\tau = (s', \overline{(k, l)})$ with $\overline{(i, j)}, \overline{(k, l)} \in PT(v(\#))$. If $p > 0$, then there must be some paths of nonzero probability in $K(v)$ from $(s, (i, j))$ to $(s', \overline{(k, l)})$, and since $K(v)$ has at most $c(H(n))a(n)$ nontrapping states, the length of the shortest path among such paths is at most $c(H(n))a(n)$. Since $\frac{1}{2}$ is the smallest nonzero transition probability of M, it follows that $p \geq 2^{-c(H(n))a(n)}$. If $\sigma = (s, \overline{(i, j)})$ with $\overline{(i, j)} \in PT(v(\#))$ and $\tau = Loop$, there must be a path Pa of nonzero probability in $K(v)$ from state $(s, (i, j))$ to some state $(s', (i', j'))$ such that there is no path of nonzero probability from $(s', (i', j'))$ to any trap state of the form $(s'', \overline{(k, l)})$ with $\overline{(k, l)} \in PT(v(\#))$. Again, if there is such a path Pa, there is one of length at most $c(H(n))a(n)$. The remaining cases are similar.

Fix an arbitrary $n \in I$. Divide $V(n)$ into M - equivalence classes by making v and v' M-equivalent if $P(v)$ and $P(v')$ are zero in exactly the same coordinates.

Let $E(n)$ be a largest M-equivalence class. Then we have

$$|E(n)| \geq \frac{|V(n)|}{2^{d(n)}}.$$

Let $d'(n)$ be the number of nonzero coordinates of $P(v)$ for $v \in E(n)$. Let $\hat{P}(v)$ be the $d'(n)$-dimensional vector of nonzero coordinates of $P(v)$. Note that $\hat{P}(v) \in [2^{-c(H(n))a(n)}, 1]^{d'(n)}$ for all $v \in E(n)$. Let $\log \hat{P}(v)$ be the componentwise log of $\hat{P}(v)$. Then $\log \hat{P}(v) \in [-c(H(n))a(n), 0]^{d'(n)}$. By dividing each coordinate interval $[-c(H(n))a(n), 0]$ into subintervals of length μ, we divide the space $[-c(H(n))a(n), 0]^{d'(n)}$ into at most $(\frac{c(H(n))a(n)}{\mu})^{d(n)}$ cells, each of size $\mu \times \mu \times \ldots \times \mu$. We want to choose μ large enough, such that the number of cells is smaller than the size of $E(n)$, that is,

$$(\frac{c(H(n))a(n)}{\mu})^{d(n)} < \frac{|V(n)|}{2^{d(n)}} \tag{3.1}$$

Concretely, we choose $\mu = 2^{-n}$. From the assumption on the rate of growth of $|V(n)|$, from the assumption that $H(n)$ is bounded by some exponentiial function in n, and since, by assumption from the contrary, $c(H(n)) = 2^{o(n^r)}$, it follows that (3.1) holds for $\mu = 2^{-n}$ with large $n \in I$. Assuming (3.1), there must be two different $v, v' \in E(n)$ such that $\log \hat{P}(v)$ and $\log \hat{P}(v')$ belong to the same cell. Therefore, if p and p' are two nonzero probabilities in the same coordinate of $P(v)$ and $P(v')$, respectively, then $|\log p - \log p'| \leq \mu$. It follows that p and p' are 2^μ-close. Therefore $P(v)$ and $P(v')$ are componentwise 2^μ-close.

For these v and v', let $u \in \sum^{(1,n)}$ be an n-chunk in Assumption (2) in th statement in the theorem. We describe two Markov chains R and R', which model the computations of M on $v[u]$ and $v'[u]$, respectively. The state space of R is

$$C(H(n)) \times (PT(v(\#)) \cup PT(u(\#))) \cup \{\text{Accept,Reject,Loop}\}.$$

Thus the number of states of R is

$$z = c(H(n))(n + 3 + n + 3) + 3 = 2c(H(n))(n + 3) + 3.$$

The state $(s, \overline{(i,j)}) \in C(H(n)) \times PT(v(\#))$ of R corresponds to M just having entered $v(\#)$ in storage state s from entrance point $\overline{(i,j)}$ of $v(\#)$, and the state $(s', \overline{(k,l)}') \in C(H(n)) \times PT(u(\#))$ of R corresponds to M just having entered $u(\#)$ in storage state s' from entrance point $\overline{(k,l)}'$ of $u(\#)$. For convenience sake, we assume that M begins to read any input tape x in the initial storage state $s_0 = (q_0, \lambda, 1)$, where q_0 is the initial state of M, by entering $x(1,1)$ from the lower edge of the cell on which $x(1,1)$ is written. Thus, the starting state of R is Initial $\overset{\triangle}{=} (s_0, \overline{(2,1)}')$. The states Accept and Reject correspond to the computations halting in the accepting state and the rejecting state, respectively, and Loop means that M has entered an infinite loop. The transition probabilities of R are obtained from the chunk probabilities of M on $u(\#)$ and $v(\#)$. For example, the transition probabilities from $(s, \overline{(i,j)})$ to $(s', \overline{(k,l)}')$ with $\overline{(i,j)} \in PT(v(\#))$ and $\overline{(k,l)}' \in PT(u(\#))$ is just $p(v, (s, \overline{(i,j)}), (s', \overline{(k,l)}))$, the transition probability from $(s', \overline{(k,l)}')$ to $(s, \overline{(i,j)})$ with $\overline{(i,j)} \in PT(v(\#))$ and $\overline{(k,l)}' \in PT(u(\#))$ is $p(u, (s', \overline{(k,l)}'), (s, \overline{(i,j)}'))$, the transition probability from $(s, \overline{(i,j)})$ to Accept is $p(v, (s, \overline{(i,j)}), \text{Accept})$, and the transition probability from $(s', \overline{(k,l)}')$ to Accept is $p(u, (s', \overline{(k,l)}'), \text{Accept})$. The states Accept, Reject and Loop are trap states. The chain R' is defined similarly, but using $v'[u]$ in place of $v[u]$.

Suppose that $v[u] \in A$ and $v'[u] \in B$, the other case being symmetric. Let $acc(v[u])$ (resp.,$acc(v'[u])$) be the probability that M accepts input $v[u]$(resp., $v'[u]$). Then, $acc(v[u])$ (resp.,$acc(v'[u])$) is exactly the probability that the Markov chain R(resp.,R') is trapped in state Accept when started in state Initial. Now $v[u] \in A$ implies $acc(v[u]) \geq 1 - \varepsilon$. Since R and R' are 2^μ-close, Lemma 3.1 implies that

$$\frac{acc(v'[u])}{acc(v[u])} \geq 2^{-2\mu z}.$$

$2^{-2\mu z}$ approaches 1 as n increases. Therefore, for Large $n \in I$, we have

$$acc(v'[u]) \geq 2^{-2\mu z}(1 - \varepsilon) > \frac{1}{2},$$

because $\varepsilon < \frac{1}{2}$. But since $v'[u] \in B$, this contradicts the assumption that M separates A and B. $\square$

As an application of Theorem 3.1, we can get the following corollary:

Corollary 3.1 *Let*
$T_1 = \{x \in \{0,1,2\}^{(2)}|\exists n \geq 1[l_1(x) = l_2(x) = 2^n \wedge x[(1,1),(2^n,n)] \in \{0,1\}^{(2)} \wedge \exists i(2 \leq i \leq n)[x[(1,1),(1,n)] = x[(i,1),(i,n)]] \wedge x[(1,n+1),(2^n,2^n)] \in \{2\}^{(2)}]\}.$
Then, $T_1 \notin 2 - PTM^s(o(\log n))$.

Proof. By using Theorem 3.1 in the previous section, we below show that $T_1 \notin 2 - PTM^s(L(n))$ for $L(n) = o(\log n)$.
For any integer $n \geq 1$, let

$$U(n) = \{ch_{(2^n,n)}(v_1,v_2)|v_1 \in \{0,1\}^{(2^n-1,n)} \wedge v_2 \in \{2\}^{(2^n,2^n-n)}\}.$$

For each $v = ch_{(2^n,n)}(v_1,v_2) \in U(n)$, let

$$contents(v) = \{w \in \{0,1\}^{(1,n)}|w = v_1[(i,1),(i,n)] \text{ for some } i \ (1 \leq i \leq 2^n - 1)\}.$$

Divide $U(n)$ into contents-equivalence classes by making v and v' *contents - equivalent* if $contents(v) = contents(v')$. There are

$$contents(n) = \binom{2^n}{1} + \binom{2^n}{2} + ... + \binom{2^n}{2^n - 1} = 2^{2^{\Omega(n)}}$$

contents - equivalence classes of the $(2^n,n)$-chunks in $U(n)$. We denote by $V(n)$ the set of all the representatives arbitrarily chosen from these $contents(n)$ contents-equivalence classes. Thus, for each $n \geq 1$, $|V(n)| = contents(n)$. Let I be an infinite set of positive integers n such that $contents(n) \geq 2^{c^n}$, where $c > 1$ is a constant. (One can easily see that there exists such a constant c.) Clearly, for each $n \in I$, $|V(n)| \geq 2^{c^n}$. It is easily seen that for every $n \in I$ and every $v, v' \in V(n)$ with $v \neq v'$, there is an n-chunk $u \in \{0,1\}^{(1,n)}$ such that

$$\text{either } \begin{cases} v[u] \in T_1 \\ v'[u] \in \overline{T_1} \end{cases} \text{ or } \begin{cases} v[u] \in \overline{T_1} \\ v'[u] \in T_1 \end{cases},$$

where for any language T, $\overline{T}$ denotes the complement of T.
Further, for each $n \in I$, $V(n)$ is a set of $(H(n),n)$-chunks, where $H(n) = 2^n$, which is bounded by some exponential in n. Thus, by Theorem 3.1, if an $L(n)$ space-bounded 2-ptm^s with error probability $\varepsilon < \frac{1}{2}$ recognizes T_1, then $L(H(n))$ can not be $o(n)$, and thus $L(n)$ can not be $o(\log n)$. This completes the proof of '$T_1 \notin 2 - PTM^s(o(\log n))$'.   □

4   2-ptm^s versus 2-stm^s

This section shows that for any $loglog\ n \leq L(n) = o(\log n)$, 2-PTM$^s(L(n)) = AM_1$2-STM$^s(L(n)) \subsetneq MA_2$2-STM$^s(L(n)))$.

Theorem 4.1 $MA_2$2-STM$^s(loglog\ n) - 2$-PTM$^s(o(\log n)) \neq \phi$.

Proof. Let T_1 be the set described in Corollary 3.1. As shown in Corollary 3.1, $T_1 \notin$ 2-PTM$^s(o(log\ n))$. We below show that $T_1 \in MA_2$2-STM$^s(loglog\ n)$.

The set T_1 is recognized by a *loglog n* space-bounded 2-stm^s M which acts as follows. Suppose that an input tape x with $l_1(x) = l_2(x) = n(n \geq 1)$ is presented to M. M first checks whether $n = 2^d$ for some integer $d > 1$. If $n \neq 2^d$ for some $d > 1$, then M rejects x. Otherwise, M marks off *loglog n* $= log\ d$ cells of the storage tape. (This is possible because *loglog n* is 2-dimensionally fully space contractible [12].)

Using these $log\ d$ cells of the storage tape, M checks whether x is well-formed, i.e., whether (i) $x[(1,1),(n,d)] \in \{0,1\}^{(2)}$, and (ii) $x[(1,d+1),(n,n)] \in \{2\}^{(2)}$. If x is not well-formed, then M rejects x. If x is well-formed, then M performs the following algorithm (this algorithm is based on the algorithm from Section 4 in [7]):

1. Nondeterministically choose some i $(2 \leq i \leq n)$, and move the input head to the segment $x[(i,1),(i,d)]$ (denoted by w_i), and interpret w_i and u as natural numbers between 0 and $2^d - 1$, where $u = x[(1,1),(1,d)]$.
2. Randomly choose a prime q with $2 \leq q \leq d^2$, compute $r_i := num(w_i)$ mod q (where $num(w_i)$ denotes an integer with binary representation w_i), and store both q and r_i in the $log\ d$ cells of the storage tape.
3. If $r_i = num(u)$ mod q, then accept x, and otherwise reject x.

If the segments w_i and u are equal, then of course $num(w_i) = num(u)$ mod q, and M accepts correctly in step 3. If $w_i \neq u$, then it could happen that $num(w_i) = num(u)$ mod q and M reaches in step 3 an accepting state, which is wrong. However, this event happens with probability that tends to 0. Indeed, since $|num(w_i) - num(u)| \leq 2^d$, it follows that $num(w_i) - num(u)$ has at most d different prime divisors. On the other hand, since the number of primes less than or equal to k is $\theta(\frac{k}{logk})$ for any positive integer k (see [6, Theorem 7]), it follows that M chooses a prime from about $d^2/2log\ d$ different primes at the beginning of step 2. So the probability that a wrong value q is chosen is at most $d/(d^2/2log\ d) = (2log\ d)/d$, which tends to 0 for large d. Obviously, M uses $O(loglog\ n)$ space, starts in a nondeterministic state, and makes only one alternation between nondeterministic and probabilistic configurations.   □

From Theorem 4.1, we have:

Corollary 4.1 *For any loglog $n \leq L(n) = o(log\ n)$,*

$$2\text{-}PTM^s(L(n)) \subsetneq MA_2\text{2-}STM^s(L(n)).$$

5   Conclusion

This section concludes this paper by posing the following open problems:

(1) For any $L(n) = o(log\ n)$ and for any positive integer $k \geq 1$,
   - MA_k2-STM$^s(L(n)) \subsetneq MA_{k+1}$2-STM$^s(L(n))$?
   - AM_k2-STM$^s(L(n)) \subsetneq AM_{k+1}$2-STM$^s(L(n))$?

(2) 2-PTM^s($o(log\ n)$) = $AM_2$2-STM^s($o(log\ n)$)?
(3) Is there a set that separates 2-PTM^s($constant$) from AM_k2-STM^s($constant$) for some $k > 1$?

References

1. A.Condon, L.Hellerstein, S.Pottle and A.Wigderson, "On the power of finite automata with both nondeterministic and probabilistic states", SIAM J.COMPUT., vol.27, no.3, pp.739-762 (1998).
2. C.Dwork and L.Stockmeyer, "Finite state verifier I:The power of interaction", J.ACM, vol.39, no.4, pp.800-828 (1992).
3. R.Freivalds, "Probabilistic two-way machines", In proceedings of the International Symposium on Mathematical Foundations of Computer Science. LNCS 118, pp.33-45 (1981).
4. R.Freivalds and M.karpinski, "Lower space bounds for randomized computation", ICALP'44, LNCS 820, pp.580-592 (1994).
5. J.Gill, "Computational complexity of probabilistic Turing machine", SIAM J.CO MPUT., vol.6, no.4, pp.675-695 (1977).
6. G.Hardy and E.Wright, "An Introduction to the Theory of Numbers', 5th ed. Oxford University Press (1979).
7. M.Liskiewicz, "Interactive proof systems with public coin : Lower space bounds and hierarchies of complexity classes", Proceedings of STACS797, LNCS 1200, pp.129-140 (1997).
8. I.I.Macarie and M.Ogihara, "Properties of probabilistic pushdown automata", Theoretical Computer Science, vol.207, pp.117-130 (1998).
9. T.Okazaki, K.Inoue, A.Ito and Y.Wang, "A note on two-dimensional probabilistic Turing machines", Information Sciences, vol.113, pp.205-220 (1999).
10. T.Okazaki, L.Zhang, K.Inoue, A.Ito and Y.Wang, "A note on two-dimensional probabilistic finite automata", Information Sciences, vol.110, pp.303-314 (1999).
11. E.Seneta, "Non-Negative Matrices and Markov Chains", 2nd Ed. Springer-Verlag, New York (1981).
12. A.Szepietowski, "Turing Machines with Sublogarithmic Space", Lecture Notes in Compputer Science 843 (1994).

Undecidability of Weak Bisimilarity
for PA-Processes

Jiří Srba[*]

BRICS[**]
Department of Computer Science
University of Aarhus
Ny Munkegade bld. 540
8000 Aarhus C, Denmark
srba@brics.dk

Abstract. We prove that the problem whether two PA-processes are weakly bisimilar is undecidable. We combine several proof techniques to provide a reduction from Post's correspondence problem to our problem: existential quantification technique, masking technique and deadlock elimination technique.

1 Introduction

The increasing interest in formal verification of concurrent systems has heightened the need for a systematic study of such systems. Of particular interest is the study of equivalence and model checking problems for classes of *infinite-state processes* [2]. To explore the decidability borders of automatic verification and to analyze in detail the situations where such verification is possible, is one of the main goals for theoretical research in this area. The primary focus of this paper is on equivalence checking problems, considering *bisimilarity* as the notion of behavioural equivalence.

The positive development in strong bisimilarity checking for many classes of infinite-state systems had led to the hope that extending the existing techniques to the case of weak bisimilarity might be a feasible step. Some of the recent results, however, contradict this hope. Opposed to the fact that strong bisimilarity is decidable between Petri nets (PN) and finite state systems [7], Jančar and Esparza proved in [6] that weak bisimilarity is undecidable. Similarly, strong bisimilarity is decidable for pushdown processes (PDA) [11], whereas weak bisimilarity is not [14]. Strong bisimilarity of Petri nets is undecidable [5], however, it is at the first level of the arithmetical hierarchy (Π_1^0-complete). On the other hand, weak bisimilarity of Petri nets lies beyond the arithmetical hierarchy [4].

In this paper we further confirm the inherent complexity of weak bisimilarity by showing its undecidability for *PA-processes*. PA (process algebra introduced

---

[*] The author is supported in part by the GACR, grant No. 201/00/0400.
[**] **B**asic **R**esearch in **C**omputer **S**cience,
Centre of the Danish National Research Foundation.

by Baeten and Weijland [1]) is a formalism which combines the parallel and sequential operator but allows neither communication nor global-state control. This makes the proof more difficult than for PDA [14] and PN [5]: the undecidability argument for PDA uses a finite-state control unit and the proof for PN relies on the possibility of communication.

Our proof of undecidability of weak bisimilarity for PA is by reduction from Post's Correspondence Problem (PCP). For a given instance of PCP we construct a pair of PA processes that are weakly bisimilar if and only if the PCP instance has a solution. We use a game-theoretic characterization of weak bisimilarity (our players are called 'attacker' and 'defender') and combine several techniques to achieve our result.

- The first technique (we call it here *existential quantification*) was first used by Jančar in [4] and explicitly formulated by Srba in [13]. It makes use of the fact that the defender in the bisimulation game has a strategy to decide on a continuation of the game in case of nondeterministic branching. This enables to encode existential quantification. In our case of weak bisimilarity it moreover provides a technique for generating arbitrarily long sequences of process constants (representing solutions of a given PCP instance).
- The second technique, used by Mayr in [8] and called the *masking technique*, deals with the following phenomenon. Assume that X is an unnormed process constant that performs an action 'a' and becomes X again. Whenever X is added via parallel composition to any process expression γ, it is capable of masking every possible occurrence of the action 'a' in γ.
- Finally, we adapt the technique of *deadlock elimination* from [12] into our context, in order to make the proofs more transparent.

Many of the infinite-state systems studied so far can be uniformly defined by means of process rewrite systems (PRS) — see Figure 1 for the PRS-hierarchy from [9]. As a result of our contribution, we can now assert that weak bisimilarity is undecidable for all systems on the third level of the PRS-hierarchy, i.e., for pushdown processes (PDA), PA-processes (PA) and Petri nets (PN).

On the other hand, the questions for the systems on the second level, namely for basic process algebra (BPA) and basic parallel processes (BPP), still remain open. The techniques used for undecidability of weak bisimilarity for PDA, PA and PN do not seem to be applicable to BPA and BPP, as these systems on the second level lack the ability of remembering global information and they do not allow to mix the sequential and parallel operator. Moreover, we think that weak bisimilarity of BPA and BPP is likely to be decidable.

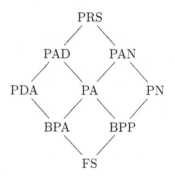

Fig. 1. PRS-hierarchy

2   Basic Definitions

Let *Const* be a set of *process constants*. The class of *process expressions* over *Const* is given by $E ::= \epsilon \mid X \mid E.E \mid E\|E$ where 'ϵ' is the *empty process*, X ranges over *Const*, '.' is the operator of *sequential composition*, and '$\|$' stands for a *parallel composition*. We do not distinguish between process expressions related by a *structural congruence*, which is the smallest congruence over process expressions such that '.' is associative, '$\|$' is associative and commutative, and 'ϵ' is a unit for '.' and '$\|$'. We shall adopt the convention that the sequential operator binds tighter than the parallel one. Thus for example $X.Y\|Z$ means $(X.Y)\|Z$.

Let *Act* be a set of *actions* such that *Act* contains a distinguished *silent* action τ. We call the elements of the set $Act \setminus \{\tau\}$ *visible* actions. A *PA process rewrite system* $((1, G)$-PRS in the terminology of [9]) is a finite set Δ of *rules* of the form $X \xrightarrow{a} E$, where $X \in Const$, $a \in Act$ and E is a process expression. Let us denote the set of actions and process constants that appear in Δ as $Act(\Delta)$ resp. $Const(\Delta)$ (note that these sets are finite).

A PA system Δ determines a *labelled transition system* where the states are process expressions over $Const(\Delta)$, and $Act(\Delta)$ is the set of labels. The *transition relation* $\longrightarrow$ is the least relation satisfying the following SOS rules (recall that '$\|$' is commutative).

$$\frac{(X \xrightarrow{a} E) \in \Delta}{X \xrightarrow{a} E} \qquad \frac{E \xrightarrow{a} E'}{E.F \xrightarrow{a} E'.F} \qquad \frac{E \xrightarrow{a} E'}{E\|F \xrightarrow{a} E'\|F}$$

As usual we extend the transition relation to the elements of Act^*. We write $E \longrightarrow^* E'$ whenever $E \xrightarrow{w} E'$ for some $w \in Act^*$ and say that E' is *reachable from* E. The notation $E \longrightarrow E'$ means that there is an $a \in Act$ such that $E \xrightarrow{a} E'$. We also write $E \xrightarrow{a}\!\!\!\!\!/\;$ if there is no E' such that $E \xrightarrow{a} E'$, and $E \not\longrightarrow$ if $E \xrightarrow{a}\!\!\!\!\!/\;$ for all $a \in Act$. By $|w|$ we denote the length of w for $w \in Act^*$, and we use $|S|$ to stand for the cardinality of a set S.

A process constant $X \in Const(\Delta)$ is called a *deadlock* iff $X \not\longrightarrow$. In the usual presentation of PA it is often assumed that Δ contains no deadlocks.

A *PA process* is a pair (P, Δ) where Δ is a PA process rewrite system and P is a process expression over $Const(\Delta)$.

Let $E \xrightarrow{\tau^*} E'$ mean that $E \xrightarrow{\tau^n} E'$ for some $n \geq 0$. A *weak transition relation* is defined as follows: $\overset{a}{\Longrightarrow} \overset{\text{def}}{=} \xrightarrow{\tau^*} \circ \xrightarrow{a} \circ \xrightarrow{\tau^*}$ if $a \in Act \setminus \{\tau\}$, and $\overset{a}{\Longrightarrow} \overset{\text{def}}{=} \xrightarrow{\tau^*}$ if $a = \tau$. As before we extend the weak transition relation to the elements of Act^* and write $E \overset{a}{\not\Longrightarrow}$ whenever there is no E' such that $E \overset{a}{\Longrightarrow} E'$.

Now we introduce the concept of *weak bisimilarity*. Let Δ be a fixed PA system. A binary relation R over process expressions is a *weak bisimulation* iff whenever $(E, F) \in R$ then for each $a \in Act(\Delta)$: if $E \xrightarrow{a} E'$ then $F \overset{a}{\Longrightarrow} F'$ for some F' such that $(E', F') \in R$; if $F \xrightarrow{a} F'$ then $E \overset{a}{\Longrightarrow} E'$ for some E' such that $(E', F') \in R$. Processes (P_1, Δ) and (P_2, Δ) are *weakly bisimilar*, and we write $(P_1, \Delta) \approx (P_2, \Delta)$, iff there is a weak bisimulation R such that $(P_1, P_2) \in R$. If Δ is clear from the context we write only $P_1 \approx P_2$.

Bisimulation equivalence has an elegant characterisation in terms of *bisimulation games* [16,15]. A bisimulation game on a pair of processes (P_1, Δ) and (P_2, Δ) is a two-player game between an 'attacker' and a 'defender'. The game is played in rounds. In each round the attacker chooses one of the processes and makes an $\xrightarrow{a}$-move for some $a \in Act(\Delta)$. The defender must respond by making an $\xRightarrow{a}$-move in the other process under the same action a. Now the game repeats, starting from the new processes. If one player cannot move, the other player wins. If the game is infinite, the defender wins. The processes (P_1, Δ) and (P_2, Δ) are weakly bisimilar iff the defender has a winning strategy (and nonbisimilar iff the attacker has a winning strategy).

The following proposition will be useful later and it simply rephrases a standard result that weak bisimilarity is a congruence w.r.t. to the parallel operator.

Proposition 1. *If the defender has a winning strategy from a pair E and F then he also has a winning strategy from $E \| \gamma$ and $F \| \gamma$ for any process expression γ.*

3   Undecidability of Weak Bisimilarity

We show that the problem whether $(P_1, \Delta) \approx (P_2, \Delta)$ for a given pair of PA processes (P_1, Δ) and (P_2, Δ) is undecidable. For technical convenience we use the power of deadlocks to achieve this result, however, at the end of this section we discuss a simple technique for deadlock elimination. Thus the undecidability result is valid even for PA without deadlocks.

Let us first define Post's correspondence problem (PCP): given a nonempty alphabet Σ and two lists $A = [u_1, \ldots, u_n]$ and $B = [v_1, \ldots, v_n]$ where $n > 0$ and $u_k, v_k \in \Sigma^+$ for all k, $1 \leq k \leq n$, the question is to decide whether the (A, B)-instance has a solution, i.e., whether there is an integer $m \geq 1$ and a sequence of indices $i_1, \ldots, i_m \in \{1, \ldots, n\}$ such that $u_{i_1} u_{i_2} \ldots u_{i_m} = v_{i_1} v_{i_2} \ldots v_{i_m}$.

According to the classical result due to Post, this problem is undecidable [10]. Let us consider an (A, B)-instance of PCP where

$$A = [u_1, \ldots, u_n] \quad \text{and} \quad B = [v_1, \ldots, v_n].$$

We construct a PA system Δ and a pair of processes (P_1, Δ) and (P_2, Δ) such that the (A, B)-instance has a solution if and only if $(P_1, \Delta) \approx (P_2, \Delta)$.

Let $\mathcal{SF}(\alpha)$ denote the set of all suffixes of a sequence $\alpha \in \Sigma^*$, i.e., $\mathcal{SF}(\alpha) \overset{\text{def}}{=} \{\alpha' \in \Sigma^* \mid \exists \alpha'' \in \Sigma^* \text{ such that } \alpha = \alpha'' \alpha'\}$. Note that $\epsilon \in \mathcal{SF}(\alpha)$ for any α. We can now define the set of process constants $Const(\Delta)$ and actions $Act(\Delta)$ by

$$Const(\Delta) \overset{\text{def}}{=} \{U^{u_k} \mid 1 \leq k \leq n\} \cup \{V^{v_k} \mid 1 \leq k \leq n\} \cup$$
$$\{T^w \mid w \in \bigcup_{k=1}^{n} \mathcal{SF}(u_k) \cup \bigcup_{k=1}^{n} \mathcal{SF}(v_k)\} \cup$$
$$\{X, X', X_1', Y, Y', Y_1, Z, C, C_1, C_2, W, D\}$$

$$Act(\Delta) \overset{\text{def}}{=} \{a \mid a \in \Sigma\} \cup \{\iota_k \mid 1 \leq k \leq n\} \cup$$
$$\{x, y, z, c_1, c_2, \tau\}.$$

Remark 1. In what follows D will be a distinguished process constant with no rules associated to it (deadlock). Hence in particular $\alpha.D.\beta \parallel \gamma \approx \alpha \parallel \gamma$ for any process expressions α, β and γ.

To make the rewrite rules introduced in this section more understandable, we define the system Δ in four stages. It is important to remark here that whenever we define the rules for some process constant $Q \in Const(\Delta)$, we always give all the rules for Q at the same stage. Our ultimate goal is to show that $(X\|C, \Delta) \approx (X'\|C, \Delta)$ if and only if the given (A, B)-instance of PCP has a solution. The first part of the system Δ is given by the following rules:

$$
\begin{array}{lll}
U^{u_k} \xrightarrow{\iota_k} \epsilon & U^{u_k} \xrightarrow{\tau} T^{u_k} & \text{for all } k \in \{1, \ldots, n\} \\
V^{v_k} \xrightarrow{\iota_k} \epsilon & V^{v_k} \xrightarrow{\tau} T^{v_k} & \text{for all } k \in \{1, \ldots, n\}
\end{array}
$$

$$
T^{aw} \xrightarrow{a} T^w \qquad T^{aw} \xrightarrow{\tau} T^w \qquad \text{for all } a \in \Sigma \text{ and } w \in \Sigma^* \text{ such that}
$$
$$
aw \in \bigcup_{k=1}^{n} \mathit{SF}(u_k) \cup \bigcup_{k=1}^{n} \mathit{SF}(v_k)
$$

$$
T^\epsilon \xrightarrow{\tau} \epsilon.
$$

This means that for a given $k \in \{1, \ldots, n\}$ the process constants U^{u_k} and V^{v_k} can perform e.g. the following transitions (or transition sequences): $U^{u_k} \xrightarrow{\iota_k} \epsilon$, $V^{v_k} \xrightarrow{\iota_k} \epsilon$, $U^{u_k} \xRightarrow{u_k} \epsilon$, $V^{v_k} \xRightarrow{v_k} \epsilon$, $U^{u_k} \xRightarrow{\tau} \epsilon$, $V^{v_k} \xRightarrow{\tau} \epsilon$. The intuition is that a solution $i_1, \ldots, i_m \in \{1, \ldots, n\}$ of the (A, B)-instance is represented by a pair of processes $U^{u_{i_1}}.U^{u_{i_2}}.\cdots.U^{u_{i_m}}$ and $V^{v_{i_1}}.V^{v_{i_2}}.\cdots.V^{v_{i_m}}$. These processes can perform the sequences of visible actions $u_{i_1} u_{i_2} \ldots u_{i_m}$ and $v_{i_1} v_{i_2} \ldots v_{i_m}$, respectively, or they can perform the actions corresponding to the indices, namely $\iota_{i_1} \iota_{i_2} \ldots \iota_{i_m}$. Moreover, since there is no global state control, the processes can produce also a combination of the actions from Σ and $\{\iota_1, \ldots, \iota_n\}$. In order to avoid this undesirable behaviour, we add (via parallel composition) a process constant C_1 or C_2 such that C_1 masks all the actions from Σ and C_2 masks all the actions testing the indices. The reason for adding Z will become clear later.

The rewrite rules for C_1, C_2 and Z are given by:

$$
C_1 \xrightarrow{a} C_1 \qquad \text{for all } a \in \Sigma
$$

$$
C_2 \xrightarrow{\iota_k} C_2 \qquad \text{for all } k \in \{1, \ldots, n\}
$$

$$
Z \xrightarrow{z} \epsilon \qquad Z \xrightarrow{\tau} D.
$$

Lemma 1. *It holds that*

$$
Z.U^{u_{i_1}}.U^{u_{i_2}}.\cdots.U^{u_{i_m}} \parallel C_1 \approx Z.V^{v_{j_1}}.V^{v_{j_2}}.\cdots.V^{v_{j_{m'}}} \parallel C_1
$$

if and only if

$$
m = m' \text{ and } i_\ell = j_\ell \text{ for all } \ell, 1 \le \ell \le m = m'.
$$

Proof. "⇒": Assume that (i) $m \neq m'$, or (ii) $m = m'$ and let ℓ, $1 \leq \ell \leq m = m'$, be the smallest number such that $i_\ell \neq j_\ell$. It is easy to show that $Z.U^{u_{i_1}}.U^{u_{i_2}}.\cdots.U^{u_{i_m}} \parallel C_1 \not\approx Z.V^{v_{j_1}}.V^{v_{j_2}}.\cdots.V^{v_{j_{m'}}} \parallel C_1$. In case (i), assuming w.l.o.g. that $m > m'$, the attacker can perform in the first process a sequence of actions $z\iota_{i_1}\iota_{i_2}\ldots\iota_{i_m}$ of length $m+1$ and the defender cannot answer by any corresponding sequence of the same length from the second process ($m > m'$). Hence the attacker wins. In case (ii), the attacker again performs the sequence $z\iota_{i_1}\iota_{i_2}\ldots\iota_{i_m}$ in the first process. The only appropriate sequence of the same length that the defender can perform in the second process is $z\iota_{j_1}\iota_{j_2}\ldots\iota_{j_{m'}}$ — obviously no τ rules can be used otherwise the defender loses (his sequence gets shorter). The attacker wins because $\iota_{i_\ell} \neq \iota_{j_\ell}$.

"⇐": We show that $Z.U^{u_{i_1}}.U^{u_{i_2}}.\cdots.U^{u_{i_m}} \parallel C_1 \approx Z.V^{v_{i_1}}.V^{v_{i_2}}.\cdots.V^{v_{i_m}} \parallel C_1$. Let $U(\ell) \stackrel{\text{def}}{=} U^{u_{i_\ell}}.U^{u_{i_{\ell+1}}}.\cdots.U^{u_{i_m}}$ and $V(\ell) \stackrel{\text{def}}{=} V^{v_{i_\ell}}.V^{v_{i_{\ell+1}}}.\cdots.V^{v_{i_m}}$ for all ℓ, $1 \leq \ell \leq m$. By definition $U(m+1) \stackrel{\text{def}}{=} \epsilon$ and $V(m+1) \stackrel{\text{def}}{=} \epsilon$. Let us consider the following relation R.

$$
\begin{aligned}
&\{ (\quad Z.U(1)\|C_1 \,,\, Z.V(1)\|C_1 \quad) \} \cup \\
&\{ (\quad D.U(1)\|C_1 \,,\, D.V(1)\|C_1 \quad) \} \cup \\
&\{ (\quad U(\ell)\|C_1 \,,\, V(\ell)\|C_1 \quad) \mid 1 \leq \ell \leq m+1 \} \cup \\
&\{ (\, T^w.U(\ell)\|C_1 \,,\, V(\ell)\|C_1 \quad) \mid 2 \leq \ell \leq m+1 \,\wedge\, w \in \mathcal{SF}(u_{\ell-1}) \} \cup \\
&\{ (\quad U(\ell)\|C_1 \,,\, T^w.V(\ell)\|C_1 \,) \mid 2 \leq \ell \leq m+1 \,\wedge\, w \in \mathcal{SF}(v_{\ell-1}) \}
\end{aligned}
$$

It is a routine exercise to check that R is a weak bisimulation. Moreover, it satisfies that $(Z.U^{u_{i_1}}.U^{u_{i_2}}.\cdots.U^{u_{i_m}} \parallel C_1, Z.V^{v_{i_1}}.V^{v_{i_2}}.\cdots.V^{v_{i_m}} \parallel C_1) \in R$. □

Lemma 2. *It holds that*

$$Z.U^{u_{i_1}}.U^{u_{i_2}}.\cdots.U^{u_{i_m}} \parallel C_2 \approx Z.V^{v_{j_1}}.V^{v_{j_2}}.\cdots.V^{v_{j_{m'}}} \parallel C_2$$

if and only if

$$u_{i_1}u_{i_2}\ldots u_{i_m} = v_{j_1}v_{j_2}\ldots v_{j_{m'}}.$$

Proof. "⇒": Let $\sigma \stackrel{\text{def}}{=} U^{u_{i_1}}.U^{u_{i_2}}.\cdots.U^{u_{i_m}}$ and $\omega \stackrel{\text{def}}{=} V^{v_{j_1}}.V^{v_{j_2}}.\cdots.V^{v_{j_{m'}}}$, and let $u \stackrel{\text{def}}{=} u_{i_1}u_{i_2}\ldots u_{i_m}$ and $v \stackrel{\text{def}}{=} v_{j_1}v_{j_2}\ldots v_{j_{m'}}$. Hence $\sigma \stackrel{u}{\Longrightarrow} \epsilon$ and $\omega \stackrel{v}{\Longrightarrow} \epsilon$, and u and v are the longest sequences (and unique ones among the sequences of the length $|u|$ resp. $|v|$) of visible actions from Σ that σ and ω can perform. From the assumption that $u \neq v$ it is easy to see that $Z.\sigma\|C_2 \not\approx Z.\omega\|C_2$.

"⇐": We show that $Z.U^{u_{i_1}}.U^{u_{i_2}}.\cdots.U^{u_{i_m}} \parallel C_2 \approx Z.V^{v_{j_1}}.V^{v_{j_2}}.\cdots.V^{v_{j_{m'}}} \parallel C_2$ assuming that $u_{i_1}u_{i_2}\ldots u_{i_m} = v_{j_1}v_{j_2}\ldots v_{j_{m'}}$. Let $\alpha \in \mathcal{SF}(u_{i_1}u_{i_2}\ldots u_{i_m}) = \mathcal{SF}(v_{j_1}v_{j_2}\ldots v_{j_{m'}})$. We define two sets $U(\alpha)$ and $V(\alpha)$. The intuition is that $U(\alpha)$ contains all the states reachable from $U^{u_{i_1}}.\cdots.U^{u_{i_m}}$ such that α is the longest sequence of visible actions from Σ that these states can perform, and similarly for $V(\alpha)$.

Let us fix the following notation: $U^{u_{m+1}}.\cdots.U^{u_m} \stackrel{\text{def}}{=} \epsilon$, $V^{v_{m'+1}}.\cdots.V^{v_{m'}} \stackrel{\text{def}}{=} \epsilon$ (here 'ϵ' stands for the empty process), and $u_{m+1}\ldots u_m \stackrel{\text{def}}{=} \epsilon$, $v_{m'+1}\ldots v_{m'} \stackrel{\text{def}}{=} \epsilon$ (here 'ϵ' means the empty sequence of actions).

$$U(\alpha) \overset{\text{def}}{=} \{U^{u_{i_\ell}}.U^{u_{i_{\ell+1}}}.\cdots.U^{u_{i_m}} \mid 1 \le \ell \le m \ \wedge \ u_{i_\ell}u_{i_{\ell+1}}\cdots u_{i_m} = \alpha\} \ \cup$$
$$\{T^w.U^{u_{i_\ell}}.U^{u_{i_{\ell+1}}}.\cdots.U^{u_{i_m}} \mid 2 \le \ell \le m+1 \ \wedge \ w \in \mathcal{SF}(u_{i_{\ell-1}}) \ \wedge$$
$$wu_{i_\ell}u_{i_{\ell+1}}\cdots u_{i_m} = \alpha\}$$

$$V(\alpha) \overset{\text{def}}{=} \{V^{v_{j_\ell}}.V^{v_{j_{\ell+1}}}.\cdots.V^{v_{j_{m'}}} \mid 1 \le \ell \le m' \ \wedge \ v_{j_\ell}v_{j_{\ell+1}}\cdots v_{j_{m'}} = \alpha\} \ \cup$$
$$\{T^w.V^{v_{j_\ell}}.V^{v_{j_{\ell+1}}}.\cdots.V^{v_{j_{m'}}} \mid 2 \le \ell \le m'+1 \ \wedge \ w \in \mathcal{SF}(v_{j_{\ell-1}}) \ \wedge$$
$$wv_{j_\ell}v_{j_{\ell+1}}\cdots v_{j_{m'}} = \alpha\}.$$

We remind the reader of the fact that $1 \le |U(\alpha)|, |V(\alpha)| \le 3$ for all α. For example if $m \ge 2$ then $U(u_{i_m}) = \{U^{u_{i_m}}, T^\epsilon.U^{u_{i_m}}, T^{u_{i_m}}\}$. Moreover, if $E \in U(\alpha)$ and $F \in V(\alpha)$ then $E \overset{\alpha}{\Longrightarrow} \epsilon$ and $F \overset{\alpha}{\Longrightarrow} \epsilon$, and α is the longest sequence of actions from Σ satisfying this property. Let us consider the following relation R where $U(1) \overset{\text{def}}{=} U^{u_{i_1}}.U^{u_{i_2}}.\cdots.U^{u_{i_m}}$, $V(1) \overset{\text{def}}{=} V^{v_{j_1}}.V^{v_{j_2}}.\cdots.V^{v_{j_{m'}}}$, and $\beta \overset{\text{def}}{=} u_{i_1}u_{i_2}\cdots u_{i_m} = v_{j_1}v_{j_2}\cdots v_{j_{m'}}$.

$$\{ \ (\ Z.U(1)\|C_2 \ , \ Z.V(1)\|C_2 \ ) \ \} \ \cup$$
$$\{ \ (\ D.U(1)\|C_2 \ , \ D.V(1)\|C_2 \ ) \ \} \ \cup$$
$$\{ \ (\qquad E\|C_2 \ , \ F\|C_2 \qquad) \mid E \in U(\alpha) \ \wedge \ F \in V(\alpha) \ \wedge \ \alpha \in \mathcal{SF}(\beta)\} \ \cup$$
$$\{ \ (\qquad\quad C_2 \ , \ C_2 \qquad\quad) \ \}$$

As in the previous lemma, it is easy to check that R is a weak bisimulation. Moreover $(Z.U^{u_{i_1}}.U^{u_{i_2}}.\cdots.U^{u_{i_m}} \ \| \ C_2, \ Z.V^{v_{j_1}}.V^{v_{j_2}}.\cdots.V^{v_{j_{m'}}} \ \| \ C_2) \in R$. $\square$

We continue with the definition of Δ by adding rules which enable the defender to generate a solution of the (A, B)-instance (if it exists).

$$\begin{array}{ll}
X \overset{x}{\longrightarrow} Y & X' \overset{x}{\longrightarrow} X_1' \\
X \overset{x}{\longrightarrow} X_1' &
\end{array}$$

$$\begin{array}{lll}
& X_1' \overset{\tau}{\longrightarrow} X_1'.V^{v_k} & \text{for all } k \in \{1, \ldots, n\} \\
& X_1' \overset{\tau}{\longrightarrow} Y'.V^{v_k} & \text{for all } k \in \{1, \ldots, n\}
\end{array}$$

$$\begin{array}{ll}
Y \overset{y}{\longrightarrow} Y_1 & Y' \overset{y}{\longrightarrow} Z \\
& Y' \overset{y}{\longrightarrow} Y_1.D
\end{array}$$

$$\begin{array}{lll}
Y_1 \overset{\tau}{\longrightarrow} Y_1.U^{u_k} & & \text{for all } k \in \{1, \ldots, n\} \\
Y_1 \overset{\tau}{\longrightarrow} Z.U^{u_k} & & \text{for all } k \in \{1, \ldots, n\}
\end{array}$$

See Figure 2 for fragments of transition systems generated by (X, Δ) and (X', Δ). The following lemma explains the purpose of the rules defined above.

Lemma 3. *Consider a bisimulation game from (X, Δ) and (X', Δ). The defender has a strategy such that after two rounds the players reach a pair of states $Z.\sigma$ and $Z.\omega$ where $\sigma = U^{u_{i_1}}.U^{u_{i_2}}.\cdots.U^{u_{i_m}}$ and $\omega = V^{v_{j_1}}.V^{v_{j_2}}.\cdots.V^{v_{j_{m'}}}$ $(m, m' \ge 1)$, and where σ and ω were chosen by the defender; or the defender wins by reaching a pair of weakly bisimilar states.*

Proof. In the first round of the bisimulation game played from (X, Δ) and (X', Δ) the attacker has only one possible move: $X \overset{x}{\longrightarrow} Y$. If the attacker

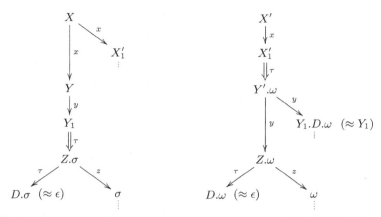

Fig. 2. Fragments of transition systems generated by (X, Δ) and (X', Δ)

plays any other move ($X \xrightarrow{x} X_1'$ or $X' \xrightarrow{x} X_1'$) then the defender can make the resulting processes syntactically equal and he wins. The defender's answer to the move $X \xrightarrow{x} Y$ is by $X' \xRightarrow{x} Y'.\omega$ for some $\omega = V^{v_{j_1}}.V^{v_{j_2}}.\cdots.V^{v_{j_{m'}}}$ such that $m' \geq 1$.

In the next round played from Y and $Y'.\omega$ the attacker is forced to continue by $Y'.\omega \xrightarrow{y} Z.\omega$. Similarly as in the first round: if the attacker chooses any other move, the defender can make the resulting processes weakly bisimilar (here we use the fact that $Y_1 \approx Y_1.D.\omega$). The defender can now choose some $\sigma = U^{u_{i_1}}.U^{u_{i_2}}.\cdots.U^{u_{i_m}}$ such that $m \geq 1$ and plays $Y \xRightarrow{y} Z.\sigma$. Hence the defender either won or he chose nonempty σ and ω and forced the attacker in two rounds to reach the pair $Z.\sigma$ and $Z.\omega$. □

We finish the definition of Δ by adding the rules:

$$C \xrightarrow{c_1} C_1 \qquad\qquad C \xrightarrow{c_2} C_2 \qquad\qquad C \xrightarrow{z} C \| W$$

$$W \xrightarrow{\tau} W.U^{u_k} \qquad\quad W \xrightarrow{\tau} W.V^{v_k} \qquad\qquad \text{for all } k \in \{1, \dots, n\}$$
$$W \xrightarrow{\tau} \epsilon.$$

The intuition is that while playing a bisimulation game from $X\|C$ and $X'\|C$, the defender can generate a solution of the (A, B)-instance by forcing the attacker to reach the states $Z.\sigma\|C$ and $Z.\omega\|C$ (see Lemma 3) such that $\sigma = U^{u_{i_1}}.U^{u_{i_2}}.\cdots.U^{u_{i_m}}$ and $\omega = V^{v_{i_1}}.V^{v_{i_2}}.\cdots.V^{v_{i_m}}$ where $i_1, \dots, i_m$ is a solution of the (A, B)-instance (if it exists). The attacker waits with using the rule $C \xrightarrow{c_1} C_1$ or $C \xrightarrow{c_2} C_2$ until the pair $Z.\sigma\|C$ and $Z.\omega\|C$ is reached and then he can check that the sequence $i_1, \dots, i_m$ is indeed a solution: from $Z.\sigma\|C_1$ and $Z.\omega\|C_1$ he checks whether the defender generated the same indices in both σ and ω, and from $Z.\sigma\|C_2$ and $Z.\omega\|C_2$ he checks whether $u_{i_1} u_{i_2} \dots u_{i_m} = v_{i_1} v_{i_2} \dots v_{i_m}$. The purpose of the rules for the process constant W is explained later.

Lemma 4. *If $(X\|C, \Delta) \approx (X'\|C, \Delta)$ then the (A, B)-instance has a solution.*

Proof. Assume that the (A, B)-instance has no solution, i.e., for every sequence of indices $i_1, \ldots, i_m \in \{1, \ldots, n\}$ where $m \geq 1$ it is the case that $u_{i_1} u_{i_2} \ldots u_{i_m} \neq v_{i_1} v_{i_2} \ldots v_{i_m}$. We show that the attacker has a winning strategy from the pair $X \| C$ and $X' \| C$. In the first round the attacker plays $X \| C \xrightarrow{x} Y \| C$. The defender can only answer by $X' \| C \xrightarrow{x} X_1' \| C$ followed by a finite number of τ actions, thus reaching a state $X_1'.\omega \| C$ or $Y'.\omega \| C$ for some ω. In the first case the attacker switches the processes and uses e.g. the rule $X_1' \xrightarrow{\tau} Y'.V^{v_1}$. Since $Y \| C \xrightarrow{\tau}\!\!\!\!\!/\,$, the defender can only stay at the state $Y \| C$. In the second case the state is already of the form $Y'.\omega \| C$.

The game now continues from the pair of states $Y \| C$ and $Y'.\omega \| C$ for some ω. The attacker chooses the move $Y'.\omega \| C \xrightarrow{y} Z.\omega \| C$. The defender has to answer by $Y \| C \xrightarrow{y} Y_1 \| C$ followed by a finite number of τ actions. This means that he can reach a state $Y_1.\sigma \| C$, or $Z.\sigma \| C$, or $D.\sigma \| C$ for some σ. The attacker wants to force the defender to reach the second possibility. We show later that if the defender reaches $D.\sigma \| C$ then he loses. Moreover, if the defender reaches $Y_1.\sigma \| C$ then the attacker can use e.g. the rule $Y_1 \xrightarrow{\tau} Z.U^{u_1}$ and the defender can only respond by staying in $Z.\omega \| C$, or by the move $Z.\omega \| C \xrightarrow{\tau} D.\omega \| C$. As we want the game to continue from $Z.\sigma \| C$ and $Z.\omega \| C$, it is enough to show that the attacker has a winning strategy from $D.\sigma \| C$ and $Z.\omega \| C$, and from $Z.\sigma \| C$ and $D.\omega \| C$. We show how the attacker wins from $D.\sigma \| C$ and $Z.\omega \| C$ (the situation from $Z.\sigma \| C$ and $D.\omega \| C$ is completely symmetric). The attacker plays in the second state: $Z.\omega \| C \xrightarrow{c_1} Z.\omega \| C_1$. The defender can only respond by $D.\sigma \| C \xrightarrow{c_1} D.\sigma \| C_1$. Now, $Z.\omega \| C_1 \xrightarrow{z} \omega \| C_1$ but $D.\sigma \| C_1 \xrightarrow{z}\!\!\!\!\!/\,$. Hence the attacker wins.

To sum up, either the attacker wins or the game continues from the pair $Z.\sigma \| C$ and $Z.\omega \| C$ for some $\sigma = U^{u_{i_1}}.U^{u_{i_2}}.\cdots.U^{u_{i_m}}$ and $\omega = V^{v_{j_1}}.V^{v_{j_2}}.\cdots.$ $V^{v_{j_{m'}}}$ where $m, m' \geq 1$. There are two cases.

- If $m = m'$ and $i_\ell = j_\ell$ for all ℓ, $1 \leq \ell \leq m = m'$, then using our assumption that the (A, B)-instance has no solution and by the fact that $m, m' \geq 1$ we get that $u_{i_1} u_{i_2} \ldots u_{i_m} \neq v_{i_1} v_{i_2} \ldots v_{i_m}$. The attacker plays $Z.\sigma \| C \xrightarrow{c_2} Z.\sigma \| C_2$ and the defender has to answer by $Z.\omega \| C \xrightarrow{c_2} Z.\omega \| C_2$ or $Z.\omega \| C \xrightarrow{c_2} D.\omega \| C_2$. From the pair $Z.\sigma \| C_2$ and $Z.\omega \| C_2$ the attacker has a winning strategy because of Lemma 2 and the attacker's strategy from the pair $Z.\sigma \| C_2$ and $D.\omega \| C_2$ is obvious: $Z.\sigma \| C_2 \xrightarrow{z} \sigma \| C_2$ but $D.\omega \| C_2 \xrightarrow{z}\!\!\!\!\!/\,$.
- If it is not the case that $m = m'$ and $i_\ell = j_\ell$ for all ℓ, $1 \leq \ell \leq m = m'$, the attacker plays $Z.\sigma \| C \xrightarrow{c_1} Z.\sigma \| C_1$ and the defender must respond by $Z.\omega \| C \xrightarrow{c_1} Z.\omega \| C_1$ or $Z.\omega \| C \xrightarrow{c_1} D.\omega \| C_1$. By Lemma 1 the attacker has a winning strategy from $Z.\sigma \| C_1$ and $Z.\omega \| C_1$. The argument for the attacker's winning strategy from $Z.\sigma \| C_1$ and $D.\omega \| C_1$ is as in the previous case. $\qquad \square$

Lemma 5. *If the (A, B)-instance has a solution then $(X \| C, \Delta) \approx (X' \| C, \Delta)$.*

Proof. Let $i_1, \ldots, i_m \in \{1, \ldots n\}$ where $m \geq 1$ be a solution of the (A, B)-instance. We show that the defender has a winning strategy from the pair $X \| C$ and $X' \| C$.

As it was already proved in Lemma 3, in the bisimulation game played from X and X' the defender can force the attacker to reach the pair $Z.\sigma$ and $Z.\omega$, or the defender has a winning strategy. In particular, the defender can make sure that the players reach the pair $Z.\sigma$ and $Z.\omega$ where σ and ω correspond to the solution of the (A, B)-instance, i.e., $\sigma = U^{u_{i_1}}.U^{u_{i_2}}.\cdots.U^{u_{i_m}}$ and $\omega = V^{v_{i_1}}.V^{v_{i_2}}.\cdots.V^{v_{i_m}}$.

The situation in this lemma, however, requires that the players start playing from $X\|C$ and $X'\|C$. We have to extend the defender's strategy by defining his responses to the attacks from the process constant C, or more generally from any context γ reachable from C (see the last part of the definition of Δ). To any attacker's move $X\|\gamma \longrightarrow X\|\gamma'$ or $X'\|\gamma \longrightarrow X'\|\gamma'$ the defender answers simply by imitating the same move in the other process. The bisimulation game then continues from the pair $X\|\gamma'$ and $X'\|\gamma'$. Since any infinite game is a winning one for the defender, the attacker must eventually use some rules for X or X'. In this case the defender uses the strategy from Lemma 3. The attacker is forced to play $X\|\gamma \xrightarrow{x} Y\|\gamma$ and the defender answers by $X'\|\gamma \xRightarrow{x} Y'.\omega\|\gamma$ where $\omega = V^{v_{i_1}}.V^{v_{i_2}}.\cdots.V^{v_{i_m}}$. From the states $Y\|\gamma$ and $Y'.\omega\|\gamma$, again the defender imitates any attacks from the context γ. Thus the attacker must eventually play $Y'.\omega\|\gamma \xrightarrow{y} Z.\omega\|\gamma$ and the defender answers by $Y\|\gamma \xRightarrow{y} Z.\sigma\|\gamma$ where $\sigma = U^{u_{i_1}}.U^{u_{i_2}}.\cdots.U^{u_{i_m}}$.

By inspecting the rules for C we can see that the context γ always contains either the process constant (i) C, (ii) C_1, or (iii) C_2. Hence γ can be written as (i) $C\|\gamma'$, (ii) $C_1\|\gamma'$, or (iii) $C_2\|\gamma'$ for some context γ'. In case (ii) the bisimulation game continues from the pair $Z.\sigma\|C_1\|\gamma'$ and $Z.\omega\|C_1\|\gamma'$, and the defender has a winning strategy because of Lemma 1 and Proposition 1. In case (iii) the bisimulation game continues from the pair $Z.\sigma\|C_2\|\gamma'$ and $Z.\omega\|C_2\|\gamma'$, and the defender has a winning strategy because of Lemma 2 and Proposition 1. It remains to demonstrate that the defender has a winning strategy also in case (i). Hence assume that the game continues from $Z.\sigma\|C\|\gamma'$ and $Z.\omega\|C\|\gamma'$. By Proposition 1 it is enough to show that $Z.\sigma\|C \approx Z.\omega\|C$. We will analyze the attacker's moves from $Z.\sigma\|C$. The arguments for the moves from $Z.\omega\|C$ are completely symmetric. The attacker has the following moves available.

$$
\begin{aligned}
&\text{(i)} \quad && Z.\sigma\|C \xrightarrow{c_1} Z.\sigma\|C_1 \\
&\text{(ii)} \quad && Z.\sigma\|C \xrightarrow{c_2} Z.\sigma\|C_2 \\
&\text{(iii)} \quad && Z.\sigma\|C \xrightarrow{z} Z.\sigma\|C\|W \\
&\text{(iv)} \quad && Z.\sigma\|C \xrightarrow{\tau} D.\sigma\|C \\
&\text{(v)} \quad && Z.\sigma\|C \xrightarrow{z} \sigma\|C
\end{aligned}
$$

In case (i) the defender answers by $Z.\omega\|C \xrightarrow{c_1} Z.\omega\|C_1$ and wins because of Lemma 1. In case (ii) the defender answers by $Z.\omega\|C \xrightarrow{c_2} Z.\omega\|C_2$ and wins because of Lemma 2. In case (iii) the defender answers by $Z.\omega\|C \xrightarrow{z} Z.\omega\|C\|W$. By Proposition 1 this case is already covered by the discussion of the defender's strategy from $Z.\sigma\|C$ and $Z.\omega\|C$. In case (iv) the defender answers by $Z.\omega\|C \xrightarrow{\tau} D.\omega\|C$ and he wins since $D.\sigma\|C \approx C \approx D.\omega\|C$. Case (v) is the only case where we need the rules for the process constant W. The defender answers by the following sequence:

$$Z.\omega\|C \xrightarrow{\tau} D.\omega\|C \xrightarrow{z} D.\omega\|C\|W \xRightarrow{\tau} D.\omega\|C\|\sigma.$$

This can be written in one step as $Z.\omega\|C \xRightarrow{z} D.\omega\|C\|\sigma$. Now the game continues from the pair $\sigma\|C$ and $D.\omega\|C\|\sigma$, however, $\sigma\|C \approx D.\omega\|C\|\sigma$. This implies that the defender has a winning strategy also in this case.   □

Theorem 1. *Weak bisimilarity of PA with deadlocks is undecidable.*

Proof. Immediately from Lemmas 4 and 5.   □

In the rest of this section we show that the presence of the deadlock D in Δ is not an essential requirement. We build upon the technique of deadlock elimination described (for the case of BPA) in [12].

Lemma 6. *There is a (polynomial time) reduction from weak bisimilarity of PA with deadlocks to weak bisimilarity of PA without deadlocks.*

Proof. Let Δ be a PA system. By $\mathcal{D}(\Delta)$ we denote the set of all process constants which have no rewrite rule in Δ, i.e., $\mathcal{D}(\Delta) = \{X \in Const(\Delta) \mid X \nrightarrow\}$. Let us consider a PA system Δ' such that $Const(\Delta') \overset{\text{def}}{=} Const(\Delta) \smallsetminus \mathcal{D}(\Delta) \cup \{D\}$ and $Act(\Delta') \overset{\text{def}}{=} Act(\Delta) \cup \{d\}$ where D is a new process constant and d is a new action. Let $\Delta' \overset{\text{def}}{=} \{X \xrightarrow{a} \overline{E} \mid (X \xrightarrow{a} E) \in \Delta\} \cup \{D \xrightarrow{d} D\}$ such that $\overline{\epsilon} \overset{\text{def}}{=} \epsilon$, $\overline{X} \overset{\text{def}}{=} X$ if $X \notin \mathcal{D}(\Delta)$, $\overline{X} \overset{\text{def}}{=} D$ if $X \in \mathcal{D}(\Delta)$, $\overline{E.F} \overset{\text{def}}{=} \overline{E}.\overline{F}$, and $\overline{E\|F} \overset{\text{def}}{=} \overline{E}\|\overline{F}$, where X is a process constant and E, F are process expressions. Obviously $\mathcal{D}(\Delta') = \emptyset$ and it is easy to verify that $(E, \Delta) \approx (F, \Delta)$ if and only if $(\overline{E}\|D, \Delta') \approx (\overline{F}\|D, \Delta')$ for any process expressions E and F.   □

Corollary 1. *Weak bisimilarity of PA (without deadlocks) is undecidable.*

4   Conclusion

We proved that weak bisimilarity of PA-processes is undecidable. In our proof we used the notion of deadlocks to make the reduction more understandable, and we also showed that the result can be easily generalized to PA without deadlocks. We took advantage of several new techniques recently developed, in particular the existential quantification technique and the masking technique.

The undecidability result of weak bisimilarity for PA contrasts to the situation of strong bisimilaririty for normed PA, which is known to be decidable in 2-NEXPTIME [3]. The problems of strong bisimilarity for unnormed PA and of weak bisimilarity for normed PA still remain open. Another question to be considered is, whether the problem of weak bisimilarity for PA is highly undecidable. In particular, we do not know whether it lies inside the arithmetical hierarchy, or whether it is beyond the hierarchy, as it is in the case of PDA (Remark 4 in [14]) and PN [4].

Acknowledgements. I would like to thank my advisor Mogens Nielsen for his kind supervision, Petr Jančar for his suggestions, and the referees for their comments.

References

1. J.C.M. Baeten and W.P. Weijland. *Process Algebra.* Number 18 in Cambridge Tracts in Theoretical Computer Science. Cambridge University Press, 1990.
2. O. Burkart, D. Caucal, F. Moller, and B. Steffen. Verification on infinite structures. In J. Bergstra, A. Ponse, and S. Smolka, editors, *Handbook of Process Algebra*, chapter 9, pages 545–623. Elsevier Science, 2001.
3. Y. Hirshfeld and M. Jerrum. Bisimulation equivalence is decidable for normed process algebra. In *Proc. of Automata, Languages and Programming, 26th International Colloquium (ICALP'99)*, volume 1644 of *LNCS*, pages 412–421. Springer-Verlag, 1999.
4. P. Jančar. High undecidability of weak bisimilarity for Petri nets. In *Proc. of Colloquium on Trees in Algebra and Programming (CAAP'95)*, volume 915 of *LNCS*, pages 349–363. Springer-Verlag, 1995.
5. P. Jančar. Undecidability of bisimilarity for Petri nets and some related problems. *Theoretical Computer Science*, 148(2):281–301, 1995.
6. P. Jančar and J. Esparza. Deciding finiteness of Petri nets up to bisimulation. In *Proc. of 23rd International Colloquium on Automata, Languages, and Programming (ICALP'96)*, volume 1099 of *LNCS*, pages 478–489. Springer-Verlag, 1996.
7. P. Jančar and F. Moller. Checking regular properties of Petri nets. In *Proc. of 6th International Conference on Concurrency Theory (CONCUR'95)*, volume 962 of *LNCS*, pages 348–362. Springer-Verlag, 1995.
8. R. Mayr. On the complexity of bisimulation problems for basic parallel processes. In *Proc. of 27st International Colloquium on Automata, Languages and Programming (ICALP'00)*, volume 1853 of *LNCS*, pages 329–341. Springer-Verlag, 2000.
9. R. Mayr. Process rewrite systems. *Information and Comp.*, 156(1):264–286, 2000.
10. E.L. Post. A variant of a recursively unsolvable problem. *Bulletion of the American Mathematical Society*, 52:264–268, 1946.
11. G. Sénizergues. Decidability of bisimulation equivalence for equational graphs of finite out-degree. In *Proc. of the 39th Annual Symposium on Foundations of Computer Science(FOCS'98)*, pages 120–129. IEEE Computer Society, 1998.
12. J. Srba. Basic process algebra with deadlocking states. *Theoretical Computer Science*, 266(1–2):605–630, 2001.
13. J. Srba. Strong bisimilarity and regularity of basic parallel processes is PSPACE-hard. In *Proc. of 19th International Symposium on Theoretical Aspects of Computer Science (STACS'02)*, volume 2285 of *LNCS*, pages 535–546. Springer-Verlag, 2002.
14. J. Srba. Undecidability of weak bisimilarity for pushdown processes. In *Proc. of 13th International Conference on Concurrency Theory (CONCUR'02)*, volume 2421 of *LNCS*, pages 579–593. Springer-Verlag, 2002.
15. C. Stirling. Local model checking games. In *Proc. of 6th International Conference on Concurrency Theory (CONCUR'95)*, volume 962 of *LNCS*, pages 1–11. Springer-Verlag, 1995.

16. W. Thomas. On the Ehrenfeucht-Fraïssé game in theoretical computer science (extended abstract). In *Proc. of 4th International Joint Conference CAAP/FASE, Theory and Practice of Software Development (TAPSOFT'93)*, volume 668 of *LNCS*, pages 559–568. Springer-Verlag, 1993.

Improved Bounds on the Number
of Automata Accepting Finite Languages

Mike Domaratzki

School of Computing, Queen's University, Kingston, ON, K7L 4S6 Canada
domaratz@cs.queensu.ca

Abstract. We improve the known bounds on the number of pairwise non-isomorphic minimal deterministic finite automata (DFAs) on n states which accept finite languages. The lower bound constructions are iterative approaches which yield recurrence relations.

1   Introduction

The study of enumeration of automata has a long history, including classic papers by Harrison [3], Radke [6], Robinson [7], Liskovets [4] and many papers by Korshunov in Russian, including the survey [5]. For a current list of references see Domaratzki *et al.* [2].

However, little work has been done on enumerating automata which accept finite languages. In this paper, we extend the work of Domaratzki *et al.* [2] by improving the upper and lower bounds on the number of finite languages accepted by deterministic finite automata (DFAs) with n states.

The author [1] has recently presented an extension to the Genocchi numbers which gives an upper bound on the number of finite languages accepted by DFAs with n states. This upper bound is not perfect, however, since it is a labeled enumeration, whereas what we seek is an unlabeled enumeration of the automata.

In the upper bound presented here, we revise the constructions of the previous work, and achieve a better upper bound. Unlike the previous work, however, the new construction does not parallel any studied sequences of numbers.

2   Definitions

A deterministic finite automaton (DFA) is a 5-tuple $M = (Q, \Sigma, \delta, q_0, F)$ where Q is a set of states, Σ is an alphabet, δ is a transition function $\delta : Q \times \Sigma \to Q$. The start state is $q_0 \in Q$ and the final states are $F \subseteq Q$. We extend δ to a function from $Q \times \Sigma^*$ to Q in the usual way. A DFA is said to be *complete* if $\delta(q, a)$ is defined for all pairs $(q, a) \in Q \times \Sigma$. In what follows, we will assume that all DFAs are complete.

A string $w \in \Sigma^*$ is accepted by a DFA $M = (Q, \Sigma, \delta, q_0, F)$ if $\delta(q_0, w) \in F$. Define the language accepted by a DFA M by $L(M) = \{w \in \Sigma^* : \delta(q_0, w) \in F\}$. A DFA $M = (Q, \Sigma, \delta, q_0, F)$ is minimal for a language L if $L(M) = L$ and for all DFAs $M' = (Q', \Sigma, \delta', q_0', F')$ with $L(M') = L$, we have $|Q| \le |Q'|$.

M. Ito and M. Toyama (Eds.): DLT 2002, LNCS 2450, pp. 209–219, 2003.

Let $[[P]]$ denote the boolean value of the expression P, i.e. $[[P]] \in \{0,1\}$ and $[[P]] = 1$ iff P is true. We use the symbol $\triangle$ to denote symmetric difference of sets. That is, for all sets S_1, S_2, we define $S_1 \triangle S_2 = (S_1 \setminus S_2) \cup (S_2 \setminus S_1)$. For the alphabet $\Sigma = \{a, b\}$, we define the morphism $\overline{\cdot} : \Sigma \to \Sigma$ by the rules $\overline{a} = b$ and $\overline{b} = a$.

3   A Lower Bound

For all $k \geq 1$, let Σ_k be an alphabet of size k, given by $\Sigma_k = \{a_1, a_2, \ldots, a_k\}$.

Definition 3.1. *Let $k \geq 1$. Let $f_k(n)$ denote the number of non-isomorphic minimal DFAs on n states over Σ_k which accept finite languages.*

The best known lower bound on $f_k(n)$ is $f_k(n) \geq 2^{n-2}((n-1)!)^{k-1}$, given by Domaratzki *et al.* [2].

We define the function $s_k(n)$ for all $k \geq 2$ as follows:

$$s_k(2) = 1$$
$$s_k(n) = 2((n-1)^k - (n-2)^k)s_k(n-1)$$

Our goal will be to establish the following lower bound:

Theorem 3.1. *For all $k \geq 2$ and $n \geq 2$, we have $f_k(n) \geq s_k(n)$.*

To do this, we will describe a set of DFAs $S_{n,k}$ with $|S_{n,k}| = s_k(n)$. Each DFA in $S_{n,k}$ will have state set $\{1, 2, \ldots, n-1, d\}$ ($d \notin \{1, 2, \ldots, n-1\}$ will represent our "dead state") and will accept a different finite language. We let Q_n denote the state set $Q_n = \{1, 2, \ldots, n-1, d\}$.

We define $S_{n,k}$ recursively as follows, beginning with $S_{2,k}$:

$$S_{2,k} = \{M = (Q_2, \Sigma_k, \delta, 1, \{1\})\}$$

with δ defined by $\delta(1, a) = \delta(d, a) = d$ for all $a \in \Sigma_k$. Thus, M accepts the language $\{\epsilon\}$. Now, let $n \geq 2$ and given the set $S_{n,k}$, define the set $S_{n+1,k}$ as follows: For each $M = (Q_n, \Sigma_k, \delta, n-1, F) \in S_{n,k}$ add a new state n. For each possible k-tuple $(i_1, i_2, \ldots, i_k) \in Q_n^k$ (subject to the condition that $i_j = n-1$ for at least one j with $1 \leq j \leq k$), we create two DFAs $M_1 = (Q_{n+1}, \Sigma_k, \delta', n, F_1)$, $M_2 = (Q_{n+1}, \Sigma_k, \delta', n, F_2) \in S_{n+1,k}$. The transition function δ' is inherited from M along with our added transitions for state n:

$$\delta'(n, a_j) = i_j$$
$$\delta'(i, a) = \delta(i, a) \quad \forall i \in Q_n, a \in \Sigma_k$$

Finally, $F_1 = F$ and $F_2 = F \cup \{n\}$. Note that by the inheritance of final states, for each $M \in S_{n,k}$, the state $1 \in F$, while $d \notin F$.

We may first state two basic lemmas about DFAs in $S_{n,k}$ for $n \geq 2$. The proofs are omitted, but will be included in a full version to appear later.

Lemma 3.1. *Let* $M = (Q_n, \Sigma_k, \delta, n-1, F) \in S_{n,k}$. *Then for each state* $i \in Q_n \setminus \{d\}$, *there exists a string* w_i *such that* $|w_i| = n-1-i$ *and* $\delta(n-1, w_i) = i$. *For* $i = d$, *there exists a string* w_d *such that* $|w_d| = n-1$ *and* $\delta(n-1, w_d) = d$.

Lemma 3.2. *Let* $M = (Q_n, \Sigma_k, \delta, n-1, F) \in S_{n,k}$. *Then for each state* $i \in Q_n \setminus \{d\}$, *there exists a string* x_i *such that* $|x_i| = i-1$ *and* $\delta(i, x_i) = 1$.

These tools will prove sufficient for establishing the following lemma:

Lemma 3.3. *Let* $M_1 \neq M_2 \in S_{n,k}$. *Then* $L(M_1) \neq L(M_2)$.

Thus, all M in $S_{n,k}$ are pairwise non-isomorphic. It remains to be shown that each M is minimal. This is established by the following lemma:

Lemma 3.4. *Let* $M = (Q_n, \Sigma_k, \delta, n-1, F) \in S_{n,k}$. *Then* M *accepts a string of length* $n-2$, *but no string of length* $n-1$ *or greater.*

Thus, we note that M is necessarily minimal, since if it were not, then a DFA M' with less than n states would accept the same language. But then M' accepts a string of length $n-2$ with less than n states. This is impossible if M' accepts a finite language.

We now established all of Theorem 3.1 except for computing the size of $S_{n,k}$. We can establish the following recurrence: Let $s_k(n) = |S_{n,k}|$. We already know that $s_k(2) = 1$ (we may also set $s_k(1) = 1$ by letting $S_{1,k}$ contain only the DFA whose language accepted is the empty set). Now, for each choice of a DFA M' in $S_{n-1,k}$, we have $(n-1)^k - (n-2)^k$ different choices for the transitions $\delta(n-1, a)$ for all $a \in \Sigma_k$ and subject to our constraints, since there are $(n-1)^k$ ways of setting the k transitions in the set Q_{n-1}, and $(n-2)^k$ ways of setting the k transitions in the set $Q_{n-1} \setminus \{n-2\}$. Further, we may assign either $n-1 \in F$ or not, giving us two more choices. This gives the recurrence:

$$s_k(n) = 2((n-1)^k - (n-2)^k)s_k(n-1) \quad \forall n \geq 3.$$

It is an easy inductive proof to show that $s_2(n) = (2n-3)!/(n-2)!$. Table 2 in Appendix A compares our lower bound to the old lower bound and the actual value for $k = 2$.

4   An Improvement

We may improve the results of Section 3 for the case $k = 2$. We set $\Sigma = \{a, b\}$. For each $n \geq 3$, we will define a set $T(n)$ of DFAs. To do this, we define $T(n, i)$ for all $0 \leq i \leq \lfloor \frac{n-3}{2} \rfloor$. The first set $T(n, 0) = S_{n,2}$ for all $n \geq 3$ (we set $T(2, 0) = S_{2,2}$ as needed). From these sets, we will define each of the sets $T(n, i)$.

For $i < \lfloor \frac{n-3}{2} \rfloor$, define $T(n, i)$ recursively as follows:
(a) for each $M \in T(n-1, i)$, we define $2((n-1)^2 - (n-2)^2) = 4n-6$ different DFAs in $T(n, i)$, as in the definition of $S_{n,2}$. Thus, let

$$M = (Q_{n-1}, \{a, b\}, \delta, n-2, F).$$

Then for each possible pair of states $(j, k) \in Q_{n-1}^2$ subject to the condition that at least one of $j, k = n - 2$, we create the DFAs

$$M' = (Q_n, \{a, b\}, \delta', n - 1, F')$$
$$M'' = (Q_n, \{a, b\}, \delta', n - 1, F'')$$

where

$$\delta'(n - 1, a) = j, \ \delta'(n - 1, b) = k,$$

and

$$\delta'(\ell, c) = \delta(\ell, c) \quad \forall c \in \{a, b\}, \ell \in Q_{n-1}.$$

Finally, $F' = F$, and $F'' = F \cup \{n - 1\}$.

(b) for each $M \in T(n - 2, i - 1)$, we define $2(2n - 5)(n - 4)$ DFAs in $T(n, i)$. Let

$$M = (Q_{n-2}, \{a, b\}, \delta, n - 3, F) \in T(n - 2, i - 1)$$

For any such M, define $D(M)$ to be the following set

$$D(M) = \{(\delta(q, a), \delta(q, b), [[q \in F]]) \ : \ q \in Q_{n-2}\}$$

Note that $D(M) \subseteq Q_{n-3}^2 \times \{0, 1\}$. Let D_n denote the set $D_n = Q_{n-3}^2 \times \{0, 1\}$. Now, $|D(M)| = n - 2$, but $|D_n| = 2(n-3)^2$. Thus, for all of the $2(n-3)^2 - (n-2) = (2n - 5)(n - 4)$ elements $(j, k, B) \in D_n - D(M)$, we define two new DFAs

$$M' = (Q_n, \{a, b\}, \delta', n - 1, F'),$$
$$M'' = (Q_n, \{a, b\}, \delta', n - 1, F'').$$

where

$$\delta'(n - 1, a) = n - 2, \ \delta'(n - 1, b) = n - 3,$$
$$\delta'(n - 2, a) = j, \qquad \delta'(n - 2, b) = k,$$

and

$$\delta'(\ell, c) = \delta(\ell, c) \quad \forall c \in \{a, b\}, \ell \in Q_{n-2}.$$

Finally, if $B = 1$ then set $F' = F \cup \{n - 2\}$ and $F'' = F \cup \{n - 2, n - 1\}$. If $B = 0$, then set $F' = F$, $F'' = F \cup \{n - 1\}$.

For $i = \lfloor \frac{n-3}{2} \rfloor$, we have two cases:

(i) If n is odd, we define $T(n, i)$ using the construction of type (b) only.

(ii) If n is even, $T(n, i)$ will be defined as usual, that is, with type (a) and (b).

We will continue to use the terms "type (a)" and "type (b)" to refer to the two different constructions used. Note the following fact:

Fact 4.1. *For all DFAs $M \in T(n, i)$ for some $n \geq 3$ and $0 \leq i \leq \lfloor (n - 3)/2 \rfloor$, $(d, d, 0), (d, d, 1) \in D(M)$. Thus, we never have $(d, d, B) in D_n \setminus D(M)$ for any $B \in \{0, 1\}$.*

To see this, note that in $M \in S_{2,2} = T(2,0)$, $D(M) = \{(d,d,0),(d,d,1)\}$. Further, note that if M is constructed from M', $D(M) \supset D(M')$.

We introduce the index i for our own use. We will see that the index i is irrelevant in the final analysis of the size of $T(n) = \cup_{i=0}^{\lfloor (n-3)/2 \rfloor} T(n,i)$. However, they will prove very useful in proving the lemmas which will establish our bound to be correct.

Lemma 4.1. *Let $M \in T(n,i)$ be arbitrary. Then for each $j \in Q_n$, there exists a string $w_j \in \{a,b\}^*$ such that $\delta(n-1, w_j) = j$. Further, if $j \neq n-1$, we may choose $w_j \neq \epsilon$.*

Lemma 4.2. *Let $M \in T(n,i)$ be arbitrary. Then for each $j \in Q_n \setminus \{d\}$ there exists a string $x_j \in \{a,b\}^*$ such that $\delta(j, x_j) = 1$.*

We now prove bounds on the length of words accepted by DFAs in $T(n,i)$. We first introduce the following notation: For any DFA M which accepts a finite language, let $m(M) = \max\{|w| : w \in L(M)\}$.

Lemma 4.3. *Let $n \geq 3$ and $0 \leq i \leq \lfloor \frac{n-3}{2} \rfloor$. Let $M = (Q_n, \{a,b\}, \delta, n-1, F) \in T(n,i)$ be constructed from $M' = (Q_{n-1}, \{a,b\}, \delta', n-2, F') \in T(n-1,i)$. Then $m(M) = m(M') + 1$. The same result also holds if $M \in T(n,i)$ is constructed from $M' = (Q_{n-2}, \{a,b\}, \delta', n-3, F') \in T(n-2, i-1)$.*

Theorem 4.2. *For all $n \geq 3$ and $0 \leq i \leq \lfloor (n-3)/2 \rfloor$, and for all $M \in T(n,i)$,*

$$m(M) = n - i - 2.$$

There are two important corollaries of Theorem 4.2:

Corollary 4.1. *Let $n \geq 3$ be fixed. For all $0 \leq i_1 < i_2 \leq \lfloor (n-3)/2 \rfloor$, $T(n,i_1) \cap T(n,i_2) = \emptyset$.*

Corollary 4.2. *Let $n \geq 3$, $0 \leq i \leq \lfloor (n-3)/2 \rfloor$. Let $M \in T(n,i)$ be arbitrary. Then for all $j \in Q_n \setminus \{n-1\}$, the following holds:*

$$\max\{|y| : \delta(j,y) \in F\} < n - i - 2$$

We now establish the minimality of each DFA.

Theorem 4.3. *Let $n \geq 1$ and $0 \leq i \leq \lfloor (n-3)/2 \rfloor$. Then for all $M \in T(n,i)$, M is minimal.*

Our final goal is to prove that each $M \in T(n,i)$ is unique:

Theorem 4.4. *Let $n \geq 3$ and $0 \leq i \leq \lfloor (n-3)/2 \rfloor$. Then for all $M_1 \neq M_2 \in T(n,i)$, $L(M_1) \neq L(M_2)$.*

The proof of Theorem 4.4 is by induction on n. To prove this, we examine the following three cases. We prove each of the three cases simultaneously, but as separate lemmas:

(i) M_1, M_2 both arrived at from elements of $T(n-1, i)$ through type (a) construction.
(ii) M_1 arrived at through an element of $T(n-1, i)$ through a type (a) construction, M_2 arrived at through an element of $T(n-2, i-1)$ through a type (b) construction.
(iii) M_1, M_2 both arrived at from elements of $T(n-2, i-1)$ through type (b) constructions.

Of these, case (i) is proved through the exact same proof idea of Lemma 3.3, so we omit it.

We will first consider case (ii). This case will be where the introduction of the index i into $T(n, i)$ will yield us the most benefit. We prove the following lemma, which involves removing states. By that we mean deleting a state s, and all transitions leading out s. Then we also delete all states (potentially none) that are unreachable after removing s. Consider then the following argument: If we take a DFA $M \in T(n, i)$ arrived at through a type (a) construction, and remove state $n - 1$, then we obviously arrive at a DFA $M' \in T(n-1, i)$. Similarly, if take a DFA $M \in T(n, i)$ arrived at through a type (b) construction and remove the states $n-1$ and $n-2$, then we will arrive at a DFA $M' \in T(n-2, i-1)$. However, what can we say about the DFA we get when we delete the states $n-1$ and $n-3$ from M? The following lemma says that we do in fact get back a DFA we have already enumerated.

This lemma gives us a reason why we fix the transitions out of state $n-1$ in a type (b) construction: $\delta(n-1, a) = n-2$ and $\delta(n-1, b) = n-3$. This is to avoid double counting. We now state our lemma, which our index i allows us to formalize easily:

Lemma 4.4. *Let $n \geq 5$ and $1 \leq i \leq \lfloor (n-3)/2 \rfloor$ and $M = (Q_n, \{a, b\}, \delta, n-1, F) \in T(n, i)$. Suppose M is constructed from a DFA $M' \in T(n-2, i-1)$. Then the DFA defined by deleting the states $n-1$ and $n-3$ and setting the initial state to $n-2$ is isomorphic to an element of $T(m, j)$ for some $m \leq n-2$ and $j \leq i-1$.*

Proof. Fix an arbitrary $n \geq 5$. The proof is then by induction on i. For $i = 1$, note that $M' \in T(n-2, 0)$. Thus, each state in $j \in Q_{n-2} \setminus \{1, d\}$ was added one at a time through type "a" construction to yield the DFA M' in $T(n-2, 0)$.

Let $\delta(n-2, a) = i_1$ and $\delta(n-2, b) = i_2$. Let $i_0 = \max\{i_1, i_2\}$ and let $c \in \{a, b\}$ be such that $\delta(n-2, c) = i_0$. Thus, there was a point where i_0 was added as the initial state as a type (a) construction. Then we let $M'' \in T(i_0+2, 0)$ be the DFA which agrees with M' on the states of Q_{i_0+1} and which satisfies $\delta(i_0 + 1, c) = i_0$ and $\delta(i_0 + 1, \bar{c}) = \min\{i_1, i_2\}$. It is clear that $M'' \in T(i_0 + 2, 0)$ and M'' is isomorphic to the DFA obtained from M by deleting $n-1$ and $n-3$. Clearly, $i_0 + 2 \leq n - 2$, so the result holds for $i = 1$.

Now, suppose the result is true for all integers less than i. Let $M \in T(n, i)$. Let $\delta(n-2, a) = i_1$, $\delta(n-2, b) = i_2$ and $i_0 = \max\{i_1, i_2\}$. There are three cases to consider when we remove the states $n-1$ and $n-3$ from M:

(i) The state i_0 was added during a type (a) construction: Then after removing $n - 1$ and $n - 3$, and all states which are then unreachable, we arrive at a DFA which is isomorphic to an element of $T(i_0 + 2, j)$ for some $j \leq i - 1$. Thus, we are done.

(ii) The state i_0 was the added initial state during a type (b) construction: This case is identical to case (i); we are again left with a DFA isomorphic to an element of $T(i_0 + 2, j)$ for some $j \leq i - 1$.

(iii) The state i_0 was the non-initial added state during a type (b) construction. Then, adding state i_0 resulted in a DFA $M' \in T(i_0+2, j)$ for some $j \leq i-1$. Note that in this case, the state $i_0 - 1$ will necessarily be unreachable after removing states $n - 1$ and $n - 3$. But now $j < i$ implies that by induction, there is some DFA $M'' \in T(m, \ell)$ which is isomorphic to the DFA obtained from removing states $i_0 + 1$ and $i_0 - 1$ from M'. Then M'' is also isomorphic to the DFA obtained from removing states $n - 1$ and $n - 3$ from M.

Thus, by induction, the result holds for all i. The lemma is proved.   □

With Lemma 4.4 in hand, we may now prove case (ii) of Theorem 4.4:

Lemma 4.5. *Let $n \geq 5$ and $0 \leq i \leq \lfloor (n - 3)/2 \rfloor$. Let $M_1, M_2 \in T(n, i)$ such that M_1 is arrived at through a type (a) construction from $M_1' \in T(n - 1, i)$ and M_2 is arrived at through a type (b) construction from $M_2' \in T(n-2, i-1)$. Then $L(M_1) \neq L(M_2)$.*

Proof. Let $M_1 = (Q_n, \{a, b\}, \delta_1, n - 1, F_1)$, $M_2 = (Q_n, \{a, b, \}, \delta_2, n - 1, F_2)$, $M_1' = (Q_{n-1}, \{a, b\}, \delta_1', n - 2, F_1')$, and $M_2' = (Q_{n-2}, \{a, b\}, \delta_2', n - 3, F_2')$. We have two cases to consider:

1. Suppose $\delta_1(n - 1, b) = n - 2$. By Theorem 4.3, the DFAs M_1' and M_2' are minimal. Since they have a different number of states, they must accept distinct languages. Namely, let $x \in L(M_1') \triangle L(M_2')$. Then note that $bx \in L(M_1) \triangle L(M_2)$.

2. Suppose $\delta_1(n - 1, a) = n - 2$. By Lemma 4.4, let $M_2'' \in T(m, j)$ be the DFA we get by deleting states $n-1$ and $n-3$ from M_2. Then $m \leq n-2$ and $j \leq i-1$. Since M_2'' has at most $n-2$ states and M_1' has $n-1$ states, by Theorem 4.3 again, the DFAs M_1' and M_2'' must accept different languages. Let $x \in L(M_1') \triangle L(M_2'')$. Then note that $ax \in L(M_1) \triangle L(M_2)$.

As one of these two cases must hold, this completes the proof.   □

Finally, we prove case (iii) of Theorem 4.4:

Lemma 4.6. *Let $n \geq 5$ and $1 \leq i \leq \lfloor (n - 3)/2 \rfloor$. Let $M_1 \neq M_2 \in T(n, i)$ such that both M_1, M_2 are constructed from $M_1', M_2' \in T(n - 2, i - 1)$ using a type (b) construction. Then $L(M_1) \neq L(M_2)$.*

Proof. The proof is by induction on n. The base case when $n = 5$ is easily verified.

1. Suppose $M_1' \neq M_2'$. Then either by induction on n or by cases (i) or (ii) of Theorem 4.4, $L(M_1') \neq L(M_2')$. Let $x \in L(M_1') \triangle L(M_2')$. Then clearly $bx \in L(M_1) \triangle L(M_2)$.

2. Suppose $M_1' = M_2'$, but $(\delta_1(n-2,a), \delta_1(n-2,b)) \neq (\delta_2(n-2,a), \delta_2(n-2,b))$. By an application of Lemma 4.4, let M_1'' and M_2'' be the DFA we get by deleting states $n-1$ and $n-3$ from M_1 and M_2, respectively. There are three simple subcases:

(a) $M_1'', M_2'' \in T(m,j)$ for the same $m \leq n-2$ and $j \leq i-1$. As the transitions leading out of the initial state (isomorphic to $n-2$) in M_1'' and M_2'' are different, $M_1'' \neq M_2''$. Thus again by induction n or by Theorem 4.4 cases (i) and (ii), as $m \leq n-2$, there is some string $x \in L(M_1'')\triangle L(M_2'')$. Then as the initial state in M_1'' and M_2'' is the same as state $n-2$ in M_1 and M_2, and $\delta_1(n-1,a) = \delta_2(n-1,a) = n-2$, then $ax \in L(M_1)\triangle L(M_2)$.

(b) $M_1'' \in T(m_1,j_1), M_2'' \in T(m_2,j_2)$ for $m_1 \neq m_2$. Since these DFAs are minimal and have different numbers of states, then there is some string $x \in L(M_1'')\triangle L(M_2'')$. Again, $ax \in L(M_1)\triangle L(M_2)$.

(c) Finally, suppose that $M_1'' \in T(m,j_1)$, $M_2'' \in T(m,j_2)$ for $j_1 \neq j_2$. Then by Theorem 4.2, there exists $x \in L(M_1'')\triangle L(M_2'')$, namely x of length $m - \min\{j_1, j_2\} - 2$.

3. Suppose $M_1' = M_2'$, and $(\delta_1(n-2,a), \delta_1(n-2,b)) = (\delta_2(n-2,a), \delta_2(n-2,b))$. Then since $M_1 \neq M_2$, we must have that $([[n-1 \in F_1]], [[n-2 \in F_1]]) \neq ([[n-1 \in F_2]], [[n-2 \in F_2]])$. Thus, we must have one of $\epsilon, a \in L(M_1)\triangle L(M_2)$. This completes the proof. $\square$

Combining the results of this section, we have established the following theorem:

Theorem 4.5. *Each DFA $M \in T(n)$ is minimal and accepts a distinct language.*

4.1   A Recurrence for the Size of $T(n)$

As we mentioned, the index i will bear no part in our estimation on the size of $T(n)$. Let $t(n) = |T(n)|$. We state the following recurrence:

Theorem 4.6. *The following recurrence holds for $t(n)$:*

$$t(n) = 2(2n-3)t(n-1) + 2(2n-5)(n-4)t(n-2) \quad \forall n \geq 5$$
$$t(n) = (2n-3)!/(n-2)! \quad otherwise.$$

We demonstrate the effect of $t(n)$ compared to the lower bound in Table 3, in Appendix A. We can show that $c_1^{n-2} \leq t(n)/s_2(n) \leq c_2^{n-2}$ for constants $c_1 \simeq 1.066946710$ and $c_2 \simeq 1.207106781$.

5   An Upper Bound

In this section, we present our upper bound argument. We first review the work of the author in [1]. Consider the following iterative construction of a DFA accepting a finite language over the alphabet $\{a,b\}$: Let $M = (Q, \{a,b\}, \delta, q_0, F)$ be our DFA, with states $Q = \{q_0, q_2, \ldots, q_{n-1}\}$. We know that if $\delta(q_i, c) = q_j$ is a transition, then $i < j$ or $i = j = n - 1$ [2]. We would like to ensure that each DFA is initially connected, that is, that for each state q_i, there is some word w_i such that $\delta(q_0, w_i) = q_i$. To this end, we ensure that each state q_i has an edge entering it. Since the states are ordered, this must mean that $\delta(q_j, a) = q_i$ or $\delta(q_j, b) = q_i$ for some $j < i$. Since our automata are deterministic, we may assign $\delta(q_j, a)$ and $\delta(q_j, b)$ only once. Thus, an upper bound is to simply count the number of ways we may do this.

For state q_1, there are at most two transitions that may enter it: $\delta(q_0, a)$ and $\delta(q_0, b)$. Thus, we can make the assignment in $\sum_{i_1=1}^{2} \binom{2}{i_1}$ ways. In general, consider having done this for the previous $j - 1$ states, with k transitions remaining unassigned: then we have $\sum_{i_j=1}^{k} \binom{k}{i_j}$ ways of assigning at least one of the unassigned transitions to enter state q_j. This suggests defining the function $A_n(i, k)$ by

$$A_n(j, k) = \sum_{i=1}^{k} \binom{k}{i} A_n(j + 1, k - i + 2).$$

Thus, $A_n(j, k)$ denotes the number of ways of defining the transitions of states $q_j, q_{j+1}, \ldots, q_{n-1}$ knowing that we have k unassigned transitions from states $q_0, q_1, \ldots, q_{j-1}$. We must only determine our terminating condition. Since all unassigned transitions after adding states $q_0, q_1, \ldots, q_{n-2}$ must necessarily enter q_{n-1}, we get that $A_n(n, k) = 1$ for all k. Under these conditions, it is proved [1] that $A_n(2, 2)$ is equivalent to the ordinary Genocchi numbers.

However, this argument does not yield a tight upper bound. Consider Figure 1: the DFAs M_1 and M_2, which are both identical after state 3, will both be counted by our upper bound. The main idea of our new upper bound is to alleviate this problem.

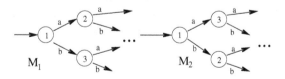

Fig. 1. Problematic DFAs M_1 and M_2.

5.1 An Improved Upper Bound

The upper bound from the last section only considered adding single states at a time. However, we may also consider adding pairs of states. This way we can control the double counting exhibited in the last section. In this section, let $M = (\{1, 2, 3, \ldots, n\}, \Sigma, \delta, 1, F)$.

Let c denote the size of our alphabet. Thus, we define a new function, call it $B_n^{(c)}(j, k)$, which will count the number of ways of defining the transitions of an n state DFA, having already defined states $1, \ldots, j - 1$ and having k extra transitions to assign to states $j, \ldots, n$. Further, the underlying graph will guaranteed to be initially connected and acyclic, and we will add states two at a time.

We let $\binom{k}{k_1 \ k_2}$ for $k_1 + k_2 \leq k$ denote the multinomial coefficient given by $k!/(k_1! k_2!(k - k_1 - k_2)!)$. The function $B_n^{(c)}(j, k)$ is defined for $n \geq 4$ by

$$B_n^{(c)}(j, k) = \sum_{k_1=1}^{k} \sum_{k_2=0}^{k-k_1} \sum_{i=1}^{c} \binom{k}{k_1 \ k_2} \binom{c}{i} B_n^{(c)}(j + 2, k - k_1 - k_2 - i + 2c)$$

$$+ \frac{1}{2} \sum_{k_1=1}^{k-1} \sum_{k_2=1}^{k-k_1} \binom{k}{k_1 \ k_2} B_n^{(c)}(j + 2, k - k_1 - k_2 + 2c)$$

$$B_n^{(c)}(n, k) = 1 \quad \forall k \geq 0$$
$$B_n^{(c)}(n - 1, k) = 2^k - 1 \quad \forall k \geq 0$$

Then we have the following theorem:

Theorem 5.1. Let $B_n^{(c)}$ be defined as above. Then $f_c(n) \leq 2^{n-2} B_n^{(c)}(2, c)$.

We may compare this upper bound for alphabets of size $c = 2$ to our old upper bounds [1], the best known lower bounds from Section 4 and the actual values [2] in Figure 4, located in Appendix A.

Acknowledgments

I must thank Kai Salomaa and Galina Jirásková, who both read this paper and made many suggestions and corrections, and the anonymous referees of DLT 2002, who made several useful comments.

References

1. Domaratzki, M.: A generalization of the Genocchi numbers with applications to enumeration of finite automata. Submitted (Preliminary version available at http://www.cs.queensu.ca.TechReports.Reports/2001-449.ps) (2001)
2. Domaratzki, M., Kisman, D., Shallit, J.: On the number of distinct languages accepted by finite automata with n states. In Dassow, J., Wotschke, D., eds.: 3rd. Intl. Workshop on Descriptional Complexity of Automata, Grammars and Related Structures, 2001. (2001) 66–78
3. Harrison, M.; A census of finite automata. Can. J. Math **17** (1965) 100–113
4. Liskovets, V.: The number of connected intial automata. Cybernetics 5 (1969) 259-262
5. Korshunov, A.: Enumeration of finite automata (In Russian). Problemy Kibernetiki **34** (1978) 5-82

6. Radke, C.: Enumeration of strongly connected sequential machines. Information and Control **8** (1965) 377-389
7. Robinson, R.: Counting strongly connected finite automata. In Alavi, Y., Chartrand, G., Lick, D., Wall, C., eds.: Graph Theory with Applications to Algorithms and Computer Science, Wiley (1985) 671-685

Appendix A: Tables

n	1	2	3	4	5	6	7
$2^{n-2}(n-1)!$	1	1	4	24	192	1920	23040
$(2n-3)!/(n-2)!$	1	1	6	60	840	15120	332640
$f_2(n)$	1	1	6	60	900	18480	487560

Fig. 2. Our lower bound, compared with the actual counts, for $k = 2$. The values of $f_2(n)$ are from Domaratzki *et al.*

n	1	2	3	4	5	6	7
$(2n-3)!/(n-2)!$	1	1	6	60	840	15120	332640
$t(n)$	1	1	6	60	900	17880	441960
$f_2(n)$	1	1	6	60	900	18480	487560

Fig. 3. Our lower bounds for an alphabet of size 2, compared with the actual counts.

n	1	2	3	4	5	6	7
$f_2(n)$	1	1	6	60	900	18480	487560
$2^{n-2}B_n^{(2)}(2,2)$	1	1	6	64	1120	26432	889216
$2^{n-2}A_n(2,2)$	1	1	6	68	1240	33168	1223264

Fig. 4. Summary of upper bounds.

Roots and Powers of Regular Languages

Sándor Horváth[1,*], Peter Leupold[2,**], and Gerhard Lischke[3,***]

[1] Department of Computer Science, Institute I of Mathematics
Eötvös Loránd University, H-1117 Budapest, Hungary
horvath@cs.elte.hu
[2] Research Group in Mathematical Linguistics
University Rovira i Virgili, E-43005 Tarragona, Spain
pl.doc@estudiants.urv.es
[3] Institute of Informatics, Faculty of Mathematics and Informatics
Friedrich Schiller University Jena, D-07743 Jena, Germany
lischke@minet.uni-jena.de

Abstract. For a set H of natural numbers, the H-power of a language L is the set of all words p^k where $p \in L$ and $k \in H$. The root of L is the set of all primitive words p such that p^n belongs to L for some $n \geq 1$. There is a strong connection between the root and the powers of a regular language L namely, the H-power of L for an arbitrary finite set H with $0, 1, 2 \notin H$ is regular if and only if the root of L is finite. If the root is infinite then the H-power for most regular sets H is context-sensitive but not context-free. The stated property is decidable.

1 Introduction

In connection with the study of properties of period and prefix languages of ω-languages, Calbrix and Nivat in [4] define the power $pow(L)$ of a language L. This is the set of all words p^k where $p \in L$ and k is a natural number. It is easy to see that for a regular language[1], its power may be a regular language too or not. Calbrix and Nivat [4] raised the problem to characterize those regular languages whose powers are also regular, and to decide the problem whether a given regular language has this property. Cachat [3] gives a partial solution to this problem showing that for a regular language L over a one-letter alphabet, it is decidable whether $pow(L)$ is regular. Also he suggests to consider as the set of exponents not only the whole set $\mathbb{N}$ of natural numbers but also an arbitrary

---

[*] The research of this author was supported by a collaboration between the HAS, Budapest, Hungary, and the JSPS, Tokyo, Japan, and also by the German-Hungarian project No. WTZ HUN 00/040.
[**] After finishing his diploma at the Friedrich Schiller University Jena he started studies in Tarragona at the 1st International PhD School in Formal Languages and Applications.
[***] Corresponding author.
[1] In the French literature the terminus *rational* is used instead of *regular*. We follow the more common usage and shall speak of regular languages.

M. Ito and M. Toyama (Eds.): DLT 2002, LNCS 2450, pp. 220–230, 2003.
© Springer-Verlag Berlin Heidelberg 2003

regular set of natural numbers. We take up this suggestion with our definition of $pow_H(L)$, see Section 2. We answer the problem of Calbrix and Nivat and the open question of Cachat for languages over any finite alphabet and almost any finite set of exponents. It turns out that the regularity of the power of a regular set L is strongly connected with the finiteness of the root of L and this in turn is connected with the structure of the Kleene term describing L.

The root of L is the set of all primitive words p such that some power p^n, for some $n \geq 1$, belongs to L. Primitive words are such words p which are not a power of any word $q \neq p$. We shall give a more formal definition in Section 2. The concept of primitive words came up with the paper [14] of Lyndon and Schützenberger, and it has received special interest in the algebraic theory of codes and formal languages (see, for instance, [16] and [10]). The relationship between the complexity of a set L and that of its root is investigated in [13]. There it is also shown that there exist regular languages the roots of which are not even context-sensitive. Thus it is quite natural that the root of a regular language L may be finite or infinite. Exactly this property is decisive for the regularity of the (finite) power of L. Therefore we split up the class REG of regular sets into two classes FR and IR of sets whose roots are finite and infinite, respectively. We show in Section 3 that FR – the class of all regular sets with a finite root – is the class of all sets which are described by Kleene terms having a special normal form. We call such a normal form to be a *root term*. IR is the class of regular sets which are not describable by a root term. We show in Section 4 that FR is closed under the power with an arbitrary finite set. In Section 5 we show that the power of any set from IR with any nonempty regular exponent containing none of the numbers 0, 1, 2, is context-sensitive butnot context-free. We also discuss the exceptions 0, 1, 2, and give some generalization. In Section 6 we discuss whether it is decidable for a regular set whether its power with any nonempty regular exponent is regular too. For the set $\mathbb{N}$ of all natural numbers as an exponent the only open case is that the language has a nonregular intersection with its root and the difference between the power of the language and the language itself has a finite root. We suppose that also in this case the regularity ofthe power of the language is decidable.

2   Preliminaries and Definitions

Even though the following is standard in the literature (see, e.g., the textbooks [15,16,7]) we briefly recall the most important notions. For our whole paper, let X be a fixed finite nonempty alphabet, and $\mathbb{N}$ be the set of all natural numbers. X^* is the free monoid generated by X or the set of all words over X. The empty word we denote by e, and $X^+ =_{Df} X^* \setminus \{e\}$. A (formal) language (over X) is a subset L of X^*, $L \subseteq X^*$. For a word $p \in X^*$, $|p|$ denotes the length of p, and for a set M, $|M|$ denotes the cardinality of M. For $k \in \mathbb{N}$, p^k denotes the concatenation of k copies of the word p, and $p^0 = e$. p^* denotes the set $\{p^k : k \in \mathbb{N}\}$, and p^*q the set $\{p^kq : k \in \mathbb{N}\}$.

Definition 1. For a language $L \subseteq X^*$ and a set $H \subseteq \mathbb{N}$, the H-*power of* L is
$pow_H(L) =_{Df} \{p^k : p \in L \wedge k \in H\}$.
The $\mathbb{N}$-*power* of L we denote by $pow(L)$.

Because of $pow_\emptyset(L) = \emptyset$, for the rest of the paper, $H \neq \emptyset$ should hold in $pow_H(L)$.

Definition 2. A word p is *primitive* iff there is no word $q \neq p$ and no number $n \in \mathbb{N}$ such that $p = q^n$.

Definition 3. The *root* of a word $p \in X^+$ is the unique primitive word q such that $p = q^n$ for some $n \in \mathbb{N}$. $\sqrt{p}$ denotes the root of p.
The *degree* of a word $p \in X^+$ is the unique natural number n such that $p = \sqrt{p}^n$. $deg(p)$ denotes the degree of p.
For a language L, $\sqrt{L} =_{Df} \{\sqrt{p} : p \in L \wedge p \neq e\}$ is the *root of* L.
$deg(L) =_{Df} \{deg(p) : p \in L \wedge p \neq e\}$ is the *degree of* L.

Definition 4. REG denotes the class of all regular sets. FR denotes the class of all regular sets L such that $\sqrt{L}$ is finite. IR denotes the class of all regular sets L such that $\sqrt{L}$ is infinite. CF and CS denote the classes of all context-free and context-sensitive languages, respectively.

Definition 5. A language L is *ultimately periodic* iff there are finitely many (not necessarily different) words $u_1, \ldots, u_k, v_1, \ldots, v_k$ such that
$L = v_1 u_1^* \cup v_2 u_2^* \cup \cdots \cup v_k u_k^*$.

This last Definition 5 is a generalization of the notion *ultimately periodic* used by Büchi [2] and Eilenberg and Schützenberger [6] (see also [3]). Transferring this definition to languages over a one-letter alphabet $X = \{a\}$ and identifying a word a^n with the natural number n, by some calculation we get:

Definition 6 ([2,6]). A set $H \subseteq \mathbb{N}$ is *ultimately periodic* iff there are natural numbers $n > 0$ and $m \geq 0$ and sets $I \subseteq \{0, \ldots, m-1\}$ and $P \subseteq \{m, \ldots, m+n-1\}$ such that $H = I \cup \{l + n \cdot r : l \in P \wedge r \in \mathbb{N}\}$.

Theorem 1 ([2,6]). *A set $H \subseteq \mathbb{N}$ is regular if and only if it is ultimately periodic.*

3   Characterizing FR and IR by Kleene Terms

In 1956 Kleene [11] introduced the calculus of regular expressions which we first want to repeat in a slightly modified way (cf. [7,17]). First we transfer our alphabet X to $X_T =_{Df} \{\mathbf{x} : x \in X\} \cup \{\mathbf{O}, (,), \circ, \vee, \langle, \rangle\}$. The *regular expressions* or *Kleene terms over* X are words over X_T which are defined recursively as follows.

(a) **O** is a Kleene term.

(b) For each $x \in X$, **x** is a Kleene term. (a) and (b) are called *elementary terms*.

(c) If t_1 and t_2 are Kleene terms, then $(t_1 \vee t_2)$ is a Kleene term, called their *sum*.

(d) If t_1 and t_2 are Kleene terms, then $(t_1 \circ t_2)$ is a Kleene term, called their *product*.

(e) If t is a Kleene term then $\langle t \rangle$ is a Kleene term, called its *iteration*.

T_X denotes the set of all Kleene terms over X.

As usual, a pair of parentheses (,) may be omitted whenever the omission will not cause any confusion and assuming that iteration has higher precedence than $\circ$ and $\vee$, and $\circ$ has higher precedence than $\vee$.

By the following function ϕ we define for each Kleene term $t \in T_X$ that set $L = \phi(t) \subseteq X^*$ which is *described by* t.

(a) $\phi(\mathbf{O}) =_{Df} \emptyset$.

(b) $\phi(\mathbf{x}) =_{Df} \{x\}$ for each $x \in X$.

(c) $\phi(t_1 \vee t_2) =_{Df} \phi(t_1) \cup \phi(t_2)$.

(d) $\phi(t_1 \circ t_2) =_{Df} \phi(t_1) \cdot \phi(t_2)$, where $L_1 \cdot L_2 =_{Df} \{pq : p \in L_1 \wedge q \in L_2\}$.

(e) $\phi(\langle t \rangle) =_{Df} \phi(t)^*$, where $L^* =_{Df} \{e\} \cup L \cup L \cdot L \cup L \cdot L \cdot L \cup \cdots = \bigcup_{i=0}^{\infty} L^i$.

It is well-known (cf. [11,7,17]) that a language is regular if and only if it may be described by a Kleene term.

Two terms describing the same set are called *equivalent*:

for $t_1, t_2 \in T_X$, $t_1 \equiv t_2 =_{Df} \phi(t_1) = \phi(t_2)$.

Now we specify special Kleene terms which we shall use to describe those regular sets whose roots are finite.

Definition 7. A *root term* is a Kleene term having the form

$$((t_{1,1} \circ \langle t_{1,2} \rangle) \vee (t_{2,1} \circ \langle t_{2,2} \rangle) \vee \cdots \vee (t_{k,1} \circ \langle t_{k,2} \rangle))$$

for a natural number k where each $t_{i,j}$ is a product of elementary terms, and for each i, either $\sqrt{\phi(t_{i,1})} = \sqrt{\phi(t_{i,2})}$ or one of the two roots is $\emptyset$. Additionally, **O** and $\langle \mathbf{O} \rangle$ are root terms.

By Definition 4, FR is the class of all regular sets having a finite root and IR = REG\FR. Now we show that both classes are characterized by special Kleene terms where the root terms characterize FR.

Theorem 2. *A regular language has a finite root if and only if it can be described by a root term, i.e.* FR $= \{\phi(t) : t$ *is a root term*$\}$.

Proof. **O** and $\langle \mathbf{O} \rangle$ are terms t such that $\sqrt{\phi(t)} = \emptyset$ is finite. Further it is obvious that $|\sqrt{\phi(t)}| \leq 1$ if t is a product of elementary terms.

If $\sqrt{\phi(t_1)} = \sqrt{\phi(t_2)} = \{p\}$ for a primitive word p or if one of the roots is empty, then $\sqrt{\phi(t_1 \circ \langle t_2 \rangle)}$ contains at most one element. Further $|\sqrt{\phi(t_1 \vee t_2)}| \leq |\sqrt{\phi(t_1)}| + |\sqrt{\phi(t_2)}|$ because of $\sqrt{\phi(t_1 \vee t_2)} = \sqrt{\phi(t_1)} \cup \sqrt{\phi(t_2)}$. It follows that each root term describes a set with finite root.

Now let L be a regular set with a finite root $\{p_1, \ldots, p_k\}$ and $e \notin L$. Then
$$L = \bigcup_{i=1}^{k} L_i \text{ where } L_i =_{Df} L \cap p_i^*,$$
and the sets L_i are pairwise disjoint. If each L_i may be described by a root term then also L. If each L_i is ultimately periodic then (because of $\sqrt{L_i} = \{p_i\}$) it may be described by a root term. If one of the L_i's is not ultimately periodic then (because of the one-one correspondence between L_i and $deg(L_i)$ in this case) also for this i, $deg(L_i)$ is not ultimately periodic. This means, by Theorem 1, that $deg(L_i)$ and therefore L_i is not regular which contradicts the regularity of L and the definition of L_i. $\square$

Corollary 1. *A regular language has an infinite root if and only if it cannot be described by a root term.*

Our next goal is to characterize those Kleene terms which are not equivalent to root terms.

We call a Kleene term a *reduced term* if it is simplified as much as possible using the following equivalences: $\langle \langle t \rangle \rangle \equiv \langle t \rangle$, $(t \vee t) \equiv t$, $(\langle \mathbf{O} \rangle \circ t) \equiv (t \circ \langle \mathbf{O} \rangle) \equiv t$, $(\mathbf{O} \circ t) \equiv (t \circ \mathbf{O}) \equiv \mathbf{O}$, $(t \vee \mathbf{O}) \equiv (\mathbf{O} \vee t) \equiv t$.

Definition 8. A *violating term* is a Kleene term having one of the forms
$$(t_1 \circ \langle t_2 \rangle) \text{ or } (\langle t_1 \rangle \circ t_2) \qquad (V1)$$
$$\text{or } \langle t_1 \vee t_2 \rangle \qquad (V2)$$
where $\sqrt{\phi(t_1)} = \sqrt{\phi(t_2)} = \{p\}$ does not hold, for any primitive word p.

Theorem 3. *A regular language has an infinite root if and only if it can be described by a reduced term which is not equivalent to a root term and which contains a violating term as a subterm.*

One direction of the theorem is trivial because of Corollary 1. An independent proof of this direction follows with the next theorem which is due to Lyndon and Schützenberger [14], for a proof see [16].

Theorem 4. *If $u^m v^n = w^k$ for nonempty words u, v, w and natural numbers $m, n, k \geq 2$ then $\sqrt{u} = \sqrt{v} = \sqrt{w}$.*

This means, u, v and w are powers of a common word. Such solutions of the equation are called *trivial solutions*.

Corollary 2. *For primitive words p and q with $p \neq q$ and numbers $i, j \geq 2$, $p^i q^j$ is primitive again.*

Corollary 3. *For primitive words p and q with $p \neq q$ there are at most two nonprimitive words in each of the languages pq^* and p^*q.*

Corollary 4. *For two nonempty words u, v with $\sqrt{u} \neq \sqrt{v}$ there are at most two nonprimitive words in each of the languages uv^* and u^*v.*

Proof of Theorem 3. Let t_{V1} and t_{V2} be violating terms of the form (V1) and (V2), respectively. Then there must exist nonempty words $u \in \phi(t_1)$ and $v \in \phi(t_2)$ such that $\sqrt{u} \neq \sqrt{v}$. Then in the case of (V1), $uv^* \subseteq \phi(t_{V1})$ or $u^*v \subseteq \phi(t_{V1})$, and because of Corollary 4 both sets have an infinite root. In the case of (V2), $\phi(t_1)^* \cdot \phi(t_2)^* \subset \phi(t_{V2})$ and it has an infinite root because of Corollary 2. Finally, $\phi(t)$ where t is a reduced term which is not equivalent to a root term and which has t_{V1} or t_{V2} as a subterm, must also have an infinite root. To show the opposite direction we have to show that each reduced Kleene term t which is not equivalent to a root term, must have a violating term as a subterm. First, it is clear that there must be an iteration in t because of otherwise $\phi(t)$ would be finite and therefore t would be equivalent to a root term. It is easy to see that each subterm of t containing an iteration either is equivalent to a subterm fulfilling the conditions of (V1) or (V2) or describes a set with a finite root. In the latter case, this subterm is equivalent to a root term. □

To illustrate the last remark in the proof, let us look at the following examples. Take $t_1 = \mathbf{b}$ and $t_2 = (\mathbf{a} \circ \mathbf{b})$. Then $(t_1 \circ \langle t_2 \rangle)$ is a violating term and a subterm of both t and t' where
$$t = (\mathbf{a} \circ (\mathbf{b} \circ \langle \mathbf{a} \circ \mathbf{b} \rangle)), \qquad t' = \langle \mathbf{a} \circ (\mathbf{b} \circ \langle \mathbf{a} \circ \mathbf{b} \rangle) \rangle.$$
It is obvious that t is equivalent to a root term. That also t' is equivalent to a root term is more generally given by the following

Fact. *Each Kleene term $\langle t_1 \circ \langle t_2 \rangle \rangle$ where $\phi(t_1) = \{w_1\}$, $\phi(t_2) = \{w_2\}$, and $\sqrt{w_1} = \sqrt{w_2} = \{p\}$, is equivalent to a root term.*

To prove this fact we consider two cases and use the following lemma which is well-known from number theory.

Lemma 1. *For any natural numbers k and l which are relatively prime there exists a number $n_0 \in \mathbb{N}$ such that each natural number $n \geq n_0$ can be represented as $n = ik + jl$ with natural numbers i, j.*

Case 1: $gcd(deg(w_1), deg(w_2)) = 1$ (*gcd* denotes the *greatest common divisor*). Then $\phi(\langle t_1 \circ \langle t_2 \rangle \rangle)$ is described as the sum of at most n_0 terms describing special words from p^* and one term describing $p^{n_0} \cdot p^*$, where n_0 is taken from Lemma 1. This is equivalent to a root term.

Case 2: $gcd(deg(w_1), deg(w_2)) = k > 1$. Then each degree of a word from $\phi(\langle t_1 \circ \langle t_2 \rangle \rangle)$ is a multiple of k and we can proceed in the same way as in Case 1 starting with p^k instead of p.

Thus $\langle t_1 \circ \langle t_2 \rangle \rangle$ is equivalent to a root term.

4   Powers of FR Sets

Now we consider powers of regular languages which have finite roots. It turns out that they have the same behavior for arbitrary finite exponents, namely that their powers are regular too with finite roots.

Theorem 5. *The class* FR *of regular sets having a finite root is closed under the power with finite sets.*

Proof. Let L be a regular language with a finite root $\{p_1, \ldots, p_k\}$ and $e \notin L$, and let $L_i =_{Df} L \cap p_i^*$ for each $i \in \{1, \ldots, k\}$ as in the proof of Theorem 2. Since $L_i \subseteq p_i^*$ and $L_i \in$ REG, L_i is isomorphic to a regular set M_i of natural numbers, namely $M_i = deg(L_i)$. For each $n \in \mathbb{N}$, $M_i \cdot n =_{Df} \{m \cdot n : m \in M_i\}$ is regular too. Therefore, for a finite set $H \subseteq \mathbb{N}$, also $\bigcup_{n \in H} M_i \cdot n$ is regular which is isomorphic to $pow_H(L_i)$. Then $pow_H(L) = \bigcup_{i=1}^{k} pow_H(L_i)$ is regular, and $\sqrt{pow_H(L)} \subseteq \{p_1, \ldots, p_k\}$ is finite. $\square$

If H is infinite then $pow_H(L)$ may be nonregular and even non-context-free. This is true even in the case of a one-letter alphabet where the root of each nonempty set (except $\{e\}$) has exactly one element, which can be seen by the following example.

Let $L = \phi(t)$ where $t = (\mathbf{a} \circ \mathbf{a} \circ \mathbf{a} \circ \langle \mathbf{a} \circ \mathbf{a} \rangle)$.

Then $L \in$ FR but $pow(L) = \{a^k : k \in \mathbb{N} \setminus \{2^m : m \geq 0\}\} \notin$ CF.

Therefore it remains a problem to characterize those regular sets L with finite roots where $pow_H(L)$ is regular for any (maybe regular) set H.

5   Powers of IR Sets

Our next goal is to show that the powers of arbitrary (not necessarily regular) languages which have infinite roots are not regular, even more, they are not even context-free. The exponent set may be an arbitrary set of natural numbers containing at least one element which is greater than 2, and the number 1 is not in the exponent set or the language has a regular root or a regular intersection with its root. To prove our theorem we need the following known results from language theory.

Lemma 2 (Pumping lemma of Bar-Hillel, Perles, Shamir [1], see also [7,15]). *For every context-free language L there exists a natural number m such that every $w \in L$ with $|w| > m$ is of the form $w_1 w_2 w_3 w_4 w_5$ where: $w_2 w_4 \neq e$, $|w_2 w_3 w_4| < m$, and $w_1 w_2^i w_3 w_4^i w_5 \in L$ for all $i \in \mathbb{N}$.*

Lemma 3 ([16]). *If two words u and u' from X^+ are cyclic permutations of one another, i.e., $u = u_1 u_2$ and $u' = u_2 u_1$ for $u_1, u_2 \in X^*$, then $deg(u) = deg(u')$, $|\sqrt{u}| = |\sqrt{u'}|$, and therefore u is primitive if and only if u' is primitive.*

Lemma 4 ([14]). *If two words $q, s \in X^+$ have powers q^i and s^j $(i, j \geq 1)$ with a common prefix of length $|q| + |s|$ then $\sqrt{q} = \sqrt{s}$.*

Theorem 6. *For every language L which has an infinite root and for every set $H \subseteq \mathbb{N}$, $pow_H(L)$ is not context-free if there is at least one number in H which is greater than 2 and one of the following conditions is true: $1 \notin H$, or at least one of $\sqrt{L}$ and $L \cap \sqrt{L}$ is regular.*

Remark. For a language L having an infinite root, the conditions $\sqrt{L}$ regular and $L \cap \sqrt{L}$ regular are independent from each other, which is shown by the following examples.
Let $L_1 =_{Df} \{ab^n ab^n : n \in \mathbb{N}\} \cup \{ab^n : n \in R\}$ where R is a nonenumerable set of natural numbers, and $L_2 =_{Df} pow_{\{2\}}(X^*)$. Then both L_1 and L_2 have an infinite root, $\sqrt{L_1}$ is regular, $L_1 \cap \sqrt{L_1}$ is nonenumerable, but $\sqrt{L_2} = Q$ (the set of all primitive words) is not regular and $L_2 \cap \sqrt{L_2} = \emptyset$ is regular.

Proof of Theorem 6. Let $L \subseteq X^*$ be a language such that $\sqrt{L}$ is infinite, and let $H \subseteq \mathbb{N}$ with $H \setminus \{0, 1, 2\} \neq \emptyset$. We define $L' =_{Df} pow_H(L)$ if $1 \notin H$. Otherwise let $L' =_{Df} pow_H(L) \setminus \sqrt{L}$ or $L' =_{Df} pow_H(L) \setminus (L \cap \sqrt{L})$ if $\sqrt{L}$ or $L \cap \sqrt{L}$ is regular, respectively.
Then, in each case, $\sqrt{L'}$ is infinite, there is no primitive word in L' and, if $pow_H(L)$ was context-free then also L' would be context-free. But we show that the latter is not true.

Assume, L' is context-free, and let $n \geq 3$ be a fixed number from H. Then there exists m according to Lemma 2, and there exists a word $z \in L'$ such that $deg(z) \geq n$ and $|\sqrt{z}| > m$. Let $p =_{Df} \sqrt{z}$ and $k =_{Df} deg(z)$. Then $|z| = k \cdot |p|$, $|p| > m$, and by Lemma 2, $z = p^k = w_1 w_2 w_3 w_4 w_5$ where $w_2 w_4 \neq e$, $|w_2 w_3 w_4| < m$, and $w_1 w_2^i w_3 w_4^i w_5 \in L'$ for each $i \in \mathbb{N}$. Especially, for $i = 0$, $x =_{Df} w_1 w_3 w_5 \in L'$, and therefore x is nonprimitive. Now let $z' =_{Df} w_5 w_1 w_2 w_3 w_4$, $q =_{Df} \sqrt{z'}$, $x' =_{Df} w_5 w_1 w_3$, and $s =_{Df} \sqrt{x'}$. Because of Lemma 3, x' is nonprimitive, $|s| \leq \frac{|x'|}{2}$, and $|q| = |p|$. We have $z' = q^k$ and $x' = q^{k-1} q'$ for some nonempty word q' with $|q'| < |q|$ (because of $|w_2 w_3 w_4| < m < |q|$). The two words z' and x' which are powers of q and s, respectively, have a common prefix q^{k-1} of length $(k-1) \cdot |q|$. $|s| \leq \frac{|x'|}{2} < \frac{k}{2} \cdot |q|$ holds and therefore $|q| + |s| \leq (k-1) \cdot |q|$ because of $k \geq 3$. It follows by Lemma 4 that $q = s$ and $x' = s^{k-1} q'$, a contradiction. $\square$

Corollary 5. *If $L \in \mathrm{IR}$, $H \setminus \{0, 1, 2\} \neq \emptyset$, and $\sqrt{pow_H(L) \setminus L}$ is infinite, then $pow_H(L)$ is not context-free.*

This is shown in the same way using $L' =_{Df} pow_H(L) \setminus L$.

Theorem 7. *For every context-sensitive set H of natural numbers and every context-sensitive language L, the H-power of L is context-sensitive.*

Proof. We sketch an algorithm which, for a fixed language $L \neq \emptyset$ and a nonempty set H of natural numbers, accepts any $u \in X^*$ if and only if $u \in pow_H(L)$ (If one of L and H is empty, then trivially $pow_H(L) = \emptyset$ is context-sensitive).

    1. If $u \in L$ and $1 \in H$ or $u = e$ and $0 \in H$ then accept.
    2. Compute $p = \sqrt{u}$ and $d = deg(u)$.
    3. For $i := 1$ to $\lfloor \frac{d}{2} \rfloor$ do
        if $p^i \in L$ and $\frac{d}{i} \in H$ then accept.
    4. Reject.

All steps of the algorithm can be done in linear bounded space if L and H are context-sensitive (for computing the root and the degree in linear bounded space see, e.g., [9]). This proves the context-sensitivity of $pow_H(L)$. $\square$

Now we consider the exceptions 0, 1, and 2 where we find out a different behavior.

Corollary 6.
    (i) *For each $L \in$ REG and $H \subseteq \{0,1\}$, $pow_H(L) \in$ REG.*
    (ii) *For each $L \in$ FR, $pow_{\{2\}}(L) \in$ FR.*
    (iii) *For each $L \in$ IR, $pow_{\{2\}}(L) \notin$ REG.*

Proof. (i) is trivial because of $pow_{\{0\}}(L) = \{e\}$, and $pow_{\{1\}}(L) = L$. (ii) follows from Theorem 5. The proof of (iii) can be done using the Pumping lemma for regular sets and Corollary 4. $\square$

It is not possible to make a general statement whether the $\{2\}$-power of a regular language with infinite root is context-free or not. This is shown by the following examples.
    Let $L_1 =_{Df} \phi(\mathbf{a} \circ \langle \mathbf{b} \rangle)$ and $L_2 =_{Df} \phi(\langle \mathbf{a} \vee \mathbf{b} \rangle)$.
Then $L_1, L_2 \in$ IR, but $L_1^{(2)} \in$ CF and $L_2^{(2)} \notin$ CF. Therefore we excluded 0, 1, 2 in our statement of Theorem 6.

Let us look now to the set $\mathbb{N}$ as set of exponents which was the starting point for the investigations in [4] and [3]. It is easy to see that $pow(L) = L$ if $L = \phi(\langle t \rangle)$ for an arbitrary Kleene term t. Therefore $pow(L) \in$ REG is possible even for $L \in$ IR. On the other hand, $pow(L_1) \notin$ CF for L_1 from our example. Also $pow(L) = L$ is not necessary for an IR set L to have the property $pow(L) \in$ REG, which is shown by the following example.
    Let $L = \phi(\mathbf{a} \vee \langle\langle \mathbf{a} \rangle \circ \mathbf{b} \circ \langle \mathbf{a} \rangle\rangle)$.
Then $L \in$ IR, $pow(L) = \{a, b\}^* \neq L$ and $pow(L) \in$ REG.

6   Decidability

One of the main (unsolved) problems of [4] and [3] was the question whether it is decidable for a regular set L whether $pow(L)$ is regular or not. If the set L is given by a regular grammar then we have some partial results. Horváth and Ito [8] have shown that it is decidable for an arbitrary regular grammar G, whether $\sqrt{L(G)}$ is finite. In the affirmative case, using Cachat's algorithm [3], it is also decidable whether $pow(L(G))$ is regular[2]. Thus for a regular grammar G and a nonempty set $H \subseteq \mathbb{N}$, we have the following procedure to decide the quality of $pow_H(L(G))$.

Let $L =_{Df} L(G)$.
1. If $H \subseteq \{0,1\}$ then $pow_H(L) \in$ REG.
2. Decide whether $\sqrt{L}$ is finite.
3. If $\sqrt{L}$ is finite then
    3.1. if H is finite then $pow_H(L) \in$ REG,
    3.2. if $H = \mathbb{N}$ then it is decidable whether $pow_H(L) \in$ REG.
4. If $\sqrt{L}$ is infinite then
    4.1. if $H \in \{\{2\}, \{0,2\}\}$ then $pow_H(L) \notin$ REG,
    4.2. if $H \setminus \{0,1,2\} \neq \emptyset$ then
        if $1 \notin H$ or $L \cap \sqrt{L}$ is regular or $pow_H(L) \setminus L$ has an infinite root
        then $pow_H(L) \notin$ CF.

It is obvious which cases are still unsolved. From the original problem of Calbrix and Nivat ($H = \mathbb{N}$) the following case is still open: Given a regular language L which has a nonregular intersection with its root, and for which $pow_H(L) \setminus L$ has a finite root – is the power of L regular? We suppose that also in this case, the regularity of $pow_H(L)$ is decidable. Therefore we formulate the

Conjecture. *For an arbitrary regular grammar G and $L = L(G)$ it is decidable whether $pow(L)$ is regular.*

References

1. Y.BAR-HILLEL, M.PERLES, E.SHAMIR, *On formal properties of simple phrase structure grammars*, Z. Phonetik Sprachwiss. Kommunikationsforsch. 14 (1961), 143–172.
2. J.R.BÜCHI, *Weak second-order arithmetic and finite automata*, Z. f. Math. Logik u. Grundl. d. Math. 6 (1960), 66–92.
3. T.CACHAT, *The power of one-letter rational languages*, in [12], 145–154.
4. H.CALBRIX, M.NIVAT, *Prefix and period languages of rational ω-languages*, in Proc. Developments in Language Theory, Magdeburg 1995, World Scientific, 1996, 341–349.
5. C.CHOFFRUT, J.KARHUMÄKI, *Combinatorics of words*, in [15], 329–438.
6. S.EILENBERG, M.P.SCHÜTZENBERGER, *Rational sets in commutative monoids*, J. of Algebra 13 (1969), 173–191.
7. J.E.HOPCROFT, J.D.ULLMAN, *Introduction to automata theory, languages, and computation*, Addison-Wesley, Reading (Mass.), 1979.
8. S.HORVÁTH, M.ITO, *Decidable and undecidable problems of primitive words, regular and context-free languages*, J. Universal Computer Science 5 (1999), 532–541.
9. S.HORVÁTH, M.KUDLEK, *On classification and decidability problems of primitive words*, PU.M.A. 6 (1995), 171–189.

---

[2] This will be published in a forthcoming paper by S.Horváth.

10. H.Jürgensen, S.Konstantinidis, *Codes*, in [15], 511–607.
11. S.C.Kleene, *Representation of events in nerve nets and finite automata*, Automata Studies, Princeton Univ. Press, Princeton (N.J.), 1956, 2–42.
12. W.Kuich, G.Rozenberg, A.Salomaa (Eds.), *Developments in Language Theory*, 5th International Conference, Wien 2001, Lecture Notes in Computer Science 2295, Springer-Verlag, Berlin-Heidelberg, 2002.
13. G.Lischke, *The root of a language and its complexity*, in [12], 272–280.
14. R.C.Lyndon, M.P.Schützenberger, *On the equation $a^M = b^N c^P$ in a free group*, Michigan Math. Journ. 9 (1962), 289–298.
15. G.Rozenberg, A.Salomaa (Eds.), *Handbook of formal languages, Vol. 1*, Springer-Verlag, Berlin-Heidelberg, 1997.
16. H.J.Shyr, *Free monoids and languages*, Hon Min Book Company, Taichung, 1991.
17. S.Yu, *Regular languages*, in [15], 41–110.

Innermost Termination
of Context-Sensitive Rewriting

Jürgen Giesl[1] and Aart Middeldorp[2],[*]

[1] LuFG Informatik II, RWTH Aachen, Ahornstr. 55, 52074 Aachen, Germany
`giesl@informatik.rwth-aachen.de`
[2] Institute of Information Sciences and Electronics, University of Tsukuba
Tsukuba 305-8573, Japan
`ami@is.tsukuba.ac.jp`

Abstract. Context-sensitive rewriting is a restriction of term rewriting used to model evaluation strategies in functional programming and in programming languages like O BJ. For example, under certain conditions termination of an O BJ program is equivalent to innermost termination of the corresponding context-sensitive rewrite system [18]. To prove termination of context-sensitive rewriting, several methods have been proposed in the literature which transform context-sensitive rewrite systems into ordinary rewrite systems such that termination of the transformed ordinary system implies termination of the original context-sensitive system. Most of these transformations are not very satisfactory when it comes to proving *innermost* termination. We investigate the relationship between termination and innermost termination of context-sensitive rewriting and we examine the applicability of the different transformations for innermost termination proofs. Finally, we present a simple transformation which is both sound and complete for innermost termination.

1 Introduction

Evaluation in functional languages is often guided by specific evaluation strategies. For example, in the program consisting of the rules

$$\mathsf{from}(x) \to x : \mathsf{from}(\mathsf{s}(x)) \quad \mathsf{nth}(0, x : y) \to x \quad \mathsf{nth}(\mathsf{s}(n), x : y) \to \mathsf{nth}(n, y)$$

the term $\mathsf{nth}(\mathsf{s}(0), \mathsf{from}(0))$ admits a finite reduction to $\mathsf{s}(0)$ as well as infinite reductions. The infinite reductions can for instance be avoided by always contracting the outermost redex. Context-sensitive rewriting [16,17] provides an alternative way of solving the non-termination problem and of dealing with infinite data objects. Here, every n-ary function symbol f is equipped with a *replacement map* $\mu(f) \subseteq \{1, \ldots, n\}$ which indicates which arguments of f may be evaluated and a contraction of a redex is allowed only if it does not take place in a forbidden argument of a function symbol somewhere above it. So by

---

[*] Partially supported by the Grant-in-Aid for Scientific Research (C)(2) 13224006 of the Ministry of Education, Culture, Sports, Science and Technology of Japan.

M. Ito and M. Toyama (Eds.): DLT 2002, LNCS 2450, pp. 231–244, 2003.

defining $\mu(:) = \{1\}$, contractions in the argument t of a term $s : t$ are forbidden. Now in the example infinite reductions are no longer possible while normal forms can still be computed. (See [20] for the relationship between normalization under ordinary rewriting and under context-sensitive rewriting.) Context-sensitive rewriting can also model the usual evaluation strategy for conditionals.

Example 1.

$$0 \leqslant y \to \text{true} \qquad\qquad \text{p}(0) \to 0$$
$$\text{s}(x) \leqslant 0 \to \text{false} \qquad\qquad \text{p}(\text{s}(x)) \to x$$
$$\text{s}(x) \leqslant \text{s}(y) \to x \leqslant y \qquad\qquad \text{if}(\text{true}, x, y) \to x$$
$$x - y \to \text{if}(x \leqslant y, 0, \text{s}(\text{p}(x) - y)) \qquad \text{if}(\text{false}, x, y) \to y$$

Because of the "$-$"-rule, this system is not terminating. But in functional languages if's first argument is evaluated first and either the second or third argument is evaluated afterwards. Again, this can easily be modeled with context-sensitive rewriting by the replacement map $\mu(\text{if}) = \{1\}$ which forbids all reductions in the arguments t_2 and t_3 of $\text{if}(t_1, t_2, t_3)$.

In programming languages like OBJ [4,5,12], the user can supply strategy annotations to control the evaluation [6,21,22]. For every n-ary symbol f, a (positive) strategy annotation is a list $\varphi(f)$ of numbers $(i_1, \ldots, i_k)$ from $\{0, 1, \ldots, n\}$. When reducing a term $f(t_1, \ldots, t_n)$ one first has to evaluate the i_1-th argument of f (if $i_1 > 0$), then one evaluates the i_2-th argument (if $i_2 > 0$), and so on, until a 0 is encountered. At this point one tries to evaluate the whole term $f(\ldots)$ at its root position. So in order to enforce the desired evaluation strategy for if in Example 1, it has to be equipped with the strategy annotation $(1, 0)$.

Context-sensitive rewriting can simulate OBJ's evaluation strategy. A strategy is *elementary* if for every defined[1] symbol f, $\varphi(f)$ contains a single occurrence of 0, at the end. Lucas [18] showed that for elementary strategies, the OBJ program terminates iff the corresponding context-sensitive rewrite system (CSRS) is *innermost* terminating[2]. Here $\mu(f) = \{i \in \varphi(f) \mid i > 0\}$. For example, the program with the rules $\text{f}(\text{a}) \to \text{f}(\text{a})$ and $\text{a} \to \text{b}$ is terminating if $\varphi(\text{f}) = (1, 0)$ and $\varphi(\text{a}) = (0)$. The corresponding CSRS with $\mu(\text{f}) = \{1\}$ is not terminating, but *innermost* terminating. Thus, to simulate OBJ evaluations with CSRSs, we have to restrict ourselves to innermost reductions where (allowed) arguments to a function are evaluated before evaluating the function.

Because of this connection to OBJ and also because innermost termination is easier to prove automatically than termination [1], it is worthwhile to investigate innermost termination of context-sensitive rewriting. (An alternative approach to prove termination of OBJ-like programs by direct induction proofs is proposed in [8].) Termination of CSRSs has been studied in a number of papers (e.g., [3,7,9,11,15,16,17,20,23]). Apart from a direct semantic characterization [23] and some recent extensions of standard termination methods for term rewriting to context-sensitive rewriting [3,15], all other proposed methods transform CSRSs into ordinary term rewrite systems (TRSs) such that termination of the transformed TRS implies termination of the original CSRS (i.e., all

---

[1] Every symbol on the root position of a left-hand side of a rule is called *defined*.

[2] The "if" direction even holds without the restriction to elementary strategies [18].

these transformations are *sound*). Direct approaches to termination analysis of CSRSs and transformational approaches both have their advantages. Techniques for proving termination of ordinary term rewriting have been studied extensively and the main advantage of the transformational approach is that in this way, all termination techniques for ordinary TRSs including future developments can be used to infer termination of CSRSs. For instance, the methods of [3,15] are unable to handle systems like Example 1.

After introducing the termination problem of context-sensitive rewriting in Section 2, in Section 3 we review the results of Lucas [18] on innermost termination of CSRSs and we show that the two transformations Θ_1 and Θ_2 of [9] are sound for innermost termination as well. Despite its soundness Θ_2 is not very useful for proving innermost termination, because termination and innermost termination coincide for the TRSs it produces. In Section 4 we show that for the class of orthogonal CSRSs, innermost termination already implies termination. This result is independent from the transformation framework and is of general interest when investigating the termination behavior of CSRSs. A consequence of this result is that for this particular class, Θ_1 is complete for innermost termination. In Section 5 we present a new transformation Θ_3 which is both sound and complete for innermost termination, for arbitrary CSRSs. Surprisingly, such a transformation can be obtained by just a small modification of Θ_1. In spite of the similarity between the two transformations, the new completeness proof is non-trivial. We make some remarks on a possible simplification of Θ_3 and on *ground* innermost termination in Section 6. In Section 7 we show that Θ_3 is equally powerful as Θ_1 when it comes to (non-innermost) termination. Due to lack of space, many proofs have been omitted. They can be found in [10].

2   Termination of Context-Sensitive Rewriting

Familiarity with the basics of term rewriting [2] is assumed. We require that every signature $\mathcal{F}$ contains a constant. A function $\mu\colon \mathcal{F} \to \mathcal{P}(\mathbb{N})$ is a *replacement map* if $\mu(f)$ is a subset of $\{1,\ldots,\text{arity}(f)\}$ for all $f \in \mathcal{F}$. A *CSRS* $(\mathcal{R},\mu)$ is a TRS $\mathcal{R}$ over a signature $\mathcal{F}$ equipped with a replacement map μ. The context-sensitive rewrite relation $\to_{\mathcal{R},\mu}$ is defined as the restriction of the usual rewrite relation $\to_{\mathcal{R}}$ to contractions of redexes at *active* positions. A position π in a term t is active if $\pi = \epsilon$ (the root position), or $t = f(t_1,\ldots,t_n)$, $\pi = i\pi'$, $i \in \mu(f)$, and π' is active in t_i. So $s \to_{\mathcal{R},\mu} t$ iff there is a rule $l \to r$ in $\mathcal{R}$, a substitution σ, and an active position π in s such that $s|_\pi = l\sigma$ and $t = s[r\sigma]_\pi$. If all active arguments of $l\sigma$ are in μ-normal form, then the reduction step is *innermost* and we write $s \xrightarrow{i}_{\mathcal{R},\mu} t$. Here a μ-normal form is a normal form w.r.t. $\to_{\mathcal{R},\mu}$. We abbreviate $\to_{\mathcal{R},\mu}$ to $\to_\mu$ and $\xrightarrow{i}_{\mathcal{R},\mu}$ to $\xrightarrow{i}_\mu$ if $\mathcal{R}$ is clear from the context. A CSRS $(\mathcal{R},\mu)$ is *left-linear* if the left-hand sides of the rewrite rules in $\mathcal{R}$ are linear terms (i.e., they do not contain multiple occurrences of the same variable). Let $l \to r$ and $l' \to r'$ be renamed versions of rewrite rules of $\mathcal{R}$ such that they have no variables in common and suppose $l|_\pi$ and l' are unifiable with most general unifier σ for some non-variable active position π in l. The pair of terms $\langle l[r']_\pi\sigma, r\sigma\rangle$ is a *critical pair* of $(\mathcal{R}, \mu)$, except when $l \to r$ and $l' \to r'$ are renamed versions of the same rewrite rule and $\pi = \epsilon$. A *non-overlapping* CSRS has no critical pairs and an *overlay* CSRS has no critical pairs with $\pi \neq \epsilon$. A CSRS is *orthogonal* if it is left-linear and non-overlapping. Notions like "termination" for a CSRS $(\mathcal{R}, \mu)$ always concern the relation $\to_\mu$ (i.e., they correspond to "μ-termination" in [17]).

To prove termination of CSRSs, several transformations from CSRSs to ordinary TRSs were suggested. We recall Giesl & Middeldorp's transformation Θ_1 [9]. It uses new unary symbols active and mark to indicate active positions in a term on the object level. If $l \to r$ is a rule in the CSRS then the transformed TRS contains the rule $\mathsf{active}(l) \to \mathsf{mark}(r)$. The symbol mark is used to traverse a term top-down in order to place the symbol active at all active positions.

Definition 2 (Θ_1). *Let $(\mathcal{R}, \mu)$ be a CSRS over a signature $\mathcal{F}$. The TRS $\mathcal{R}^1_\mu$ over the signature $\mathcal{F}_1 = \mathcal{F} \cup \{\mathsf{active}, \mathsf{mark}\}$ has the following rules:*

$$(\sharp) \qquad \mathsf{active}(l) \to \mathsf{mark}(r) \qquad\qquad \forall l \to r \in \mathcal{R}$$
$$(\sharp) \quad \mathsf{mark}(f(x_1, \ldots, x_n)) \to \mathsf{active}(f([x_1]^f, \ldots, [x_n]^f)) \qquad \forall f \in \mathcal{F}$$
$$\mathsf{active}(x) \to x$$

Here $[x_i]^f = \mathsf{mark}(x_i)$ if $i \in \mu(f)$ and $[x_i]^f = x_i$ otherwise. The transformation $(\mathcal{R}, \mu) \mapsto \mathcal{R}^1_\mu$ is denoted by Θ_1 and we shorten $\to_{\mathcal{R}^1_\mu}$ to $\to_1$.

Because every infinite reduction of a term t in the original CSRS would correspond to an infinite reduction of $\mathsf{mark}(t)$ in the transformed TRS, Θ_1 is *sound* for termination: Termination of the transformed TRS implies termination of the original CSRS. The second transformation Θ_2 of [9,11], Θ_L of Lucas [16], Θ_Z of Zantema [23], and Θ_{FR} of Ferreira & Ribeiro [7] are also sound for termination[3]. However, only Θ_2 is *complete*, i.e., the other transformations do not transform every terminating CSRS into a terminating TRS. Nevertheless, Θ_2 does not render the other transformations superfluous, since in practice, termination of $\Theta_2(\mathcal{R}, \mu)$ can be harder to show than termination of the TRSs resulting from the other transformations.

Example 3 ([9]). The non-terminating TRS $\mathcal{R} = \{f(b, c, x) \to f(x, x, x), d \to b, d \to c\}$ demonstrates the incompleteness of Θ_1. If $\mu(f) = \{3\}$ then the CSRS is terminating because the cyclic reduction of $f(b, c, d)$ to $f(d, d, d)$ and further to $f(b, c, d)$ cannot be done, as one would have to reduce the first and second argument of f. However, the TRS $\mathcal{R}^1_\mu$

$$\mathsf{active}(f(b, c, x)) \to \mathsf{mark}(f(x, x, x)) \qquad \mathsf{mark}(f(x, y, z)) \to \mathsf{active}(f(x, y, \mathsf{mark}(z)))$$
$$\mathsf{active}(d) \to \mathsf{mark}(b) \qquad\qquad\qquad \mathsf{mark}(b) \to \mathsf{active}(b)$$
$$\mathsf{active}(d) \to \mathsf{mark}(c) \qquad\qquad\qquad \mathsf{mark}(c) \to \mathsf{active}(c)$$
$$\mathsf{active}(x) \to x \qquad\qquad\qquad\qquad \mathsf{mark}(d) \to \mathsf{active}(d)$$

---

[3] The interested reader is referred to [11] for definitions and a comparison of these transformations.

is not terminating:

$$\text{m ark}(f(b, c, d)) \to_1 \text{active}(f(b, c, m\, \text{ark}(d))) \to_1 \text{active}(f(b, c, \text{active}(d)))$$
$$\to_1 \text{m ark}(f(\text{active}(d), \text{active}(d), \text{active}(d))) \to_1^+ \text{m ark}(f(m\, \text{ark}(b), m\, \text{ark}(c), d))$$
$$\to_1^+ \text{m ark}(f(\text{active}(b), \text{active}(c), d)) \to_1^+ \text{m ark}(f(b, c, d))$$

Note that in the third step the 'active' subterm active(d) is copied to the first and second argument positions of f, which are inactive according to $\mu(f)$. This can only happen if the reduction step is non-innermost.

3   Innermost Termination of Context-Sensitive Rewriting

Now we examine the usefulness of the five transformations for *innermost* termination of CSRSs. Lucas [18] showed that Θ_L and Θ_Z are unsound[4] for innermost termination, i.e., innermost termination of the transformed TRS does not imply innermost termination of the original CSRS. The example showing the latter ([18, Example 12]) also demonstrates that Θ_{FR} is unsound for innermost termination. Moreover, none of these transformations is complete for innermost termination. The following new result shows that Θ_1 is sound for innermost termination[5].

Theorem 4. *Let $(\mathcal{R}, \mu)$ be a CSRS. If $\mathcal{R}_\mu^1$ is innermost terminating then $(\mathcal{R}, \mu)$ is innermost terminating.*

Proof. Let $\mathcal{F}$ be the signature of $\mathcal{R}$ and let c be an arbitrary constant in $\mathcal{F}$. We show that every innermost reduction step $s \xrightarrow{i}_\mu t$ in $(\mathcal{R}, \mu)$ corresponds to an innermost reduction sequence $\text{mark}(s\theta) \downarrow_{\mathcal{M}} \xrightarrow{i}_1^+ \text{mark}(t\theta) \downarrow_{\mathcal{M}}$ in $\mathcal{R}_\mu^1$. Here $\mathcal{M}$ consists of all rules in $\mathcal{R}_\mu^1$ of the form

$$\text{mark}(f(x_1, \ldots, x_n)) \to \text{active}(f([x_1]^f, \ldots, [x_n]^f))$$

and θ is the substitution that maps all variables to c [6]. Note that since $\mathcal{M}$ is confluent and terminating, every term u has a unique $\mathcal{M}$-normal form $u \downarrow_{\mathcal{M}}$. First we show $\text{mark}(u\theta) \downarrow_{\mathcal{M}} \xrightarrow{i}_1^* \text{active}(u\theta)$ by induction on $u \in \mathcal{T}(\mathcal{F}, \mathcal{V})$. If u is a variable then $u\theta = c$ and thus $\text{mark}(u\theta) \downarrow_{\mathcal{M}} = \text{active}(u\theta)$. If $u = f(u_1, \ldots, u_n)$ then $\text{mark}(u\theta) \downarrow_{\mathcal{M}} = \text{active}(f(u_1', \ldots, u_n'))$ with $u_i' = \text{mark}(u_i\theta) \downarrow_{\mathcal{M}}$ if $i \in \mu(f)$ and $u_i' = u_i\theta$ if $i \notin \mu(f)$. Let $i \in \mu(f)$. The induction hypothesis yields $u_i' \xrightarrow{i}_1^*$ active$(u_i\theta)$. Since $u_i\theta$ is an $\mathcal{R}_\mu^1$-normal form, active$(u_i\theta) \xrightarrow{i}_1 u_i\theta$ and thus $u_i' \xrightarrow{i}_1^*$ $u_i\theta$. It follows that $\text{mark}(u\theta) \downarrow_{\mathcal{M}} \xrightarrow{i}_1^*$ active$(f(u_1\theta, \ldots, u_n\theta)) = \text{active}(u\theta)$.

---

[4] Θ_L is sound for the subclass of left-linear CSRSs with the property that all function symbols in the left-hand sides are on active positions [18].

[5] The same claim is made in [18, Theorem 11]. However, Lucas only proved the soundness of Θ_1 and Θ_2 for *ground* innermost termination (cf. Section 6) and later claimed that Θ_1 and Θ_2 are unsound for innermost termination [19].

[6] It is interesting to note that the instantiated context-sensitive reduction step $s\theta \to_\mu t\theta$ need not be innermost.

Now let π be the position of the redex contracted in the reduction step $s \xrightarrow{i}_\mu t$. We prove the lemma by induction on π. If $\pi = \epsilon$ then $s \to t$ and thus also $s\theta \to t\theta$ is an instance of a rule in $\mathcal{R}$. We have $\mathrm{mark}(s\theta){\downarrow}_{\mathcal{M}} \xrightarrow{i}_1^* \mathrm{active}(s\theta)$ by the above observation. Moreover, $\mathrm{active}(s\theta) \xrightarrow{i}_1 \mathrm{mark}(t\theta)$ since $\mathrm{active}(s\theta) \to \mathrm{mark}(t\theta)$ is an instance of a rule in $\mathcal{R}_\mu^1$. We also have $\mathrm{mark}(t\theta) \xrightarrow{i}_1^* \mathrm{mark}(t\theta){\downarrow}_{\mathcal{M}}$. Combining all reductions yields $\mathrm{mark}(s\theta){\downarrow}_{\mathcal{M}} \xrightarrow{i}_1^+ \mathrm{mark}(t\theta){\downarrow}_{\mathcal{M}}$.

If $\pi = i\pi'$ then $s = f(s_1, \ldots, s_i, \ldots, s_n)$ and $t = f(s_1, \ldots, t_i, \ldots, s_n)$ with $s_i \xrightarrow{i}_\mu t_i$. Note that we have $i \in \mu(f)$ due to the definition of context-sensitive rewriting. For $1 \leqslant j \leqslant n$ define $s_j' = \mathrm{mark}(s_j\theta){\downarrow}_{\mathcal{M}}$ if $j \in \mu(f)$ and $s_j' = s_j\theta$ if $j \notin \mu(f)$. The induction hypothesis yields $s_i' \xrightarrow{i}_1^+ \mathrm{mark}(t_i\theta){\downarrow}_{\mathcal{M}}$. The result follows since $\mathrm{mark}(s\theta){\downarrow}_{\mathcal{M}} = \mathrm{active}(f(s_1', \ldots, s_i', \ldots, s_n'))$ and $\mathrm{mark}(t\theta){\downarrow}_{\mathcal{M}} = \mathrm{active}(f(s_1', \ldots, \mathrm{mark}(t_i\theta){\downarrow}_{\mathcal{M}}, \ldots, s_n'))$. $\square$

Not surprisingly, Θ_1 is incomplete for innermost termination.

Example 5 ([18]). Consider $(\mathcal{R}, \mu)$ with $\mathcal{R} = \{\mathsf{f(a)} \to \mathsf{f(a)}, \mathsf{a} \to \mathsf{b}\}$ and $\mu(\mathsf{f}) = \{1\}$. The CSRS $(\mathcal{R}, \mu)$ is innermost terminating but $\mathcal{R}_\mu^1$

$$
\begin{array}{ll}
\mathsf{active(f(a))} \to \mathsf{mark(f(a))} & \mathsf{mark(f(}x\mathsf{))} \to \mathsf{active(f(mark(}x\mathsf{)))} \\
\mathsf{active(a)} \to \mathsf{mark(b)} & \mathsf{mark(a)} \to \mathsf{active(a)} \\
\mathsf{active(}x\mathsf{)} \to x & \mathsf{mark(b)} \to \mathsf{active(b)}
\end{array}
$$

is not an innermost terminating TRS: $\mathsf{active(f(a))} \xrightarrow{i}_1 \mathsf{mark(f(a))} \xrightarrow{i}_1 \mathsf{active(f(mark(a)))} \xrightarrow{i}_1 \mathsf{active(f(active(a)))} \xrightarrow{i}_1 \mathsf{active(f(a))}$. Note that applying the rule $\mathsf{active(a)} \to \mathsf{mark(b)}$ instead of $\mathsf{active(}x\mathsf{)} \to x$ in the fourth step would break the cycle. So the rule $\mathsf{active(}x\mathsf{)} \to x$ can delete innermost redexes, causing non-innermost active redexes of the underlying CSRS to become innermost. We come back to this in Section 5.

Transformation Θ_2 is sound for innermost termination as well. However, it is also incomplete and (in contrast to Θ_1) rather useless for innermost termination. These observations are consequences of the following new result. In particular, Θ_2 cannot prove innermost termination of non-terminating CSRSs.

Theorem 6. *Let $(\mathcal{R}, \mu)$ be a CSRS. The TRS $\mathcal{R}_\mu^2$ resulting from transformation Θ_2 is innermost terminating iff it is terminating.*

Soundness of Θ_2 for innermost termination is an immediate consequence of Theorem 6 and the soundness of Θ_2 for termination.

So Θ_1 is the only sound and useful transformation for innermost termination of CSRSs so far. The next theorem shows that it is complete for an important subclass of CSRSs. More precisely, while in general termination of a CSRS $(\mathcal{R}, \mu)$ does not imply termination of the transformed TRS $\mathcal{R}_\mu^1$ (as demonstrated by Example 3), it at least implies *innermost* termination of $\mathcal{R}_\mu^1$.

Theorem 7. *Let $(\mathcal{R}, \mu)$ be a CSRS. If $(\mathcal{R}, \mu)$ is terminating then $\mathcal{R}_\mu^1$ is innermost terminating.*

Theorem 7 implies that for subclasses of CSRSs where innermost termination is equivalent to termination, Θ_1 is complete for innermost termination. In the next section we show that this subclass contains all orthogonal systems (e.g., CSRSs like Example 1).

4   Termination versus Innermost Termination

There are two motivations for studying innermost termination of CSRSs. First, innermost context-sensitive rewriting models evaluation in OBJ and thus, innermost termination analysis of CSRSs can be used for termination proofs of OBJ programs. But second, innermost termination analysis of CSRSs can also be helpful for (non-innermost) termination proofs of CSRSs. This is similar to the situation with ordinary TRSs: Proving innermost termination is much easier than proving termination, cf. [1]. There are classes of TRSs where innermost termination already implies termination and therefore for such systems, one should rather use innermost termination techniques for investigating their termination behavior.

In order to use a corresponding approach for context-sensitive rewriting, in this section we examine the connection between termination and innermost termination for CSRSs. In general, termination implies innermost termination, but not vice versa as demonstrated by Example 5. For ordinary TRSs, being *non-overlapping* suffices to ensure that innermost termination is equivalent to termination [13]. Unfortunately, as noted by Lucas [19], this criterion cannot be extended to CSRSs. However, we show the new result that the desired equivalence between innermost and full termination at least holds for *orthogonal* CSRSs. Thus, this includes all CSRSs which correspond to typical functional programs like Example 1. Theorem 9 states that for such systems we only have to prove innermost termination in order to verify their termination.

In order to prove the theorem, we need some preliminaries. For non-overlapping CSRSs $(\mathcal{R}, \mu)$ the relation $\xrightarrow{i}_\mu$ is confluent. Hence, for every term s there is at most one μ-normal form reachable by innermost reductions. We call this term the *innermost μ-normal form* of s and denote it by $s\downarrow^i_\mu$. Now for any term s, let $\nabla(s)$ be the set of those terms which result from repeatedly replacing subterms of s by their innermost μ-normal form (if it exists). Here, one may also consider subterms on inactive positions. However, the replacement must go "from the inside to the outside" (i.e., after replacing at position π one cannot replace at positions below π any more). Moreover, one may only perform replacements on such positions π where the original term $s|_\pi$ is terminating.

Definition 8. *Let $(\mathcal{R}, \mu)$ be a non-overlapping CSRS. For any term s we define non-empty sets $\nabla(s)$ and $\nabla'(s)$ as follows. If s is terminating, then $\nabla(s) = \nabla'(s) \cup \{u\downarrow^i_\mu \mid u \in \nabla'(s) \text{ is innermost terminating}\}$. Otherwise, we have $\nabla(s) = \nabla'(s)$. Moreover, $\nabla'(s) = \{f(u_1, \ldots, u_n) \mid u_i \in \nabla(s_i)\}$ if $s = f(s_1, \ldots, s_n)$ and $\nabla'(s) = \{s\}$ if s is a variable.*

Theorem 9. *An orthogonal CSRS* $(\mathcal{R}, \mu)$ *is terminating iff it is* innermost *terminating.*

Proof. The "only if" direction is trivial. We prove the "if" direction. Let $s \to_\mu t$ where the contracted redex is either terminating or a minimal non-terminating term (i.e., all proper subterms of the redex on active positions are terminating). We prove the following statements for all innermost terminating $s' \in \nabla(s)$:

(1) There exists a $t' \in \nabla(t)$ such that $s' \xrightarrow{i}{}^*_\mu t'$.
(2) If the contracted redex in $s \to_\mu t$ is not terminating, then there even exists a $t' \in \nabla(t)$ such that $s' \xrightarrow{i}{}^+_\mu t'$.

With (1) and (2) one can prove the theorem: If $(\mathcal{R}, \mu)$ is not terminating, then there is an infinite reduction $s_0 \to_\mu s_1 \to_\mu \ldots$ in which only terminating or minimal non-terminating redexes are contracted. Assume that $(\mathcal{R}, \mu)$ is innermost terminating. Then all $\nabla(s_i)$ contain only innermost terminating terms and since $s_0 \in \nabla(s_0)$, we can construct an infinite innermost reduction $s_0 \xrightarrow{i}{}^*_\mu t_1 \xrightarrow{i}{}^*_\mu t_2 \xrightarrow{i}{}^*_\mu \ldots$ with $t_i \in \nabla(s_i)$. However, since the reduction contains infinitely many steps of type (2), this gives rise to an infinite innermost reduction, contradicting our assumption.

Now we prove (1) and (2) by structural induction on s. Since $s \to_\mu t$, s must have the form $f(s_1, \ldots, s_n)$. We first regard the case where $s \to_\mu t$ is not a root reduction step. Then we have $t = f(s_1, \ldots, t_i, \ldots, s_n)$ with $s_i \to_\mu t_i$ for some $i \in \mu(f)$. Let $s' \in \nabla(s)$ be innermost terminating. First, let $s' = f(u_1, \ldots, u_n)$ with $u_j \in \nabla(s_j)$ for all j. Because $i \in \mu(f)$, u_i is innermost terminating. Hence by the induction hypothesis, $u_i \in \nabla(s_i)$ implies that there exists a $v_i \in \nabla(t_i)$ such that $u_i \xrightarrow{i}{}^*_\mu v_i$. Therefore, we also have $s' = f(u_1, \ldots, u_i, \ldots, u_n) \xrightarrow{i}{}^*_\mu f(u_1, \ldots, v_i, \ldots, u_n) \in \nabla(t)$. Moreover, if the contracted redex in $s \to_\mu t$ and hence, in $s_i \to_\mu t_i$ is not terminating, then by the induction hypothesis we even have $u_i \xrightarrow{i}{}^+_\mu v_i$ and therefore $s' \xrightarrow{i}{}^+_\mu f(u_1, \ldots, v_i, \ldots, u_n) \in \nabla(t)$.

Now let $s' = f(u_1, \ldots, u_n)\!\downarrow^i_\mu$ with $u_j \in \nabla(s_j)$ for all j. Hence, s is terminating and thus, we only have to prove (1). As before, there is a $v_i \in \nabla(t_i)$ such that $u_i \xrightarrow{i}{}^*_\mu v_i$ and $f(u_1, \ldots, v_i, \ldots, u_n) \in \nabla(t)$. Since innermost reduction is confluent, we have $s' = f(u_1, \ldots, u_i, \ldots, u_n)\!\downarrow^i_\mu = f(u_1, \ldots, v_i, \ldots, u_n)\!\downarrow^i_\mu \in \nabla(t)$, since t inherits termination from s.

Finally, we regard the case where $s = f(s_1, \ldots, s_n)$ and $s \to_\mu t$ is a root reduction step. Hence, there must be a rule $l \to r \in \mathcal{R}$ with $l = f(l_1, \ldots, l_n)$ and a substitution σ such that $s_i = l_i\sigma$ and $t = r\sigma$. First let $s' = f(u_1, \ldots, u_n)$ with $u_i \in \nabla(s_i)$ for all i. Since $(\mathcal{R}, \mu)$ is orthogonal and since $s_i = l_i\sigma$, there must be a substitution σ' such that $u_i = l_i\sigma'$ for all i [7]. Because s' is innermost terminating, $x\sigma'$ must also be innermost terminating for all variables x which occur on active positions of l. Let σ'' be the substitution where $x\sigma'' = x\sigma'\!\downarrow^i_\mu$ for all x in active positions of l and $x\sigma'' = x\sigma'$ for all other x. Then we have the innermost reduction $s' = f(l_1\sigma', \ldots, l_n\sigma') \xrightarrow{i}{}^*_\mu f(l_1\sigma'', \ldots, l_n\sigma'') \xrightarrow{i}_\mu r\sigma''$. We claim that $r\sigma'' \in \nabla(t) = \nabla(r\sigma)$. To this end, it suffices to show that $x\sigma'' \in \nabla(x\sigma)$

---

[7] A formal proof of this observation can be found in [10].

for all variables x in r, because in the construction of ∇ arbitrary subterms q can be replaced by terms from $\nabla(q)$. Each variable x occurs in some l_i and we have $l_i\sigma' \in \nabla(l_i\sigma)$. It follows that $x\sigma' \in \nabla(x\sigma)$ for all variables x.[7] If x is on an inactive position of l, then $x\sigma'' = x\sigma' \in \nabla(x\sigma)$. If x is on an active position of l, then $x\sigma'' = x\sigma'\downarrow^i_\mu \in \nabla(x\sigma)$, since $x\sigma'$ is innermost terminating and because in this case, $x\sigma$ is terminating due to the fact that s is either a terminating or a *minimal* non-terminating term.

Now let $s' = f(u_1, \ldots, u_n)\downarrow^i_\mu$ with $u_i \in \nabla(s_i)$ for all i. Hence, s is terminating and thus we only have to prove (1). As before, $u_i = l_i\sigma'$ and $f(l_1\sigma', \ldots, l_n\sigma') \xrightarrow{i}{}^*_\mu$ $f(l_1\sigma'', \ldots, l_n\sigma'') \xrightarrow{i}_\mu r\sigma''$ with $r\sigma'' \in \nabla(t)$. Since innermost reduction is confluent and t inherits termination from s, $s' = f(u_1, \ldots, u_n)\downarrow^i_\mu = r\sigma''\downarrow^i_\mu \in \nabla(t)$.   □

Very recently, Gramlich and Lucas [14] showed that termination and innermost termination coincide for locally confluent overlay CSRSs with the additionally property that variables that occur at an active position in a left-hand side l of a rewrite rule $l \to r$ do not occur at inactive positions in l or r. The latter condition is quite restrictive, e.g., it is not satisfied by the CSRS of Example 1.

5   A Sound and Complete Transformation

In Section 3 we showed that the existing transformations are incomplete for innermost termination and that only Θ_1 and Θ_2 are sound. Because of Theorem 6, Θ_2 cannot distinguish innermost termination from termination. So when developing a sound and complete transformation for innermost termination, we take Θ_1 as starting point. As observed in Example 5, we must make sure that in innermost reductions, rules of the form $\mathsf{active}(l) \to \mathsf{mark}(r)$ get preference over the rule $\mathsf{active}(x) \to x$, because then this counterexample no longer works. Hence, we modify the rule $\mathsf{active}(x) \to x$ such that the innermost reduction strategy ensures that $\mathsf{active}(l) \to \mathsf{mark}(r)$ is applied with higher preference. In the modification, $\mathsf{active}(l) \to \mathsf{mark}(r)$ no longer overlaps with the *root position* of $\mathsf{active}(x) \to x$, but with a non-root position of the new modified rule(s).

Definition 10 (Θ_3). *Let $(\mathcal{R}, \mu)$ be a CSRS over a signature $\mathcal{F}$. The TRS $\mathcal{R}^3_\mu$ over the signature $\mathcal{F}_1$ consists of all ($\sharp$)-marked rewrite rules of $\mathcal{R}^1_\mu$ together with the rewrite rules*

$$f(x_1, \ldots, \mathsf{active}(x_i), \ldots, x_n) \to f(x_1, \ldots, x_i, \ldots, x_n)$$
(♭) $$f(x_1, \ldots, \mathsf{mark}(x_i), \ldots, x_n) \to f(x_1, \ldots, x_i, \ldots, x_n)$$

for all $f \in \mathcal{F}$ and $1 \leqslant i \leqslant \mathrm{arity}(f)$. We denote the transformation $(\mathcal{R}, \mu) \mapsto \mathcal{R}^3_\mu$ by Θ_3 and we abbreviate $\to_{\mathcal{R}^3_\mu}$ to $\to_3$ and $\xrightarrow{i}_{\mathcal{R}^3_\mu}$ to $\xrightarrow{i}_3$.

For the CSRS $(\mathcal{R}, \mu)$ of Example 5, $\mathcal{R}^3_\mu$ differs from $\mathcal{R}^1_\mu$ in two respects: $\mathsf{active}(x) \to x$ is replaced by $\mathsf{f}(\mathsf{active}(x)) \to \mathsf{f}(x)$ and moreover, the rule $\mathsf{f}(\mathsf{mark}(x)) \to \mathsf{f}(x)$ is added. As a consequence, the cycle $\mathsf{active}(\mathsf{f}(a)) \xrightarrow{i}{}^+ \mathsf{active}(\mathsf{f}(a))$ can no longer be obtained with $\mathcal{R}^3_\mu$, since $\mathsf{active}(\mathsf{f}(\mathsf{active}(a))) \to \mathsf{active}(\mathsf{f}(a))$ is not an *innermost* rewrite step in $\mathcal{R}^3_\mu$. Indeed, $\mathcal{R}^3_\mu$ is innermost terminating and in general, Θ_3 is sound and complete for innermost termination.

Theorem 11. *A CSRS $(\mathcal{R}, \mu)$ is innermost terminating iff $\mathcal{R}_\mu^3$ is innermost terminating.*

In [10] we show for several CSRSs $(\mathcal{R}, \mu)$ including Example 1 how innermost termination of $\mathcal{R}_\mu^3$ can be proved with *dependency pairs* [1].

With the new rules $f(x_1, \ldots, \mathsf{active}(x_i), \ldots, x_n) \to f(x_1, \ldots, x_n)$ we can remove almost every active-symbol, compensating to a large extent the lack of the rule $\mathsf{active}(x) \to x$. The ($\flat$)-marked rules can never be used in an innermost reduction if x_i is instantiated to a non-variable term from $\mathcal{T}(\mathcal{F}, \mathcal{V})$. However, they are required if x_i is instantiated by a variable or by terms containing the symbols mark and active. As a matter of fact, the transformation without the ($\flat$)-marked rules is neither sound nor complete for innermost termination (see [10] for counterexamples).

6   Ground Innermost Termination

Unlike for termination, to conclude innermost termination it is not sufficient to prove that all ground terms are innermost terminating.

Example 12. The TRS $\{\mathsf{f}(\mathsf{f}(x)) \to \mathsf{f}(\mathsf{f}(x)), \mathsf{f}(\mathsf{a}) \to \mathsf{a}\}$ is not innermost terminating but ground innermost terminating over the signature $\{\mathsf{f}, \mathsf{a}\}$, i.e., all ground terms permit only finite innermost reductions.

It is well known that innermost termination of a TRS $\mathcal{R}$ over a signature $\mathcal{F}$ is equivalent to ground innermost termination of $\mathcal{R}$ over the signature $\mathcal{F} \cup \{\mathsf{c}, \mathsf{h}\}$ where c is a fresh constant and h is a fresh unary function symbol. The reason is that a term t with the variables $x_1, \ldots, x_n$ starts an infinite innermost reduction iff the ground term $t\sigma$ starts an infinite innermost reduction where $\sigma(x_i) = \mathsf{h}^i(\mathsf{c})$. So the fresh symbols c and h are needed to create arbitrarily many different ground terms (in order to handle non-linear rewrite rules). A similar correspondence holds for innermost *context-sensitive* reductions with $\mu(\mathsf{h}) = \varnothing$ or $\mu(\mathsf{h}) = \{1\}$.

The following result states that Θ_1 and Θ_2 cannot distinguish ground innermost termination from innermost termination and thus they are sound but incomplete for ground innermost termination as well.

Theorem 13. *Let $(\mathcal{R}, \mu)$ be a CSRS. For $i \in \{1, 2\}$, the TRS $\mathcal{R}_\mu^i$ is ground innermost terminating iff it is innermost terminating.*

On the other hand, the proof of Theorem 11 can easily be adapted to show that Θ_3 is sound and complete for ground innermost termination.

Theorem 14. *A CSRS $(\mathcal{R}, \mu)$ is ground innermost terminating iff $\mathcal{R}_\mu^3$ is ground innermost terminating.*

One might think that the ($\flat$)-marked rules in Definition 10 are not needed to obtain a sound and complete transformation for ground innermost termination. While soundness is easily proved, completeness does *not* hold, as shown in the following example.

Example 15. Consider the (ground) innermost terminating CSRS $(\mathcal{R}, \mu)$ with $\mathcal{R} = \{f(x, x) \rightarrow b, g(f(x, y)) \rightarrow g(f(y, y))\}$ and $\mu(f) = \mu(g) = \{1\}$. The transformed TRS without the $(\flat)$-marked rules however is not ground innermost terminating as can be seen from the following cycle, with $t = \text{mark}(\text{active}(b))$:

$$\text{mark}(g(f(t, t))) \xrightarrow{i}^{+} \text{active}(g(\text{active}(f(\text{mark}(t), t))))$$
$$\xrightarrow{i} \text{active}(g(f(\text{mark}(t), t))) \xrightarrow{i} \text{mark}(g(f(t, t)))$$

As explained above, a transformation that is sound for ground innermost termination can also be used for innermost termination analysis by adding fresh function symbols to the signature. However, for completeness the situation is different. Here, it is desirable that the transformation is not only complete for ground, but also for full innermost termination. The reason is that while there do exist techniques to analyze ground innermost termination [8], the best-known technique for automated innermost termination analysis [1] really checks full (non-ground) innermost termination of TRSs. A complete transformation for innermost termination transforms every innermost terminating CSRS into an innermost terminating TRS and hence, innermost termination of this TRS can potentially be checked by every technique for innermost termination analysis of ordinary TRSs. But if the transformed TRS is only ground innermost terminating, (full) innermost termination analysis techniques for TRSs cannot be applied successfully.

7   Conclusion and Comparison

Figure 1 contains a summary of the soundness and completeness results covered in the preceding sections. The negative results for ground innermost termination for Θ_{L}, Θ_{Z}, and Θ_{FR} are shown by the same examples used to demonstrate the corresponding results for innermost termination, cf. the first paragraph of Section 3. Of the existing transformations, only Θ_1 and Θ_2 from [9] are sound for innermost termination of CSRSs. We showed that Θ_2 is not very useful for proving innermost termination, but that termination of a CSRS $(\mathcal{R}, \mu)$ already implies innermost termination of $\Theta_1(\mathcal{R}, \mu)$. So for classes of CSRSs where termination and innermost termination are equivalent, Θ_1 is already complete for

	termination		ground innermost termination		innermost termination	
	sound	complete	sound	complete	sound	complete
Θ_{L}	✓	✗	✗	✗	✗	✗
Θ_{Z}	✓	✗	✗	✗	✗	✗
Θ_{FR}	✓	✗	✗	✗	✗	✗
Θ_1	✓	✗	✓	✗	✓	✗
Θ_2	✓	✓	✓	✗	✓	✗
Θ_3	✓	✗	✓	✓	✓	✓

Fig. 1. Summary.

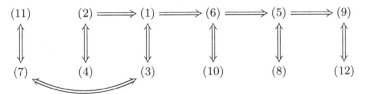

Fig. 2. Comparison.

innermost termination. We proved that this equivalence holds for the class of orthogonal CSRSs. While in general Θ_1 is still incomplete, we developed a new transformation Θ_3 which is sound and complete for innermost termination. The results on termination for Θ_3 follow from Theorem 16 below.

In order to assess the power of our transformations, Figure 2 illustrates the relationship between the following twelve properties ($i = 1, 2, 3$):

(1)	$(\mathcal{R}, \mu)$ is terminating
(5)	$(\mathcal{R}, \mu)$ is innermost terminating
(9)	$(\mathcal{R}, \mu)$ is ground innermost terminating

(1+i)	$\mathcal{R}^i_\mu$ is terminating
(5+i)	$\mathcal{R}^i_\mu$ is innermost terminating
(9+i)	$\mathcal{R}^i_\mu$ is ground innermost terminating

Implication (2) $\Rightarrow$ (1) is the soundness of transformation Θ_1 for termination [9], implication (1) $\Rightarrow$ (6) is Theorem 7, implication (6) $\Rightarrow$ (5) is Theorem 4, and implication (5) $\Rightarrow$ (9) is trivial.

Equivalence (1) $\Leftrightarrow$ (3) is the soundness and completeness of Θ_2 for termination [9], equivalence (3) $\Leftrightarrow$ (7) is Theorem 6, equivalence (5) $\Leftrightarrow$ (8) is Theorem 11, equivalences (10) $\Leftrightarrow$ (6) and (11) $\Leftrightarrow$ (7) are Theorem 13, and equivalence (9) $\Leftrightarrow$ (12) is Theorem 14. Equivalence (2) $\Leftrightarrow$ (4) means that Θ_1 and Θ_3 are equally powerful when proving termination. This may not come as a surprise, but the proof is surprisingly difficult.

Theorem 16. *Let $(\mathcal{R}, \mu)$ be a CSRS. The TRS $\mathcal{R}^3_\mu$ is terminating iff $\mathcal{R}^1_\mu$ is terminating.*

None of the missing implications in Figure 2 hold, except those that follow by transitivity: (1) $\not\Rightarrow$ (2) and (5) $\not\Rightarrow$ (6) are the incompleteness of Θ_1 for termination (Example 3) and innermost termination (Example 5). Moreover, (6) $\not\Rightarrow$ (1) follows by using $\mu(f) = \{1, 2, 3\}$ in Example 3 and (9) $\not\Rightarrow$ (5) follows from Example 12 with $\mu(f) = \{1\}$.

To conclude, with our new transformation Θ_3, innermost termination of context-sensitive rewriting can be reduced to innermost termination of ordinary rewriting. Moreover, for orthogonal CSRSs innermost termination already suffices for termination. So for such systems, innermost termination of the transformed TRS even implies termination of the CSRS. Hence, our result now enables the use of powerful methods for *innermost* termination analysis of TRSs for (innermost) termination of context-sensitive rewriting.

Acknowledgments

We thank Salvador Lucas and the anonymous referees for many helpful remarks.

References

1. T. Arts and J. Giesl. Termination of term rewriting using dependency pairs. *Theoretical Computer Science*, 236:133–178, 2000.
2. F. Baader and T. Nipkow. *Term Rewriting and All That.* Cambr. Univ. Pr., 1998.
3. C. Borralleras, S. Lucas, and A. Rubio. Recursive path orderings can be context-sensitive. In *Proc. 18th CADE*, volume 2392 of *LNAI*, pages 314–331, 2002.
4. M. Clavel, S. Eker, P. Lincoln, and J. Meseguer. Principles of Maude. In *Proc. 1st WRLA*, volume 4 of *ENTCS*, 1996.
5. R. Diaconescu and K. Futatsugi. CafeOBJ *Report*, volume 6 of *AMAST Series in Computing*. World Scientific, 1998.
6. S. Eker. Term rewriting with operator evaluation strategies. In *Proc. 2nd WRLA*, volume 15 of *ENTCS*, pages 1–20, 1998.
7. M.C.F. Ferreira and A.L. Ribeiro. Context-sensitive AC-rewriting. In *Proc. 10th RTA*, volume 1631 of *LNCS*, pages 173–187, 1999.
8. O. Fissore, I. Gnaedig, and H. Kirchner. Induction for termination with local strategies. In *Proc. 4th International Workshop on Strategies in Automated Deduction*, volume 58 of *ENTCS*, 2001.
9. J. Giesl and A. Middeldorp. Transforming context-sensitive rewrite systems. In *Proc. 10th RTA*, volume 1631 of *LNCS*, pages 271–285, 1999.
10. J. Giesl and A. Middeldorp. Innermost termination of context-sensitive rewriting. Technical Report AIB-2002-04[8], RWTH Aachen, Germany, 2002.
11. J. Giesl and A. Middeldorp. Transformation techniques for context-sensitive rewrite systems. Technical Report AIB-2002-02[8], RWTH Aachen, Germany, 2002. Extended version of [9].
12. J. Goguen, T. Winkler, J. Meseguer, K. Futatsugi, and J.-P. Jouannaud. Introducing OBJ. In J. Goguen and G. Malcolm, editors, *Software Engineering with OBJ: algebraic specification in action*. Kluwer, 2000.
13. B. Gramlich. Abstract relations between restricted termination and confluence properties of rewrite systems. *Fundamenta Informaticae*, 24:3–23, 1995.
14. B. Gramlich and S. Lucas. Modular termination of context-sensitive rewriting. In *Proc. 4th PPDP*, pages 50–61. ACM Press, 2002.
15. B. Gramlich and S. Lucas. Simple termination of context-sensitive rewriting. In *Proc. 3rd Workshop on Rule-Based Programming*, pages 29–41. ACM Press, 2002.
16. S. Lucas. Termination of context-sensitive rewriting by rewriting. In *Proc. 23rd ICALP*, volume 1099 of *LNCS*, pages 122–133, 1996.
17. S. Lucas. Context-sensitive computations in functional and functional logic programs. *Journal of Functional and Logic Programming*, 1:1–61, 1998.
18. S. Lucas. Termination of rewriting with strategy annotations. In *Proc. 8th LPAR*, volume 2250 of *LNAI*, pages 669–684, 2001.
19. S. Lucas, 2001–2002. Personal communication.
20. S. Lucas. Termination of (canonical) context-sensitive rewriting. In *Proc. 13th RTA*, volume 2378 of *LNCS*, pages 296–310, 2002.

---

[8] Available from http://aib.informatik.rwth-aachen.de.

21. T. Nagaya. *Reduction Strategies for Term Rewriting Systems.* PhD thesis, School of Information Science, Japan Advanced Institute of Science and Technology, 1999.

22. M. Nakamura and K. Ogata. The evaluation strategy for head normal form with and without on-demand flags. In *Proc. 3rd WRLA*, volume 36 of *ENTCS*, 2001.

23. H. Zantema. Termination of context-sensitive rewriting. In *Proc. 8th RTA*, volume 1232 of *LNCS*, pages 172–186, 1997.

A Unique Structure
of Two-Generated Binary Equality Sets

Štěpán Holub[*]

Charles University, Prague
Sokolovská 83, 186 75 Praha 8, Czech Republic
holub@karlin.mff.cuni.cz
and
Turku Center for Computer Science, Turku, Finland

Abstract. Let L be the equality set of two distinct injective morphisms g and h, and let L be generated by at least two words. Recently it was proved ([2]) that such an L is generated by two words and g and h can be chosen marked from both sides. We use this result to show that L is of the form $\{a^i b, ba^i\}^*$, with $i \geq 1$.

1 Introduction

Binary equality sets are the simplest non-trivial equality languages. Nevertheless, their precise description is still not known. They were for the first time studied extensively by K. Čulík II and J. Karhumäki in [3]. There the authors indicate that the only existing binary equality sets of rank two have the form $\{a^i b, ba^i\}^*$, but avoid to state this as a conjecture. Instead, they conjectured that in non-periodic cases (periodic cases being easy to deal with) the equality set is generated by at most two words. This statement was partially proved by A. Ehrenfeucht, J. Karhumäki and G. Rozenberg ([4]) leaving open the possibility of an infinitely generated equality set of the form $(\alpha \gamma^* \beta)^*$. The result is a corollary of the proof that the binary Post Correspondence Problem is decidable, previously achieved by the same authors ([1]). The mentioned possibility, contradicting the original conjecture, was excluded recently in [2], where we prove a stronger statement: the two words generating the equality set start (end resp.) with different letters. Especially, this means that the equality set belongs to a pair of morphisms marked from both sides. In the present paper we therefore investigate such morphisms and show that their equality set can be generated by two words only if it is of the form $\{a^i b, ba^i\}^*$. This yields the complete characterization of binary equality sets generated by more than one word.

The paper at hand is actually an exercise in combinatorial analysis. After preliminaries (Section 2) we present some auxiliary lemmas based mostly on the primitivity of a word (Section 3). In the fourth section some general results concerning our morphisms are obtained. In Section 5 special cases are dealt with.

[*] Work supported by institutional grant MSM 113200007

M. Ito and M. Toyama (Eds.): DLT 2002, LNCS 2450, pp. 245–257, 2003.
© Springer-Verlag Berlin Heidelberg 2003

2   Assumptions and Definitions

We first fix our notation.

By A we denote the binary alphabet $\{a, b\}$. The empty word is denoted by ε.

The set of all prefixes of u is denoted by $\mathrm{pref}(u)$. A prefix v of u is *proper* if $v \neq \varepsilon$ and $v \neq u$. Similarly *proper suffix* is defined. The set of all suffixes of u is denoted by $\mathrm{suff}(u)$. The first (the last resp.) letter of a non-empty word u is denoted by $\mathrm{pref}_1(u)$ ($\mathrm{suff}_1(u)$ resp.). A word v is called a *factor* of u if there exist words $w, w' \in A^*$ such that $u = w v w'$. A factor is said to be proper if and only if both w and w' are non-empty. If $v \in \mathrm{pref}(u)$ or $u \in \mathrm{pref}(v)$, we say that u and v are *comparable*. If $uv = w$ we also write $u = wv^{-1}$ and $v = u^{-1}w$. A word w is called an *overlap* of u and v if $w \in \mathrm{suff}(u) \cap \mathrm{pref}(v)$, or $w \in \mathrm{suff}(v) \cap \mathrm{pref}(u)$.

By a factor (prefix, suffix resp.) of a language we mean a factor (prefix, suffix resp.) of any of its elements.

Let $g, h : A^* \rightarrow A^*$ be binary morphisms. Their *equality set* is defined by

$$\mathrm{Eq}(g, h) = \{u \in A^* |\ g(u) = h(u)\}.$$

The choice of A as the target alphabet does not harm generality, since any alphabet can be encoded by two letters.

A binary morphism g is said to be *marked* if and only if $\mathrm{pref}_1(g(a)) \neq \mathrm{pref}_1(g(b))$. If, moreover, $\mathrm{suff}_1(g(a)) \neq \mathrm{suff}_1(g(b))$, we say that g is *marked from both sides*. Similarly we say that two non-empty words x and y are marked from both sides, if and only if $\mathrm{pref}_1(x) \neq \mathrm{pref}_1(y)$ and $\mathrm{suff}_1(x) \neq \mathrm{suff}_1(y)$.

We say that g is periodic, if words $g(a)$ and $g(b)$ commute, i.e., they have the same primitive root.

It is easy to verify that the set $\mathrm{Eq}(g, h)$ is a free submonoid of A^* generated by the set of its minimal elements

$$\mathrm{eq}(g, h) = \mathrm{Eq}(g, h) \setminus (\mathrm{Eq}(g, h) \setminus \{\varepsilon\})^2 \setminus \{\varepsilon\}.$$

If $g \neq h$, and u and v are non-empty elements of $\mathrm{Eq}(g, h)$ then $\mathrm{rat}(u) = \mathrm{rat}(v)$. This follows easily from the length agreement of g and h on elements of their equality set.

The following is known about the structure of $\mathrm{Eq}(g, h)$ (see [2]).

Theorem 1. *Let g and h be non-periodic binary morphisms. Then*

$$\mathrm{Eq}(h, g) = \{\alpha, \beta\}^*$$

for some (possibly empty) words $\alpha, \beta \in A^*$. If α and β are both non-empty then they are marked from both sides. Moreover, there are binary morphisms g' and h' marked from both sides, such that $\mathrm{Eq}(g, h) = \mathrm{Eq}(g', h')$.

In this paper we investigate binary morphisms $g, h : A^* \rightarrow A^*$, $g \neq h$, whose equality set is generated by two non-empty words α and β. The symmetry of the letters a and b, the words α and β, and of the morphisms g and h allow to suppose that $h(b)$ is the longest of the four considered images, and the first letter of α as well as of $g(\alpha)$ is a. Therefore, by Theorem 1, we can adopt the following assumptions without loss of generality.

Conditions 2

- $|g(a)| > |h(a)|$
- $|g(b)| < |h(b)|$
- $|h(b)| \geq |g(a)|$
- $\mathrm{pref}_1(g(a)) = \mathrm{pref}_1(h(a)) = a$
- $\mathrm{pref}_1(g(b)) = \mathrm{pref}_1(h(b)) = b$
- $\mathrm{suff}_1(g(a)) = \mathrm{suff}_1(h(a)) \neq \mathrm{suff}_1(g(b)) = \mathrm{suff}_1(h(b))$
- $\mathrm{pref}_1(\alpha) = a$
- $\mathrm{pref}_1(\beta) = b$
- $\mathrm{suff}_1(\alpha) \neq \mathrm{suff}_1(\beta)$

We are going to prove the following

Theorem 3. *Let $g, h : A^* \to A^*$ be binary morphisms, such that $\mathrm{eq} = \{\alpha, \beta\}$, satisfying Conditions 2. Then there is a positive integer i such that $\alpha = a^i b$ and $\beta = ba^i$.*

Since $g \neq h$, both α and β contain both letters a and b. Note that the difference between the letters a and b is only given by the condition $|h(b)| \geq |g(a)|$. Therefore if $|h(b)| = |g(a)|$ then $i = 1$.

Throughout the paper k, k', l and l' will be positive integers such that

- $a^k b$ is a prefix of α,
- $b^l a$ is a prefix of β,
- $ba^{k'}$ and $ab^{l'}$ are elements of $\mathrm{suff}\{\alpha, \beta\}$.

3   Auxiliary Lemmas

In this section we present several auxiliary lemmas. The proofs are easy and we omit them. We also omit well-known characterization of conjugate words and the Periodicity Lemma.

The following Lemma is a consequence of the fact that two words generate a free semigroup if and only if they do not commute.

Lemma 1. *Let all words $g(a)$, $g(b)$, $h(a)$ and $h(b)$ be generated by words x and y, which do not commute. Define morphism $\pi : A^* \to A^*$ by $\pi(a) = x$ and $\pi(b) = y$. Then π is injective and $\mathrm{Eq}(g, h) = \mathrm{Eq}(\pi^{-1} \circ g, \pi^{-1} \circ h)$.*

It is the well known fact that a primitive word p is not a proper factor of pp. This implies the following list of claims.

Lemma 2. *Let swp be a factor of w^+. Then s is a suffix, and p a prefix of w^+.*

Lemma 3. *Let x and y be words marked from both sides. Let u be a factor of $(xy)^+$. Then any overlap of u and $xyyx$ is strictly shorter than $|xy|$.*

Lemma 4. *Let x and y be words marked from both sides. Let u be a word with a prefix (suffix resp.) xyx. Let w be a word such that*

$$w \in \mathrm{pref}(u) \cap \mathrm{suff}(xyyx) \qquad (w \in \mathrm{suff}(u) \cap \mathrm{pref}(xyyx) \ resp.)$$

Then w is strictly shorter than $|xy|$.

Lemma 5. *Let x and y be words marked from both sides. Then xyx is not a factor of $xyyx$, and $xyyx$ is not a factor of $\{xy, xyx\}^+$.*

Lemma 6. *Let x and y be words marked from both sides. Let u and v be non-empty words such that*

- *$yx \in \mathrm{pref}(u)$, $xy \in \mathrm{suff}(u)$,*
- *$v \in \{xy, xyx\}^+$,*
- *$|v| > |u|$.*

Then v is not a factor of u^+.

4   General Considerations

Lemma 7. *The words $g(a)$ and $h(a)$ ($g(b)$ and $h(b)$ resp.) do not commute.*

Proof. Suppose $g(a) = t^i$ and $h(a) = t^j$, with $i > j \geq 1$. Then the maximal element of t^+, which is a prefix of $g(\alpha\beta)$, is $t^{i \cdot k}$. On the other hand, $t^{j \cdot k}$ is the maximal element of t^+, which is a prefix of $h(\alpha\beta)$. This is a contradiction with $g(\alpha\beta) = h(\alpha\beta)$. Similarly for $g(b)$ and $h(b)$.

Lemma 8. *The word $g(b)^l$ ($h(a)^k$ resp.) is a prefix of $h(b)$ ($g(a)$ resp.). Similarly, $g(b)^{l'}$ ($h(a)^{k'}$ resp.) is a suffix of $h(b)$ ($g(b)$ resp.)*

Proof. Suppose, on the contrary, that $h(b)$ is a prefix of $g(b)^l$. Since $g(b)$ is a suffix of $h(b)$, the words $g(b)$ and $h(b)$ commute, a contradiction with Lemma 7. The rest is analogical.

We list some situations, which typically occure if $\mathrm{Eq}(g, h)$ is not empty.

Conditions 4

(A) *There is a proper suffix s of $g(b)$, such that $h(b)$ is a prefix of $sg(a)^+$.*
(B) *There is a proper prefix p of $g(b)$, such that $h(b)$ is a suffix of $g(a)^+p$.*
(C) *The word $h(b)$ is a factor of $g(a)^+$.*
(D) *There is a non-empty suffix u of $g(a)^+$ and a non-empty prefix v of $g(a)^+$, such that $ug(b)v = h(b)$.*

Lemma 9. *Let ub be a prefix of βα.*

(i) *If $|g(u)| < |h(u)|$ or $|g(u)| > |h(ub)|$ then condition (C) or (A) holds.*
(ii) *If $|g(ub)| > |h(ub)|$ or $|g(ub)| < |h(u)|$ then condition (C) or (B) holds.*

Proof. We first introduce some terminology. Let $m = |\alpha\beta|_b$ and $w = g(\alpha\beta) = h(\alpha\beta)$. Each letter b is mapped by g to a factor $g(b)$ of w, and by h to a factor $h(b)$ of w. The factor of w, which is an image of the i-th occurrence of letter b, with $1 \le i \le m$, will be called the i-th $g(b)$-factor of w. Similarly we define the i-th $h(b)$-factor of w. We shall consider the position of $g(b)$-factors with respect to corresponding $h(b)$-factors.

(i) Let $|g(u)| < |h(u)|$ and let u be the longest prefix of $\beta\alpha$ satisfying the assumption. Put $i = |ub|_b$. By assumption, the i-th $g(b)$-factor of w does not start within the i-th $h(b)$-factor. If the $(i+1)$-th $g(b)$-factor starts there, then $|g(u')| < |h(u')|$ for the prefix u' of $\alpha\beta$ such that $|u'b|_b = i + 1$. But we supposed that u is the longest possible. Therefore no $g(b)$-factor starts within the i-th $h(b)$-factor and the claim follows. Similarly for the shortest possible u, if $|g(u)| > |h(ub)|$.
(ii) The proof is analogical.

Corollary 1.

(i) *If $l > 1$ or $l' > 1$ then either conditions (A) and (B) hold, or condition (C) holds.*
(ii) *If none of conditions (A), (B), (C) and (D) holds, then eq$(g, h) = \{a^i b, ba^i\}$, $i \ge 1$.*

Proof.

(i) Let $l > 1$ and put $u = b$. Then $|g(u)| < |h(u)|$ and, by Lemma 8, also $|g(ub)| < |h(u)|$. The statement now follows from Lemma 9. Similarly if $l' > 1$.
(ii) It is not difficult to deduce, by Lemma 9, that if none of the conditions holds, all letters b in $\alpha\beta$ must be starting or ending. Therefore there are only two letters b in $\{\alpha, \beta\}$. Since the case $\{ba^i b, a^j\}$ implies $g = h$, we are left with $\{a^i b, ba^j\}$. The length agreement yields $i = j$.

Lemma 10. *Let x and y be words such that xy is primitive, and*

$$g(a) \in (xy)^* x, \qquad\qquad h(a) \in (xy)^* x,$$
$$g(b) \in (yx)^* y, \qquad\qquad h(b) \in (yx)^* y.$$

Then eq$(g, h) = \{ab, ba\}$.

Proof. Let $w = g(u) = h(u)$. By Lemma 1, we can suppose $x = a$ and $y = b$.

Let u be an element of Eq(g, h). Suppose that aa is a factor of u and $u = u_1 a a u_2$, where aa is not a factor of $u_1 a$. The word $g(u_1 a)a$ is the shortest prefix of $g(u)$ ending with aa. Similarly $h(u_1a)a$ is the shortest prefix of $h(u)$ of that form. Thus $g(u_1a) = h(u_1a)$.

This implies that aa is a factor of neither α nor β. In the same way we can show that neither α nor β contains bb as a factor. Thus either $\alpha \in (ab)^+a$ and $\beta \in (ba)^+b$, or $\alpha = ab$ and $\beta = ba$. The first possibility is excluded by the length argument.

The previous lemma has the following modification.

Lemma 11. *Let xy be a primitive word, with $x, y \in A^+$, such that*

$$g(a) \in (xy)^+x,$$
$$g(b) \in (yx)^+y, \qquad\qquad h(b) \in (yx)^+y.$$

Then eq$(g, h) = \{ab, ba\}$.

Proof. By Lemma 10, it is enough to show $h(a)$ is in $(xy)^*x$. The assumptions imply that x and y are marked from both sides.

1. Suppose ab is a prefix of α. Then $h(a)yx$ is a prefix of $(xy)^+$ and therefore $h(a) \in (xy)^*x$.
2. Suppose, on the other hand, that aa is a prefix of α. Then, by Lemma 8, the word $yxxy$ is either a factor of $h(b)$, or $h(b)$ is a factor of $yxxy$, or the two words have an overlap of length at least $|xy|$. This is a contradiction with Lemma 5 or Lemma 3.

5   Cases

The main *principium divisionis* is whether the word $g(ab)$ is longer or shorter than the word $h(b)$.

Case 1. $|g(ba)| \leq |h(b)|$.

The point of this case is to prove the following

Claim 1
$$g(b^l a) \in \operatorname{pref}(h(b)) \quad and \quad g(ab^{l'}) \in \operatorname{suff}(h(b)).$$

Proof. It is enough to prove $|g(b^l a)| \leq |h(b)|$ and $|g(ab^{l'})| \leq |h(b)|$.

Proceed by contradiction, and suppose, by symmetry, $|g(b^l a)| > |h(b)|$. Since $|g(ba)| < |h(b)|$, $l \geq 2$. Therefore the word $g(b^l a)$ is a prefix of $h(b)g(b)^{l-1}$ and there is a word u and a proper prefix q of $g(b)$, such that

$$h(b) = g(b)^l u, \qquad\qquad g(a) = ug(b)^i q,$$

with $0 \leq i \leq l - 2$.

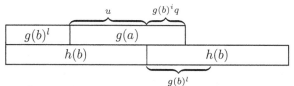

Suppose that $b^l ab$ is a prefix of β. Then $g(b)^i qg(b)$ is a factor of $g(b)^l$, and Lemma 2 yields that q is a suffix of $g(b)^+$, a contradiction. Therefore $b^l aa$ is a prefix of β.

By Corollary 1 we have to consider two possibilities.

1. Suppose $h(b)$ is a factor of $g(a)^+$. Let t be the primitive root of $g(a)$ and let $v_1 \in \text{suff}(t)$ and $v_2 \in \text{pref}(t)$ be words such that $h(b) \in (v_1 t^* v_2)$. Since $g(b)^i qt \in \text{suff}(t^+)$ is comparable with $h(b)$, it is also comparable with $v_1 t$, and the primitivity of t yields that $g(b)^i q \in v_1 t^*$. Therefore $h(bb)$ is a prefix of $g(b)^l t^+$. Similarly we deduce that $h(bb)$ is a suffix of $t^+ g(b)^{l'}$. Hence, by primitivity of t, $h(bbb) = g(b)^l t^m g(b)^{l'}$, for some positive integer m. From

$$|t| + |g(b)| \le |g(a)| + |g(b)| < |h(b)|,$$
$$3 \cdot |h(b)| = (l + l') \cdot |g(b)| + m \cdot |t|$$

it is not difficult to deduce that either

$$(l + l') \cdot |g(b)| > |g(b)| + |h(b)|,$$

or

$$m \cdot |t| > |t| + |h(b)|.$$

This implies, by Periodicity Lemma, that either $g(b)$ or t commutes with $h(b)$. We thus get a contradiction with Lemma 7 or with $\text{pref}_1(h(b)) \ne \text{pref}_1(g(a))$.

2. Suppose now that $h(b)$ is a prefix of $sg(a)^+$ and a suffix of $g(a)^+ p$, with a proper suffix s and a proper prefix p of $g(b)$. Lemma 2 and g is marked from both sides imply that

$$|h(b)| < |s| + |p| + |g(a)|.$$

	$h(b)$		
	s	$g(a)$	$g(a)^+$
$g(a)^+$	$g(a)$	p	

Therefore there are words x and y such that xy is primitive, $g(a) \in (xy)^+ x$, yx is a prefix and xy a suffix of $g(b)$. Then $xyyx$ occurs on the edge of $h(b)h(b)$ and it is easy to derive a contradiction with Lemma 5 or Lemma 4.

$g(b)^l$	$g(a)$	$g(a)$	
$h(b)$	xy	yx	$h(b)$

It is now straightforward to see that

Claim 2 *None of conditions* (A), (B) *and* (C) *holds.*

Proof.

1. If $h(b)$ is a factor of $g(a)^+$, then, by Claim 1, $g(b)^l g(a)$ is a factor of $g(a)^+$. This is a contradiction with Lemma 2 and g being marked from both sides.
2. Let (A) hold, and $h(b)$ be a prefix of $sg(a)^+$ for some proper suffix s of $g(b)$. Lemma 2 implies that $s^{-1} g(b)^l$ is a suffix of $g(a)^+$, a contradiction. Similarly for condition (B).

Lemma 9 now implies that $l = l' = 1$, and $h(b)$ is a prefix of $g(b)g(a)^+$ and a suffix of $g(a)^+ g(b)$. It is slightly more complicated to see that

Claim 3 *The condition* (D) *does not hold.*

Proof. In this proof p_i (s_i resp.) will always denote a proper prefix (a proper suffix resp.) of $g(a)$.

Suppose that (D) holds. We have

$$h(b) = g(b)g(a)^{m_1}p_1 = s_2 g(a)^{m_2} g(b) = s_3 g(a)^{m_3} g(b)g(a)^{m_4}p_4,$$

with $m_1, m_2, m_3, m_4 \geq 0$. Since $g(a)^{m_3} r$ is a factor of $g(a)^+$ for a non-empty prefix r of $g(b)$, Lemma 2 and g is marked imply that $m_3 = 0$. The mirrored consideration yields $m_4 = 0$.

Hence $|h(b)| < |g(b)| + 2 \cdot |g(a)|$, and therefore $m_1 = m_2 = 1$. We can write

$$h(b) = g(b)g(a)p_1 \tag{1}$$
$$h(b) = s_2 g(a)g(b) \tag{2}$$
$$h(b) = s_3 g(b)p_4, \tag{3}$$

$g(b)$		$g(a)$	p_3	p_1
s_2	$g(a)$		$g(b)$	
s_3		$g(b)$	p_4	

where $|s_2| < |s_3|$ and $|p_1| < |p_4|$. From (1) and (3) we deduce $p_4 = p_3 p_1$ and

$$g(b)g(a) = s_3 g(b)p_3,$$

with $s_3 p_3 = g(a)$. Hence

$$g(b)p_3 s_3 = s_3 g(b)p_3,$$

and words $g(b)p_3$ and s_3 have a common primitive root, say t. Let $t = t_1 t_2$ be a factorization of t such that

$$g(b) = (t_1 t_2)^{i_1} t_1, \qquad p_3 = t_2(t_1 t_2)^{i_2}, \qquad s_3 = (t_1 t_2)^j.$$

with $i_1, i_2 \geq 0$, $j \geq 1$. Then also

$$g(a) = p_3 s_3 = (t_2 t_1)^{i_2 + j} t_2,$$
$$g(b)g(a) = (t_1 t_2)^{i_1 + i_2 + j + 1},$$
$$g(a)g(b) = (t_2 t_1)^{i_1 + i_2 + j + 1}.$$

From (2) and (1) it follows that $s_2(t_2 t_1)$ is a prefix of $g(b)g(a)$ and thus

$$s_2 = (t_1 t_2)^{i_3} t_1, \qquad h(b) = s_2 g(a)g(b) = (t_1 t_2)^{i_1 + i_2 + i_3 + j + 1} t_1,$$

with $i_3 \geq 0$. The equality (3) gives

$$p_4 = (t_1 t_2)^{i_2 + i_3 + 1}$$

and, since p_4 is a prefix of $g(a)$, the words t_1 and t_2 commute. Therefore also $g(a)$ and $g(b)$ commute, a contradiction.

Corollary 1 together with the above claims now yields $\mathrm{Eq}(g,h) = \{a^i b, b a^i\}^*$.

Case 2. $|g(ab)| > |h(b)|$

We first consider a special situation:

Lemma 12. *If* $k = k' = l = l' = 1$, *then* $\mathrm{eq}(g,h) = \{ab, ba\}$.

Proof. If $|g(ab)| = |h(ab)|$, we are through. Suppose that $|g(ab)| > |h(ab)|$. The case $|g(ab)| < |h(ab)|$ is analogical. Assumptions now imply that

$$g(ab) = h(ab)v \tag{4}$$

for some non-empty word v. Since $h(b)$ is a suffix of $g(ab)$, there is a word u such that

$$uh(b) = h(b)v.$$

Let xy be a primitive word such that x is non-empty and

$$u = (yx)^i, \qquad\qquad v = (xy)^i, \qquad\qquad h(b) = (yx)^j y,$$

with $i \geq 1$ and $j \geq 0$. From $|h(ab)| > |g(a)|$ and from (4) we deduce $|g(b)| > |v|$. Since $g(b)$ is both prefix and suffix of $h(b)$, primitivity of xy yields $g(b) = (yx)^{j_1} y$, $j_1 \geq 1$. We have

$$g(a) = h(a)(yx)^{i+j-j_1} = (xy)^{i+j-j_1} h(a).$$

Therefore, by characterization of conjugates, $h(a), g(a) \in (xy)^* x$ and we are through by Lemma 10.

Subcase 1 $(l + l')|g(b)| \geq |h(b)|$

If $(l + l')|g(b)| = |h(b)|$ then $g(b)$ and $h(b)$ commute, a contradiction with Lemma 7.

If $(l + l' - 1)|g(b)| \geq |h(b)|$, then $g(b)$ and $h(b)$ again commute, by Periodicity Lemma.

Therefore $|h(b)| + |g(b)| > (l + l')|g(b)| > |h(b)|$. This implies that there exists a primitive word xy, with $x, y \in A^+$, such that

$$g(b) = (yx)^i y, \qquad\qquad h(b) = ((yx)^i y)^{l-1} (yx)^m y ((yx)^i y)^{l'-1},$$

with $i \geq 1$, and $i < m \leq 2i$. The factor $(yx)^m y$ in the expression of $h(b)$ represents the overlapping occurrences of $g(b)$.

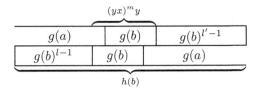

Then also

$$(xy)^{m-i}((yx)^iy)^{l'-1} \in \text{pref}(g(a)), \qquad ((yx)^iy)^{l-1}(yx)^{m-i} \in \text{suff}(g(a)). \qquad (5)$$

Note that x and y are marked from both sides.

1. Suppose first that either $l > 1$ or $l' > 1$, and apply Corollary 1. By Lemma 6, the word $h(b)$ is not a factor of $g(a)^+$. Therefore there is a proper suffix s of $g(b)$, such that $h(b)$ is a prefix of $sg(a)^+$, and

$$s^{-1}g(b)^l x \quad \text{is a prefix of} \quad g(a)^+. \qquad (6)$$

The prefix yx of $h(b)$ is also a prefix of $sxy \in \text{pref}(sg(a))$, which is a suffix of $(yx)^{i+1}y$. This implies that $s = (yx)^{i_1}y$, with $i > i_1 \geq 0$. Therefore

$$(yx)^{i_1+m-i}y((yx)^iy)^{l'-1} \in \text{pref}(h(b)). \qquad (7)$$

We shall show that $l' \leq l$ and $i_1 = 2i - m$.

1.1. Suppose that $l = 1$ and $l' > 1$. By (5), the word $(yx)^iy(xy)^{m-i}y$ is the shortest prefix of $h(b)$ ending with xyy. From (7) we get another expression of this word, namely $(yx)^{i_1+m-i}yy$. This implies $m = i_1 + m - i$. Therefore $i_1 = i$, a contradiction with s being proper prefix of $g(b)$.

1.2. If $l > 1$, the shortest prefix of $h(b)$ ending with xyy is $(yx)^iyy$, and, as above, we deduce $i = i_1 + m - i$, in accordance with the claim. Thus both $((yx)^iy)^{l'}$ and $((yx)^iy)^l x$ are prefixes of $h(b)$, and l' is at most l.

Mirror considerations yield $l \leq l'$ and thus $l = l'$. From (6) we now conclude that

$$(xy)^{m-i}y((yx)^iy)^{l-1}x \quad \text{is a prefix of} \quad g(a).$$

It follows that the word

$$g(b)^l(xy)^{m-i}((yx)^iy)^{l-1}x$$

is a prefix of $h(b)^l$, and x is a prefix of $h(b)$, a contradiction.

2. Suppose then that either $k > 1$ or $k' > 1$. By symmetry, let $k > 1$. We shall use the fact that $g(aa)$ contains a factor $yxxy$. By Lemma 6, the word $h(b)$ is not a factor of $g(a)^+$. Therefore $g(a)^k$ is a factor of $h(a^kb)$. Since $|h(a^k)| < |g(a)|$, we get a contradiction with Lemma 4 or Lemma 5.

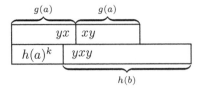

We have shown that any possibility, except $k = k' = l = l'$, is contradictory, and can use Lemma 12.

Subcase 2 $(l + l')|g(b)| < |h(b)|$

We have

$$g(a)g(b)^{l'}u = vg(b)^l g(a) = vh(b)u = vg(b)^l wg(b)^l u,$$

with $u, v, w \in A^+$. The word w is both a prefix and a suffix of $g(a)$, and

$$g(a) = vg(b)^l w = wg(b)^{l'} u.$$

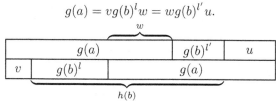

Thus there is a primitive word xy, such that x and y are marked from both sides, and

$$w = (xy)^j x, \qquad g(b)^{l'} u = (yx)^i, \qquad vg(b)^l = (xy)^i,$$

with $i \geq 1$, $j \geq 0$. We have

$$g(a) = (xy)^{i+j} x$$

and $h(b)$ is a factor of $(yx)^+$. We first note the following direct consequence of Lemma 3.

Claim 4 If $k > 1$, or $k' > 1$, or (C) holds, then $h(b)$ is a factor of $yxxy$.

1. Let first $|g(b)| \geq |y|$. Then yxy is a factor of $h(b)$. Claim 4 and Lemma 5 imply that (C) does not hold, and $k = k' = 1$.
   Suppose $l > 1$. Then $u = g(b)^{l-1}q$, with a prefix q of $g(b)$, and $g(b)^l q$ is a factor of $(xy)^+$. By Corollary 1, the condition (A) holds. Consequently, the word $g(b)^l xy$ is a prefix of $s(xy)^+$ for some suffix s of $g(b)$. The primitivity of xy implies that $s^{-1}g(b)^l$ is in $(xy)^+$, especially xy is a suffix of $g(b)^l$. By Lemma 2, the word q is a prefix of $(xy)^+$, a contradiction. Similarly if $l' > 1$.
2. Let now $|g(b)| < |y|$.
2.1. Suppose $l > 1$. Then the word u is a prefix of $h(b)$ and consequently yx is a prefix of $g(b)^{l'+l}$. Since $g(b)$ is a suffix of y, Lemma 2 yields that x is a prefix of $g(b)^+$, a contradiction. Similarly if $l' > 1$.
2.2. Suppose now $l = l' = 1$ and $k > 1$. Claim 4 implies $|h(b)| \leq |yxxy|$ and $|h(b)| \geq |g(a)|$ yields $i + j = 1$. Thus $i = 1$ and $j = 0$, and from

$$2|g(b)| + |x| = |h(b)| \geq |g(a)| = 2|x| + |y|$$

we deduce

$$|x| + |y| \leq 2|g(b)|. \tag{8}$$

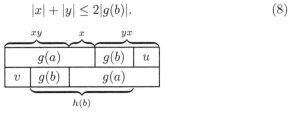

The word $g(b)$ is a prefix and a suffix of y. The inequality (8) implies that there exist a primitive word $x_1 y_1$, with $y_1 \in A^+$, $x_1 \in A^*$, and integers $1 \leq i_1 \leq j_1$ such that

$$y = (y_1 x_1)^{i_1 + j_1} y_1,$$

$$g(b) = (y_1 x_1)^{j_1} y_1. \tag{9}$$

Therefore

$$h(b) = (y_1 x_1)^{j_1} y_1 \, x \, (y_1 x_1)^{j_1} y_1$$

and Claim 4 now yields

$$u_1 (y_1 x_1)^{j_1} y_1 \, x \, (y_1 x_1)^{j_1} y_1 v_1 = (y_1 x_1)^{i_1 + j_1} y_1 \, xx \, (y_1 x_1)^{i_1 + j_1} y_1, \tag{10}$$

for some words u_1 and v_1.

Note that x and y_1 are marked from both sides, and thus, by Lemma 5, the word $y_1 x y_1$ is not a factor of $y_1 xx y_1$. By symmetry of (10) and by Lemma 4, we can suppose

$$|u_1 (y_1 x_1)^{j_1} y_1| < |(y_1 x_1)^{i_1 + j_1}|. \tag{11}$$

Consider the factor $x_1 y_1 x y_1 x_1$ in (10). The primitivity of its prefix $x_1 y_1$ yields that the word $x y_1 x_1$ is a prefix of $(x_1 y_1)^m x$, $m \geq 1$. This implies $x = (x_1 y_1)^{m'} x_1$, $m' \geq 0$. From (9) we have $yx = (y_1 x_1)^{i_1 + j_1 + m' + 1}$, a contradiction with the primitivity of xy.

We are left with $l = l' = k = k' = 1$, and Lemma 12 concludes the proof.

Conclusion

We have shown that any binary equality set of rank two is of the form $\{a^i b, ba^i\}^*$ (up to the symmetry of the letters a and b). On the other hand, for any $i \geq 1$ it is not difficult to find morphisms g and h with $\mathrm{Eq}(g,h) = \{a^i b, ba^i\}^*$. A precise description of such morphisms can be found in [3], p. 365 n.

Our proof depends on Theorem 1. Independently of [2] it works for a pair of morphisms marked from both sides.

We believe that a similar approach can be also used for a complete characterization of binary equality sets generated by just one word.

References

1. A. Ehrenfeucht, J. Karhumäki, and G. Rozenberg, *The (generalized) Post correspondence problem with lists consisting of two words is decidable*, Theoret. Comput. Sci. **21** (1982), no. 2, 119–144.
2. Š. Holub, *Binary equality sets are generated by two words*, J. Algebra, DOI: 10.1016/S0021-8693(02)00534-3, in press.
3. K. Culik II and J. Karhumäki, *On the equality sets for homomorphisms on free monoids with two generators*, RAIRO Theor. Informatics **14** (1980), 349–369.
4. A. Ehrenfeucht, J. Karhumäki and G. Rozenberg, *On binary equality sets and a solution to the test set conjecture in the binary case*, J.Algebra **85** (1983), 76–85.

On Deterministic Finite Automata
and Syntactic Monoid Size

Markus Holzer and Barbara König

Institut für Informatik, Technische Universität München,
Arcisstraße 21, D-80290 München, Germany
{holzer,koenigb}@informatik.tu-muenchen.de

Abstract. We investigate the relationship between regular languages and syntactic monoid size. In particular, we consider the transformation monoids of n-state (minimal) deterministic finite automata. We show tight upper bounds on the syntactic monoid size, proving that an n-state deterministic finite automaton with singleton input alphabet (input alphabet with at least three letters, respectively) induces a linear (n^n, respectively) size syntactic monoid. In the case of two letter input alphabet, we can show a lower bound of $n^n - \binom{n}{\ell}\ell!n^k - \binom{n}{\ell}k^k\ell^\ell$, for some natural numbers k and ℓ close to $\frac{n}{2}$, for the size of the syntactic monoid of a language accepted by an n-state deterministic finite automaton. This induces a family of deterministic finite automata such that the fraction of the size of the induced syntactic monoid and n^n tends to 1 as n goes to infinity.

1 Introduction

Regular languages and their implementations have received more and more attention in recent years due to the many new applications of finite automata and regular expressions in object-oriented modeling, programming languages and other practical areas of computer science. In recent years, quite a few software systems for manipulating formal language objects, with an emphasis on regular-language objects, have been developed. Examples include AMoRE, Automata, FIRE Engine, FSA, Grail, and INTEX [1,10]. These applications and implementations of regular languages motivate the study of descriptive complexity of regular languages. A very well accepted and studied measure of descriptional complexity for regular languages is the size, i.e., number of states, of deterministic finite automata.

Besides machine oriented characterization of regular languages, they also obey several algebraic characterizations. It is a consequence of Kleene's theorem [3], that a language $L \subseteq \Sigma^*$ is regular if and only if there exists a finite monoid M, a morphism $\varphi : \Sigma^* \to M$, and a finite subset $N \subseteq M$ such that $L = \varphi^{-1}(N)$. The monoid M is said to recognize L. The syntactic monoid of L is the smallest monoid recognizing the language under consideration. It is uniquely defined up to isomorphism and is induced by the syntactic congruence $\sim_L$ defined over Σ^* by $v_1 \sim_L v_2$ if and only if for every $u, w \in \Sigma^*$ we have

M. Ito and M. Toyama (Eds.): DLT 2002, LNCS 2450, pp. 258–269, 2003.

$uv_1w \in L \iff uv_2w \in L$. The syntactic monoid of L is the quotient monoid $M(L) = \Sigma^* / \sim_L$. In this paper we propose the size of the syntactic monoid as a natural measure of descriptive complexity for regular languages and study the relationship between automata and monoid size in more detail.

In most cases, we show tight upper bounds on the syntactic monoid, proving that there are languages accepted by n-state deterministic finite automata whose syntactic monoid has a certain size. It is easy to see that for unary regular languages the size is linear, while that for regular languages over an input alphabet with at least three letters is maximal, i.e., n^n. The challenging part is to determine the size of the syntactic monoid for regular languages over a binary alphabet. The trivial lower and upper bounds are $n!$—induced by the two generators of S_n—and $n^n - n! + g(n)$, respectively, where $g(n)$ denotes Landau's function [4,5,6], which equals the maximal order of all permutations in S_n. Compared to the trivial lower bound, where $\lim_{n \to \infty} \frac{n!}{n^n} = 0$, we can do much better, since we present binary regular languages whose syntactic monoid is at least $n^n - \binom{n}{\ell}\ell!n^k - \binom{n}{\ell}k^k\ell^\ell$, for some natural numbers k and ℓ close to $\frac{n}{2}$, and the fraction of this number (for appropriate k and ℓ) and n^n tends to 1 as n goes to infinity.

The paper is organized as follows. In the next section we introduce the necessary notations. Then in Section 3 we prove the easy cases on syntactic monoid size and devote Section 4 to the study of binary regular languages. Finally, we summarize our results and state some open problems.

2   Definitions

We assume the reader to be familiar with the basic notions of formal language theory and semigroup theory, as contained in [2] and [8]. In this paper we are dealing with regular languages and their syntactic monoid. A *semigroup* is a nonempty set S equipped with an associative binary operation, i.e., $(\alpha\beta)\gamma = \alpha(\beta\gamma)$ for all $\alpha, \beta, \gamma \in S$. The semigroup S is called a *monoid* if it contains an identity element *id*. If E is a set, then we denote by $T(E)$ the monoid of functions from E into E together with the composition of functions. We read composition from left to right, i.e., first α, then β. Because of this convention, it is natural to write the argument i of a function to the left: $(i)\alpha\beta = ((i)\alpha)\beta$. The image of a function α in $T(E)$ is defined as $img(\alpha) = \{ (i)\alpha \mid i \in E \}$ and the kernel of α is the equivalence relation $\equiv$, which is induced by $i \equiv j$ if and only if $(i)\alpha = (j)\alpha$. In particular, if $E = \{1, \ldots, n\}$, we simply write T_n for the monoid $T(E)$. The monoid of all permutations over n elements is denoted by S_n and trivially is a sub-monoid of T_n.

A deterministic finite automaton is a 5-tuple $A = (Q, \Sigma, \delta, q_0, F)$, where Q is the finite set of states, Σ is a finite alphabet, $\delta : Q \times \Sigma \to Q$ denotes the transition function, $q_0 \in Q$ is the initial state, and $F \subseteq Q$ is the set of final states. Observe, that a deterministic finite automaton is complete by definition. As usual, δ is extended to act on $Q \times \Sigma^*$ by $\delta(q, \lambda) = q$ and $\delta(q, aw) = \delta(\delta(q, a), w)$ for $q \in Q$, $a \in \Sigma$, and $w \in \Sigma^*$, where λ denotes the empty word of length zero. Unless otherwise stated, we assume that $Q = \{1, \ldots, n\}$ for some $n \in \mathbb{N}$. The language accepted by the deterministic finite automaton A is defined as

$$L(A) = \{\, w \in \Sigma^* \mid \delta(q_0, w) \in F \,\}.$$

The family of regular languages is the set of all languages which are accepted by some deterministic finite automaton.

In order to compute the syntactic monoid of a language it is convenient to consider the transition monoid induced by a finite automaton. Let $A = (Q, \Sigma, \delta, q_0, F)$ be a deterministic finite automaton. Naturally, each word $w \in \Sigma^*$ defines a function from Q into Q. The monoid generated by all these functions thus defined, where w varies over Σ^*, is a sub-monoid of $T(Q)$; it is the transition monoid $M(A)$ of the automaton A. Clearly, $M(A)$ is generated by the functions defined by the letters of the alphabet and we have a canonical morphism $\Sigma^* \to M(A)$. The intrinsic relationship between the transition monoid $M(A)$ and the syntactic monoid of the language $L(A)$ is as follows: The transition monoid of the minimal deterministic finite automata is isomorphic to $M(L)$. This allows the computation of $M(L)$ in a convenient way.

3   Syntactic Semigroup Size – The Easy Cases

We start our investigation on syntactic monoid size with two easy cases, which mostly follow from results from the literature. We state these results for completeness only. Firstly, we consider unary regular languages, where we can profit from the following result on monogenic (sub)semigroups, which can be found in [2].

Theorem 1. *Let α be an element of a semigroup S. Then either all powers of α are distinct and the monogenic sub-semigroup $\langle \alpha \rangle := \{\, \alpha^i \mid i \geq 1 \,\}$ of S is isomorphic to the semigroup $(\mathbb{N}, +)$ of the natural numbers under addition, or there exists positive integers m and r such that $\alpha^m = \alpha^{m+r}$ and $\langle \alpha \rangle = \{\alpha, \alpha^2, \ldots, \alpha^{m+r-1}\}$. Here m is called the index and r the period of α.*

Then we can estimate the syntactic monoid size of regular languages over a unary input alphabet as follows:

Theorem 2. *Let A be an n-state deterministic finite automaton with a unary input alphabet. Then a monoid of size n is sufficient and necessary in the worst case to recognize the language $L(A)$.*

Proof. Observe, that the transition graph of a deterministic finite automaton A with unary input alphabet consists of a path, which starts from the initial state, followed by a cycle of one or more states. Assume that m is the number of states of the path starting from the initial state, and r the number of states in the cycle. Then $n = m + r$ and A, by appropriately numbering the states, induces the mapping

$$\alpha = \begin{pmatrix} 1 & 2 & \ldots & m & m+1 & \ldots & m+r-1 & m+r \\ 2 & 3 & \ldots & m+1 & m+2 & \ldots & m+r & m+1 \end{pmatrix}$$

of the semigroup T_n. It is a routine matter to verify that α has index m and period r. Hence by Theorem 1 the semigroup generated by α equals the $n-1$ element set $\{\alpha, \alpha^2, \ldots, \alpha^{m+r-1}\}$. This shows the upper bound n on the monoid size, since the neutral element has to be taken into consideration, too. On the other hand, if A was chosen to be a minimal deterministic finite automaton then the induced transformation monoid equals $\{id\} \cup \{\alpha, \alpha^2, \ldots, \alpha^{m+r-1}\}$ by our previous investigation. Therefore, n is also a lower bound for the maximal syntactic monoid size. $\qquad\square$

In the remainder of this section we consider regular languages over an input alphabet with at least three letters. Obviously, for all n, the elements

$$\alpha = \begin{pmatrix} 1 & 2 & \ldots & n-1 & n \\ 2 & 3 & \ldots & n & 1 \end{pmatrix}, \quad \beta = \begin{pmatrix} 1 & 2 & 3 & \ldots & n \\ 2 & 1 & 3 & \ldots & n \end{pmatrix}, \quad \text{and} \quad \gamma = \begin{pmatrix} 1 & 2 & \ldots & n-1 & n \\ 1 & 2 & \ldots & n-1 & 1 \end{pmatrix}$$

of T_n form a complete basis of T_n, i.e., they generate all of the monoid T_n. In particular, if $n=2$ we find that $\alpha = \beta$, and thus two elements suffice for the generation of T_2, while for $n=1$ trivially one element is enough to generate all of T_1—here $\alpha = \beta = \gamma$ holds. Thus we have shown the following theorem:

Theorem 3. *Let A be an n-state deterministic finite automaton with input alphabet Σ. Then a monoid of size n^n is sufficient and necessary in the worst case to recognize the language $L(A)$ if either (i) $n=1$, or (ii) $n=2$ and $|\Sigma| \geq 2$, or (iii) $n \geq 3$ and $|\Sigma| \geq 3$.*

Proof. The upper bound n^n is trivial. From the above given generators α, β, and γ we define the deterministic finite automata $A = (\{Q, \{a, b, c\}, \delta, 1, F)$, where $Q = \{1, \ldots, n\}$, $F = \{n\}$ and $\delta(i, a) = (i)\alpha$, $\delta(i, b) = (i)\beta$, and $\delta(i, c) = (i)\gamma$. It remains to prove that A is minimal. In order to show this, it is sufficient to verify that all states of A are reachable and lie in different equivalence classes. The reachability claim is easy to see, since for every state $i \in Q$ we have $\delta(1, a^{i-1}) = i$ and the latter claim follows since for $i, j \in Q$ with $i < j$ we find $\delta(i, a^{n-j}) = i + (n-j) \notin F$, since $i + (n-j) < n$, and $\delta(j, a^{n-j}) = n \in F$. Thus, i and j are not in the same equivalence class. $\qquad\square$

The question arises, whether the above given theorem can be improved with respect to the alphabet size. By easy calculations one observes, that for $n=2$ this is not the case, since a unary language will only induce a syntactic monoid of size 2, due to Theorem 2. For $n \geq 3$ the following completeness theorem for unary functions given in [9], shows that an improvement is also not possible. The completeness result reads as follows.

Theorem 4. *Assume $n \geq 3$. Then three elements of T_n generate all functions of T_n if and only if two of them generate the symmetric group S_n and the third has kernel size $n-1$. Moreover, no less than three elements generate all functions from T_n.*

Thus, it remains to classify the syntactic monoid size of binary languages in general, which is done in the remaining part of the paper.

4   Syntactic Semigroup Size – A More Complicated Case

In this section we consider binary languages and the size of their syntactic monoid in more detail. Compared to the previous section here we are only able to prove a trivial upper and a non-matching lower bound on the syntactic monoid size for languages accepted by n-state deterministic finite automata.

The outline of this section is as follows: First we define a subset of T_n by some easy properties, verify that it is a semigroup and that it is generated by two generators only. Then, we argue that there is a minimal deterministic finite automaton, the transition monoid of which equals the defined semigroup and finally, we determine a lower bound of the semigroup size. The advantage of the explicit definition of the semigroup is that we don't have to go into some tedious analysis of the Green's relations if the semigroup would be given by generators only. The subset of T_n we are interested in, is defined as follows:

Definition 1. *Let $n \geq 2$ such that $n = k + \ell$ for some natural numbers k and ℓ. Furthermore, let $\alpha = (1\,2\,\ldots\,k)(k+1\,k+2\,\ldots\,n)$ be a permutation of S_n consisting of two cycles. We define $U_{k,\ell}$ as a subset of T_n as follows: A transformation γ is an element of $U_{k,\ell}$ if and only if*

1. *there exists a natural number $m \in \mathbb{N}$ such that $\gamma = \alpha^m$ or*
2. *the transformation γ satisfies that*
   (a) *there exist $i \in \{1, \ldots, k\}$ and $j \in \{k+1, \ldots, n\}$ such that $(i)\gamma = (j)\gamma$ and*
   (b) *there exists $h \in \{k+1, \ldots, n\}$ such that $h \notin img(\gamma)$.*

The intuition behind choosing this specific semigroup $U_{k,\ell}$ is the following: We intend to generate it with two transformations, one being the permutation α, the other a non-bijective transformation β. Since β is non-bijective there are at least two indices i, j such that $(i)\beta = (j)\beta$. By applying a multiple of α before applying β the number of index pairs which may be mapped to the same image can be increased. If the permutation is one cycle of the form $(1\,2\,\ldots\,n)$ the number of pairs is only n, whereas in the case of the α above which consists of two cycles whose lengths do not have a non-trivial common divisor, there are $k\ell$ possible pairs to choose from. And if k and ℓ are chosen close to $\frac{n}{2}$, then $k\ell > n$. Next we have to show that $U_{k,\ell}$ is indeed a semigroup.

Lemma 1. *The set $U_{k,\ell}$ is closed under composition and is therefore a (transformation) semigroup.*

Proof. Let $\gamma_1, \gamma_2 \in U_{k,\ell}$ be two transformations. We show that $\gamma_1\gamma_2$ is also an element of $U_{k,\ell}$. We have to distinguish the following four cases:

1. The transformation γ_1 is of the form α^{m_1} and the transformation γ_2 is of the form α^{m_2} for some $m_1, m_2 \geq 1$. Then clearly $\gamma_1\gamma_2 = \alpha^{m_1+m_2}$ is an element of $U_{k,\ell}$.

2. Let $\gamma_1 = \alpha^m$, for some $m \geq 1$, and γ_2 satisfies the second condition of Definition 1, i.e., there are indices $i \in \{1, \ldots, k\}$ and $h, j \in \{k+1, \ldots, k+\ell\}$ such that $(i)\gamma_2 = (j)\gamma_2$ and $h \notin img(\gamma_2)$.
The element h also fails to be a member of $img(\gamma_1\gamma_2)$. Furthermore, because of the nature of α it holds that $i' = (i)\gamma_1^{-1} \in \{1, \ldots, k\}$ and $j' = (j)\gamma_1^{-1} \in \{k+1, \ldots, k+\ell\}$. And it holds that $(i')\gamma_1\gamma_2 = (i)\gamma_2 = (j)\gamma_2 = (j')\gamma_1\gamma_2$. Therefore $\gamma_1\gamma_2$ satisfies also the second condition of Definition 1.

3. Assume that $\gamma_2 = \alpha^m$, for some $m \geq 1$, and γ_1 satisfies the second condition of Definition 1, i.e., there are indices $i \in \{1, \ldots, k\}$ and $h, j \in \{k+1, \ldots, k+\ell\}$ such that $(i)\gamma_1 = (j)\gamma_1$ and $h \notin img(\gamma_1)$.
It obviously holds that $(i)\gamma_1\gamma_2 = (j)\gamma_1\gamma_2$. And since $\gamma_2 = \alpha^m$ and the permutation α maps elements of its second cycle only to other elements of the second cycle, it holds that $h' = (h)\gamma_2 \notin img(\gamma_1\gamma_2)$, since otherwise $h = (h')\gamma_2^{-1}$ would be in the image of γ_1 which is a contradiction.

4. Finally, let γ_1 and γ_2 both satisfy the second condition of Definition 1. Then there are indices $i_1, i_2, \in \{1, \ldots, k\}$ and $h_1, h_2, j_1, j_2 \in \{k+1, \ldots, k+\ell\}$ such that $(i_r)\gamma_r = (j_r)\gamma_r$ and $h_r \notin img(\gamma_r)$ for $1 \leq r \leq 2$.
By setting $i = i_1$, $j = j_1$, and $h = h_2$, it is easy to see that $\gamma_1\gamma_2$ satisfies also the second part of Definition 1. $\square$

Before we can prove that $U_{k,\ell}$ is generated by two elements of $T_{k+\ell}$ we need some result, which constitutes how to find a complete basis for the symmetric group S_n. The below given result was shown in [7].

Theorem 5. *Given a non-identical element α in S_n, then there exists β such both generate the symmetric group S_n, provided that it is not the case that $n = 4$ and α is one of the three permutations $(1\,2)(3\,4)$, $(1\,3)(2\,4)$, and $(1\,4)(2\,3)$.*

Now we are ready for the proof that two elements are enough to generate all of $U_{k,\ell}$, provided that k and ℓ obey some nice properties.

Theorem 6. *Let $k, \ell \in \mathbb{N}$ be two natural numbers with $k < \ell$ and $\gcd\{k, \ell\} = 1$, and set $n = k+\ell$. The semigroup $U_{k,\ell}$ can be generated with two elements of T_n, where one element is the permutation $\alpha = (1\,2\,\ldots\,k)(k+1\,k+2\,\ldots\,n)$ and the other is an element β of kernel size $n - 1$.*

Proof. The first generator of $U_{k,\ell}$ is the permutation α of Definition 1. Now set $\pi_1 = (1\,2\,\ldots\,k)$, which will be considered as a permutation in S_{n-1}. Since π_1 is not the identity and not an element of the listed exceptions, then according to Theorem 5, there exists a permutation π_2 such that π_1 and π_2 generate S_{n-1}. Now define the second generator β of $U_{k,\ell}$ as follows: Let $(i)\beta = (i)\pi_2$ whenever $1 \leq i \leq n - 1$ and $(n)\beta = (1)\pi_2$. Hence β has kernel size $(k + \ell) - 1 = n - 1$.

We will first show that α and β generate at most the transformations specified in Definition 1. Let γ therefore be an element generated by α and β. If no β was used in the generation of γ, then $\gamma = \alpha^m$, for some natural number m. Otherwise $\gamma = \alpha^m\beta\gamma'$ for some natural number m (possibly $m = 0$) and some transformation γ'. By definition $(1)\beta = (n)\beta$. We set $i = (1)\alpha^{-m}$ and $j =$

$(n)\alpha^{-m}$. Since the element 1 is located in the first cycle of α and the element n is located in the second cycle of α it follows that $i \in \{1, \ldots, k\}$ and $j \in \{k + 1, \ldots, n\}$. Furthermore, $(i)\gamma = (i)\alpha^m\beta\gamma' = (1)\beta\gamma' = (n)\beta\gamma' = (j)\alpha^m\beta\gamma = (j)\gamma$. On the other hand γ can be written as $\gamma = \gamma''\beta\alpha^r$, for some $r \geq 0$. Since n is not in the image of β, the same is true for the image of $\gamma''\beta$. This implies that $h = (n)\alpha^r$ is not in the image of γ and since n is an element of the second cycle of α, this implies $h \in \{k + 1, \ldots, n\}$.

Conversely, we show that α and β generate at least the transformations specified in Definition 1. Clearly transformations of the form $\gamma = \alpha^m$, for some $m \geq 1$, can be generated easily. Now let γ be a transformation such that $(i)\gamma = (j)\gamma$ and $h \notin img(\gamma)$ for $i \in \{1, \ldots, k\}$ and $h, j \in \{k+1, \ldots, n\}$. Since k and ℓ do not have a common divisor, the cycles of α can be "turned" independently and therefore there exists a natural number $r \in \{1, \ldots, k\ell\}$ such that $(i)\alpha^r = 1$ and $(j)\alpha^r = n$. And there exists a number p such that $\alpha^p = (1\,2\,\ldots\,k)$. Furthermore there exists a number s such that $(n)\alpha^s = h$.

We are now looking for a transformation γ' such that $\gamma = \alpha^r\beta\gamma'\alpha^s$ and γ' can be generated from α and β. This condition can be rewritten to $\gamma\alpha^{-s} = \alpha^r\beta\gamma'$. Both transformation $\gamma\alpha^{-s}$ and $\alpha^r\beta$ do not have the element n in their image. So it suffices to show that for every transformation δ on $\{1, \ldots, n-1\}$ we can generate a transformation γ' on $\{1, \ldots, n\}$ such that $\gamma'|_{\{1,\ldots,n-1\}} = \delta$. Observe, that the transformations α^p and β (see the definition of β) act as permutations on the set $\{1, \ldots, n-1\}$ and their restrictions to this set are generators of S_{n-1}.

We can also generate the transformation η that maps $(1)\eta = (2)\eta = 1$ and is the identity on $\{3, \ldots, n-1\}$. This can be done by first creating a transformation with the same kernel as η. The kernel of β partitions the set $\{1, \ldots, n\}$ into $\{\{1, n\}, \{2\}, \ldots, \{n-1\}\}$. We can now construct a transformation σ that acts as a permutation on $\{1, \ldots, n-1\}$ and that maps $(2)\beta \mapsto n-1$ and $(1)\beta \mapsto k$. Therefore the transformation $\beta\sigma\alpha$ maps 2 to n, and 1 to itself, and has the same kernel as β. Consequently the transformation $\beta\sigma\alpha\beta$ has the kernel $\{\{1, 2, n\}, \{3\}, \ldots, \{n-1\}\}$ and all its images are contained in $\{1, \ldots, n-1\}$. Therefore there exists a permutation σ' that acts on $\{1, \ldots, n-1\}$ and for which $\beta\sigma\alpha\beta\sigma' = \eta$. Since this gives us three generators for T_{n-1}, it is clear that with these three transformations α^p, β, and η we can construct a transformation γ' such that $\gamma'|_{\{1,\ldots,n-1\}} = \delta$ for every transformation $\delta \in S_{n-1}$. $\square$

Before we continue our investigations estimating the size of $U_{k,\ell}$, we show that $U_{k,\ell}$ is in fact a syntactic monoid of a regular language accepted by some n-state deterministic finite automaton.

Theorem 7. *Let $k, \ell \in \mathbb{N}$ be two natural numbers with $k < \ell$ and $\gcd\{k, \ell\} = 1$, and set $n = k + \ell$. Then there is an n-state minimal deterministic finite automaton A with binary input alphabet the transition monoid of which equals $U_{k,\ell}$. Hence, $U_{k,\ell}$ is the syntactic monoid of $L(A)$.*

Proof. By Theorem 6 the semigroup $U_{k,\ell}$ is generated the permutation $\alpha = (1\,2\,\ldots\,k)(k+1\,k+2\,\ldots\,n)$ and by an element β of kernel size $n-1$. Define the deterministic finite automaton $A = (Q, \{a, b\}, \delta, 1, F)$, where $Q = \{1, \ldots, n\}$,

$F = \{k, n\}$, and $\delta(i, a) = (i)\alpha$ and $\delta(i, b) = (i)\beta$ for all $i \in Q$. In order to show that $U_{k,\ell}$ is the syntactic monoid of $L(A)$ we have to prove that all states are reachable and belong to different equivalence classes. For reachability we argue as follows: Obviously, the transition monoid of A equals $U_{k,\ell}$ by construction. Thus, all states are reachable since $U_{k,\ell}$ contains all constant functions. For the second claim we distinguish three cases:

1. Let $i, j \in \{1, \ldots, k\}$ with $i < j$. Then $\delta(i, a^{k-j}) \notin F$ and $\delta(j, a^{k-j}) = k \in F$. Thus, states i and j are inequivalent.
2. Let $i, j \in \{k+1, \ldots, n\}$ with $i < j$. Then a similar argumentation as above shows that both states are not equivalent.
3. Finally, let $i \in \{1, \ldots, k\}$ and $j \in \{k+1, \ldots, n\}$. Here we can not exclude that $k - i = n - j$. Nevertheless, since $\gcd\{k, \ell\} = 1$ it follows in that case that $\delta(i, a^{k-i}a^k) = k$ and $k \in F$, while $\delta(j, a^{k-i}a^k) \notin F$. This implies that both states are inequivalent, too.

This completes our proof and shows that A is a minimal deterministic finite automaton. Hence, A's transition monoid equals the syntactic monoid of $L(A)$. □

In order to determine the size of $U_{k,\ell}$ the following lemma, relating size and number of colourings of a particular graph, is very useful in the sequel.

Lemma 2. *Let $n = k + \ell$ for some naturals numbers k and ℓ satisfying $k < \ell$ and $\gcd\{k, \ell\} = 1$. Denote the complete bipartite graph with two independent sets C and D having k and ℓ nodes, respectively, by $K_{k,\ell}$. Then*

$$|U_{k,\ell}| = k\ell + N,$$

where N is the number of invalid colourings of $K_{k,\ell}$ with colours from $\{1, \ldots, n\}$, such that at least one colour from the set $\{k+1, \ldots, n\}$ is missing.

Proof. We assume, without loss of generality, that $V = \{1, \ldots, n\}$ is the set of nodes of $K_{k,\ell}$ and that $C = \{1, \ldots, k\}$ and $D = \{k+1, \ldots, n\}$. Thus every (valid or invalid) colouring of $K_{k,\ell}$ can be considered as a transformation of T_n and *vice versa*. It is rather straightforward to see that the transformations of $U_{k,\ell}$ satisfying the second part of Definition 1 coincide exactly with the invalid colourings of $K_{k,\ell}$, where at least one colour from the set $\{k+1, \ldots, n\}$ is missing. □

Now we are ready to estimate the size of $U_{k,\ell}$ and prove some asymptotics for particular values of k and ℓ.

Theorem 8. *Assume $n \geq 3$. Let $n = k + \ell$ for some natural numbers k and ℓ obeying $k < \ell$ and $\gcd\{k, \ell\} = 1$. Then*

$$|U_{k,\ell}| \geq n^n - \binom{n}{\ell}\ell! n^k - \binom{n}{\ell}k^k \ell^\ell.$$

Moreover, for every n there exists $k(n)$ and $\ell(n)$ satisfying the above properties and $\ell(n) - k(n) \leq 4$, such that

$$\lim_{n \to \infty} \frac{|U_{k(n),\ell(n)}|}{n^n} = 1.$$

Proof. By our previous investigation on the relationship between the size of $U_{k,\ell}$ and the number of (in)valid colourings of the complete bipartite graph $K_{k,\ell}$ we have

$$U_{k,\ell} \supseteq T_n - \underbrace{\{\, \gamma \in T_n \mid \{k+1, \dots, n\} \subseteq img(\gamma)\,\}}_{A}$$
$$- \underbrace{\{\, \gamma \in T_n \mid \gamma \text{ is a valid colouring of the graph } K_{k,\ell}\,\}}_{B}.$$

This is also due to the fact that every permutation is a valid colouring of $K_{k,\ell}$.

Thus, in order to determine $|U_{k,\ell}|$ it is sufficient to estimate the size of A and B. We over-estimate both sets in the forthcoming. Let

$$A' = \{\, (\gamma, a_1, \dots, a_\ell) \mid \gamma \in A \text{ and } \gamma(a_i) = k + i, \text{ for } 1 \le a_i \le n\,\}.$$

It is easy to see that that $|A| \le |A'|$ and furthermore $|A'| = \binom{n}{\ell} \ell! n^k = n^{\underline{\ell}} n^k$ where $n^{\underline{\ell}} = n(n-1) \cdots (n-\ell+1)$ denotes the falling factorial. We first choose the values of the a_i, then assign a different element of $\{k+1, \dots, k+\ell\}$ to each of them and finally assign an arbitrary element to each of the remaining k pre-images. For B we argue as follows: Let

$$B' = \{\, (\gamma, X, Y) \mid \gamma \in B, \ X \uplus Y = \{1, \dots, n\}, \ |X| = k, \ |Y| = \ell,$$
$$(\{1, \dots, k\})\gamma \subseteq X, \text{ and } (\{k+1, \dots, n\})\gamma \subseteq Y\,\}.$$

One observes, that $|B| \le |B'|$ and furthermore $|B'| = \binom{n}{\ell} k^k \ell^\ell$, since we first choose the elements of Y (which gives us automatically the elements of X), then we assign a colour from X to the nodes in $\{1, \dots, k\}$, and afterwards we assign a colour from Y to the nodes in $\{k+1, \dots, k+\ell\}$. This shows that

$$|U_{k,\ell}| \ge n^n - \binom{n}{\ell} \ell! n^k - \binom{n}{\ell} k^k \ell^\ell.$$

For the asymptotic result, we first show the following claim: Assume $n \ge 3$. Then there exists $k, \ell \in \mathbb{N}$ such that $n = k + \ell$, $\ell - k \le 4$, and $\gcd\{k, \ell\} = 1$.

We argue as follows: Whenever $n = 2m + 1$ then set $k = m$ and $\ell = m + 1$. If n is even, we have to distinguish the following two cases: Either $n = 4m$, then we can set $k = 2m - 1$ and $\ell = 2m + 1$, both can not be divided by 2 and since $\ell - k = 2$ there is no other candidate for a common divisor. If $n = 4m + 2$, then we can set $k = 2m - 1$ and $\ell = 2m + 3$. Since $\ell - k = 4$, the only candidates for common divisors are 2 and 4, but clearly k and ℓ are not divisible by any of them. This proves the existence of some k and ℓ, which are close to $\frac{n}{2}$.

Then the asymptotic result is seen by using Stirling's approximation for the factorials, proving that both $\frac{|A|}{n^n}$ and $\frac{|B|}{n^n}$ converge to 0 whenever n goes to infinity. Let A' and B' be the sets defined above. We obtain

$$\frac{|A|}{n^n} \le \frac{|A'|}{n^n} = \frac{n^{\underline{\ell}} n^k}{n^n} = \frac{n!}{\ell! n^{n-k}} = \frac{1}{n^{n-k}} \cdot \frac{\sqrt{2\pi n} \left(\frac{n}{e}\right)^n (1 + \Theta(\frac{1}{n}))}{\sqrt{2\pi \ell} \left(\frac{\ell}{e}\right)^\ell (1 + \Theta(\frac{1}{\ell}))}$$

$$= \sqrt{\frac{n}{\ell}} \left(\frac{n}{e}\right)^k \frac{1}{\ell^\ell} \frac{1 + \Theta(\frac{1}{n})}{1 + \Theta(\frac{1}{\ell})}$$

Since both k and ℓ are close to $\frac{n}{2}$ as shown above, we can infer that the last factor converges to 1 whenever n goes to infinity. Furthermore, since $k \leq \frac{n}{2} \leq \ell$, it follows that

$$\sqrt{\frac{n}{\ell}} \left(\frac{n}{e}\right)^k \frac{1}{\ell^\ell} \leq \sqrt{2} \left(\frac{n}{e}\right)^{\frac{n}{2}} \frac{1}{\left(\frac{n}{2}\right)^{\frac{n}{2}}} \leq \sqrt{2} \left(\frac{2}{e}\right)^{\frac{n}{2}}.$$

Thus, the last term obviously converges to 0 whenever n goes to infinity. For the fraction of $|B|$ and n^n we do similar. We find

$$\frac{|B|}{n^n} \leq \frac{|B'|}{n^n} = \frac{1}{n^n} \frac{n!}{k!\ell!} k^k \ell^\ell$$

$$= \frac{1}{n^n} \frac{\sqrt{2\pi n} \left(\frac{n}{e}\right)^n}{\sqrt{2\pi k} \left(\frac{k}{e}\right)^k \sqrt{2\pi \ell} \left(\frac{\ell}{e}\right)^\ell} k^k \ell^\ell \frac{1 + \Theta(\frac{1}{n})}{(1 + \Theta(\frac{1}{k}))(1 + \Theta(\frac{1}{\ell}))}$$

$$= \frac{1}{\sqrt{2\pi}} \sqrt{\frac{n}{k\ell}} \frac{1 + \Theta(\frac{1}{n})}{(1 + \Theta(\frac{1}{k}))(1 + \Theta(\frac{1}{\ell}))}$$

Again the last factor converges to 1. Now it holds that

$$\sqrt{\frac{n}{k\ell}} = \sqrt{\frac{n}{kn - k^2}} = \frac{1}{\sqrt{k - \frac{k^2}{n}}} \overset{\frac{3n}{8} \leq k \leq \frac{n}{2}}{\leq} \frac{1}{\sqrt{\frac{3n}{8} - \frac{n}{4}}} = \sqrt{\frac{8}{n}},$$

where $\frac{3n}{8} \leq k \leq \frac{n}{2}$ follows for large enough n, since k and $\frac{n}{2}$ differ only by a constant. And the last term converges to 0 whenever n goes to infinity. This proves the second statement of our result. □

Now we come to the main result of this section. Recall that $g(n)$ denotes Landau's function [4,5,6], which gives the size of the maximal subgroup of S_n which can be generated by one generator.

Theorem 9. *Assume $n \geq 3$ and let A be a n-state deterministic finite automata. Then a monoid of size $n^n - n! + g(n)$ is sufficient to recognize the language $L(A)$ and a monoid of size*

$$n^n - \binom{n}{\ell} \ell! n^k - \binom{n}{\ell} k^k \ell^\ell,$$

where $n = k + \ell$, $\ell - k \leq 4$, and $\gcd\{k, \ell\} = 1$ for some natural numbers k and ℓ, is necessary in the worst case.

Proof. The upper bound $n^n - n! + g(n)$ is immediate, since we assume that only one of the two generators is a permutation and the lower bound follows by Theorems 7 and 8. □

Moreover, we obtain the following corollary, which we state without proof:

Corollary 1. *There is a sequence $L_1, L_2, \dots$ of binary regular languages such that*

$$\lim_{n \to \infty} \frac{|M(L_i)|}{n^n} = 1,$$

and each L_i is accepted by a minimal deterministic finite automaton with exactly n states. □

5   Conclusions

We have studied the relationship between the size of a deterministic finite automaton A and the size of the syntactic monoid, which is necessary to recognize the language $L(A)$.

In most cases, we were able to prove tight upper bounds. The only exception are binary regular languages were we have presented a non-matching upper and lower bound. We summarize some computed values on the size of some of the semigroups (monoids) involved in Table 1.

There, the number $max(n)$ denotes the size of the maximal transformation semigroup (monoid) with two generators, which might not coincide with the size of some $U_{k,\ell}$. A table entry with a question mark indicates that the precise value is not known and thus is a conjecture. The generators for the groups with 24, 176, 2110, and 32262 elements all contain a single cycle permutation $(1\,2\,\ldots\,n)$. However, already for $n = 7$, the case where one of the generators is the cycle is beat by our semigroup $U_{k,\ell}$.

It remains to tighten the bound on the syntactic monoid size on two generators in future research. To understand the very nature of this question it seems to be very important, to precisely characterize the maximal size transformation

Table 1. Sizes of some investigated semigroups.

n	$\|S_n\| = n!$	k	ℓ	$\|U_{k,\ell}\|$	$max(n)$	$n^n - n! + g(n)$	$\|T_n\| = n^n$
3	6	1	2	13	24	24	27
4	24	1	3	133	176	236	256
5	120	2	3	1857	2110	3011	3125
		1	4	1753			
6	720	1	5	27311	32262 (?)	45942	46656
7	5040	3	4	607285	610871 (?)	818515	823543
		2	5	610871			
		1	6	492637			
8	40320	3	5	13492007	13492007 (?)	16736911	16777216
		1	7	10153599			
9	362880	4	5	323534045	323534045 (?)	387057629	387420489
		2	7	306605039			
		1	8	236102993			
10	3628800	3	7	8678434171	8678434171 (?)	9996371230	10000000000
		1	9	6122529199			
11	39916800	5	6	256163207631	258206892349 (?)	285271753841	285311670611
		4	7	258206892349			
		3	8	251856907425			
		2	9	231326367879			
		1	10	175275382621			

semigroup on two generators, in a similar way as the generators for T_n and S_n are characterized in Theorems 4 and 5. We conjecture, that for every $n \geq 7$, there exists natural numbers k and ℓ with $n = k + \ell$ such that the semigroup $U_{k,\ell}$ is maximal under all two generator transformation semigroups (monoids).

Acknowledgments

Thanks to J. M. Howie and J.-E. Pin for some fruitful discussions on the subject.

References

1. J.-M. Champarnaud, D. Maurel, and D. Ziadi, editors. *Automata Implementation, Proceedings of the 3rd International Workshop on Implementing Automata*, number 1660 in LNCS, Rouen, France, September 1998. Springer.
2. J. M. Howie. *An Introduction to Semigroup Theory*, volume 7 of *L. M. S. Monographs*. Academic Press, 1976.
3. S. C. Kleene. Representation of events in nerve nets and finite automata. In C. E. Shannon and J. McCarthy, editors, *Automata studies*, volume 34 of *Annals of mathematics studies*, pages 2–42. Princeton University Press, 1956.
4. E. Landau. Über die Maximalordnung der Permutationen gegebenen Grades. *Archiv der Mathematik und Physik*, 3:92–103, 1903.
5. J.-L. Nicolas. Sur l'ordre maximum d'un élément dans le groupe s_n des permutations. *Acta Arithmetica*, 14:315–332, 1968.
6. J.-L. Nicolas. Ordre maximum d'un élément du groupe de permutations et highly composite numbers. *Bulletin of the Mathematical Society France*, 97:129–191, 1969.
7. S. Piccard. *Sur les bases du groupe symétrique et les couples de substitutions qui engendrent un groupe régulier*. Librairie Vuibert, Paris, 1946.
8. J.-E. Pin. *Varieties of formal languages*. North Oxford, 1986.
9. A. Salomaa. On the composition of functions of several variables ranging over a finite set. *Annales Universitatis Turkuensis*, 41, 1960. Series AI.
10. D. Wood and S. Yu, editors. *Automata Implementation, Proceedings of the 2nd International Workshop on Implementing Automata*, number 1436 in LNCS, London, Canada, September 1997. Springer.

An Inverse Automata Algorithm
for Recognizing 2-Collapsing Words[*]

Dmitry S. Ananichev[1], Alessandra Cherubini[2], and Mikhail V. Volkov[1]

[1] Department of Mathematics and Mechanics,
Ural State University, 620083 Ekaterinburg, RUSSIA
{Dmitry.Ananichev,Mikhail.Volkov}@usu.ru
[2] Dipartimento di Matematica "Francesco Brioschi"
Politecnico di Milano
20133 Milano, ITALIA
aleche@mate.polimi.it

Abstract. A word w over a finite alphabet Σ is *n-collapsing* if for an arbitrary DFA $\mathscr{A} = \langle Q, \Sigma, \delta \rangle$, the inequality $|\delta(Q, w)| \leq |Q| - n$ holds provided that $|\delta(Q, u)| \leq |Q| - n$ for some word $u \in \Sigma^+$ (depending on $\mathscr{A}$). We give a new algorithm to test whether a word w is 2-collapsing. In contrast to our previous group-theoretic algorithm, the present algorithm is of a geometric nature, and if the word $w \in \Sigma^*$ is not 2-collapsing, it directly produces a DFA $\mathscr{A}_w = \langle Q, \Sigma, \delta \rangle$ such that $|Q| < \max\{|w|, 4\}$, $|\delta(Q, u)| \leq |Q| - 2$ for some word $u \in \Sigma^*$, but $|\delta(Q, w)| \geq |Q| - 1$.

1 Background and Motivation

Let $\mathscr{A} = \langle Q, \Sigma, \delta \rangle$ be a deterministic finite automaton (DFA, for short) with the state set Q, the input alphabet Σ, and the transition function $\delta : Q \times \Sigma \to Q$. The action of the letters in Σ on the states in Q defined via δ extends in a natural way to an action of the words in the free Σ-generated monoid Σ^*; the latter action is still denoted by δ. For any $w \in \Sigma^*$, we set $Q.w = \{\delta(q, w) \mid q \in Q\}$.

Let n be a positive integer. A DFA $\mathscr{A} = \langle Q, \Sigma, \delta \rangle$ is said to be *n-compressible* if there is a word $w \in \Sigma^*$ such that $|Q.w| \leq |Q| - n$. The word w is then called *n-compressing with respect to* $\mathscr{A}$. These notions (under varying names) have been around for some time mainly in connection with *Pin's conjecture* [9, 10] which in our terminology can be formulated as follows: for each n-compressible

[*] This work was completed during the third-named author's visit to Politecnico di Milano which was supported by Fondo di Ricerca di Ateneo. The second-named author acknowledges support from MIUR Progetto Cofinanziato 2001–2002 "Linguaggi formali e automi: teoria e applicazioni" while the first and the last named authors acknowledge support from Russian Education Ministry, grant E00-1.0-92, Russian Basic Research Foundation, grant 01-01-00258, and the INTAS (through Network project 99-1224 "Combinatorial and Geometric Theory of Groups and Semigroups and its Applications to Computer Science").

automaton $\mathscr{A}$, there exists a word which is n-compressing with respect to $\mathscr{A}$ and has length n^2. Even though this particular conjecture has been recently disproved by J. Kari [5] (who came up with a surprising counter example in the case $n = 4$), the area—in which such easily looking problems turn out to be so difficult to solve—remains rather vivid and reveals some interesting connections with algebra, language theory and combinatorics.

The notion that plays a central role in the present paper may be thought of as a 'black-box' version of the notion of an n-compressing word. Namely, we say that a word $w \in \Sigma^*$ is n-*collapsing* if w is n-compressing with respect to every n-compressible DFA whose input alphabet is Σ. In other terms, a word $w \in \Sigma^*$ is n-collapsing if for a DFA $\mathscr{A} = \langle Q, \Sigma, \delta \rangle$, we have $|Q.w| \leq |Q| - n$ whenever $\mathscr{A}$ is n-compressible. Thus, such a word is a 'universal tester' whose action on the state set of an arbitrary DFA with a fixed input alphabet exposes whether or not the automaton is n-compressible.

The very first problem related to n-collapsing words is of course the question of whether such words exist for every n and over every finite alphabet Σ. This question (which is by no means obvious provided that both n and $|\Sigma|$ are greater than 1) was solved in the positive by Sauer and Stone [11, Theorem 3.3] who arguably were the first to introduce words with this property (*the property* Δ_k in the terminology of [11]). While Sauer and Stone considered n-collapsing words in a purely combinatorial environment, tight relations between this notion and certain problems of automata theory (such as Pin's conjecture, for instance) were observed somewhat later by Margolis, Pin and the third-named author who extracted another existence proof from their approach [8, Theorem 2] and also found some bounds on the length of the shortest n-collapsing word over a given alphabet, see [8, Theorems 5 and 11] where words which we call n-collapsing here are said to *witness for deficiency* n.

As the existence has been established, the next crucial step is to master, for each positive integer n, an algorithm that recognizes if a given word is n-collapsing. In an earlier paper [1], we solved this problem for the first non-trivial case $n = 2$. The solution proposed in [1] was based on a reduction of the initial problem to a question concerning finitely generated subgroups of free groups which can be efficiently solved by certain classic methods of combinatorial group theory. It is fair to say that the resulting algorithm was quite complex and difficult to analyze. Moreover, for a problem concerning finite automata, it would be rather natural to have an answer formulated in the language of finite automata. This is in fact the aim of the present paper.

In order to achieve our aim, we have applied the inverse automata approach to studying finitely generated subgroups of free groups as originated by Stallings in a seminal paper [12] and developed in [6] (see also a detailed and rather elementary presentation in [4]). In this approach, departing from any given finite set V of group words, one builds a finite inverse automaton which can be shown to depend only on the subgroup H generated by V and not on the set V itself. The automaton can be then used to provide elegant solutions for many natural questions concerning H including those that have played a crucial role for the algorithm from [1] that recognizes if a given word $w \in \Sigma^*$ is 2-collapsing. However, rather than passing first from the word w to a finite family of finitely generated subgroups of free groups as in [1] and then from the subgroups to corresponding inverse automata in the style of [6], now we have managed to find a 'shortcut' and to directly produce from w a finite bunch of inverse automata that controls the property of being 2-collapsing. In the case when w is not 2-collapsing but contains each word of length 2 over Σ as a factor, at least one of these inverse automata can be completed to a 2-compressible DFA $\mathscr{A}_w = \langle Q, \Sigma, \delta \rangle$ such that $|Q| < \max\{|w|, 4\}$ and $|Q.w| \geq |Q| - 1$. This feature is a considerable advantage in comparison with the algorithm from [1]. In particular, it allows us to show that the language of all 2-collapsing words over a fixed alphabet Σ is context-sensitive.

Our algorithm is presented in Sect. 2. Because of length limitations, we have only commented on how its proof can be deduced from the main results of [1]. A complete proof will be published elsewhere. In Sect. 3 we briefly discuss some computational complexity issues and a language-theoretical corollary.

2   The Algorithm

In the sequel, we call our algorithm C2C (Checking 2-Collapsability). For the purpose of time and space complexity analysis, it is necessary to have a rigorous description of C2C. On the other hand, describing C2C in a kind of pseudocode would perhaps be too technical and would create unnecessary difficulties for those readers who rather want to understand how the algorithm works than to calculate how many steps it takes. As a compromise, we have chosen to present C2C as a collection of precisely specified steps whose descriptions are interwoven with less formal comments.

The input of the algorithm C2C is a word w over a finite alphabet Σ. The algorithm outputs "YES" if the word w is 2-collapsing; otherwise it returns a 2-compressible DFA $\mathscr{A}_w = \langle Q, \Sigma, \delta \rangle$ such that $|Q| \leq 2|w|$ and $|\delta(Q, w)| \geq |Q| - 1$.

The algorithm starts with a preparatory step whose aim is to isolate some easy cases.

Step 0: Check if the word w contains each word of length 2 over Σ as a factor. If some word u of length 2 does not appear as a factor of w, then take as $\mathscr{A}_w$ the minimal automaton of the language $L = \Sigma^* u \Sigma^*$ and stop. If all words of length 2 appear as factors of w, then pass to Step 1 provided that $|\Sigma| > 1$, and in the case $|\Sigma| = 1$ output "YES" and stop.

This step is based on the following elementary observation from [8] (see the proof of Theorem 5 there):

Lemma 1. *Let u be a word over Σ and $|u| = n$. Then the minimal automaton $\mathscr{A}(L)$ of the language $L = \Sigma^* u \Sigma^*$ has $n + 1$ states and is n-compressible. A word $v \in \Sigma^*$ is n-compressing with respect to $\mathscr{A}(L)$ if and only if it belongs to L. In particular, every n-collapsing word over Σ belongs to L.*   □

Thus, for the rest of the algorithm we may assume that $|\Sigma| > 1$ and that the word w contains each word of length 2 over Σ as a factor.

Now the main loop of the algorithm begins. Its parameter is an arbitrary ordered pair (Υ, Π) of non-empty subsets of the alphabet Σ such that Υ and Π form a partition of Σ, that is, $\Upsilon \cup \Pi = \Sigma$, $\Upsilon \cap \Pi = \varnothing$. We fix such a pair and perform the next steps of the algorithm. If on some step we obtain the desired DFA $\mathscr{A}_w$, then we stop; if not, then we choose another pair and repeat. If after running the main loop for each of $2^{|\Sigma|} - 2$ pairs no automaton $\mathscr{A}_w$ is produced, then the word w is 2-collapsing, so the algorithm returns "YES" and stops.

Thus, from now on we assume that a pair (Υ, Π) of non-empty sets forming a partition of the alphabet Σ is fixed.

Step 1: Factor the word w as an alternating product of words over Υ and over Π:

$$w = u_0 p_1 u_1 \cdots u_{m-1} p_m u_m, \tag{1}$$

where $u_0, u_m \in \Upsilon^*$, $u_1, \ldots, u_{m-1} \in \Upsilon^+$, $p_1, \ldots, p_m \in \Pi^+$ and m is a positive integer.

We say that the factor p_i of the decomposition (1) is an *inner segment* of the word w if both the 'neighbors' u_{i-1} and u_i of p_i are non-empty. We denote the set of all inner segments of w by S.

At this stage, the algorithm branches into two loops. This branching corresponds to a classification of 2-compressible automata suggested in [1, Section 1]. We say that a 2-compressible automaton $\mathscr{A}$ is *proper* if no word of length 2 is 2-compressing with respect to $\mathscr{A}$. The following observation is obvious:

Lemma 2. *If $\mathscr{A} = \langle Q, \Sigma, \delta \rangle$ is a proper 2-compressible automaton and $a \in \Sigma$, then either $Q.a = Q$ or $|Q.a| = |Q| - 1$ and $Q.a = Q.a^2$.* $\square$

In the first case we call a a *permutation letter with respect to $\mathscr{A}$* and in the second case we say that a is a *just-non-permutation letter with respect to $\mathscr{A}$*). For each just-non-permutation letter $a \in \Sigma$, there exists a unique state in $Q \setminus Q.a$ which we call *the exception state of a* and denote by e_a. Further, since $Q.a = Q.a^2$, there exists a unique *doubling state* $d_a \in Q.a$ such that $\delta(d_a, a) = \delta(e_a, a)$. Given a proper 2-compressible automaton $\mathscr{A}$, let

$$\Pi[\mathscr{A}] = \{a \in \Sigma \mid a \text{ is a permutation letter with respect to } \mathscr{A}\},$$

$$\Upsilon[\mathscr{A}] = \{a \in \Sigma \mid a \text{ is a just-non-permutation letter with respect to } \mathscr{A}\}.$$

Clearly, $\Upsilon[\mathscr{A}] \neq \varnothing$, and it is observed in [1, Lemma 1.5] that also $\Pi[\mathscr{A}] \neq \varnothing$. Thus, if there exists a 2-compressing automaton $\mathscr{A}_w$ such that the word w under inspection is not 2-compressing with respect to it, then $\mathscr{A}_w$ must be proper (since w contains each word of length 2 over Σ as a factor), and running over all pairs (Υ, Π) of non-empty sets forming a partition of Σ, we sooner or later encounter the partition $(\Upsilon[\mathscr{A}_w], \Pi[\mathscr{A}_w])$.

We have proved in [1, Proposition 1.4] that if $\mathscr{A}$ is a proper 2-compressible automaton, then either there exists a state e such that $e_a = e$ for all $a \in \Upsilon[\mathscr{A}]$, or $\{d_b, e_b\} = \{d_c, e_c\}$ for all $b, c \in \Upsilon[\mathscr{A}]$. In the first case we say that the automaton $\mathscr{A}$ is a *mono automaton*, and if no common exception state for all $a \in \Upsilon[\mathscr{A}]$ exists, we call $\mathscr{A}$ a *stereo automaton*. In the two branches of the algorithm C2C we search for the automaton $\mathscr{A}_w$ among mono respectively stereo automata. We label steps of the first/second branch by adding an "a"/a "b" to the step numbers.

The parameter of the first inner loop is an arbitrary subset P of the set S of all inner factors of the word w under inspection. We fix such a subset and perform the next steps of the algorithm. If they result in constructing the automaton $\mathscr{A}_w$ which we are looking for, we stop; if not, we choose another subset and repeat. If after running the loop for each of $2^{|S|}$ subsets no automaton $\mathscr{A}_w$ is produced, we pass to the second branch of the algorithm.

It should be mentioned that, while the number of the repetitions of the main loop does not depend on the length of the input word w (it depends only on the size of the alphabet Σ), the number of the repetitions of the first inner loop (as well as of the second one) depends on $|w|$. It can be calculated that for sufficiently long words w and for the pairs (Υ, Π) with $|\Pi| > 1$, the number $|S|$ of inner segments is of the magnitude $O(|w|)$ whence the first inner loop is to be repeated an exponential (of $|w|$) number of times. Thus, by the construction our algorithm is not polynomial; we shall see, however, that in some cases it leads to a polynomial time procedure for deciding if a given word is 2-collapsing.

We proceed with presenting the steps of the first inner loop. Recall that we assume that a subset $P \subseteq S$ is fixed. For each letter $a \in \Upsilon$, define S_a to be the set of all $p_i \in S$ which are followed by a in w. (We note that the sets S_a for different letters $a \in \Upsilon$ need not be disjoint as the same word from Π^+ may appear several times as a factor in (1) preceding different letters $a \in \Upsilon$.) Let $P_a = S_a \setminus P$. In [1, Proposition 2.1] we have shown the property of w being 2-compressing with respect to all mono automata is controlled by a certain collection of subgroups of the free group $FG(\Pi)$ over the alphabet Π; namely, to each $P \subset S$ we assign the subgroup H_P generated by the set $P \cup \bigcup_{a \in \Upsilon} P_a \cdot P_a^{-1}$. In our present approach we substitute the subgroup H_P by an automaton which accepts H_P (as the set of reduced words in $FG(\Pi)$).

Step 2a: Construct the following 'flower' automaton $\mathscr{F}$ with labels in Π. It has a distinguished state e, and for each word $p_i \in P$, there is a 'petal' that spells out p_i in a path from e to e. This means that if $p_i = b_{i1} b_{i2} \cdots b_{in_i}$ with $b_{ij} \in \Pi$, then one adds $n_i - 1$ new states $q_{i1}, q_{i2}, \ldots, q_{in_i - 1}$ and draws an arrow labeled b_{i1} from e to q_{i1}, an arrow labeled b_{i2} from q_{i1} to q_{i2}, $\ldots$, an arrow labeled b_{in_i} from $q_{in_i - 1}$ to e. Of course, if $|p_i| = 1$, then the corresponding 'petal' reduces to a loop on e labeled by the only letter in p_i. Besides that $\mathscr{F}$ has a distinguished state d_a for every letter $a \in \Upsilon$, and for each word $p_j \in P_a$, there is a path from e to d_a that spells out p_j in the way described above.

Generally speaking, the automaton $\mathscr{F}$ is incomplete and non-deterministic. Clearly, the time and the space needed to construct $\mathscr{F}$ (given the sets P and

P_a) are linear as functions of $|w|$. Observe that if one declares the state e to be the unique initial and terminal state of $\mathscr{F}$ and assumes that for every arrow $q \to q'$ in $\mathscr{F}$ labeled b, there is the opposite arrow $q' \to q$ labeled b^{-1}, then the set of reduced words accepted by $\mathscr{F}$ coincides with the subgroup H_P defined above.

On the next step we fold the flower automaton. Let us precisely define this operation on automata.

Suppose we have a (non-deterministic incomplete) finite automaton $\mathscr{B} = \langle Q, \Xi, \delta \rangle$ in which there is a state $q^{\sharp} \in Q$ such that for some letter $b \in \Xi$, the set $R = \delta(q^{\sharp}, b)$ consists of more than one element. By the *direct folding of* $\mathscr{B}$ *at* $q^{\sharp}$ we mean the automaton $\mathscr{B}' = \langle Q', \Xi, \delta' \rangle$ in which all states of the set R are collapsed to a single new state r, say, (so that $Q' = (Q \setminus R) \cup \{r\}$) and the transition function $\delta' : Q' \times \Xi \to Q'$ is defined as follows:

$$\delta'(q, c) = \begin{cases} \delta(q, c) & \text{if } \delta(q,c) \cap R = \varnothing, \\ (\delta(q, c) \setminus R) \cup \{r\} & \text{otherwise} \end{cases} \quad \text{for } q \in Q \setminus R;$$

$$\delta'(r, c) = \begin{cases} \bigcup_{q \in R} \delta(q, c) & \text{if } \left(\bigcup_{q \in R} \delta(q, c) \right) \cap R = \varnothing, \\ \left(\bigcup_{q \in R} \delta(q, c) \setminus R \right) \cup \{r\} & \text{otherwise.} \end{cases}$$

This somewhat bulky formal definition has in fact a very transparent geometric meaning shown on Fig. 1 on the left:

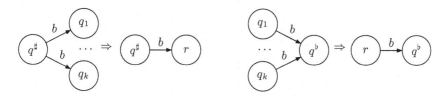

Fig. 1. Direct and reverse folding

The second kind of folding arises when in the automaton $\mathscr{B}$ there exists a state $q^{\flat} \in Q$ such that for some letter $b \in \Xi$ and for some non-singleton subset $R \subseteq Q$, $R.b = \{q^{\flat}\}$. In this situation we first perform the direct folding at $q^{\flat}$ in *the dual automaton* of $\mathscr{B}$ (that is, in the automaton obtained from $\mathscr{B}$ by changing the direction of all arrows) and then take the dual of the folded automaton. The result is referred to as the *reverse folding of* $\mathscr{B}$ *at* $q^{\flat}$. The geometric meaning of this operation should be also clear, see the right part in Fig. 1.

Since each folding decreases the number of states, repeatedly folding a given automaton $\mathscr{B} = \langle Q, \Xi, \delta \rangle$, we eventually arrive at an automaton $\mathscr{I}$ to which no further folding (direct or reverse) can be applied. Clearly, this means that letters of Ξ act on the state set of the automaton $\mathscr{I}$ as partial one-to-one mappings. Automata with the latter property are called *inverse*. It follows from [12] that the inverse automaton $\mathscr{I}$ does not depend on the order of edges of $\mathscr{B}$ chosen to fold.

Thus, the next step of our algorithm is

Step 3a: Fold the flower automaton $\mathscr{F}$ to an inverse automaton $\mathscr{I}$.

It is known (see [12, 6, 4]) that the automaton $\mathscr{I}$ inherits from $\mathscr{F}$ the property to recognize the subgroup H_P: if $\bar{e}$ is the image in $\mathscr{I}$ of the distinguished state e of the flower automaton $\mathscr{F}$, then the set of reduced words that label a path from $\bar{e}$ to $\bar{e}$ coincides with H_P. (Of course, we retain the convention that for every arrow $q \rightarrow q'$ labeled b, there is the opposite arrow $q' \rightarrow q$ labeled b^{-1}.) Moreover, properties of the subgroup H_P can be easily read from the automaton $\mathscr{I}$. According to [1, Proposition 2.1], we can produce a 2-compressible automaton with respect to which the word w fails to be 2-compressing whenever the index of H_P in the free group $FG(\Pi)$ is greater than or equal to 2 and in the latter case $P_a = \varnothing$ for some letter $a \in \Upsilon$. Using [6, Theorems 5.1 and 5.3], we conclude that in the language of the automaton $\mathscr{I}$ this amounts to saying that either $\mathscr{I}$ is not complete (this corresponds to the case when the index of H_P in $FG(\Pi)$ is infinite) or $\mathscr{I}$ is complete and the number of states in the connected component C of $\bar{e}$ is greater than or equal to 2 and in the latter case there is a state beyond C. Thus, if $\mathscr{I}$ satisfies the condition just formulated, we proceed with the next step; otherwise (that is, if C has only one state or if $\mathscr{I}$ is complete, connected and has exactly 2 states) we must choose a subset $P \subseteq S$ not yet examined and return to Step 2a.

Step 4a: Complete the automaton $\mathscr{I}$ to a permutation automaton $\mathscr{I}'$ over Π in the following way.

If $\mathscr{I}$ has exactly 2 states, say, q and r, then we add a new state s and define the action of each letter $b \in \Pi$ as follows:

- if the action of b on both q and r is defined, then let b fix s;
- if b sends one of the states $x \in \{q, r\}$ to the other state $y \in \{q, r\}$ while its action on the latter is undefined, then let b send y to s and s to x;
- if b fixes one of the states $x \in \{q, r\}$ while its action on the other state $y \in \{q, r\}$ is undefined, then let b interchange y and s;
- if the action of b on neither q nor r is defined, then let b send q to r, r to s, and s to q.

If $\mathscr{I}$ has more than 2 states, then extend the action of each letter in Π to a permutation of the state set of $\mathscr{I}$ in an arbitrary way (such an extension is always possible since the letters act as partial one-to-one mappings).

On the final steps of the first branch we define the action of letters in Υ on the state set of the permutation automaton $\mathscr{I}'$ thus producing a DFA with the input alphabet Σ. This DFA is the desired mono automaton $\mathscr{A}_w$ which exposes that the word w under examination is not 2-collapsing.

If the connected component C' of the state $\bar{e}$ in $\mathscr{I}'$ has at least 3 states, the DFA $\mathscr{A}_w$ is constructed as follows:

Step 5a: For each letter $a \in \Upsilon$ let a fix all states of $\mathscr{J}'$ except $\bar{e}$ and send $\bar{e}$ to the image $\bar{d}_a$ of the distinguished state d_a of $\mathscr{F}$ provided that $\bar{e} \neq \bar{d}_a$ and to some state in $C' \setminus \{\bar{e}\}$ (which can be chosen arbitrarily) if $\bar{e} = \bar{d}_a$.

The situation when the connected component C' of $\bar{e}$ has less than 3 states is only possible if C' has exactly 2 states and the permutation automaton $\mathscr{J}'$ has also another connected component. Obviously, neither foldings nor completions can split a connected automaton into several components so this means that the flower automaton $\mathscr{F}$ is not connected. It follows from the construction of $\mathscr{F}$ that this is only possible provided that for some letter $a \in \Upsilon$, the set $P_a = S_a \setminus P$ happens to be empty, and therefore, the distinguished state d_a is isolated in $\mathscr{F}$ and remains so in $\mathscr{J}'$ (which means that it is fixed by all letters in Π). Then the following construction applies:

Step 6a: Let all letters in Υ fix all states of $\mathscr{J}'$ except $\bar{e}$ and let a send $\bar{e}$ to d_a while all other letters in Υ should send $\bar{e}$ to the only state in $C' \setminus \{\bar{e}\}$.

Figure 2 below schematically presents the connected component of $\bar{e}$ in the automaton $\mathscr{A}_w$ produced on Step 6a. Here Π_1 is some non-empty subset of Π and q stands for the only state in $C' \setminus \{\bar{e}\}$.

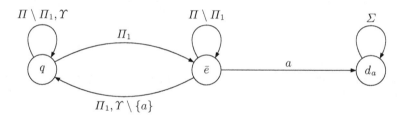

Fig. 2. A part of automaton $\mathscr{A}_w$ for the case when $|C'| = 2$

Thus, the first inner loop of C2C is completed. Since the underlying ideas of the second inner loop are very similar, we restrict our comments to a minimum.

First of all, since now we are searching for the automaton $\mathscr{A}_w$ among stereo automata, the loop only arises when $|\Upsilon| > 1$. Provided that this is the case, partition the set Υ in two non-empty subsets Υ_1 and Υ_2 and denote by S_k the set of all inner segments $p_i \in S$ that follow a letter from Υ_k, $k = 1, 2$. Now take arbitrary subsets $P_{11} \subseteq S_1$ and $P_{22} \subseteq S_2$ and define $P_{12} = S_1 \setminus P_{11}$ and $P_{21} = S_2 \setminus P_{22}$. The pair (P_{11}, P_{22}) as well as the partition (Υ_1, Υ_2) are the parameters of the second inner loop. By [1, Proposition 2.2] the property of the word w being 2-compressing with respect to all stereo automata is determined by certain properties of subgroups $H_{(\Upsilon_1, P_{11}, P_{22})}$ of $FG(\Pi)$ assigned to these parameters; the subgroup $H_{(\Upsilon_1, P_{11}, P_{22})}$ is generated by the set

$$P_{11} \cup P_{12} \cdot P_{21} \cup P_{12} \cdot P_{12}^{-1} \cup P_{12} \cdot P_{22} \cdot P_{12}^{-1}.$$

As in the first loop, we construct an automaton that accepts this subgroup (as the set of reduced words in $FG(\Pi)$), fold the automaton and finally express the desired property of the subgroup in the language of the resulting inverse automaton.

Thus, assume that we have selected a partition (Υ_1, Υ_2) and a pair (P_{11}, P_{22}).

Step 2b: Construct the following 'bi-flower' automaton $\mathscr{BF}$ with labels in Π. It has two distinguished states x_1 and x_2. For each word $p_i \in P_{kk}$, there is a 'petal' that spells out p_i in a loop from x_k to x_k, $k = 1, 2$. Besides that, for each word $p_j \in P_{k\ell}$ where $\ell = 3 - k$, there is a path from x_k to x_ℓ that spells out p_j.

Step 3b: Fold the automaton $\mathscr{BF}$ to an inverse automaton $\mathscr{J}$.

Let $\bar{x}_k$ be the image of the distinguished state x_k in $\mathscr{J}$. If $\bar{x}_1 = \bar{x}_2$ or if the automaton $\mathscr{J}$ is complete and has only two states, then we must choose another tuple of parameters (Υ_1, Υ_2), (P_{11}, P_{22}) not yet examined and return to Step 2b. Otherwise we proceed with the next steps.

Step 4b: Complete the automaton $\mathscr{J}$ to a permutation automaton $\mathscr{J}'$ over Π in the same way as in Step 4a.

Step 5b: Let all letters in Υ_1 fix all states of $\mathscr{J}'$ except $\bar{x}_1$ and send $\bar{x}_1$ to $\bar{x}_2$, while all letters in Υ_2 should fix all states of $\mathscr{J}'$ except $\bar{x}_2$ and send $\bar{x}_2$ to $\bar{x}_1$.

The resulting stereo automaton is the desired DFA $\mathscr{A}_w$ showing that the word w is not 2-collapsing.

This completes the description of the algorithm C2C.

3   Applications and Refinements

From the description of our algorithm it should be clear that if a word $w \in \Sigma^*$ is not 2-collapsing but contains all factors of length 2, then the automaton $\mathscr{A}_w$ output by the algorithm has less than $|w|$ states. This immediately leads to a non-deterministic linear space and polynomial time algorithm recognizing the language of words which are not 2-collapsing: the algorithm simply makes a guess consisting of a DFA $\mathscr{A} = \langle Q, \Sigma, \delta \rangle$ with $|Q| < |w|$ and then verifies that $\mathscr{A}$ is 2-compressible (this can be easily done in low polynomial time) and that w is not 2-compressing with respect to $\mathscr{A}$. By classical results of formal language theory (cf. [7, Sections 2.4 and 2.5]) this implies that the language of 2-collapsing words over a fixed finite alphabet Σ is context-sensitive. We note that in [2] it was proved that for $|\Sigma| > 1$ this language is not regular, and now we conjecture that it is not context-free.

In some cases we can improve the idea of the above non-deterministic algorithm by predicting more precisely what automata should be taken as guesses. For instance, if $|\Sigma| = 2$, then the algorithm C2C reduces to its first inner loop and the output of C2C may be either a 3-state automaton of the type shown on

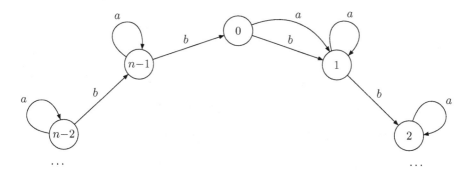

Fig. 3. The automaton $\mathscr{C}_n$

Fig. 2, or one of the automata $\mathscr{C}_n$ (see Fig. 3) discovered in Černý's pioneering paper [3], with $n < |w|$.

Thus, we have only to check if the given word w is 2-compressing with respect to $\leq 2|w|$ automata with $< |w|$ states each, and this clearly can be done in deterministic polynomial time. In general, however, we do not know yet if there is a polynomial time procedure for recognizing 2-collapsing words over any larger alphabet.

Acknowledgments

Fruitful discussions with Stuart Margolis as well as useful comments of anonymous referees are gratefully acknowledged. We also thank Ilya Petrov who has implemented our algorithm and discovered with its help many rather surprising examples of 2-collapsing words. For instance, he has found the following 2-collapsing word of length 21:

$$aba \cdot acc \cdot bab \cdot bac \cdot bab \cdot cac \cdot bcb.$$

It is worth mentioning that our previous shortest example of a 2-collapsing word over 3 letters (constructed by hand, see [2, Proposition 5] and also [1, Example 2.1]) was of length 27.

References

1. D. S. Ananichev, A. Cherubini, and M. V. Volkov. Image reducing words and subgroups of free groups. *Theor. Comput. Sci.*, (to appear).
2. D. S. Ananichev and M. V. Volkov. Collapsing words vs. synchronizing words. In W. Kuich, G. Rozenberg, A. Salomaa (eds.), *Developments in Language Theory* [*Lect. Notes Comput. Sci.* **2295**], Springer-Verlag, Berlin-Heidelberg-N.Y., 2002, 166–174.

3. J. Černý. Poznámka k homogénnym eksperimentom s konecnými avtomatami. *Mat.-Fyz. Cas. Slovensk. Akad. Vied.* **14** (1964) 208–216 [in Slovak].
4. I. Kapovich and A. Myasnikov. Stallings foldings and subgroups of free groups, *J. Algebra* **248** (2002) 608–668.
5. J. Kari. A counter example to a conjecture concerning synchronizing words in finite automata. *EATCS Bull.* **73** (2001) 146.
6. S. Margolis and J. Meakin. Free inverse monoids and graph immersions. *Internat. J. Algebra and Computation* **3** (1993) 79–99.
7. A. Mateescu and A. Salomaa. Aspects of classical language theory. In G. Rozenberg, A. Salomaa (eds.), *Handbook of Formal Languages, Vol. I. Word. Language, Grammar*, Springer-Verlag, Berlin-Heidelberg-N.Y., 1997, 175–251.
8. S. Margolis, J.-E. Pin, and M. V. Volkov. Words guaranteeing minimal image. In M. Ito (ed.), *Proc. III Internat. Colloq. on Words, Languages and Combinatorics*, World Scientific, Singapore, 2003 (to appear).
9. J.-E. Pin. *Le Problème de la Synchronisation. Contribution à l'Étude de la Conjecture de Černý*, Thèse de 3éme cycle, Paris, 1978 [in French].
10. J.-E. Pin. Sur les mots synchronisants dans un automate fini. *Elektronische Informationverarbeitung und Kybernetik* **14** (1978) 283–289 [in French].
11. N. Sauer and M. G. Stone. Composing functions to reduce image size. *Ars Combinatoria* **31** (1991) 171–176.
12. J. Stallings. Topology of finite graphs. *Inv. Math.* **71** (1971) 551–565.

Appendix: An Illustrating Example

In order to show some highlights of the algorithm C2C, we show some steps of its work on the following word of length 27:

$$V_{27} = abcb^2c^2bcac^2a^2caba^2b^2(abc)^2.$$

Obviously, the word contains each word of length 2 over the alphabet $\{a, b, c\}$ as factor so we bypass Step 0. On Step 1 we first fix a partition of the alphabet; let it be the partition $(\{a\}, \{b, c\})$. The decomposition of the word V_{27} with respect to this partition is

$$V_{27} = a \cdot bcb^2c^2bc \cdot a \cdot c^2 \cdot a^2 \cdot c \cdot a \cdot b \cdot a^2 \cdot b^2 \cdot a \cdot bc \cdot a \cdot bc$$

and the corresponding set S of the inner segments consists of the words b, c, b^2, bc, c^2, and bcb^2c^2bc.

On Step 2a we have to choose a subset $P \subset S$ and to construct the corresponding flower automaton. Suppose that P is chosen to be the subset $\{b, b^2, c^2, bcb^2c^2bc\}$ whence $P_a = S_a \setminus P = \{c, bc\}$. Applying the construction described on Step 2a we arrive at the following automaton:

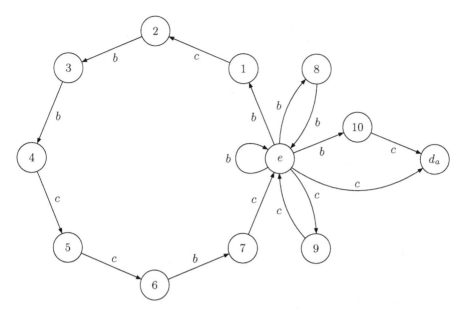

Fig. 4. The flower automaton for the word V_{27}

On Step 3a we fold the automaton $\mathscr{F}$ to an inverse automaton. One readily checks that folding the flower automaton shown on Fig. 4 stops after few steps yielding the inverse automaton $\mathscr{J}$ shown on Fig. 5 below. (For the states of $\mathscr{J}$ we have chosen as labels the corresponding state sets of the initial flower automaton.)

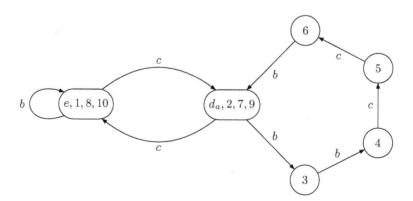

Fig. 5. The result of folding of the automaton shown on Fig. 4

Since the automaton $\mathscr{J}$ is connected and has more than two states, we proceed with Step 4a which aim is to complete $\mathscr{J}$ to a permutation automaton. Figure 6 presents one of the possible completions of $\mathscr{J}$.

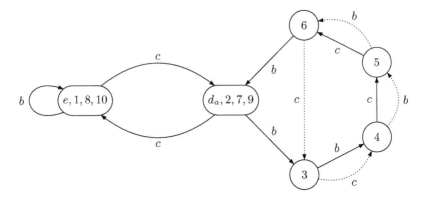

Fig. 6. A permutation completion of the automaton shown on Fig. 5

Since the automaton on Fig. 6 is connected and has more than 3 states, the construction of Step 5a applies to it yielding the following DFA:

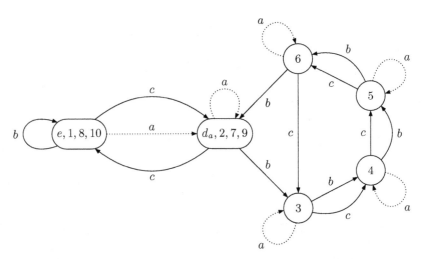

Fig. 7. A DFA proving that the word V_{27} is not 2-collapsing

One readily sees that this DFA is 2-compressible: the word $acba$ sends 3 states ($\bar{e} = \{e, 1, 8, 10\}$, $\bar{d_a} = \{d_a, 2, 7, 9\}$, and 5) to the state $\bar{d_a}$. On the other hand, one can check that the word V_{27} is not 2-compressing with respect to this automaton.

It is of some interest that a small "shift" of V_{27} produces a 2-collapsing word, namely, the word

$$W_{27} = abc^2b^2cbca^2c^2acab^2a^2b(abc)^2.$$

(It is the word of length 27 mentioned at the end of this paper.)

Efficient Algorithm for Checking Multiplicity Equivalence for the Finite $\mathbf{Z} - \Sigma^*$-Automata

Kostyantyn Archangelsky

Kiev National University, cybernetics dept,P.O.Box 129, 04070 Kiev, Ukraine
akv@zeos.net

Abstract. We represent a new fast algorithm for checking multiplicity equivalence for finite $\mathbf{Z} - \Sigma^*$ - automata. The classical algorithm of Eilenberg [1]is exponential one. On the other hand, for the finite deterministic automata an analogous Aho - Hopcroft - Ullman [2] algorithm has "almost" linear time complexity. Such a big gap leads to the idea of the existence of an algorithm which is more faster than an exponential one. Hunt and Stearns [3] announced the existence of polynomial algorithm for the $\mathbf{Z} - \Sigma^*$ - automata of finite degree of ambiguity only. Tzeng [4] created $O(n^4)$ algorithm for the general case. Diekert [5] informed us Tzeng's algorithm can be implemented in $O(n^3)$ using traingular matrices. Any way, we propose a new $O(n^3)$ algorithm (for the general case). Our algorithm utilizes recent results of Siberian mathematical school (Gerasimov [6], Valitckas [7]) about the structure of rings and does not share common ideas with Eilenberg [1] and Tzeng [4].

Let $\mathbf{Z}$ be the set of integers, $\Sigma = \{\sigma_1, \sigma_2, \ldots, \sigma_m,\}$ be a finite alphabet, ε - an empty word, $\emptyset$- an empty set, $\mathbf{Z} < \Sigma^* >$ - a set of all noncommutative polynomials with integer coefficients over Σ, $\mathbf{Z} < \Sigma >$ — a set of all linear polynomials, $\overline{\emptyset}$ - a zero row, $\overline{\emptyset}'$ - a zero column, $\overline{\overline{\emptyset}}$ - a zero matrix over $\mathbf{Z} < \Sigma^* >$. Define $\mathbf{Z} - \Sigma^*$ - automata as a tuple $\mathfrak{B} = (Q, E, \mu, I, F)$, where $Q = \{q_1, q_2, \ldots, q_n\}$ is a finite set of states, $E \subseteq Q \times \Sigma \times Q$ is a finite set of transitions, $\mu : E \to \mathbf{Z}$ is a multipicity function, I is a set of initial states and F is a set of final states. Thus $\mathbf{Z} - \Sigma^*$ - automata $\mathfrak{B}$ defines a mapping $\mathfrak{B} : \Sigma^* \to \mathbf{Z}$. Two $\mathbf{Z} - \Sigma^*$ - automata are multiplicitly equivalent, if they define the same mapping. Transition matrix B, initial row A and finial column C will be defined in a usual way:

$$b_{ij} = \sum_{\substack{1 \le k \le m \\ (q_i, \sigma_k, q_j) \in E}} \mu(q_i, \sigma_k, q_j)\sigma_k, \quad i, j = \overline{1, n}$$

$$a_i = \begin{cases} \varepsilon, & \text{if } q_i \in I \\ \emptyset, & \text{else} \end{cases} \quad i = \overline{1, n}$$

$$c_i = \begin{cases} \varepsilon, & \text{if } q_i \in F \\ \emptyset, & \text{else} \end{cases} \quad i = \overline{1, n}$$

For two $\mathbf{Z} - \Sigma^*$ - automata $\mathfrak{B}_1$, $\mathfrak{B}_2$ consider a $\mathbf{Z} - \Sigma^*$ - automata $\mathfrak{B}$ (resulting automata) with following matrix B, row A and column C.

M. Ito and M. Toyama (Eds.): DLT 2002, LNCS 2450, pp. 283–289, 2003.
© Springer-Verlag Berlin Heidelberg 2003

A

B 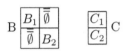 C

Thus, $\mathfrak{B}_1$ is equivalent to $\mathfrak{B}_2$ iff AB^NC is a zero matrix for all $N = 0, 1, 2, \dots$. We will prove that this will be true iff there exist such matrices $\tilde{A}, \tilde{B}, \tilde{C}$ (over $\mathbf{Z} < \Sigma >$) of *smaller* dimension, based on the same principle: $\tilde{A}\tilde{B}^N\tilde{C} = \bar{\bar{\emptyset}}$ for all $N = 0, 1, 2, \dots$.

Theorem 1. *If nonzero* $(k \times n)$ — *matrix* A *and nonzero* $(n \times l)$ — *matrix* C *(both over* $\mathbf{Z} < \Sigma >$*) are such that* AC *is a zero* $(k \times l)$ — *matrix, then there exists a* $(n \times n)$ — *matrix* T *over* $\mathbf{Z}$, *invertable over* $\mathbf{Z}$, *such that one of the following two statements holds:*

1) matrix $\tilde{A} = AT$ *has* $n - s_1$ $(1 \le s_1 \le n)$ *final zero columns and* s_1 *initial linear independent columns over* $\mathbf{Z} < \Sigma >$; *matrix* $\tilde{C} = T^{-1}C$ *has* s_2 $(1 \le s_1 \le s_2 \le n)$ *initial zero rows and* $n - s_2$ *final linear independent rows over* $\mathbf{Z} < \Sigma >$.

2) At least one of the matrices from $\tilde{A}, \tilde{C}$ *is a zero one.*

Proof: Let E be a unity matrix, U_{ij} be a matrix with 1 at (i,j) position and zeroes at others (both matrices are over $\mathbf{Z}$ of dimension $n \times n$). We will do equivalent transformations with columns of matrix A and rows of matrix C without changing the dimensions of matrices and their product AC:

(I) Exchange the places of the i-th and j-th columns of matrix A and simultaneously the i-th and j-th rows of matrix C. This transformation is in fact a multiplication of matrix A from the right by a $(n \times n)$-matrix $T_1 = E - U_{ii} - U_{jj} + U_{ij} + U_{ji}$ and matrix C from the left by matrix $T_1^{-1} = T_1$.

(II) Adding in matrix A to the j-th column its i-th column multiplied by z $(z \in \mathbf{Z})$; simultaneous subtracting in matrix C the j-th row multiplied by z of the i-th row. This transformation is in fact a multiplication of matrix A from the right by $(n \times n)$-matrix $T_2 = E + zU_{ij}$ and matrix C from the left by matrix $T_2^{-1} = E - zU_{ij}$.

The result of such transformations will be the matrices $\tilde{A} = AT_{i_1} \dots T_{i_h}$ and $\tilde{C} = T_{i_h}^{-1} \dots T_{i_1}^{-1}C$, i.e., $\tilde{A}\tilde{C} = AC = \bar{\bar{\emptyset}}$.

Let us start from the matrix A. Chose the first upper nonzero row, let it be the i-th row. Let $\bar{a}'_{j_1}, \dots, \bar{a}'_{j_s}$ $(1 \le s \le n)$ be a set of all columns with nonzero coefficients near the letter $\sigma_1 \in \Sigma$ in the i-th row. We will subtract the j_1-th column, multiplied by correspondent coefficient from columns j_2-th,...,j_s-th in order to place a zero near σ_1 in the entries $a_{ij_2}, \dots, a_{ij_s}$. (Simultaneous transformations with coefficients of the rest of σ_i, $i \ge 2$, as well as with rows of matrix C are not essential for our consideration).

Let us change this column with the 1-st. In the matrix obtained there will be no nonzero coefficients near σ_1 in the i-th row, except in the first column. Let us move to the next row containing at least one nonzero coefficient near σ_1 and perform steps analogous on the columns from 2-nd to n-th. After the annihilation of all coefficients near σ_1 in the final row and columns $t+1, t+2, \dots, n$ let us

repeat all the above actions for the matrix consisting of columns t+1, t+2, ...,
n with respect to the next letter of alphabet σ_2 and so on.

All nonzero columns of the obtained matrix $\tilde{A}$ became linear independent
over $\mathbf{Z} < \Sigma >$ (and even over $\mathbf{Z} < \Sigma^* >$). If there are no zero columns in $\tilde{A}$
then $\tilde{C}$ became a zero one.

If no, we assume that $\tilde{A}$ consists of columns $(\overline{a}'_1, \ldots, \overline{a}'_s, \overline{\emptyset}', \ldots, \overline{\emptyset}')$ $\overline{a}'_i, i = \overline{1, s}$
being nonzero and linear independent. If there is at least one nonzero element
in the first row of $\tilde{C}$, for example c_{1j}, then in the j-th column of $\tilde{A}\tilde{C}$ there is
a nonzero element too. If we go on, we will see that initial s rows of matrix $\tilde{C}$
are zero ones. Unfortunately, rows $\overline{c}_{s+1}, \ldots \overline{c}_n$ are not necessary linear independent. In order to make them linear independent let us use the above algorithm,
starting from the lower row. This transformations will not disrupt the linear in-
dependency of columns $\overline{a}'_1, \ldots, \overline{a}'_s$ because only zero columns from the $(s+1)$-th
to the n-th in matrix $\tilde{A}$ will be added, subtracted, exchanged. ∘

(Gerasimov [6] proved that there exist such a matrix T over $\mathbf{Z} < \Sigma >$,
Valitckas [7] said T may be over $\mathbf{Z}$.)

We will call *right-up* $(k \times l)$ *-cell* of $(n \times n)$ -matrix $\{a_{ij}\}$ a nonempty submatrix
$\{a_{ij}\}_{j=n-l+1,n}^{i=\overline{1,k}}$.

Theorem 2. *If for all* $N = 0, 1, 2, \ldots$ *for the* $(k \times n)$ *- matrix* A, $(n \times n)$ *-matrix*
B, $(n \times l)$ *- matrix* C, *all are over* $\mathbf{Z} < \Sigma >$, $AB^N C$ *is a zero* $(k \times l)$ *-matrix,*
then there exists a $(n \times n)$ *- matrix* T *over* $\mathbf{Z}$, *invertable over* $\mathbf{Z}$, *such that:*
either a right-up cell of every matrix $(T^{-1}BT)^N$ *is a zero one, or there is a zero*
matrix among AT *and* $T^{-1}C$.

Proof: Matrix AC is a zero $(k \times l)$-matrix, then we may use the transformation
matrix T from Theorem 1. Denote $\tilde{A} = AT, \tilde{B} = T^{-1}BT, \tilde{C} = T^{-1}C$. Initial s_1
columns of $(k \times n)$ - matrix $\tilde{A}$ are linear independent over $\mathbf{Z} < \Sigma >$ so that at
least initial s_1 rows of $(n \times l)$ - matrix $\tilde{B}^N \tilde{C}$ are zero ones. It follows that the
product of initial s_1 rows of matrix $\tilde{B}^N$ to matrix $\tilde{C}$ is a zero $(s_1 \times l)$-matrix.
Because final $(n - s_2)$ rows of matrix $\tilde{C}$ are linear independent over $\mathbf{Z} < \Sigma >$,
then at least all elements of the right-up $s_1 \times (n - s_2)$ - cell of matrix $\tilde{B}^N$ are
zeroes. ∘

Theorem 3. *Let B be a $(n \times n)$ - matrix over $\mathbf{Z} < \Sigma >$ of the following structure*
*$(B_0, B_1, B_2, B_3$ are matrices of corresponding dimensions, * denotes a matrix to*
be ignored):

$$
k\left\{
\begin{array}{|c|c|c|}
\hline
B_0 & B_1 & \overline{\overline{\emptyset}} \\
\hline
* & B_2 & B_3 \\
\hline
* & * & * \\
\hline
\end{array}
\right\} l
$$

then matrix B^N has a right-up zero $(k \times l)$ - cell for all $N = 0, 1, 2, \ldots$. iff
$B_1 B_2^N B_3$ is a zero $(k \times l)$ - matrix for all $N = 0, 1, 2, \ldots$.

Proof. Let's introduce following designations for cells in matrix B^N:

We will prove by induction on $N \geq 2$:

(i) $R_N = P_0 B_1 + P_1 B_1 B_2 + \ldots + P_{N-2} B_1 B_2^{N-2} + + B_1 B_2^{N-1}$

(ii) $B_1 B_2^{N-2} B_3 = \overline{\overline{\emptyset}}$

(P_i, $\overline{i = 0, \ N - 2}$ are several matrices over $\mathbf{Z} < \Sigma^* >$ of corresponding dimensions).

Base. For $N = 2$ we have

$$B^2 = \begin{array}{|c|c|c|} \hline B_0 & B_1 & \overline{\overline{\emptyset}} \\ \hline * & B_2 & B_3 \\ \hline * & * & * \\ \hline \end{array} \cdot \begin{array}{|c|c|c|} \hline B_0 & B_1 & \overline{\overline{\emptyset}} \\ \hline * & B_2 & B_3 \\ \hline * & * & * \\ \hline \end{array} = \begin{array}{|c|c|c|} \hline * & B_0 B_1 + B_1 B_2 & B_1 B_3 \\ \hline * & * & * \\ \hline * & * & * \\ \hline \end{array}$$

So,

(i) $R_2 = B_0 B_1 + B_1 B_2$ is true ($P_0 = B_0$)

(ii) $B_1 B_3 = \overline{\overline{\emptyset}}$ is true.

Hypothesis.

$$B^{N+1} = B^N B = \begin{array}{|c|c|c|} \hline Q_N & R_N & \overline{\overline{\emptyset}} \\ \hline * & * & * \\ \hline * & * & * \\ \hline \end{array} \cdot \begin{array}{|c|c|c|} \hline B_0 & B_1 & \overline{\overline{\emptyset}} \\ \hline * & B_2 & B_3 \\ \hline * & * & * \\ \hline \end{array} = \begin{array}{|c|c|c|} \hline * & Q_N B_1 + R_N B_2 & R_N B_3 \\ \hline * & * & * \\ \hline * & * & * \\ \hline \end{array}$$

So,

(i) $R_{N+1} = Q_N B_1 + R_N B_2 = Q_N B_1 + P_0 B_1 B_2 + P_1 B_1 B_2^2 +$
$+ \ldots + P_{N-2} B_1 B_2^{N-1} + B_1 B_2^N$ is true.

(ii) $\overline{\overline{\emptyset}} = R_N B_3 = P_0 B_1 B_3 + P_1 B_1 B_2 B_3 + \ldots + P_{N-2} B_1 B_2^{N-2} B_3 + + B_1 B_2^{N-1} B_3$
$= B_1 B_2^{N-1} B_3$ is true. ∘

Let us consider an example of application of our algorithm for two $\mathbf{N} - \{\mathbf{a}, \mathbf{b}\}^*$ — automata $\mathfrak{B}_1$ and $\mathfrak{B}_2$ (Eilenberg [1])

Both these automata define the same mapping $\mu : \{\mathbf{a}, \mathbf{b}\}^* \to \mathbf{N}$ as follows ($|\beta|$ denotes the length of the word β):

$$\mu(\alpha) = \sum_{a\beta \ \text{is a suffix of} \ \alpha} 2^{|\beta|}.$$

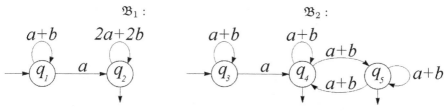

Construct the resulting automata $\mathfrak{B}$ (empty cells contain $\emptyset$):

	1	2	3	4	5
A_0	ε		$-\varepsilon$		

B_0

$a+b$	a			
	$2a+2b$			
		$a+b$	a	
			$a+b$	$a+b$
			$a+b$	$a+b$

C_0

ε
ε
ε

Step 0. Check the A_0C_0 (zero.) Columns of A_0 (in fact not columns, but cells only) are linear dependent.

Step 1. Add the 1-st column of A_0 and B_0 to the 3-rd column; subtract the 3-rd row of B_0 and C_0 from the 1-st. Rows of C_0 (in fact not rows, but cells only) are still linear dependent.

Subtract the 5-th row of B_0 and C_0 from the 2-nd; add the 2-nd column of A_0 and B_0 to the 5-th.

Subtract the 5-th row of B_0 and C_0 from the 4-th; add the 4-th column of A_0 and B_0 to the 5-th.

	1	2	3	4	5
	ε				

	1	2	3	4	5
1	$a+b$	a		$-a$	
2		$2a+2b$		$-a-b$	
3			$a+b$	a	a
4					
5				$a+b$	$2a+2b$

(with column vector, last cell ε)

Step 2. Now columns of A_0 and rows of C_0 are linear independent. According to Theorem 3 let us turn to the following matrices:

A_1

	2	3	4
	a		$-a$

	2	3	4
2	$2a+2b$		$-a-b$
3		$a+b$	a
4			

B_1

C_1

a

Step 3. Check A_1C_1 (zero.)

Step 4. Add the 2-nd column of A_1 and B_1 to the 4-th; subtract the 4-th row of B_1 and C_1 from the 2-nd. Exchange the places of the 3-rd and 4-th rows of B_1 and C_1; exchange the places of the 3-rd and 4-th columns of A_1 and B_1

$$\begin{array}{ccc} 2 & 3 & 4 \end{array}$$

a		

	2	3	4
2	$2a+2b$	$a+b$	
3			$a+b$
4		a	$a+b$

a

Step 5. Now columns of A_1 and rows of C_1 are linear independent. According to Theorem 3 turn to the following matrices:

A_2 $\boxed{a+b}$

$\boxed{}$ B_2 $\boxed{a+b}$ C_2

Step 6. Since $B_2 = \bar{\bar{\emptyset}}$, we have $A_2 B_2^N C_2 = \bar{\bar{\emptyset}}$ for all $N = 1, 2, \dots$ so $A_0 B_0^N C_0 = \bar{\bar{\emptyset}}$ for all $N = 0, 1, 2 \dots$.

Algorithm 1.

    INPUT: $(k \times n)$—matrix A_0, $(n \times n)$—matrix B_0, $(n \times l)$—matrix C_0

    begin $A := A_0$; $B := B_0$; $C := C_0$;

        if $AC \neq \bar{\bar{\emptyset}}$ **then OUTPUT "No"**;

        while $AC = \bar{\bar{\emptyset}}$ **do**

            begin

            if $(A = \bar{\bar{\emptyset}}$ **or** $B = \bar{\bar{\emptyset}}$ **or** $C = \bar{\bar{\emptyset}})$ **then OUTPUT "Yes"**;

            transform A to the kind of Theorem 1;

            $(B, C$ are being transformed simultaneously)

            transform C to the kind of Theorem 1;

            $(A, B$ are being transformed simultaneously)

            $A := \{b_{ij}\}_{j=\overline{s_1+1, s_2}}^{i=\overline{1, s}}$;

            $B := \{b_{ij}\}_{j=\overline{s_1+1, s_2}}^{i=\overline{s_1+1, s_2}}$;

            $C := \{b_{ij}\}_{j=\overline{s_2+1, n}}^{i=\overline{s_1+1, s_2}}$;

            end

        if $AC \neq \bar{\bar{\emptyset}}$ **then**

                if $B = \bar{\bar{\emptyset}}$ **then OUTPUT "Yes"**

                    else OUTPUT "No"

    end.

Note. We have not been discussed the case when $s_1 = s_2$ in this paper. Matrix B disintegrates into 4 not 9 parts in this case. It simplifies much more considerations above.

Theorem 4. *Algorithm 1 has $O(n^3)$ time complexity.*

Proof. Suppose after $i - 1$ steps of while-do-cycle the dimensions of the current matrice A, B, C are $k_i \times n_i, n_i \times n_i, n_i \times l_i$ correspondingly. It is easy to count that the next i-th step of the cycle of Algorithm 1 works no more $f(k_i, n_i, l_i) = 1/6|\Sigma|^2(12n_i^2(k_i + l_i) + n_i(27k_i + 27l_i + 12k_il_i - 3k_i^2 - 3l_i^2) + (-k_i^3 + 7k_i - 3k_i^2l_i + 18k_il_i - l_i^3 + 7l_i - 3l_i^2k_i))$ elementary operations of adding and multiplying of integer coefficients of the entries of matrices. Suppose a cycle has been working t times. So we have to find a maximum of the function $F(k_1, ..., k_t; n_1, ..., n_t; l_1, ..., l_t) = \sum_{1 \le i \le t} f(k_i, n_i, l_i)$ in the domain

$$\begin{cases} k_1 = 1, n_1 = n, l_1 = 1 \\ 0 \le k_i, n_i, l_i \le n, i = \overline{2, t} \\ k_{i+1} + n_{i+1} + l_{i+1} = n_i, i = \overline{1, t-1} \end{cases}$$

Exact solution of this problem is available but too extensive. To do the estimation note that the abovementioned equation implies

$$\sum_{1 \le i \le t} (k_i + l_i) \le n + 2,$$

so

$$F(k_1, ..., k_t; n_1, ..., n_t; l_1, ..., l_t) \le 83/6|\Sigma|^2(n + 2)^3.$$

On the other hand,

$$F(1, n/3; n, n/3; 1, n/3) = 1/6|\Sigma|^2(7/9n^3 + 287/9n^2 + 582/9n + 24)$$

so Algorithm 1 has exactly $O(n^3)$ time complexity. ∘

Open Problem. Our considerations are not acceptable to the case of finite multitape automata: problem of polynomial algorithm for the checking equivalence for the deterministic multitape automata is still open.

References

1. Eilenberg, S.: Automata, Languages, and Machines, Vol. A. Acad. Press (1974)
2. Aho,A., Hopcroft,J.E., Ullman,J.D.:The Design and Analysis of Computer Algorithms. Reading, Mass. (1976)
3. Hunt,H.B., Stearns,R.E.:On the Complexity of Equivalence, Nonlinear Algebra, and Optimization on Rings, Semirings, and Lattices. (Extended Abstract), Technical Report 86-23, Computer Science Dept, State University of NY at Albany (1986)
4. Tzeng,W.-G.:A Polynomial-time Algorithm for the Equivalence of Probabilistic Automata. FOCS (1989)
5. Diekert,V., Stuttgart University, private E-mail, December 12, (2001)
6. Gerasimov,V.N.:Localizations in the Associative Rings. Siberian Mathematical Journal, Vol.XXIII, N 6(1982) 36-54 (in Russian).
7. Valitckas,A.I., Institute of Mathematics, Siberian Division of Russian Academy, private communication (1992).

Some Remarks on Asynchronous Automata[*]

Balázs Imreh

Department of Informatics, University of Szeged,
Árpád tér 2, H-6720 Szeged, Hungary

Abstract. This paper is concerned with the class of all asynchronous automata. We characterize isomorphically complete systems for this class with respect to the α_i-products and present a description of languages accepted by asynchronous automata.

1 Introduction

An automaton is asynchronous if for every input sign and state the next state is stable for the input sign considered. Asynchronous automata were studied in different aspects. We mention here only some papers. In [3] and [4], the authors deal with the decomposition of an arbitrary automaton into a serial composition of two ones having fewer states than the original automaton and one of them being asynchronous. A subclass of the asynchronous automata, the class of commutative asynchronous automata is studied in [10] and [11]. In [11], it is shown that every commutative asynchronous automaton can be embedded isomorphically into a quasi-direct power of a suitable two-state automaton. Moreover, the exact bound for the maximal lengths of minimum-length directing words of n states directable commutative asynchronous automata, generated by one element, is presented. In [10], languages recognized by commutative asynchronous automata are described.

This paper is organized as follows. After the preliminaries of Section 2, we characterize isomorphically complete systems for the class of the asynchronous automata with respect to the α_i-products in Section 3. It turns out that this class is large enough in such a sense that there is no finite isomorphically complete system for it with respect to any of the α_i-products. In Section 4, we present a characterization of the languages recognized by asynchronous automata.

2 Preliminaries

Let X be a finite nonempty alphabet. The set of all finite words over X is denoted by X^* and $X^+ = X^* \setminus \{\lambda\}$, where λ denotes the empty word of X^*. In particular, we write x^+ for $\{x\}^+$.

By an *automaton* we mean a system $\mathbf{A} = (A, X)$, where A is a finite nonempty set of *states*, X is a finite nonvoid set of the *input symbols*, and every $x \in X$ is

---

[*] This work has been supported by the Hungarian National Foundation for Science Research, Grant T037258

M. Ito and M. Toyama (Eds.): DLT 2002, LNCS 2450, pp. 290–296, 2003.

realized as a unary operation $x^{\mathbf{A}} : A \rightarrow A$. For any word $w = x_1 \ldots x_s \in X^*$, $w^{\mathbf{A}} : A \rightarrow A$ is defined as the composition of the mappings $x_1^{\mathbf{A}}, \ldots, x_s^{\mathbf{A}}$. If $\mathbf{A}$ is known from the context, we simply write aw for $aw^{\mathbf{A}}$. An automaton $\mathbf{A} = (A, X)$ is *asynchronous* if for every $a \in A$ and $x \in X$, $axx = ax$ is valid.

By the definition above, an automaton can be considered as a unoid. Therefore, such notions as subautomata, isomorphisms, embeddings, can be defined in the usual way (see *e.g.* [1] or [7]).

A *recognizer* is a system $\mathcal{A} = (\mathbf{A}, a_0, F)$ which consists of an automaton $\mathbf{A} = (A, X)$, an *initial state* $a_0 \in A$, and a set $F(\subseteq A)$ of *final states*. The language *recognized* by $\mathcal{A}$ is

$$L(\mathcal{A}) = \{w : w \in X^* \text{ and } a_0 w^{\mathbf{A}} \in F\}.$$

It is also said that $L(\mathcal{A})$ is *recognizable* by the automaton $\mathbf{A}$.

Now, let $\mathcal{A} = (\mathbf{A}, a_0, F)$ be a recognizer. $\mathcal{A}$ is said to be *connected* if for each state $a \in A$, there exists a word $w \in X^*$ such that $a_0 w = a$ holds. Moreover, two states $a, b \in A$ are *equivalent*, denoted by $a \sim b$, if

$$aw \in F \text{ if and only if } bw \in F$$

is valid, for all $w \in X^*$. It is easy to see that $\sim$ is a congruence relation of $\mathbf{A}$. The recognizer $\mathcal{A}$ is *reduced* if $a \sim b$ implies $a = b$, for all $a, b \in A$. Furthermore, the recognizer $\mathcal{A}$ is *minimal* if no recognizer with fewer states recognizes the language $L(\mathcal{A})$. The following statement is well-known.

Lemma 1. *A recognizer $\mathcal{A}$ is minimal if and only if it is connected and reduced.*

We now recall the notion of the α_i-products (see *e.g.* [5], [6]). This product family is a natural generalization of the serial connection or cascade product of automata.

Throughout this paper, let i be an arbitrarily fixed nonnegative integer. Consider the automata $\mathbf{A} = (A, X)$, $\mathbf{A}_j = (A_j, X_j)$, $j = 1, \ldots, m$, and let Φ be a family of the following feedback functions

$$\varphi_j : A_1 \times \cdots \times A_{j+i-1} \times X \rightarrow X_j, \quad j = 1, \ldots, m.$$

It is said that $\mathbf{A}$ is the α_i-*product* of $\mathbf{A}_j$, $j = 1, \ldots, m$, if the conditions below are satisfied:

(1) $A = \prod_{j=1}^m A_j$,

(2) for every $(a_1, \ldots, a_m) \in A$ and $x \in X$,

$$(a_1, \ldots, a_m)x^{\mathbf{A}} = (a_1 x_1^{\mathbf{A}_1}, \ldots, a_m x_m^{\mathbf{A}_m})$$

is valid where $x_j = \varphi_j(a_1, \ldots, a_{j+i-1}, x)$, for all $j \in \{1, \ldots, m\}$.

For the α_i-product introduced above, we use the notation

$$\mathbf{A} = \prod_{j=1}^m \mathbf{A}_j(X, \Phi).$$

In particular, if an α_i-product consists of one factor, *i.e.*, $m = 1$, then we speak of α_i-*product with a single factor*.

The next statement is obvious.

Lemma 2. *Let* **A**, **B**, *and* **C** *be automata. If* **A** *can be embedded into an* α_0-*product of* **B** *with a single factor and* **B** *can be embedded into an* α_i-*product of* **C** *with a single factor, then* **A** *can be embedded into an* α_i-*product of* **C** *with a single factor.*

Let $\mathcal{N}$ be an arbitrary class and $\mathcal{M}$ be a system of automata. It is said that $\mathcal{M}$ is *isomorphically complete* for $\mathcal{N}$ with respect to the α_i-product if for any automaton $\mathbf{B} \in \mathcal{N}$, there are automata $\mathbf{A}_j \in \mathcal{M}$, $j = 1, \ldots, m$, such that $\mathbf{B}$ can be embedded into an α_i-product of the automata $\mathbf{A}_j$, $j = 1, \ldots, m$.

3   Isomorphic Representation

In this section, we characterize the isomorphically complete systems for the class of the asynchronous automata with respect to the α_i-products. First we recall some particular asynchronous automata introduced in [8]. For any positive integer n, let $[n]$ denote the set $\{1, \ldots, n\}$ and $\mathrm{Part}([n])$ the set of all partitions of $[n]$. For every partition $\pi = \{J_1, \ldots, J_k\} \in \mathrm{Part}([n])$, let

$$X_\pi^{(n)} = \{\{(J_1, j_1), \ldots, (J_k, j_k)\} : j_t \in J_t, \ t = 1, \ldots, k\}.$$

Moreover, let

$$X^{(n)} = \bigcup \{X_\pi^{(n)} : \pi \in \mathrm{Part}([n])\}.$$

Let us define now the automaton $\mathbf{R}^{(n)} = ([n], X^{(n)})$ follows. For every $j \in [n]$ and $\{(J_1, j_1), \ldots, (J_k, j_k)\} \in X^{(n)}$, let

$$j\{(J_1, j_1), \ldots, (J_k, j_k)\}^{\mathbf{R}^{(n)}} = j_l \ \text{if} \ j \in J_l$$

for some $1 \le l \le k$. Since $\{J_1, \ldots, J_k\}$ is a partition, j_l is determined uniquely. On the other hand, by the definition of $X^{(n)}$, $j_l \in J_l$, which implies that $\mathbf{R}^{(n)}$ is an asynchronous automaton. The next statement shows a useful property of $\mathbf{R}^{(n)}$.

Lemma 3. *Every asynchronous automaton of n states is isomorphic to an* α_0-*product of* $\mathbf{R}^{(n)}$ *with a single factor.*

Proof. Let $\mathbf{A} = (A, X)$ be an arbitrary asynchronous automaton of n states. Without loss of generality, we may assume that $A = \{a_1, \ldots, a_n\}$. Now, we define a mapping φ of X into $X^{(n)}$. For this reason, let $x \in X$ be an arbitrary input sign of $\mathbf{A}$. Let us define the relation $\sim_x$ on A as follows. For every pair $a, b \in A$, let $a \sim_x b$ if $ax^{\mathbf{A}} = bx^{\mathbf{A}}$. It is easy to check that the relation $\sim_x$ is reflexive, symmetric, and transitive, thus, $\sim_x$ is an equivalence relation on A. Let $\{A_1, \ldots, A_k\}$ be the partition belonging to the relation $\sim_x$. Observe that each equivalence class A_j contains one and only one state a_{s_j} such that $A_j x^{\mathbf{A}} = \{a_{s_j}\}$. For each equivalence classes A_j let us consider the pair (J_j, s_j), where J_j consists of the indices of the elements of A_j. Then obviously, $\{J_1, \ldots, J_k\}$ is a partition of $[n]$. Let us denote this partition of $[n]$ by π. Since $s_j \in J_j$, $j = 1, \ldots, k$, the set $\{(J_1, s_1), \ldots, (J_k, s_k)\}$ is contained in $X_\pi^{(n)}$. Therefore, the mapping φ defined by

$$\varphi : x \to \{(J_1, s_1), \ldots, (J_k, s_k)\}$$

is a mapping of X into $X^{(n)}$. Let us define now the α_0-product $\mathbf{B} = \mathbf{R}^{(n)}(X, \varphi)$ with a single factor. It is easy to see that $\mathbf{A}$ is isomorphic to $\mathbf{B}$ under the isomorphism μ, where $\mu : a_j \to j$, $j = 1, \ldots, n$.

Now, by using the automata $\mathbf{R}^{(n)}$, $n = 1, 2, \ldots$, and the idea of [9], we can characterize the isomorphically complete systems for the class of the asynchronous automata with respect to the α_i-products. Our characterization is based on the following statement.

Theorem 1. *If $\mathbf{R}^{(n)}$ can be embedded into an α_i-product $\prod_{j=1}^{k}(\mathbf{A}_j(X^{(n)}, \varphi)$, then $\mathbf{R}^{\lfloor n^{1/i^*} \rfloor}$ can be embedded into an α_i-product of $\mathbf{A}_s$ with a single factor for some $s \in \{1, \ldots, k\}$, where $i^* = i$ if $i > 0$ and $i^* = 1$ if $i = 0$.*

Proof. If $n = 1$ or $k = 1$, then the statement is obvious. Now, let us suppose that $n > 1$ and $k > 1$, moreover, $\mathbf{R}^{(n)}$ can be embedded into an α_i-product $\mathbf{B} = \prod_{j=1}^{k} \mathbf{A}_j(X^{(n)}, \varphi)$. Let $\mu : t \to (a_{t1}, \ldots, a_{tk})$, $t \in [n]$, be a suitable isomorphism. Without loss of generality, we may assume that there are integers $u \neq v \in [n]$ such that $a_{u1} \neq a_{v1}$. Indeed, in the opposite case, we could embed $\mathbf{R}^{(n)}$ into an α_i-product of the automata $\mathbf{A}_2, \ldots, \mathbf{A}_k$.

As a first step, we prove that the tuples $(a_{t1}, \ldots, a_{ti^*})$, $t \in [n]$, are pairwise different. Contrary, let us suppose that there are integers $p \neq q \in [n]$ such that $a_{ps} = a_{qs}$, $s = 1, \ldots, i^*$. For every $t \in [n] \setminus \{q\}$, let y_t be the following input sign of $\mathbf{R}^{(n)}$: $y_t = \{(\{q\}, q), ([n] \setminus \{q\}, t)\}$. Obviously, $p y_t^{\mathbf{R}^{(n)}} = t$ and $q y_t^{\mathbf{R}^{(n)}} = q$, for all $t \in [n] \setminus \{q\}$. Since μ is an isomorphism, we have that for every $t \in [n] \setminus \{q\}$,

$$(a_{t1}, \ldots, a_{tk}) = t\mu = (p y_t^{\mathbf{R}^{(n)}})\mu = (p\mu)y_t^{\mathbf{B}} = (a_{p1}x_1^{\mathbf{A}_1}, \ldots, a_{pk}x_k^{\mathbf{A}_k}),$$

where $x_1 = \varphi_1(a_{p1}, \ldots, a_{pi^*}, y_t)$ if $i > 0$ and $x_1 = \varphi_1(y_t)$ if $i = 0$. Therefore, $a_{t1} = a_{p1}x_1^{\mathbf{A}_1}$.

Similarly,

$$(a_{q1}, \ldots, a_{qk}) = q\mu = (q y_t^{\mathbf{R}^{(n)}})\mu = (q\mu)y_t^{\mathbf{B}} = (a_{q1}\bar{x}_1^{\mathbf{A}_1}, \ldots, a_{qk}\bar{x}_k^{\mathbf{A}_k}),$$

where $\bar{x}_1 = \varphi_1(a_{q1}, \ldots, a_{qi^*}, y_t)$ if $i > 0$ and $\bar{x}_1 = \varphi_1(y_t)$ if $i = 0$. Therefore, $a_{q1} = a_{q1}\bar{x}_1^{\mathbf{A}_1}$.

Now, let us observe that from our assumption on p and q it follows that $x_1 = \bar{x}_1$ and $a_{p1} = a_{q1}$. This results in $a_{t1} = a_{q1}$, for all $t \in [n] \setminus \{q\}$ which contradicts our assumption on u and v. Consequently, the elements $(a_{t1}, \ldots, a_{ti^*})$, $t \in [n]$, are pairwise different.

Based on the fact above it is easy to prove that $\mathbf{R}^{(n)}$ can be embedded into an α_i-product of the automata $\mathbf{A}_1, \ldots, \mathbf{A}_{i^*}$. Then there exists at least one integer $s \in \{1, \ldots, i^*\}$ such that the number of the pairwise different elements among $a_{1s}, \ldots, a_{ns}$ is not less than $v = \lfloor n^{1/i^*} \rfloor$, implying that $\mathbf{R}^{\lfloor n^{1/i^*} \rfloor}$ can be embedded into an α_i-product of $\mathbf{A}_s$ with a single factor. This ends the proof of Theorem 1.

Now, we are ready to characterize the isomorphically complete systems with respect to the α_i-products.

Theorem 2. *A system $\mathcal{M}$ of automata is isomorphically complete for the class of the asynchronous automata with respect to the α_i-product if and only if for every positive integer n, there exists an automaton $\mathbf{A}_n \in \mathcal{M}$ such that the automaton $\mathbf{R}^{(n)}$ can be embedded into an α_i-product of $\mathbf{A}_n$ with a single factor.*

Proof. The necessity of the condition follows from Theorem 1. The sufficiency can be easily obtained by Lemmas 2 and 3.

By Theorem 2, we get the following observation.

Corollary 1. There is no finite isomorphically complete system for the class of the asynchronous automata with respect to any of the α_i-products.

4   Recognizable Languages

A recognizer $\mathcal{A} = (\mathbf{A}, a_0, F)$ is called *asynchronous recognizer* if $\mathbf{A}$ is an asynchronous automaton. A language L is called *asynchronous language* if it can be recognized by an asynchronous recognizer. Let us denote the class of all asynchronous languages over X by $\mathcal{L}$. Now, we have the following description of the languages in $\mathcal{L}$.

Theorem 3. *For every regular language $L \subseteq X^*$, $L \in \mathcal{L}$ if and only if, for every word $w = x_1 \ldots x_m \in X^*$,*

(i) *if $w \in L$, then $x_1^+ \ldots x_m^+ \subseteq L$,*
(ii) *if $w \notin L$, then $x_1^+ \ldots x_m^+ \cap L = \emptyset$.*

Proof. If $L = \emptyset$ or $L = \{\lambda\}$, then the statement is obviously true. Now, let us suppose that L contains at least one word different from the empty word.

To prove the necessity of the conditions, let us suppose that $L \in \mathcal{L}$. Then L is recognized by an asynchronous recognizer $\mathcal{A} = (\mathbf{A}, a_0, F)$. Let $w \in X^*$ be an arbitrary word.

If $w = \lambda$, then (i) and (ii) hold obviously. Now, let us suppose that $w = x_1 \ldots x_m \in L$ with $m \geq 1$. Since $w \in L$, there is a sequence of not necessarily distinct states $a_0, a_1, \ldots, a_m$ of $\mathbf{A}$ such that $a_{j-1} x_j = a_j, j = 1, \ldots, m$, moreover, $a_m \in F$. Since $\mathbf{A}$ is an asynchronous automaton, we have $a_{j-1} x_j = a_{j-1} x_j^k = a_j$, for all positive integer k and $j \in \{1, \ldots, m\}$. This yields that $a_0 x_1^{k_1} x_2^{k_2} \ldots x_m^{k_m} = a_m$, where $k_1, \ldots, k_m$ are arbitrary positive integers. This implies the inclusion $x_1^+ \ldots x_m^+ \subseteq L$, and hence, (i) is valid.

To prove (ii) let us suppose that $w \notin L$ and $x_1^+ \ldots x_m^+ \cap L \neq \emptyset$. Then there are positive integers $k_1, \ldots, k_m$ such that $x_1^{k_1} \ldots x_m^{k_m} \in L$. Then $a_0 x_1^{k_1} \ldots x_m^{k_m} = a'$, where $a' \in F$. For every $j \in \{1, \ldots, m\}$, let us denote the state $a_{j-1} x_j$ by a_j. Since $\mathbf{A}$ is asynchronous, $a_j = a_{j-1} x_j = a_{j-1} x_j^{k_j}$ is valid for all $j = 1, \ldots, m$. This yields that $a_0 x_1 \ldots x_m = a'$, and therefore, $w = x_1 \ldots x_m \in L$ which is a contradiction.

In order to prove sufficiency, let us suppose that $L \subseteq X^*$ satisfies conditions (i) and (ii). Since L is regular language, there exists a minimal recognizer $\mathcal{A} = (\mathbf{A}, a_0, F)$ such that L is recognized by $\mathcal{A}$. Now, we prove that $\mathcal{A}$ is an asynchronous recognizer, *i.e.*, $\mathbf{A}$ is an asynchronous automaton. For this purpose, let $a \in A$ and $x \in X$ be arbitrary state and input sign, respectively. Since $\mathcal{A}$ is reduced, $\mathbf{A}$ is connected. This yields that there exists a word $p \in X^*$ such that $a_0 p = a$. Let us denote ax by b and axx by c. We prove that $b = ax = axx = c$. Let $q \in X^*$ be an arbitrary word. We shall prove that $bq \in F$ if and only if $cq \in F$. This equivalence implies $b = c$ since $\mathcal{A}$ is a minimal recognizer. We distinguish the following cases.

Case 1. Let us suppose that $q = \lambda$. If $b \in F$, then $px \in L$, but then $pxx \in L$ by (i), and therefore, $a_0 pxx = c$ is also contained in F. If $c \in F$, then $pxx \in L$. Now, if $b \notin F$, then $px \notin L$, but then, $pxx \notin L$ by (ii) which is a contradiction. Hence, $bq = b \in F$ if and only if $cq = c \in F$.

Case 2. Let us assume that $q = x_1 \ldots x_m$ for some integer $m \geq 1$. If $bq \in F$, then $pxq \in L$. Now, by (i), we obtain that this inclusion implies $pxxq \in L$ which results in $cq \in F$. If $cq \in F$, then $pxxq \in L$. We claim that $a_0 pxq \in F$ as well. If it is not so, then $pxq \notin L$, and by (ii), we get that $pxxq \notin L$ which is a contradiction.

From Theorem 3 the next observation follows immediately.

Corollary 2. Every asynchronous language different from the languages $\emptyset$, $\{\lambda\}$ is infinite.

Obviously, the direct product of asynchronous automata is also an asynchronous automaton, and hence, we have the following assertion.

Corollary 3. The class of the asynchronous languages $\mathcal{L}$ is closed under finite union and intersection, and the operation of complementation.

Regarding the operations concatenation and iteration, let us consider the following example.

Let $L = L_1 \cup L_2 = \{x^{k_1} y^{k_2} : 1 \leq k_1 \ \& \ 1 \leq k_2\} \cup \{y^{l_1} x^{l_2} : 1 \leq l_1 \ \& \ 1 \leq l_2\}$. It can be easily seen that both L_1 and L_2 are contained in $\mathcal{L}$, namely, they can be recognized by asynchronous recognizers of four states. Then, by Corollary 3, $L \in \mathcal{L}$ is also true. Now, let us consider the language LL. LL is a regular language since L is regular. Now, by using Theorem 3, it can be proved that $LL \notin \mathcal{L}$. Indeed, $xyx \notin LL$, but $xy^2 x \in LL$, and therefore, LL does not satisfies condition (ii) which results in $LL \notin \mathcal{L}$. This example leads to the following observation.

Corollary 4. The class of asynchronous languages is not closed under concatenation and iteration.

It is worth noting that the asynchronous languages form a variety in the sense of [12] although they form neither a +-variety nor a *-variety in the sense of [2].

References

1. Burris, S., Sankappanavar, H. P.: *A Course in Universal Algebra*. Springer-Verlag, New York Berlin (1981)
2. Eilenberg, S.: *Automata, languages, and machines, Volume B*. Academic Press, New York (1976)
3. Gerace, G. B., Gestri, G.: Decomposition of Synchronous Sequential Machines into Synchronous and Asynchronous Submachines. *Information and Control* **11** (1968) 568-591
4. Gerace, G. B., Gestri, G.: Decomposition of Synchronous Machine into an Asynchronous Submachine driving a Synchronous One. *Information and Control* **12** (1968) 538-548
5. Gécseg, F.: Composition of automata. In: Proceedings of the 2nd Colloquium on Automata, Languages and Programming, Saarbrücken, LNCS, Vol. 14. Springer-Verlag (1974) 351-363
6. Gécseg, F.: *Products of Automata*. Springer-Verlag, Berlin Heidelberg New York Tokyo (1986)
7. Grätzer, G.: *Universal Algebra*. 2nd edn. Springer-Verlag, New York Berlin Heidelberg Tokyo (1979)
8. Imreh, B.: On completeness with respect to the quasi-direct product. In: Conference on Automata, Languages and Mathematical Systems, Salgótarján (1984) 83-89
9. Imreh, B.: On α_i-products of automata. *Acta Cybernetica* **3** (1978) 301-307
10. Imreh, B., Ito, M., Pukler, A.: A note on the languages recognized by commutative asynchronous automata. *RIMS Kokyuroku* **1166** (2000) 95-99
11. Imreh, B., Ito, M., Pukler, A.: On commutative asynchronous automata. In: Proceedings of the Third International Colloquium on Words, Languages, and Combinatorics, Kyoto 2000, to appear
12. Steinby, M.: Classifying Regular Languages by Their Syntactic Algebras. In: Karhumäki, J., Maurer, H., Rozenberg, G. (eds.): *Results and Trends in Theoretical Computer Science*. LNCS Vol. 812. Springer Verlag, Berlin Heidelberg New York (1994) 396-409

Tiling Systems over Infinite Pictures and Their Acceptance Conditions

Jan-Henrik Altenbernd, Wolfgang Thomas, and Stefan Wöhrle

Lehrstuhl für Informatik VII
RWTH Aachen, 52056 Aachen, Germany
{altenbernd,thomas,woehrle}@i7.informatik.rwth-aachen.de

Abstract. Languages of infinite two-dimensional words (ω-pictures) are studied in the automata theoretic setting of tiling systems. We show that a hierarchy of acceptance conditions as known from the theory of ω-languages can be established also over pictures. Since the usual pumping arguments fail, new proof techniques are necessary. Finally, we show that (unlike the case of ω-languages) none of the considered acceptance conditions leads to a class of infinitary picture languages which is closed under complementation.

1 Introduction

In the theory of automata over infinite words, many types of acceptance conditions have been studied, such as Büchi and Muller acceptance. In the framework of nondeterministic automata, three kinds of acceptance conditions have been singled out to which all other standard conditions can be reduced [11, 9, 2]: Referring to a nondeterministic automaton $\mathcal{A}$, an ω-word α is

1. A-accepted if some complete run of $\mathcal{A}$ on α exists,
2. E-accepted if some complete run of $\mathcal{A}$ on α exists, reaching a state in a given set F of final states,
3. Büchi accepted if some complete run of $\mathcal{A}$ on α exists, reaching infinitely often a state in a given set F of final states.

It is well-known that these acceptance conditions lead to a strict hierarchy of three classes of ω-languages in the listed order.

The purpose of the present paper is to study these acceptance conditions over two-dimensional infinite words, i. e. labeled ω-grids or "infinite pictures". We use a model of "nondeterministic automaton" which was introduced under the name "tiling system" in [4] (see also the survey [3]). While the notion of run of a tiling system on a given infinite picture is natural, there are several versions of using the above acceptance conditions; for example one may refer to the occurrence of states on arbitrary picture positions, or one only considers the diagonal positions. As a preparatory step, we give a reduction to the latter case and thus use only the diagonal positions for visits of final states.

M. Ito and M. Toyama (Eds.): DLT 2002, LNCS 2450, pp. 297–306, 2003.

The first main result says that over infinite pictures we obtain the same hierarchy of languages as mentioned above for ω-languages. Whereas in the case of ω-words one can use simple state repetition arguments for the separation proofs, we need different arguments here, combining König's Lemma with certain boundedness conditions.

In the second part of the paper we show that the class of Büchi recognizable infinitary picture languages is not closed under complementation. We use a recursion theoretic result on infinitely branching infinite trees, namely that such trees with only finite branches constitute a set which is not in the Borel class Σ^1_1. From this we easily obtain a picture language that is not Σ^1_1 and thus not Büchi recognizable: One uses pictures which consist of a code of an infinitely branching finite-branch tree in the first row and which otherwise contain dummy symbols. The hard part of the nonclosure result on complementation is to show the Büchi recognizability of pictures which code infinite-branch trees. For this, it is necessary to implement precise comparisons between an infinity of segments of the first row, using only finitely many states. It turns out that this is possible by using the "work space" $\omega \times \omega$.

This nonclosure proof should be compared with a corresponding result of Kaminski and Pinter [6] where a kind of Büchi acceptance is used over arbitrary acyclic graphs; in that case one has much more freedom to construct counterexamples and thus does not obtain the nonclosure result over pictures.

A related work on tiling problems over $\omega \times \omega$ appeared in [5]. There "dominoes" are placed on the ω-picture in a non-overlapping tiling, the information flow being realized by matching the colors of the domino boundaries. This difference is not essential; however in [5] only the unlabeled ω-picture is considered, and no picture languages are associated with the domino systems.

The paper is structured as follows: In Section 2 tiling systems and their acceptance conditions are introduced, and we show how picture languages can be defined in monadic second-order logic. Section 3 contains the hierachy results mentioned above, and in Section 4 we show that the class of Büchi-recognizable picture languages is not closed under complementation.

2   Tiling Systems over Infinite Pictures

2.1   Preliminaries

Let Σ be a finite alphabet and $\hat{\Sigma} = \Sigma \uplus \{\#\}$. An ω-picture over Σ is a function $p : \omega^2 \to \hat{\Sigma}$ such that $p(i,0) = p(0,i) = \#$ for all $i \geq 0$ and $p(i,j) \in \Sigma$ for $i,j > 0$. (So we use $\# \notin \Sigma$ as a border marking of pictures.) $\Sigma^{\omega,\omega}$ is the set of all ω-pictures over Σ. An ω-picture language L is a subset of $\Sigma^{\omega,\omega}$.

We call the restriction $p \restriction_{\{0,\ldots,m\} \times \{0,\ldots,n\}}$ to an initial segment of a picture p an (m,n)-prefix picture or just a prefix of p. If $n = m$ such a prefix is called a square. A k-extension with Σ of an (m,n)-prefix p is an $(m+k, n+k)$-prefix p' such that $p' \restriction_{\{0,\ldots,m\} \times \{0,\ldots,n\}} = p$ and the vertices of p' outside p are labeled with Σ. Analogously we define an ω-extension with Σ of p. We write $p' = p \cdot^k \Sigma$ or $p' = p \cdot^\omega \Sigma$ if p' is a k-extension, respectively an ω-extension, of p with Σ.

We denote by $S_v = \{((i,j),(i+1,j)) \mid i,j \in \omega\}$ and $S_h = \{((i,j),(i,j+1)) \mid i,j \in \omega\}$ the vertical, respectively horizontal successor relation on ω^2.

A *path* in a picture p is a sequence $\pi = (v_0, v_1, v_2, \ldots)$ of vertices such that $(v_i, v_{i+1}) \in S_v \cup S_h$ for all $i \geq 0$. If $v_0 = (0,0)$ we call π an *initial path*, and if $\pi = (v_0, \ldots, v_n)$ is finite we call π a path from v_0 to v_n. A vertex v_1 is *beyond* a vertex v_0 ($v_1 > v_0$) if there is a path from v_0 to v_1.

The *origin* of a picture p is the vertex $(0,0)$, the only "corner" of p. The *diagonal* of a picture p is the set of vertices $\mathrm{Di}(p) = \{(i,i) \mid i \in \omega\}$.

2.2   Tiling Systems

A *tiling system* is a tuple $\mathcal{A} = (Q, \Sigma, \Delta, Acc)$ consisting of a finite set Q of states, a finite alphabet Σ, a finite set $\Delta \subseteq (\hat{\Sigma} \times Q)^4$ of *tiles*, and an acceptance component Acc (which may be a subset of Q or of 2^Q).

Tiles will be denoted by $\left(\begin{smallmatrix} a_1,q_1 & a_2,q_2 \\ a_3,q_3 & a_4,q_4 \end{smallmatrix} \right)$ with $a_i \in \hat{\Sigma}$ and $q_i \in Q$, and in general, over an alphabet Γ, by $\left(\begin{smallmatrix} \gamma_1 & \gamma_2 \\ \gamma_3 & \gamma_4 \end{smallmatrix} \right)$ with $\gamma_i \in \Gamma$. To indicate a combination of tiles we write $\left(\begin{smallmatrix} \gamma_1 & \gamma_2 \\ \gamma_3 & \gamma_4 \end{smallmatrix} \right) \circ \left(\begin{smallmatrix} \gamma_1' & \gamma_2' \\ \gamma_3' & \gamma_4' \end{smallmatrix} \right)$ for $\left(\begin{smallmatrix} (\gamma_1,\gamma_1') & (\gamma_2,\gamma_2') \\ (\gamma_3,\gamma_3') & (\gamma_4,\gamma_4') \end{smallmatrix} \right)$.

A *run* of a tiling system $\mathcal{A} = (Q, \Sigma, \Delta, Acc)$ on a picture p is a mapping $\rho : \omega^2 \to Q$ such that for all $i,j \in \omega$ with $p(i,j) = a_{i,j}$ and $\rho(i,j) = q_{i,j}$ we have $\left(\begin{smallmatrix} a_{i,j} & a_{i,j+1} \\ a_{i+1,j} & a_{i+1,j+1} \end{smallmatrix} \right) \circ \left(\begin{smallmatrix} q_{i,j} & q_{i,j+1} \\ q_{i+1,j} & q_{i+1,j+1} \end{smallmatrix} \right) \in \Delta$. One can view a run of $\mathcal{A}$ on p as an additional labeling of p with states from Q such that every 2×2 segment of the thus labeled picture is a tile in Δ.

2.3   Acceptance Conditions

We consider different acceptance conditions for tiling systems, all of them similar to the well-known ones from ω-automata over words. First consider the case where the acceptance component is a set $F \subseteq Q$ of states. A tiling system $\mathcal{A} = (Q, \Sigma, \Delta, F)$

- A-accepts p if there is a run ρ of $\mathcal{A}$ on p such that $\rho(v) \in F$ for all $v \in \omega^2$,
- E-accepts p if there is a run ρ of $\mathcal{A}$ on p such that $\rho(v) \in F$ for at least one $v \in \omega^2$,
- Büchi accepts p if there is a run ρ of $\mathcal{A}$ on p such that $\rho(v) \in F$ for infinitely many $v \in \omega^2$,
- co-Büchi accepts p if there is a run ρ of $\mathcal{A}$ on p such that $\rho(v) \in F$ for all but finitely many $v \in \omega^2$.

There are other natural acceptance conditions referring to an acceptance component Acc which is a set $\mathcal{F} \subseteq 2^Q$. We denote by $\mathrm{Oc}(\rho)$ the set of states occurring in ρ, and by $\mathrm{In}(\rho)$ the set of states occurring infinitely often in ρ. A tiling system $\mathcal{A}$ *Staiger-Wagner-accepts* (*Muller-accepts*) a picture p if $\mathrm{Oc}(\rho) \in \mathcal{F}$ ($\mathrm{In}(\rho) \in \mathcal{F}$) for some run ρ of $\mathcal{A}$ on p.

For any acceptance condition C, we say that a picture language L is C-*recognizable* if some tiling system $\mathcal{A}$ C-accepts precisely the pictures in L.

In the conditions above we consider a run on the whole picture and therefore call them *global acceptance conditions*. To emphasize this, we speak of picture languages which are globally A-recognizable, globally E-recognizable, etc. However, for our proofs it is convenient to look at more restricted conditions.

A tiling system $\mathcal{A} = (Q, \Sigma, \Delta, F)$ accepts a picture p with an *A-condition on the diagonal* if there is a run ρ of $\mathcal{A}$ on p such that $\rho(v) \in F$ for all $v \in \mathrm{Di}(p)$. Similarly we can define diagonal versions of all other acceptance conditions above by replacing $v \in \omega^2$ with $v \in \mathrm{Di}(p)$.

Theorem 1. *For every acceptance condition C from above, an ω-picture language L is globally recognizable with condition C if, and only if, L is recognizable with C on the diagonal.* □

The proof of this Theorem uses, independent of the acceptance condition, a powerset construction where at vertex (i, i) on the diagonal all states occurring at positions (i, j) and (j, i) for $j \leq i$ are collected. Due to space restrictions we have to skip the details here.

Using the same idea, we can reprove the standard simulation results for nondeterministic ω-automata over pictures, namely that nondeterministic Muller recognizability reduces to nondeterministic Büchi recognizability, and that nondeterministic co-Büchi and Staiger-Wagner recognizability reduce to E-recognizability.

Again, using a classical (product-) construction, we obtain:

Proposition 1. *For every acceptance condition C from above, the class of C-recognizable ω-picture languages is closed under union and intersection.* □

An alternative proof uses the results of the following subsection.

2.4 Monadic Definability

Similarly as for ω-words and for finite pictures, one verifies that every Büchi recognizable picture language can be defined by an existential monadic second order sentence (a Σ_1^1-formula). Here we view pictures p as relational structures over the signature $\{S_v, S_h, \leq_v, \leq_h, (P_a)_{a \in \Sigma}\}$ with universe ω^2, where S_v, S_h are interpreted as the usual vertical and horizontal successor relations, and $\leq_v$ and $\leq_h$ as the corresponding linear orderings. We write $u < v$ if v is beyond u. $P_a v$ holds for a vertex $v \in \omega^2$ iff $p(v) = a$.

Proposition 2. *Let $\mathcal{A} = (Q, \Sigma, \Delta, \mathrm{Acc})$ be a tiling system. The ω-picture language recognized by $\mathcal{A}$ with any of the acceptance conditions of Section 2.3 can be defined by an existential monadic second order sentence φ.*

Proof: The desired Σ_1^1-sentence φ describes (over any given picture p) that there is a successful run of $\mathcal{A}$ on p. For the details we refer only to Büchi acceptance, the other cases are easy variations.

Let $Q := \{1,\ldots,k\}$. We code states assigned to the p-vertices by disjoint subsets $Q_1,\ldots,Q_k$ of ω^2, Q_i containing the vertices where state i is assumed. The formula has to assure that the states are distributed in the picture according to the transition relation Δ. Furthermore we have to express that a state from F is assumed infinitely often. This can be done by expressing that there is an infinite sequence of vertices, all of them labeled with the same state from F, which strictly increases with respect to $\leq$.

The following sentence φ describes the existence of a successful run of $\mathcal{A}$ on p:

$$\exists Q_1 \ldots Q_k \exists R \quad \forall x \bigvee_{1 \leq i \leq k} \left(Q_i x \wedge \bigwedge_{j \neq i} \neg Q_j x \right) \tag{1}$$

$$\wedge \; \forall x_1 \ldots x_4 \Big(S_h x_1 x_2 \wedge S_h x_3 x_4 \wedge S_v x_1 x_3 \wedge S_v x_2 x_4 \rightarrow$$

$$\bigvee_{\left(\begin{smallmatrix} a_1,q_1 & a_2,q_2 \\ a_3,q_3 & a_4,q_4 \end{smallmatrix} \right) \in \Delta} \bigwedge_{1 \leq i \leq 4} P_{a_i} x_i \wedge Q_{q_i} x_i \Big) \tag{2}$$

$$\wedge \bigvee_{i \in F} \forall x \big(Rx \rightarrow Q_i x \big) \tag{3}$$

$$\wedge \; \forall x \in R \; \exists y \in R \; \exists z \big(\neg z = x \wedge x \leq_h z \wedge z \leq_v y \big)$$

$\square$

3   Hierarchy Results

Let us start with some natural examples of picture languages. Over the alphabet $\{a, b\}$, define

- $L_0 = \{b\}^{\omega,\omega}$ as the set of pictures carrying solely label b,
- L_1 as the set of all pictures containing at least one a,
- L_2 as the set of all pictures containing a infinitely often.

It is clear that L_0 is A-recognizable (by a tiling system which allows to cover only b-labeled vertices), that L_1 is E-recognizable (by a tiling system that assumes a final state precisely at the a-labeled vertices), and that L_2 is Büchi recognizable (by the same tiling system applied with the Büchi condition). Let us show that L_1 is not A-recognizable and L_2 not Büchi recognizable:

Theorem 2. *(a) L_1 is not A-recognizable.*
(b) L_2 is not E-recognizable.

Proof: (a): Assume that $\mathcal{A} = (Q, \Sigma, \Delta, F)$ A-recognizes L_1. Consider the pictures p_i which are labeled b on the (i, i)-prefix, and a everywhere else. Each p_i is A-accepted by $\mathcal{A}$. Let ρ_i be the partial accepting run on the (i, i)-prefix of p_i.

For the partial runs of $\mathcal{A}$ on the b-labeled (i, i)-prefixes we use the extension relation: ρ' on the b-labeled (j, j)-prefix extends ρ on the b-labeled (i, i)-prefix if $j > i$ and the restriction of ρ' to the (i, i)-square is ρ.

Via the extension relation, the partial runs ρ are arranged in a finitely branching tree, where the empty run represents the root and on level i all possible runs on the b-labeled (i, i)-square are collected. (Note that for each such run on level i there are only finitely many possible extensions on level $i + 1$.)

By assumption the tree is infinite (use the runs ρ_i from above). So by König's Lemma there is an infinite path. It determines a run of $\mathcal{A}$ on the infinite picture which is labeled b everywhere. Contradiction.

(b): Assume $\mathcal{A} = (Q, \Sigma, \Delta, F)$ E-recognizes L_2. Two cases must be considered:

Case 1: There is a square prefix p allowing a run ρ of $\mathcal{A}$ with final state on p, and p can be extended to infinitely many square prefixes p' which beyond p are labeled solely with b such that ρ can be extended to a run ρ' on p'. Then, by König's Lemma, $\mathcal{A}$ E-accepts the picture consisting of p and b-labeled vertices everywhere else. Contradiction.

Case 2: For any square prefix p allowing a run ρ of $\mathcal{A}$ with final state on p, there exist only finitely many extensions of p solely by b-labeled vertices to square prefixes p' such that ρ can be extended to a run ρ' on p'.

Let us consider one such square prefix p_0. For every possible run ρ with final state on p_0, let p_0^ρ be the largest possible square extension of p_0 only with b, such that ρ can be extended to a run ρ' on p_0^ρ.

Choose p_0' as the largest of all p_0^ρ and let p_1 be the extension of p_0' to a larger square in which extra columns and rows are added: first a row and column labeled b, then a row and column labeled a. (Note that a run assuming a final state on p_0 cannot be extended to any run on p_1.)

We repeat this procedure with p_1. If there is no run with final state on p_1, we can directly set p_2 as the extension to the next larger square with an extra column and row of a-labeled vertices; otherwise, we apply the construction mentioned above to obtain p_2.

By iteration, we get a sequence $p_0, p_1, p_2, \ldots$ of square prefixes such that

- p_{i+1} is a square extension of p_i containing a's in the last column and row,
- there is no run of $\mathcal{A}$ on p_{i+1} with final state on p_i.

In the limit (the unique common extension of all p_i) we obtain an ω-picture p with infinitely many occurrences of a which does not admit a run with an occurrence of a final state. Contradiction. $\qquad\square$

It is instructive to compare the proofs above with the corresponding arguments over ω-words. Regarding the set of ω-words over $\{a, b\}$ with at least one letter a, one refutes A-recognizability as follows: Assume an ω-automaton $\mathcal{A}$ with n states A-recognizes the language. Then it accepts $b^n a b^\omega$ and assumes a loop before the a (note that up to a already $n + 1$ states are visited), allowing also to accept b^ω. A similar repetition argument applies to the set of ω-words with infinitely many a's. If it is E-recognized by $\mathcal{A}$ with n states, then it will accept $(b^n a)^\omega$ by a visit to a final state after a prefix $(b^n a)^i b^n$; and again via a loop in the last b-segment it will also accept $(b^n a)^i b^\omega$. Over pictures these simple constructions of runs from loops cannot be copied.

The situation for deterministic tiling systems is much easier. We mention these systems here only shortly. A tiling system is called *deterministic* if on any picture it allows at most one tile covering the origin, the state assigned to position $(i+1, j+1)$ is uniquely determined by the states at positions $(i, j), (i+1, j), (i, j+1)$, and the states at the border positions $(i+1, 0)$ and $(0, j+1)$ are determined by the state $(i, 0)$, respectively $(0, j)$. The classical Landweber hierachy (see [7, 10]) of ω-languages is defined using deterministic ω-automata with the acceptance conditions from Section 2.3. The hierarchy proofs carry over without essential change to pictures, so we do not enter the details for deterministic tiling systems.

4   The Complementation Problem

It is easy (following the pattern of the well-known proofs over ω-words) to verify that the classes of A-recognizable and of E-recognizable picture languages are not closed under complement. Over ω-words, the Büchi recognizable languages are closed under complement. We show here that this result fails over pictures. For this purpose we use a well-known result of recursion theory, namely that the codes of finite-path trees form a set which is not Σ_1^1.

The ω-*trees* considered in this context are (possibly) infinitely branching. For technical purposes it is convenient to work with a coding of trees where the nodes are represented by nonempty sequences of positive integers, the sequence (1) representing the root, and a sequence $(1, i_2, \ldots, i_k)$ representing a node on level $k-1$. We do not require that the sons $(1, i_2, \ldots, i_k, j)$ of a node $(1, i_2, \ldots, i_k)$ have j-values which form an initial segment of the positive integers. So we identify an ω-tree with a nonempty prefix-closed set of sequences $(1, i_2, \ldots, i_k)$ with positive integers i_j. An ω-tree is called *finite-path* if all paths from the root are finite.

In a natural way, we use a unary coding (over the alphabet $\{1\}$) of numbers and code an ω-tree t by an ω-word over $\{1, \$\}$, taking $\$$ as a separation marker. A node $(1, i_2, \ldots, i_k)$ is encoded as the finite word $1\$1^{i_2}\$\ldots\$1^{i_k}\$\$$. The tree t itself is encoded by a concatenation of the encodings of all its nodes, with the restriction that the encoding of any given node must be preceded by the encoding of its father. In addition, we begin the whole encoding with an extra $\$\$$. Finite trees are encoded by ω-words by repeating the encoding of a node infinitely often.

Let T_1 be the set of ω-pictures over the alphabet $\{0, 1, \$\}$ which contain a code of an ω-tree in the first row and which are labeled 0 on the remaining positions. Let T_2 be the subset of T_1 of those pictures where the coded tree contains an infinite path.

Theorem 3. *The class of Büchi recognizable ω-picture languages is not closed under complement. In particular, $T_2 \subseteq \{0, 1, \$\}^{\omega, \omega}$ is Büchi recognizable, but its complement is not.*

Proof: We use a standard result of recursion theory (see, e. g. [8, Sect. 16.3, Thm. XX]), saying that the set FT of finite-path trees is Π_1^1-complete. Thus it is not Σ_1^1-definable in second-order arithmetic. This implies in particular that the ω-picture language $T_1 \setminus T_2$ containing the corresponding tree codes is not definable by a monadic Σ_1^1-sentence as introduced in Section 2.3, and hence by Proposition 2 is not Büchi recognizable.

In the next Lemma we show that T_1 and T_2 are Büchi recognizable. Assuming that this class of picture languages is closed under complement, we get that $\{0, 1, \$\}^{\omega, \omega} \setminus T_2$ is Büchi recognizable. Hence by Proposition 1 the set $T_1 \setminus T_2$ of pictures encoding finite-path trees, which is $(\{0, 1, \$\}^{\omega, \omega} \setminus T_2) \cap T_1$, would be Büchi recognizable, too. Contradiction.   □

Lemma 1. *(a) The language $T_1 \subseteq \{0, 1, \$\}^{\omega, \omega}$ of all ω-pictures encoding an ω-tree is Büchi recognizable.*

(b) The language $T_2 \subseteq \{0, 1, \$\}^{\omega, \omega}$ of all ω-pictures encoding an ω-tree with an infinite path is Büchi recognizable.

Proof: *(a)*: The idea for a tiling system which will Büchi recognize T_1 is to check increasing prefixes of a tree-encoding for correctness. An accepting run divides a picture into horizontal slices, one for every node. On every slice we test whether the father of the corresponding node, which will be guessed nondeterministically, has already been listed before. The tiles will not allow a node to be skipped from this procedure and the acceptance condition will require that these checks succeed infinitely often.

Every state of the tiling system will consist of three components. Since we will always have to know the encoding on the first row anywhere in the picture, we spend the first state component for the vertical propagation of the letters of this row.

The second and the third component of the states are used to check whether the father of the current node was guessed correctly. A typical accepting run checking 1$1111$$ and 1$1111$111$$ is sketched in Figure 1, where only the third component of a corresponding state is shown. The correct beginning of an encoding including the root node has to be checked separately using another set of states.

Let us outline the behaviour of $\mathcal{A}$ and the interpretation of each state. In the following we will use the term "labeling" for the third state component only.

The starting point of each comparison is the line consisting of 0,1 and 2 only. State 1 marks the end of the tree node checked last, everything to the left is labeled 0 (already checked), everything to the right is labeled 2 (to be checked).

In the next line we start the comparison of the node encoded just to the right of the vertex labeled 1. For this, four signals are used, two diagonal ones from the beginning of the encoding of the father (state 3) and of the son (state 5), and two vertical ones (4 and 6) marking their respective ends.

From every vertex (i, j) labeled 3 a signal to the right is initiated. This signal forwards the first state component of the state at (i, j) to the right using the second state component until it reaches a vertex $(i, j + k)$ labeled 5. The first state component of $(i, j + k)$ has to be the same as the second state component

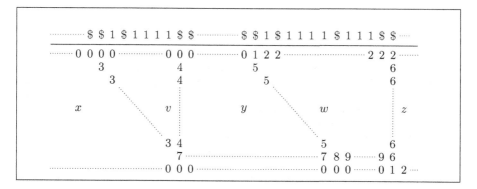

Fig. 1. Accepting tiling of father and son

of the arriving signal for the run to be continued. This is the case only if vertices $(1, j)$ and $(1, j + k)$ are labeled with the same symbol.

We mark the vertex where signals 3 and 4 meet by 7, which is forwarded to the right to mark at its meeting point with signal 5 (labeled 8) the end of the enconding of the father contained in the encoding of the son. We use label 9 to check that henceforth only 1's appear in the encoding of the son. Finally, the vertex below the one where signals 9 and 6 meet is labeled 1, indicating that the node whose encoding ends here has been successfully verified. The labels x, v, y, w, z are used to distinguish the vertices contained in the fields bordered by signals 0-9.

(b): To accept the language T_2 of ω-trees with an infinite path we modify the tiling system above slightly by adding another two state components. The first one is used to indicate (using state $*$) that the corresponding node is the last one which has already been verified and which is on the infinite path to be checked. The second new component is used to forward this mark to the right when we verify the encoding of the son of this node on the infinite path. Once this son has been verified, it becomes the only node labeled $*$ in the first new component for the next slices below the current one.

The tiling system will still ensure that there is a run on a picture p if, and only if, p encodes an ω-tree. The acceptance condition now requires that the consistency check for nodes whose father is labeled $*$ in the first new component succeeds infinitely often.   □

5   Conclusion

In this paper we have isolated those aspects of acceptance of ω-pictures by tiling systems which differ from the theory of ω-languages. This concerns the proofs (but not the results) in the comparison of acceptance conditions. For the class of Büchi recognizable picture languages we showed the nonclosure under complementation.

Among the many questions raised by this research we mention the following: Find (or disprove the existence of) decision procedures which test Büchi recognizable picture languages for E-, respectively A-recognizability. Compare the tiling system acceptance with an acceptance of pictures row by row, using an automaton model over ordinal words of length ω^2 (see [1]). Finally, it would be nice to have elegant characterizations of the A- and E-recognizable picture languages which do not use the obvious restrictions in Part (3) of the formula in Proposition 2.

References

1. Y. Choueka. Finite automata, definable sets, and regular expressions over ω^n-tapes. *Computer and System Sciences*, 17:81–97, 1978.
2. J. Engelfriet and H. Hoogeboom. X-automata on ω-words. *Theoretical Computer Science*, 110:1–51, 1993.
3. D. Giammarresi and A. Restivo. Two-dimensional languages. In *Handbook of Formal Languages*, vol. III, pages 215–267. 1997.
4. D. Giammarresi, A. Restivo, S. Seibert, and W. Thomas. Monadic second-order logic over rectangular pictures and recognizability by tiling systems. *Information and Computation*, 125(1):32–45, 1996.
5. D. Harel. Recurring dominoes: Making the highly undecidable highly understandable. *Annals of Discrete Mathematics*, 24:51–72, 1985.
6. M. Kaminski and S. Pinter. Finite automata on directed graphs. *Computer and System Sciences*, 44:425–446, 1992.
7. L. Landweber. Decision problems for ω-automata. *Mathematical Systems Theory*, 3:376–384, 1969.
8. H. Rogers. *Theory of Recursive Functions and Effective Computability*. McGraw-Hill. New York, 1967.
9. L. Staiger. Research in the theory of ω-languages. *Information Processing and Cybernetics EIK*, 23:415–439, 1987.
10. W. Thomas. Automata on infinite objects. In *Handbook of Theoretical Computer Science*, vol. B, pages 133–192. Amsterdam, 1990.
11. K. Wagner. On ω-regular sets. *Information and Control*, 43(2):123–177, 1979.

The Average Lengths of the Factors
of the Standard Factorization of Lyndon Words

Frédérique Bassino, Julien Clément, and Cyril Nicaud

Institut Gaspard Monge
Unversité de Marne-la-Vallée
77454 Marne-la-Vallée Cedex 2 - France
{bassino,clementj,nicaud}@univ-mlv.fr

Abstract. A non-empty word w of $\{a, b\}^*$ is a Lyndon word if and only if it is strictly smaller for the lexicographical order than any of its proper suffixes. Such a word w is either a letter or admits a standard factorization uv where v is its smallest proper suffix. For any Lyndon word v, we show that the set of Lyndon words having v as right factor of the standard factorization is rational and compute explicitly the associated generating function. Next we establish that, for the uniform distribution over the Lyndon words of length n, the average length of the right factor v of the standard factorization is asymptotically equivalent to $3n/4$. Finally we present algorithms on Lyndon words derived from our work together with experimental results.

1 Introduction

Given a totally ordered alphabet A, a *Lyndon word* is a word that is strictly smaller, for the lexicographical order, than any of its conjugates (*i.e.*, all words obtained by a circular permutation on the letters). Lyndon words were introduced by Lyndon [14] under the name of "standard lexicographic sequences" in order to give a base for the free Lie algebra over A; the standard factorization plays a central role in this framework (see [12], [16], [17]).

One of the basic properties of the set of Lyndon words is that every word is uniquely factorizable as a non increasing product of Lyndon words. As there exists a bijection between Lyndon words over an alphabet of cardinality k and irreducible polynomials over $\mathbb{F}_k$ [10], lot of results are known about this factorization: the average number of factors, the average length of the longest factor [8] and of the shortest [15].

Several algorithms deal with Lyndon words. Duval gives in [6] an algorithm that computes, in linear time, the factorization of a word into Lyndon words; he also presents in [7] an algorithm for generating all Lyndon word up to a given length in lexicographic order. This algorithm runs in a constant average time (see [3]).

In Section 2, we define more formally Lyndon words and give some enumerative properties of these sets of words. Then we introduce the standard factorization of a Lyndon word w which is the unique couple of Lyndon words u, v such that $w = uv$ and v is of maximal length.

M. Ito and M. Toyama (Eds.): DLT 2002, LNCS 2450, pp. 307–318, 2003.

In Section 3, we study the set of Lyndon words of $\{a, b\}^*$ having a given right factor in their standard factorization and prove that it is a rational language. We also compute its associated generating function. But as the set of Lyndon words is not context-free [1], we are not able to directly derive asymptotic properties from these generating functions. Consequently in Section 4 we give the average length of the standard right factors and present the combinatorial construction on which is based the proof. Probabilistic techniques and results from analytic combinatorics (see [9]) are used in order to establish this result. Section 5 is devoted to algorithms and experimental results.

The results contained in this paper constitute a first step in the study of the average behavior of the binary Lyndon trees obtained from Lyndon words by a recursive application of the standard factorization.

2   Preliminary

We denote A^* the free monoid over the alphabet $A = \{a, b\}$ obtained by all finite concatenations of elements of A. The length $|w|$ of a word w is the number of the letters w is product of, $|w|_a$ is the number of occurrences of the letter a in w. We consider the lexicographical order $<$ over all non-empty words of A^* defined by the extension of the order $a < b$ over A.

We record two properties of this order

(i) For any word w of A^*, $u < v$ if and only if $wu < wv$.
(ii) Let $x, y \in A^*$ be two words such that $x < y$. If x is not a prefix of y then for every $x', y' \in A^*$ we have $xx' < yy'$.

By definition, a *Lyndon word* is a primitive word (*i.e*, it is not a power of another word) that is minimal, for the lexicographical order, in its conjugate class (*i.e*, the set of all words obtained by a circular permutation). The set of Lyndon words of length n is denoted by $\mathcal{L}_n$ and $\mathcal{L} = \cup_n \mathcal{L}_n$.

$$\mathcal{L} = \{a, b, ab, aab, abb, aaab, aabb, abbb,$$
$$aaaab, aaabb, aabab, aabbb, ababb, abbbb, \dots\}$$

Equivalently, $w \in \mathcal{L}$ if and only if

$$\forall u, v \in A^+, \quad w = uv \Rightarrow w < vu.$$

A non-empty word is a Lyndon word if and only if it is strictly smaller than any of its proper suffixes.

Proposition 1. *A word $w \in A^+$ is a Lyndon word if and only if either $w \in A$ or $w = uv$ with $u, v \in \mathcal{L}$, $u < v$.*

Theorem 1 (Lyndon). *Any word $w \in A^+$ can be written uniquely as a non-increasing product of Lyndon words:*

$$w = l_1 l_2 \dots l_n, \quad l_i \in \mathcal{L}, \quad l_1 \geq l_2 \geq \dots \geq l_n.$$

Moreover, l_n is the smallest suffix of w.

The number $\text{Card}(\mathcal{L}_n)$ of Lyndon words of length n over A (see [12]) is

$$\text{Card}(\mathcal{L}_n) = \frac{1}{n} \sum_{d \mid n} \mu(d) \, \text{Card}(A)^{n/d},$$

where μ is the Moebius function defined on $\mathbb{N} \setminus \{0\}$ by $\mu(1) = 1$, $\mu(n) = (-1)^i$ if n is the product of i distinct primes and $\mu(n) = 0$ otherwise.

When $\text{Card}(A) = 2$, we obtain the following estimate

$$\text{Card}(\mathcal{L}_n) = \frac{2^n}{n} \left(1 + O\left(2^{-n/2}\right)\right).$$

Definition 1 (Standard factorization). *For $w \in \mathcal{L} \setminus A$ a Lyndon word not reduced to a letter, the pair (u, v), $u, v \in \mathcal{L}$ such that $w = uv$ and v of maximal length is called the standard factorization. The words u and v are called the* left factor *and* right factor *of the standard factorization.*

Equivalently, the right factor v of the standard factorization of a Lyndon word w which is not reduced to a letter can be defined as the smallest proper suffix of w.

Examples.

$$aaabaab = aaab \cdot aab, \qquad aaababb = a \cdot aababb, \qquad aabaabb = aab \cdot aabb.$$

3   Counting Lyndon Words with a Given Right Factor

In this section, we prove that the set of Lyndon words with a given right factor in their standard factorization is a rational language and compute its generating function. The techniques used in the following basically come from combinatorics on words.

Let $w = vab^i$ be a word containing one a and ending with a sequence of b. The word $R(w) = vb$ is the *reduced word* of w.

For any Lyndon word v, we define the set

$$\mathcal{X}_v = \{v_0 = v, v_1 = R(v), v_2 = R^2(v), \ldots, v_k = R^k(v)\}.$$

where $k = |v|_a$ is the number of occurrences of a in v. Note that $\text{Card}(\mathcal{X}_v) = |v|_a + 1$ and $v_k = b$.

Examples.

1. $v = aabab$: $\mathcal{X}_{aabab} = \{aabab, aabb, ab, b\}$.
2. $v = a$: $\mathcal{X}_a = \{a, b\}$.
3. $v = b$: $\mathcal{X}_b = \{b\}$.

By construction, v is the smallest element of $\mathcal{X}_v^+$ for the lexicographical order.

Lemma 1. *Every word $x \in \mathcal{X}_v$ is a Lyndon word.*

Proof. If $v = a$, then $\mathcal{X}_v = \{a, b\}$, else any element of $\mathcal{X}_v$ ends by a b. In this case, if $x \notin \mathcal{L}$, there exists a decomposition $x = x_1 x_2 b$ such that $x_2 b x_1 \leq x_1 x_2 b$ and $x_1 \neq \varepsilon$. Thus $x_2 a$ is not a left factor of $x_1 x_2 b$ and $x_2 a < x_1 x_2 a$. By construction of $\mathcal{X}_v$, as $x \neq v$, there exists a word w such that $v = x_1 x_2 a w$. We get that $x_2 a w x_1 < x_1 x_2 a w$. This is impossible since $v \in \mathcal{L}$.

A *code* C over A^* is a set of non-empty words such any word w of A^* can be written in at most one way as a product of elements of C. A set of words is *prefix* if none of its elements is the prefix of another one. Such a set is a code, called a *prefix code*. A code C is said to be *circular* if any word of A^* written along a circle admits at most one decomposition as product of words of C. These codes can be characterized as the bases of very pure monoids, *i.e.*, if $w^n \in C^*$ then $w \in C^*$. For a general reference about codes, see [2].

Proposition 2. *The set $\mathcal{X}_v$ is a prefix circular code.*

Proof. If $x, y \in \mathcal{X}_v$ with $|x| < |y|$, then, by construction of $\mathcal{X}_v$, $x > y$. So x is not a left factor of y and $\mathcal{X}_v$ is a prefix code.

Moreover, for every $n \geq 1$, if w is a word such that $w^n \in \mathcal{X}_v^*$ then $w \in \mathcal{X}_v^*$. Indeed if $w \notin \mathcal{X}_v^*$, then either w is a proper prefix of a word of $\mathcal{X}_v$ or w has a prefix in $\mathcal{X}_v^*$. If w is a proper prefix of a word of $\mathcal{X}_v$, it is a prefix of v and it is strictly smaller than any word of $\mathcal{X}_v$. As $w^n \in \mathcal{X}_v^*$, w or one of its prefix is a suffix of a word of $\mathcal{X}_v$. But all elements of $\mathcal{X}_v$ are Lyndon words greater than v, so their suffixes are strictly greater than v and w can not be a prefix of a word of $\mathcal{X}_v$.

Now if $w = w_1 w_2$ where w_1 is the longest prefix of w in $\mathcal{X}_v^+$, then w_2 is a non-empty prefix of a word $\mathcal{X}_v$, so w_2 is strictly smaller than any word of $\mathcal{X}_v$. As $w^n \in \mathcal{X}_v^*$, w_2 or one of its prefix is a suffix of a word of $\mathcal{X}_v$, but all elements of $\mathcal{X}_v$ are Lyndon words greater than v, so their suffixes are strictly greater than v and w can not have a prefix in $\mathcal{X}_v^+$.

As a conclusion, since $\mathcal{X}_v$ is a code and for every $n \geq 1$, if $w^n \in \mathcal{X}_v^*$ then $w^n \in \mathcal{X}_v^*$, $\mathcal{X}_v$ is circular code.

Proposition 3. *Let $l \in \mathcal{L}$ be a Lyndon word, $l \geq v$ if and only if $l \in \mathcal{X}_v^+$.*

Proof. If $l \geq v$, let l_1 be the longest prefix of l which belongs to $\mathcal{X}_v^*$, and l_2 such that $l = l_1 l_2$. If $l_2 \neq \varepsilon$, we have the inequality $l_2 l_1 > l \geq v$, thus $l_2 l_1 > v$. The word v is not a prefix of l_2 since l_2 has no prefix in $\mathcal{X}_v$, hence we have $l_2 = l_2' b l_2''$ and $v = l_2' a v''$. Then, by construction of $\mathcal{X}_v$, $l_2' b \in \mathcal{X}_v$ which is impossible. Thus $l_2 = \varepsilon$ and $l \in \mathcal{X}_v^+$.

Conversely, if $l \in \mathcal{X}_v^+$, as a product of words greater than v, $l \geq v$.

Theorem 2. *Let $v \in \mathcal{L}$ and $w \in A^*$. Then awv is a Lyndon word with $aw \cdot v$ as standard factorization if and only if $w \in \mathcal{X}_v^* \setminus (a^{-1}\mathcal{X}_v)\mathcal{X}_v^*$. Hence the set $\mathcal{F}_v$ of Lyndon words having v as right standard factor is a rational language.*

Proof. Assume that awv is a Lyndon word and its standard factorization is $aw \cdot v$. By Theorem 1, wv can be written uniquely as

$$wv = l_1 l_2 \ldots l_n, \quad l_i \in \mathcal{L}, \quad l_1 \geq l_2 \geq \cdots \geq l_n.$$

As v is the smallest (for the lexicographical order) suffix of awv, and consequently of wv, we get $l_n = v$; if $w = \varepsilon$, then $n = 1$, else $n \geq 2$ and for $1 \leq i \leq n-1, l_i \geq v$. Thus, $w \in \mathcal{X}_v^*$.

Moreover if $w \in (a^{-1}\mathcal{X}_v)\mathcal{X}_v^*$, then $aw \in \mathcal{X}_v^+ \cap \mathcal{L}$. Hence $aw \geq v$ which is contradictory with the definition of the standard factorization. So $w \in \mathcal{X}_v^* \setminus (a^{-1}\mathcal{X}_v)\mathcal{X}_v^*$.

Conversely, if $w \in \mathcal{X}_v^* \setminus (a^{-1}\mathcal{X}_v)\mathcal{X}_v^*$, then

$$w = x_1 x_2 \ldots x_n, \quad x_i \in \mathcal{X}_v \quad \text{and} \quad aw \notin \mathcal{X}_v^+.$$

From Proposition 1, the product ll' of two Lyndon words such that $l < l'$ is a Lyndon word. Replacing as much as possible $x_i x_{i+1}$ by their product when $x_i < x_{i+1}$, w can be rewritten as

$$w = y_1 y_2 \ldots y_m, \quad y_i \in \mathcal{X}_v^+ \cap \mathcal{L}, \quad y_1 \geq y_2 \geq \cdots \geq y_m.$$

As $aw \notin \mathcal{X}_v^+$, for any integer $1 \leq i \leq m, ay_1 \ldots y_i \notin \mathcal{X}_v^+$.

Now we prove by induction that $aw \in \mathcal{L}$. As $y_1 \in \mathcal{L}$ and $a < y_1$, $ay_1 \in \mathcal{L}$. Suppose that $ay_1 \ldots y_i \in \mathcal{L}$. Then, as $y_{i+1} \in \mathcal{L} \cap \mathcal{X}_v^+$, and $ay_1 \ldots y_i \in \mathcal{L} \setminus \mathcal{X}_v^+$, from Proposition 3, we get $ay_1 \ldots y_i < v \leq y_{i+1}$. Hence $ay_1 \ldots y_{i+1} \in \mathcal{L}$. So, $aw \in \mathcal{L}$.

As $aw \in \mathcal{L} \setminus \mathcal{X}_v^+$, $aw < v$ and $awv \in \mathcal{L}$. Setting $v = y_{m+1}$, we have

$$wv = y_1 y_2 \ldots y_m y_{m+1}, \quad y_i \in \mathcal{X}_v^* \cap \mathcal{L}, \quad y_1 \geq y_2 \geq \cdots \geq y_{m+1}.$$

Moreover any proper suffix s of awv is a suffix of wv and can be written as $s = y_i' y_{i+1} \ldots y_{m+1}$ where y_i' is a suffix of y_i. As $y_i \in \mathcal{L}$, $y_i' \geq y_i$. As $y_i \in \mathcal{X}_v^+$, $y_i \geq v$ and thus $s \geq v$. Thus, v is the smallest suffix of awv and $aw \cdot v$ is the standard factorization of the Lyndon word awv.

Finally as the set of rational languages is closed by complementation, concatenation, Kleene star operation and left quotient, for any Lyndon word v, the set $\mathcal{F}_v$ of Lyndon words having v right standard factor is a rational language.

Remark 1. The proof of Theorem 2 leads to a linear algorithm that computes the right factor of a Lyndon word using the fact that the factorization of Theorem 1 can be achieved in linear time and space (by an algorithm of Duval [6], see Section 5).

We define the generating functions $X_v(z)$ of $\mathcal{X}_v$ and $X_v^*(z)$ of $\mathcal{X}_v^*$:

$$X_v(z) = \sum_{w \in \mathcal{X}_v} z^{|w|} \quad \text{and} \quad X_v^*(z) = \sum_{w \in \mathcal{X}_v^*} z^{|w|}.$$

As the set $\mathcal{X}_v$ is a code, the elements of $\mathcal{X}_v^*$ are sequences of elements of $\mathcal{X}_v$ (see [9]):

$$X_v^*(z) = \frac{1}{1 - X_v(z)}.$$

Denote by $F_v(z) = \sum_{x \in \mathcal{F}_v} z^{|x|}$ the generating function of the set

$$\mathcal{F}_v = \{awv \in \mathcal{L} \mid aw \cdot v \text{ is the standard factorization}\}.$$

Theorem 3. *Let v be a Lyndon word. The generating function of the set $\mathcal{F}_v$ of Lyndon words having a right standard factor v can be written*

$$F_v(z) = z^{|v|} \left(1 + \frac{2z - 1}{1 - X_v(z)} \right).$$

Proof. First of all, note that any Lyndon word of $\{a, b\}^*$ which is not a letter ends with the letter b, so $F_a(z) = 0$. And as $\mathcal{X}_a = \{a, b\}$, the formula given for $F_v(z)$ holds for $v = a$.

Assume that $v \neq a$. From Theorem 2, $F_v(z)$ can be written as

$$F_v(z) = z^{|av|} \sum_{w \in \mathcal{X}_v^* \setminus a^{-1} \mathcal{X}_v^+} z^{|w|}.$$

In order to transform this combinatorial description involving $\mathcal{X}_v^* \setminus a^{-1} \mathcal{X}_v^+$ into an enumerative formula of the generating function $F_v(z)$, we prove first that $a^{-1} \mathcal{X}_v^+ \subset \mathcal{X}_v^*$ and, next that the set $a^{-1} \mathcal{X}_v^+$ can be described as a disjoint union of rational sets.

If $x \in \mathcal{X}_v \setminus \{b\}$, then x is greater than v and as x is a Lyndon word, its proper suffixes are strictly greater than v; consequently, writing $a^{-1}x$ as a non-increasing sequence of Lyndon word $l_1, \ldots, l_m$, we get, since $l_m \geq v$, that for all i, l_i is greater than v. Consequently from Proposition 3, for all i, $l_i \in \mathcal{X}_v^+$ and as a product of elements of $\mathcal{X}_v^+$, $a^{-1}x \in \mathcal{X}_v^+$. Therefore $a^{-1}(\mathcal{X}_v \setminus \{b\}) \mathcal{X}_v^* \subset \mathcal{X}_v^*$.

Moreover if $x_1, x_2 \in \mathcal{X}_v$ and $x_1 \neq x_2$, as $\mathcal{X}_v$ is a prefix code,

$$a^{-1}x_1 \mathcal{X}_v^* \cap a^{-1}x_2 \mathcal{X}_v^* = \emptyset.$$

Thus $a^{-1}(\mathcal{X}_v \setminus \{b\}) \mathcal{X}_v^*$ is the disjoint union of the sets $(a^{-1}x_i) \mathcal{X}_v^*$ when x_i ranges over $\mathcal{X}_v \setminus \{b\}$. Consequently the generating function of the set $\mathcal{F}_v$ of Lyndon words having v as right factor satisfies

$$F_v(z) = z^{|v|+1} \frac{1 - \frac{X_v(z) - z}{z}}{1 - X_v(z)}$$

and finally the announced equality.

Note that the function $F_v(z)$ is rational for any Lyndon word v. But the right standard factor runs over the set of Lyndon words which is not context-free [1]. Therefore in order to study the average length of the factors in the standard factorization of Lyndon words, we adopt another point of view.

4   Average Length of the Standard Factors

Making use of probabilistic techniques and of results from analytic combinatorics (see [9]), we are able to establish the following result.

Theorem 4. *The average length for the uniform distribution over the Lyndon words of length n of the right factor of the standard factorization is asymptotically*

$$\frac{3n}{4}\left(1 + O\left(\frac{\log^3 n}{n}\right)\right).$$

Remark 2. The error term comes from successive approximations at different steps of the proof and, for this reason, it is probably overestimated (see experimental results in Section 5).

First we partition the set $\mathcal{L}_n$ of Lyndon words of length n in the two following subsets: $a\mathcal{L}_{n-1}$ and $\mathcal{L}'_n = \mathcal{L}_n \setminus a\mathcal{L}_{n-1}$.

Note that $a\mathcal{L}_{n-1} \subset \mathcal{L}_n$ (that is, if w is a Lyndon word then aw is also a Lyndon word). Moreover if $w \in a\mathcal{L}_{n-1}$, the standard factorization is $w = a \cdot v$ with $v \in \mathcal{L}_{n-1}$. As

$$\mathrm{Card}\,(\mathcal{L}_{n-1}) = \frac{2^{n-1}}{n-1}\left(1 + O\left(2^{-n/2}\right)\right),$$

the contribution of the set $a\mathcal{L}_{n-1}$ to the mean value of the length of the right factor is

$$(n-1) \times \frac{\mathrm{Card}\,(a\mathcal{L}_{n-1})}{\mathrm{Card}\,(\mathcal{L}_n)} = \frac{n}{2}\left(1 + O\left(\frac{1}{n}\right)\right).$$

The study of the standard factorization of the words of $\mathcal{L}'_n$ requires a careful analysis.

Proposition 4. *The contribution of the set $\mathcal{L}'_n$ to the mean value of the length of right factor is*

$$\frac{n}{4}\left(1 + O\left(\frac{\log^3 n}{n}\right)\right).$$

This proposition basically asserts that in average for the uniform distribution over $\mathcal{L}'_n$, the length of the right factor is asymptotically $n/2$.

Proof. (sketch) Denote by **right** the paramater associated to the length of the right factor of a Lyndon word.

The idea is to build a transformation φ, which is a bijection on a set $\mathcal{D}_n \subset \mathcal{L}'_n$, such that the sum of the lengths of standard right factors of w and $\varphi(w)$, both belonging to $\mathcal{D}_n$, is about $|w|$ the length of w. Indeed with such a relation we can compute the contribution of $\mathcal{D}_n$ to the expectation of the parameter **right**. Then if the contribution of $\mathcal{L}'_n \setminus \mathcal{D}_n$ to the parameter **right** is negligible we are able to conclude for the expectation of the parameter **right**.

It remains to construct such a bijection φ and determine a "good" set $\mathcal{D}_n$. This is done in the following way: assume that w is a Lyndon word in $\mathcal{L}'_n = \mathcal{L}_n \setminus a\mathcal{L}_{n-1}$. Denote by k the length of the first run of a's of the standard right factor. We partition the set $\mathcal{L}_n \setminus a\mathcal{L}_{n-1}$ in two sets depending on the factorization. Indeed the standard factorization of a word w can only be one of the following factorizations

$$w = a^{k+1}b\,u \ \cdot \ a^k b\,v \text{ (first kind)}$$
$$w \ = a^k b\,u \ \cdot \ a^k b\,v \text{ (second kind)}.$$

This means that the left factor of a Lyndon word w can only begin by $a^{k+1}b$ or $a^k b$ when we know that the right factor begin by $a^k b$ (otherwise w cannot be in $\mathcal{L}_n \setminus a\mathcal{L}_{n-1}$). Let us fix a integer parameter $\lambda \in \mathbb{Z}^+$. Then the words u, v of $\mathcal{X}^*_k$ can be uniquely written as $u = u'u''$ and $v = v'v''$ where u' and v' are the smallest prefixes of u and v of length greater than λ and ending by a b (there is always such a symbol b if these words are not empty since then u and v end with a b). When $|u|, |v| \geq \lambda$ we define $\varphi(w)$ for a word $w = a^k b\,u \ \cdot \ a^k b\,v$ (resp. $w = a^{k+1}b\,u \ \cdot \ a^k b\,v$) by

$$\varphi(w) = a^k b\,u'v''a^k b\,v'u'' \qquad (\text{resp. } a^{k+1}b\,u'v''a^k b\,v'u'').$$

For example, considering $w = a^k b\,abb \ \cdot \ a^k b\,baab\,a^k b\,bbb$, if we choose $\lambda = 2$ and so $u' = ab, u'' = b, v' = baab, v'' = a^k b\,bbb$ then we get

$$\varphi(w) = a^k b\,aaba^k b\,bbb \ \cdot \ a^k b\,ab \in \mathcal{L}'_n.$$

Here $|w| = |\varphi(w)| = 3k+13$ and the length of the right standard factor are $2k+9$ and $k + 3$ respectively.

Some words give hints of what we must be careful about if we want the application φ to be well defined and the word $\varphi(w)$ to be a Lyndon word of $\mathcal{L}'_n$.

- First of all the parameter λ must be greater or equal to 1. So the longest runs of a's have to be separated by non-empty words. Otherwise, for example, if $w = a^k b \ \cdot \ a^k b\,b$, then $u = \varepsilon$ is the empty word. The application exchanging u and v gives a words which is no longer a Lyndon word.
- If $w = a^k b\,ab \ \cdot \ a^k b\,abb$, then $u = ab$ and it is a prefix of v. For any choice of λ, $\varphi(w)$ is not a Lyndon word. So the longest runs of a's have to be separated by words having distinct prefixes to ensure that $\varphi(w)$ is a Lyndon word.
- If $w = a^k b\,bab \ \cdot \ a^k b\,bbab$, then if we choose $\lambda = 1$, we get $\varphi(w) = a^k b\,bbaba^k b\,bab \notin \mathcal{L}$ (since $u' = v'$ and $u'v'' = bbab > v'u'' = bab$). Thus we have to take care, when we apply the transformation that $\varphi(w)$ is still smaller than its proper suffixes (this is ensured if $u' \neq v'$).

The application φ and the set $\mathcal{D}_n$ are dependent and to suit the "good" conditions they are implicitly determined by the following constraints:

1. The function φ is an involution on $\mathcal{D}_n$: $\varphi(\varphi(w)) = w$.

2. The standard factorization of $\varphi(w)$ for $w \in \mathcal{D}_n$ is

$$\varphi(w) = a^{k+1}b\,u'v'' \cdot a^k b\,v'u'' \text{ (first kind)}$$
$$\varphi(w) = a^k b\,u'u'' \cdot a^k b\,v'v'' \text{ (second kind)}.$$

3. The lengths of right factors of w and $\varphi(w)$ satisfy

$$|\text{right}(w)| + |\text{right}(\varphi(w))| = |w|\,(1 + o(|w|)).$$

4. The set $\mathcal{D}_n$ "captures" most of the set $\mathcal{L}'_n$ in an asymptotic way when n grows to ∞, that is

$$\frac{\text{Card}(\mathcal{D}_n)}{\text{Card}(\mathcal{L}'_n)} = 1 - o(1).$$

Most of these conditions are related to the properties of the longest runs of a's. More precisely a word w of $\mathcal{L}'_n$ can be written as

$$w = a^{k+1}b\,w_1 a^k b\,w_2 \cdots a^k b\,w_m$$
$$\text{or } w = a^k b\,w_1 a^k b\,w_2 \cdots a^k b\,w_m.$$

In order to construct φ and the domain $\mathcal{D}_n$ we study the distribution of the length k of the longest runs, the number m of these longest runs and the properties of the separating words w_i.

This analysis is based on probabilistic techniques and results from analytic combinatorics (generating functions, bootstrapping method, asymptotic estimation of the coefficients of a series, ...). Main references on these subjects are [9] and [11].

5   Algorithms and Experimental Results

In this section we give an algorithm to generate random Lyndon words of a given length n and use it to establish some experimental results about the length of the right factor in the standard factorization.

Our algorithms use Duval's algorithm [6], which computes in linear time the decomposition of a word into decreasing Lyndon words (see Theorem 1). So we assume that we have a algorithm named Duval(u) which produce the Lyndon words $l_1 \geq l_2 \geq \cdots \geq l_k$ such that

$$u = l_1 l_2 \ldots l_k.$$

Let the function Duval(string u, int k, array pos) be the function which computes the Lyndon decomposition of u by storing in an array pos of size k the positions of the factors.

There exists an algorithm SmallestConjugate(u), proposed by Booth [13] and [4], that computes the smallest conjugate a random Lyndon word of length n in linear time. We use it to make a reject algorithm which is efficient to generate randomly a Lyndon word of length n:

```
RandomLyndonWord(n)                // return a random Lyndon word
string u, v;
  do
    u = RandomWord(n);             // u is a random word of A^n
    v = SmallestConjugate(u);      // v is the smallest conjugate of u
  until (length(v) == n);          // v is primitive
  return v;
```
The algorithm `RandomLyndonWord` computes uniformly a Lyndon word.

Lemma 2. *The average complexity of* `RandomLyndonWord(n)` *is linear.*

Proof. Each execution of the do ... until loop is done in linear time. The condition is not satisfied when u is a conjugate of a power v^p with $p > 1$. This happens with probability $O(\frac{n}{2^{n/2}})$. Thus the loop is executed a bounded number of times in the average.

Lemma 3. *Let* $l = au$ *be a Lyndon word of length greater or equal to 2 starting with a letter* a. *Let* $l_1 \ldots l_k$ *be the Lyndon factorization of* u. *The right factor of* l *in its standard factorization is* l_k.

Proof. By Theorem 1, l_k is the smallest suffix of u, thus it is the smallest proper suffix of l.

The algorithm to compute the right factor of a Lyndon word l such that $|l| \geq 2$ is the following:

```
RightFactor(string l[1..n])
array pos;
int k;
  pos = Duval(l[2..n], k, pos);
  // omit the first letter a and apply Duval's algorithm
  return l[pos[k]..n];   // return the last factor
```

This algorithm is linear in time since Duval's algorithm is linear.
Figures 1 and 2 present some experimental results obtained with our algorithms.

6   Open Problem

The results obtained in this paper are only the first step toward the average case-analysis of the Lyndon tree. The Lyndon tree $T(w)$ of a Lyndon word w is recursively built in the following way

- if w is a letter, then $T(w)$ is a external node labeled by the letter.
- otherwise, $T(w)$ is an internal node having $T(u)$ and $T(v)$ as children where the standard factorization of w is $u \cdot v$.

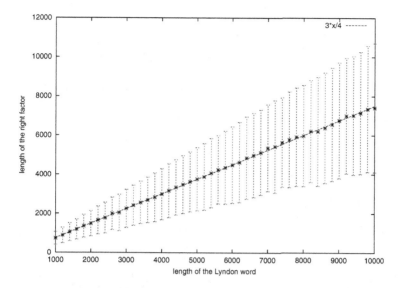

Fig. 1. Average length of the right factor of random Lyndon words with lengths from 1,000 to 10,000. Each plot is computed with 1,000 words. The error bars represents the standard deviation.

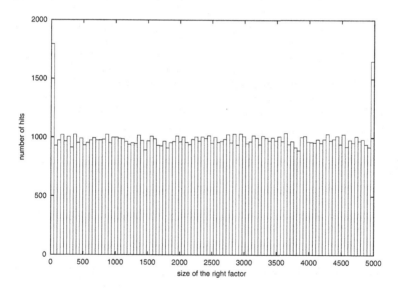

Fig. 2. Distribution of the length of the right factor. We generated 100,000 random Lyndon words of length 5,000.

This structure encodes a nonassociative operation, either a commutator in the free group [5], or a Lie bracketing [12]; both constructions leads to bases of the free Lie algebra.

In order to study the height of the tree obtained from a Lyndon word by successive standard factorizations, it would be very interesting to get more precise informations about the distribution of the right factors of words of $\mathcal{L}'_n$. Fig. 2 hints a very strong equi-repartition property of the length of the right factor over this set. This suggests a very particular subdivision process at each node of the factorization tree which needs further investigations.

References

1. J. Berstel and L. Boasson. The set of lyndon words is not context-free. *Bull. Eur. Assoc. Theor. Comput. Sci. EATCS*, 63:139–140, 1997.
2. J. Berstel and D. Perrin. *Theory of codes*. Academic Press, 1985.
3. J. Berstel and M. Pocchiola. Average cost of Duval's algorithm for generating Lyndon words. *Theoret. Comput. Sci.*, 132(1-2):415–425, 1994.
4. K. S. Booth. Lexicographically least circular substrings. *Inform. Process. Lett.*, 10(4-5):240–242, 1980.
5. K.T. Chen, R.H. Fox, and R.C. Lyndon. Free differential calculus IV: The quotient groups of the lower central series. *Ann. Math.*, 58:81–95, 1958.
6. J.-P. Duval. Factorizing words over an ordered alphabet. *Journal of Algorithms*, 4:363–381, 1983.
7. J.-P. Duval. Génération d'une section des classes de conjugaison et arbre des mots de Lyndon de longueur bornée. *Theoret. Comput. Sci.*, 4:363–381, 1988.
8. P. Flajolet, X. Gourdon, and D. Panario. The complete analysis of a polynomial factorization algorithm over finite fields. *Journal of Algorithms*, 40:37–81, 2001.
9. P. Flajolet and R. Sedgewick. Analytic combinatorics–symbolic combinatorics. Book in preparation, 2002. (Individual chapters are available as INRIA Research reports at http://www.algo.inria.fr/flajolet/publist.html).
10. S. Golomb. Irreducible polynomials, synchronizing codes, primitive necklaces and cyclotomic algebra. In *Proc. Conf Combinatorial Math. and Its Appl.*, pages 358–370, Chapel Hill, 1969. Univ. of North Carolina Press.
11. D. E. Knuth. The average time for carry propagation. *Indagationes Mathematicae*, 40:238–242, 1978.
12. M. Lothaire. *Combinatorics on Words*, volume 17 of *Encyclopedia of mathematics and its applications*. Addison-Wesley, 1983.
13. M. Lothaire. *Applied Combinatorics on Words*. Cambridge University Press, 2003. in preparation, chapters available at http://www-igm.univ-mlv.fr/~berstel/Lothaire.
14. R. C. Lyndon. On Burnside problem I. *Trans. American Math. Soc.*, 77:202–215, 1954.
15. D. Panario and B. Richmond. Smallest components in decomposable structures: exp-log class. *Algorithmica*, 29:205–226, 2001.
16. C. Reutenauer. *Free Lie algebras*. Oxford University Press, 1993.
17. F. Ruskey and J. Sawada. Generating Lyndon brackets: a basis for the n-th homogeneous component of the free Lie algebra. *Journal of Algorithms*, (to appear). Available at http://www.cs.uvic.ca/ fruskey/Publications/.

Circular Words Avoiding Patterns

James D. Currie* and D. Sean Fitzpatrick**

Department of Mathematics and Statistics
University of Winnipeg
Winnipeg, Manitoba R3B 2E9, Canada
currie@uwinnipeg.ca, dfitzpat@io.uwinnipeg.ca

Abstract. We introduce the study of circular words avoiding patterns. We prove that there are circular binary cube-free words of every length and present several open problems regarding circular words avoiding more general patterns.

Keywords: Combinatorics on words, words avoiding patterns, Thue-Morse word

1 Introduction

The study of words avoiding patterns can be traced to the beginning of the twentieth century, when Thue showed that there are infinite sequences over $\{a, b\}$ not containing any overlaps, and infinite sequences over $\{a, b, c\}$ not containing any squares [9].

Later Erdos initiated the search for words avoiding abelian squares [5], while Dejean generalized Thue's work to repetitions with fractional exponents[4]. Later Zimin[10], and independently Bean et al.[2] introduced the general study of words avoiding patterns.

A word w **encounters** pattern p if w has a subword which is the image of p under a non-erasing morphism. Otherwise we say that w **avoids** p. For example, the word $mississippi$ encounters $p = aabccb$, since $ississippi = h(aabccb)$, where $h(a) = iss$, $h(b) = i$, $h(c) = p$. However, $mississippi$ avoids xxx.

A pattern p is k-**avoidable** if for $\Sigma = \{1, 2, 3, \ldots, k\}$ there are arbitrarily long strings over Σ avoiding p. We say that p is **avoidable** if it is k-avoidable for some finite k. In [2,10] an algorithm for testing whether w is avoidable is given. However, it is not known whether k-avoidability is decidable.

When studying words avoiding patterns, it is usual to study infinite words (sequences) rather than arbitrarily long finite words. As usual in mathematics, the introduction of infinity smoothes out certain irregularities which disfigure finite words. Nevertheless, even in an infinite word, structural peculiarities can take place at the beginning of the word. For this reason, it is sometimes convenient to study bi-sequences. A different way to avoid "ugliness" at the endpoints of a finite word is to study **circular words**.

---

* The author's research was supported by an NSERC operating grant.
** The author's research was supported by an NSERC Undergraduate Research Award.

M. Ito and M. Toyama (Eds.): DLT 2002, LNCS 2450, pp. 319–325, 2003.
© Springer-Verlag Berlin Heidelberg 2003

We can conceive of a circular word as a labelling of the vertices of C_n, the cycle on n vertices. Removing an edge from C_n leaves us with a path on n vertices, and reading the labels along the path gives a word of length n. This is analogous to cutting a necklace of beads. (See Figure 1.) If each way of 'cutting' a circular word gives a word avoiding pattern p, then the circular word avoids pattern p. Thue showed that there are binary circular overlap-free words of length n for exactly those n of the form 2^m or $3 \cdot 2^m$ for some integer m.

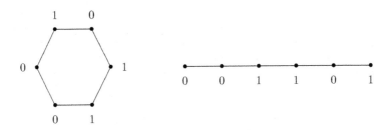

Fig. 1. An overlap-free circular word.

Although ordinary, linear words have been extensively studied, the same cannot be said for circular words. Basic open problems remain unanswered. Recently it was shown [3] that circular square-free words of length n exist over a three-letter alphabet for every $n \geq 18$. This contrasts with binary overlap-free words, which Thue showed were exponentially sparse. Again, Thue's result contrasts with binary circular cube-free words, since we prove the following:

Main Theorem: *Let n be a natural number. There is a binary circular cube-free word of length n.*

2   Preliminaries

If u and v are words, we write $u \leq v$ (equivalently, $v \geq u$) if u is a subword of v, that is, if $v = xuy$ for words x and y. Sometimes u is a subword of v in more than one way, that is, if $v = x_1 u y_1 = x_2 u y_2$ where $x_1 \neq x_2$. In this case we say that there is more than one occurrence of u in v. We write $u \leq_p v$ (equivalently, $v \geq_p u$) if u is a prefix of v, that is, if $v = uy$ for some word y. Again, we write $u \leq_s v$ (equivalently, $v \geq_s u$) if u is a suffix of v, that is, if $v = xu$ for some word x. If w is a word, we denote by $|w|$ its length, that is, the number of letters in w. Thus $|01101001| = 8$.

A **cube** is a non-empty word of the form xxx for some word x. An **overlap** is a non-empty word of the form XXx where X is a word, and x the first letter of X. A word w is **cube-free (overlap-free)** if none of its subwords are non-empty cubes (overlaps). Word v is a **conjugate** of word w if there are words x and y such that $w = xy$ and $v = yx$. We say that w is a **circular cube-free (overlap-free) word** if all of its conjugates are cube-free (overlap-free).

Example 2.1. The set of conjugates of word 001101 is

$$\{001101, 011010, 110100, 101001, 010011, 100110\}.$$

Each of these is cube-free, so 001101 is a circular cube-free word. On the other hand, 0101101 is cube-free, but its conjugate 0101011 starts with the cube 010101. Thus 0101101 is not a circular cube-free word.

Let w be a word, $w = w_1 w_2 \ldots w_n$ where the w_i are letters. We say that w is **periodic** if for some k we have $w_i = w_{i+k}$, $i = 1, 2, \ldots, n - k$. We call k a **period** of w. The **exponent of** w is $|w|/k$, where k is the shortest period of w. For example, the exponent of 01010 is $5/2$. Squares, overlaps and cubes are periodic.

Our constructions deal with binary words, i.e. strings over $\{0, 1\}$. If w is a binary word we denote by $\overline{w}$ the binary complement of w, obtained from w by replacing 0's with 1's and vice versa. For example, $\overline{01101001} = 10010110$.

3   A Few Properties of the Thue-Morse Word

The Thue-Morse word t is defined to be $t = h^\omega(0) = \lim_{n \to \infty} h^n(0)$, where $h : \{0, 1\}^* \to \{0, 1\}^*$ is the substitution generated by $h(0) = 01$, $h(1) = 10$. Thus

$$t = 0110100110010110100101100110100 1 \cdots$$

The Thue-Morse word has been extensively studied. (See [6,8,9] for example.) We use the following facts about it:

1. Word t is overlap-free.
2. If w is a subword of t then so is $\overline{w}$. (The set of subwords of t is closed under binary complementation.)
3. None of 00100, 01010 or 11011 is a subword of t.

Lemma 3.1. *Let n be an integer, $n \geq 2$. Word t contains a subword of the form $0u1$ with $|0u1| = n$.*

Proof: Since t is overlap-free, it is not periodic, whence $n - 1$ is not a period of t. Thus t contains a subword of length n for which initial letter and final letter are different. To finish the proof, we note that the set of subwords of t is closed under binary complementation.□

Lemma 3.2. *Suppose that $u \leq xxx$ and $|u| \leq |x|$. Then is more than one occurrence of u in xxx.*

Proof: Since $u \leq xxx$ and $|u| \leq |x|$ we must have $u \leq xx$. Since there is more than one occurrence of xx in xxx, there is more than one occurrence of u in xxx.□

Lemma 3.3. *Let $u \le t$. Let $v \in \{0,1\}^*$. Suppose that uv contains a cube. Then for some prefix w of v, word uw ends in a cube xxx with $|x| \le |w|$.*

Proof: Let w be the shortest prefix of v such that uw contains a cube; possibly $w = v$. It follows that uw ends in xxx for some non-empty word x. We claim that $|x| \le |w|$; if $|x| > |w|$, write $x = x_0 y w$, where x_0 is the first letter of x. then $t \ge u \ge xxx_0$, an overlap. This is impossible, since t is overlap-free.□

Similarly one proves the following two lemmas:

Lemma 3.4. *Let $v \le t$. Let $u \in \{0,1\}^*$. Suppose that uv contains a cube. For some suffix w of u, word wv begins in a cube xxx with $|x| \le |w|$.*

Lemma 3.5. *Let $v \le t$. Let $u'', u' \in \{0,1\}^*$. Suppose that $u'' vu' = xxx$. Then $|x| \le |u'| + |u''|$.*

4   Cube-Free Words

Lemma 4.1. *Let $u_2 = 1010010$. Let $v = h^2(0\nu)$ where $0\nu \le t$. Then $u_2 v$ is cube-free.*

Proof: Word v commences 0110. Also, $v = h^2(0\nu) \le h^2(t) = t$. For each non-empty suffix w of u_2, let p_w be the prefix of wv with period at most $|w|$ and maximal exponent. Here are the w and p_w:

w	v	p_w	exponent of p_w
0	0110⋯	00	2
10	0110⋯	1	1
010	0110⋯	01001	5/3
0010	0110⋯	001001	2
10010	0110⋯	1001001	7/3
010010	0110⋯	01001001	8/3
1010010	0110⋯	1010	2

Each p_w has exponent less than 3; thus, no wv has a prefix $p = xxx$ where $|x| \le |w|$. It follows from Lemma 3.4 that $u_2 v$ is cube-free.□

The following lemma is proved similarly.

Lemma 4.2. *Let $v_2 = 010010$. Let $u = h^2(v1)$ where $v1 \le t$. Then uv_2 is cube-free.*

Lemma 4.3. *Let $g_2 = 010010$. Let $u = h^2(v1)$, $v = h^2(0\nu)$, where $v1, 0\nu \le t$. Then $ug_2 v$ is cube-free.*

Proof: Word u ends in 1001, while v commences 0110. Suppose for the sake of getting a contradiction that $xxx \le ug_2 v$. Because $g_2 = v_2$ and $1g_2 = u_2$, it follows from Lemmas 4.1 and 4.2 that $1g_2v$ and ug_2 are cube-free. We may thus assume $xxx = v''g_2u'$ where $01 \leq_s v'' \leq_s v$, $0 \leq_p u' \leq_p u$. This means that $01g_20 = 010100100 \leq xxx$.

It follows that word $e_2 := 01010 \leq_p 01g_20$ is a subword of xxx. This overlap is not a subword of t. Since e_2 is not a subword of t, and u, v are subwords of t, we conclude that e_2 occurs exactly once in ug_2v, and hence exactly once in

xxx. By the contrapositive of Lemma 3.2, we must have $|x| < |e_2| = 5$. Since xxx has period $|x|$, we conclude that subword 010100100 of xxx must have period 1, 2, 3 or 4. However, this is not the case, giving a contradiction.

We conclude that ug_2v is cube-free.$\square$

Theorem 4.4. Let $u = h^2(0v1)$ where $0v1 \leq t$. Then ug_2 is circular cube-free.

Proof: Word u begins with 0110 and ends with 1001. Suppose that some conjugate of ug_2 contains a cube xxx. By Lemma 4.3 xxx cannot be a subword of g_2u or of ug_2.

Case 1: Suppose that $xxx = g''ug'$ where $g_2 = g'gg''$, some g'', $g' \neq \epsilon$. By Lemma 3.5, $|x| \leq |g''| + |g'|$. Since $3|x| = |xxx| = |g''ug'| \geq 1 + |0110| + |1001| + 1 = 10$, we have $|x| \geq 4$.

Suppose that $|x| = 4$. Notice that $|u| = |h^2(0v1)|$ is a multiple of 4. Since $|u| < |xxx| = 12$, we have $u = 01101001$. This means that $|g''| + |g'| = 12 - |u| = 4$, implying $|g''|, |g'| \leq 3$. Thus $xxx = g''ug' \leq 010\ 01101001\ 010$. This is quickly seen to be impossible.

If $|x| = 5$, then $|xxx| = 15$, so that $|u| = 15 - |g''| - |g'| \geq 9$. Since $|u| = |h^2(0v1)|$ is a multiple of 4, we conclude that $|u| \geq 12$. However, now $5 = |x| \leq |g''| + |g'| = 15 - |u| \leq 3$, a contradiction.

If $|x| = 6$, then $xxx = 18$, so that $|u| = 18 - |g''| - |g'| \geq 12$. However, now $6 = |x| \leq |g''| + |g'| = 18 - |u| \leq 6$. This forces $|g''| + |g'| = 6$, so that $g_2 = g'g''$.

If $|g'| \geq 3$, then $ug' \geq e_2 = 01010$. As earlier, e_2 can appear at most once in $g''ug'$. However, $|e_2| = 5 < 6 = |x|$, so that e_2 appears twice by Lemma 3.2, a contradiction. We conclude that $|g'| \leq 2$. If $|g''| \geq 4$, then $f_2 := 00100$ is a subword of xxx. Since f_2 is not a subword of t, word f_2 occurs at most once in xxx. Again $|f_2| = 5 < 6 = |x|$, giving a contradiction. We conclude that $|g''| \leq 3$. Now, however, $6 = |g'| + |g''| \leq 2 + 3$, a contradiction.

Case 2: Suppose that $xxx = u''g_2u'$ where $u \geq_p u' \neq \epsilon$, $u \geq_s u'' \neq \epsilon$. Since $0 \leq_p u'$, $f_2 = 00100 \leq g_2u' \leq xxx$. Word f_2 occurs at most once in xxx. By Lemma 3.2, $|x| < |f_2| = 5$. This implies that subword $1g_20$ of xxx must have period 1, 2, 3 or 4. Inspection shows that none of these periods occurs, which is a contradiction.

We conclude that ug_2v is circular cube-free.$\square$

Define words u_i, v_i, e_i, f_i and g_i, $i = 0, \ldots, 3$ as follows:

i	u_i	v_i	e_i	f_i	g_i
0	010011	001101	00100	11011	0011
1	010		0 00100	00100	0
2	1010010	010010	01010	00100	010010
3	1010		010 01010	01010	010

Notice that $|g_i| \equiv i \pmod 4$ for each i.

Theorem 4.5. *Let* $u = h^2(0v1)$ *where* $0v1 \leq t$. *Then* ug_i *is circular cube-free, for* $i = 0, 1, 2, 3$.

Proof: We have seen the proof of this theorem in the case that $i = 2$. The other cases are proved analogously, using u_i, v_i, e_i, f_i and g_i in place of u_2, v_2, e_2, f_2 and g_2, and using the corresponding analogs of Lemmas 4.1, 4.2 and 4.3.□

Main Theorem: *Let* n *be a positive integer. There is a binary circular cube-free word of length* n.

Proof: The result is true for $n < 14$. Suppose $n \geq 14$, and suppose that $n \equiv i \pmod 4$. Let $m = (n - |g_i|)/4 \geq 2$. By Lemma 3.1, t contains a subword $0v1$ of length m. Let $u = h^2(0v1)$. By Theorem 4.5 the word ug_i is circular cube-free; finally, $|ug_i| = 4m + |g_i| = n$□.

5   Discussion and Open Problems

The following conjecture is due to V. Linek [7]

Conjecture 5.1. *Let* n *be a positive integer. There is a binary circular cube-free subword of* t *of length* n.

Computer search shows that such subwords certainly exist for $n \leq 200$.

Conjecture 5.2. *If* p *is* k*-avoidable, then there are arbitrarily long circular words on* k *letters avoiding* p.

If p is a pattern then any word of length at most $|p| - 1$ is a circular p-free word. Even the following weak version of Conjecture 5.2 is open, as far as we know:

Conjecture 5.3. *If* p *is* k*-avoidable, then there is a circular word of length* $|p|$ *on* k *letters avoiding* p.

The number of cube-free binary words of length n, and the number of square-free ternary words of length n grow exponentially as a function of n. On the other hand, the number of overlap-free binary words of length n grows only polynomially.

Conjecture 5.4. *Let* p *be* k*-avoidable.*

1. *If the number of words on k letters avoiding p of length n grows exponentially with n, then for some N_0 there are circular p-free words on k letters for every length $n > N_0$.*

2. *If the number of words on k letters avoiding p of length n grows only polynomially with n, the possible lengths for circular p-free words on k letters has density 0 in the set of positive integers.*

Remark 5.5. The set of circular words over $\{0, 1, 2, 3\}$ avoiding $w_\Delta = abxbcycazbawcb$ [1] should provide evidence for this conjecture.

References

1. Kirby A. Baker, George. F. McNulty & Walter Taylor, Growth problems for avoidable words, Theoret. Comput. Sci. **69** (1989), no. 3, 319–345; MR **91f**:68109.

2. Dwight R. Bean, Andrzej Ehrenfeucht & George McNulty, Avoidable Patterns in Strings of Symbols, *Pacific J. Math.* **85** (1979), 261–294.

3. James D. Currie, There are ternary circular square-free words of length n for $n \geq 18$, *Elec. J. Comb.* **9(1)** N10.

4. Françoise Dejean, Sur un théorème de Thue, J. Combin. Theory Ser. A **13** (1972), 90–99.

5. Paul Erdös, Some unsolved problems, Magyar Tud. Akad. Mat. Kutato. Int. Kozl. **6** (1961), 221–254.

6. Earl D. Fife, Binary sequences which contain no BBb, Trans. Amer. Math. Soc. **261** (1980), 115–136; MR **82a**:05034

7. Vaclav Linek, Personal communication.

8. Marston Morse & Gustav A. Hedlund, Symbolic dynamics I, II, Amer. J. Math. **60** (1938), 815–866; **62** (1940), 1–42; MR **1**, 123d.

9. Axel Thue, Über unendliche Zeichenreihen, Norske Vid. Selsk. Skr. I. Mat. Nat. Kl. Christiana (1912), 1–67.

10. A. Zimin, Blocking sets of terms, Mat. Sb. (N.S.) **119 (161)** (1982); Math. USSR Sbornik **47** (1984), 353–364.

Safety Verification for Two-Way Finite Automata with Monotonic Counters*

Oscar H. Ibarra[1,**], Zhe Dang[2], and Zhi-Wei Sun[3]

[1] Department of Computer Science
University of California
Santa Barbara, CA 93106, USA
[2] School of Electrical Engineering and Computer Science
Washington State University
Pullman, WA 99164, USA
[3] Department of Mathematics
Nanjing University
Nanjing 210093, China

Abstract. We look at a model of a two-way nondeterministic finite automaton augmented with monotonic counters operating on inputs of the form $a_1^{i_1}...a_n^{i_n}$ for some fixed n and distinct symbols $a_1, ..., a_n$, where $i_1, ..., i_n$ are nonnegative integers. Our results concern the following Presburger safety verification problem: Given a machine M, a state q, and a Presburger relation E over counter values, is there $(i_1, ..., i_n)$ such that M, when started in its initial state on the left end of the input $a_1^{i_1}...a_n^{i_n}$ with all counters initially zero, reaches some configuration where the state is q and the counter values satisfy E? We give positive and negative results for different variations and generalizations of the model (e.g., augmenting the model with reversal-bounded counters, discrete clocks, etc.). In particular, we settle an open problem in [10].

1   Introduction

Recently, there has been significant progress in automated verification techniques for finite-state systems. One such technique, called model-checking [5,6,22,20,23], explores the state space of a finite-state system and checks that a desired temporal property is satisfied. In recent years, model-checkers like SMV [20] and SPIN [16] have been successful in some industrial-level applications. Successes in finite-state model-checking have inspired researchers to develop model-checking techniques for infinite-state systems (e.g., arithmetic programs that contain integer variables and parameters).

However, for infinite-state systems, a fundamental problem - the decidability problem - should be addressed before any attempts to develop automatic verification procedures. It is well known that in general, it is not possible to

---

* The work by Oscar H. Ibarra has been supported in part by NSF Grants IIS-0101134 and CCR02-08595.

** Corresponding author (`ibarra@cs.ucsb.edu`).

M. Ito and M. Toyama (Eds.): DLT 2002, LNCS 2450, pp. 326–338, 2003.

automatically verify whether an arithmetic program with two integer variables (i.e., a Minsky machine) is going to halt [21]. Therefore, a key aspect of the research on infinite-state system verification is to identify what kinds of practically important infinite-state models are decidable with respect to a particular form of properties.

Counter machines are considered a natural model for specifying reactive systems containing integer variables. They have also been found to have a close relationship to other popular models of infinite-state systems, such as timed automata [1]. Since general counter machines like Minsky machines are undecidable for verification of any nontrivial property, studying various restricted models of counter machines may help researchers to obtain new decidable models of infinite-state systems and identify the boundary between decidability and undecidability.

In this paper, we study a model of a two-way nondeterministic finite automaton augmented with monotonic counters operating on inputs of the form $a_1^{i_1}...a_n^{i_n}$ for some fixed n and distinct symbols $a_1, ..., a_n$, where $i_1, ..., i_n$ are nonnegative integers. Our results concern the following Presburger safety verification problem: Given a machine M, a state q, and a Presburger relation E over counter values, is there $(i_1, ..., i_n)$ such that M, when started in its initial state on the left end of the input $a_1^{i_1}...a_n^{i_n}$ with all counters initially zero, reaches some configuration where the state is q and the counter values satisfy E? We give positive and negative results for different variations and generalizations of the model (e.g., augmenting the model with reversal-bounded counters, discrete clocks, etc.). In particular, we settle an open problem in [10]. In the sequel, we refer to the tuples of counter values that can be reached by M (on all inputs) as the reachable set at state q, or simply reachability set when q is understood.

Some of our decidable results are quite powerful in the sense that the underlying machines may have nonsemilinear (i.e., non Presburger) reachability sets. This is in contrast to most existing decidable results in model-checking infinite-state systems (e.g., pushdown systems [4]) that essentially have semilinear reachability sets. For instance, consider a nondeterministic transition system M that is equipped with a number of monotonic counters $\boldsymbol{X} = X_1, ..., X_k$ and a number of nonnegative integer parameterized constants $A_1, ..., A_n$. M has a finite number of control states. On a transition from one control state to another, M is able to increment the counter synchronously (i.e., $\boldsymbol{X} := \boldsymbol{X} + \delta$ that increments every X_i by the same amount δ, where $\delta = 0, 1$, or some A_j), test the truth value of a Presburger formula on $\boldsymbol{X}$ and $A_1, \ldots, A_n$, or reset a counter to 0. M is a form of a generalized timed automaton (with discrete clocks, Presburger enabling conditions, and parameterized durations) [10] in disguise. We further require that each counter is reset for at most a fixed number of times (such as 10). We show in this paper that the safety verification problem is decidable for such M. It is easy to see that M, with these reset-bounded counters, can exhibit a reachability set satisfying $A_j | X_i$, which is not Presburger.

2   Preliminaries

Let $\mathbf{N}$ be the set of nonnegative integers and $c \in \mathbf{N}$. A c-counter machine is a two-way nondeterministic finite automaton with input endmarkers (two-way NFA) augmented with c counters, each of which can be incremented by 1, decremented by 1, and tested for zero. We assume, w.l.o.g., that each counter can only store a nonnegative integer, since the sign can be stored in the states. If r is a nonnegative integer, let 2NCM(c,r) denote the class of c-counter machines where each counter is r reversal-bounded, i.e., it makes at most r alternations between nondecreasing nonincreasing modes in any computation; e.g., a counter whose values change according to the pattern 0 1 1 2 3 4 $\underline{4}$ $\underline{3}$ 2 1 $\underline{0}$ $\underline{1}$ $\underline{1}$ $\underline{0}$ is 3-reversal, where the reversals are underlined. For convenience, we sometimes refer to a machine in the class as a 2NCM(c,r). Note that the counters in a 2NCM(c,0) are monotonic (i.e., nondecreasing). A 2NCM(c,r) is *finite-crossing* if there is a positive integer c such that in any computation, the input head crosses the boundary between any two adjacent cells of the input no more than c times. Note that a 1-crossing 2NCM(c,r) is a one-way nondeterministic finite automaton augmented with c r-reversal counters. 2NCM(c) will denote the class of c-counter machines whose counters are r-reversal bounded for some given r. For deterministic machines, we use 'D' in place of 'N'. If M is a machine, $L(M)$ denotes the language it accepts.

We will need the following result from [17].

Theorem 1. *There is a* fixed r *such that the emptiness problem for 2DCM(2,r) over bounded languages is undecidable.*

We will also need the following results.

Theorem 2. *The emptiness problem is decidable for the following machine classes: (a) 2DCM(1) [18], (b) 2NCM(1) over a bounded language [8], (c) 2NCM(c) over a unary alphabet (i.e., over a bounded language on 1 letter) for every c [18], and (d) finite-crossing 2NCM(c) for every c [17,15].*

Let Y be a finite set of variables over integers. For all integers a_y, with $y \in Y$, b and c (with $b > 0$), $\sum_{y \in Y} a_y y < c$ is an *atomic linear relation* on Y and $\sum_{y \in Y} a_y y \equiv_b c$ is a *linear congruence* on Y. A *linear relation* on Y is a Boolean combination (using $\neg$ and $\wedge$) of atomic linear relations on Y. A *Presburger formula* on Y is the Boolean combination of atomic linear relations on Y and linear congruences on Y. A set P is *Presburger-definable* if there exists a Presburger formula $\mathcal{F}$ on Y such that P is exactly the set of the solutions for Y that make $\mathcal{F}$ true. It is well known that Presburger formulas are closed under quantification.

Let k be a positive integer. A subset S of $\mathbf{N}^k$ is a *linear set* if there exist vectors $v_0, v_1, \ldots, v_t$ in $\mathbf{N}^k$ such that $S = \{v \mid v = v_0 + a_1 v_1 + \ldots + a_t v_t, \forall 1 \leq i \leq t, a_i \in \mathbf{N}\}$. S is a *semilinear set* if it is a finite union of linear sets. It is known that S is a semilinear set iff S is Presburger-definable [14].

Let Σ be an alphabet consisting of k symbols $a_1, \ldots, a_k$. For each string (word) w in Σ^*, we define the *Parikh map* $p(w)$ of w as follows: $p(w) = (i_1, \ldots, i_k)$, where i_j is the number of occurrences of a_j in w. If L is a subset of Σ^*, the *Parikh map* of L is defined by $p(L) = \{p(w) \mid w \in L\}$. L is a *semilinear language* if its Parikh map $p(L)$ is a semilinear set. Obviously, if the languages accepted by machines in a class C are effectively semilinear (that is, for each M in C, the semilinear set $p(L(M))$ can be effectively computed), then the emptiness problem for C is decidable. We will need the following theorem from [17]:

Theorem 3. *Let M be a finite-crossing 2NCM(c). Then $p(L(M))$ is a semilinear set effectively computable from M.*

Note the result above is not true for machines that are not finite-crossing. For example, a 2DCM(1,1) can recognize the language $\{0^i 1^j \mid i \text{ divides } j\}$, which is not semilinear.

3   Machines with Monotonic Counters over Bounded Languages

3.1   One-Way Input

Let M be a nondeterministic one-way finite automaton augmented with a number of reversal-bounded counters. An input to M is of the form $a_1^{i_1}...a_n^{i_n}$ for some fixed n and distinct symbols $a_1, ..., a_n$, where $i_1, ..., i_n$ are nonnegative integers. It is known that the Presburger safety verification problem for these machines is decidable [17], even when *one* of the counters is unrestricted in that it can change mode from nondecreasing to nonincreasing and vice-versa an unbounded number of times. The machine M can further be generalized by allowing some of the reversal-bounded counters to reset to zero. Thus M now has one unrestricted counter, some "reset counters", and some reversal-bounded counters. There is no fixed bound on the number of times a reset counter can reset (but between two resets, the counter is reversal-bounded). Reset counters are quite useful in modeling clocks in real-time systems [1]. If each counter is monotonic (i.e., 0-reversal-bounded) between two resets, the problem is decidable [7]. (The decidability holds even when the machine is allowed to have discrete clocks which operate and can be tested as in a timed automaton [1].) For the general case, the aforementioned machines with at least four reset counters is undecidable for the problem. This can be shown by a reduction to two-counter machines, noticing that an unrestricted counter can be simulated by two reset counters, each of which is 1-reversal-bounded between two resets.

Other generalizations of the model which allows linear constraints as counter tests have also been shown decidable for the problem [19].

3.2   Two-Way Input

We now study the more complicated model when the input to the machine is two-way. So now M is a nondeterministic two-way finite automaton augmented with monotonic counters $C_1, ..., C_k$ operating on an input of the form $a_1^{i_1}...a_n^{i_n}$, with left and right endmarkers. There is an interesting model equivalent to this model.

Let M' be a nondeterministic machine with 1 unrestricted counter U and k monotonic counters $C_1, ..., C_k$. M' has *no* input tape. There are n parameterized constants $A_1, ..., A_n$. Initially, all the counters are zero. During the computation, M' can check if the unrestricted counter U is equal to 0 or equal to A_i ($i = 1, .., n$). Moreover, U is not allowed to exceed $max(A_1, ..., A_n)$.

M' is equivalent to M. An input $a_1^{i_1}...a_n^{i_n}$ (with left and right endmarkers) to M corresponds to the parameterized constants, by interpreting $i_1, (i_1 + i_2), ..., (i_1 + i_2 + ... + i_n)$ to be the values of $A_1, A_2, ..., A_n$, respectively. This correspondence assumes that $A_1 \leq A_2 ... \leq A_n$. However, since there are only a finite number of these orderings, this assumption can be made without loss of generality. Clearly, M starting on the left endmarker corresponds to the unrestricted counter U of M' starting at zero. M moving right (left) on the input corresponds to M' incrementing (decrementing) U. M' can determine if M is on the boundary between a_i and a_{i+1} by checking if U is equal to A_i. In view of the equivalence of the two models, will use them interchangeably in the sequel.

Let E be a Presburger relation on the values $c_1, ..,c_k$ of the monotonic counters and q be a state of M. We say that M *satisfies* E if there is $(i_1, ..., i_n)$ such that M, when started in its initial state on the left end of the input $a_1^{i_1}...a_n^{i_n}$ with all counters initially zero, reaches some configuration where the state is q and the counter values satisfy E. Note that we do not assume that M necessarily halts. So, in fact, a configuration satisfying E may be an intermediate configuration in some computation. Also, there may be several configurations satisfying E. In the paper, we usually do not explicitly specify the state q when E is satisfied, since its specification is usually obvious from construction.

An *atomic equality relation* is a relation $c_i = c_j$, $i \neq j$. A relation E is an *equality relation* if it is a conjunction of atomic equality relations and no c_i is involved in more than one atomic relation in E. We denote such a relation by E_e.

Theorem 4. *Let M be as specified above. Let E and E_e be a Presburger relation and an equality relation, respectively.*

1. *There is a fixed k such that it is undecidable whether M with k monotonic counters satisfies E_e.*
2. *When $n = 1$ (i.e., there is only one parameterized constant A_1), it is decidable whether M satisfies E.*
3. *When M is deterministic and always halts, it is decidable whether M satisfies E.*

Proof. We first prove Part 1. From Theorem 1, there is a fixed r such the emptiness problem for 2DCM(2,r) over bounded languages is undecidable. Let M be such an automaton. Clearly, we can convert M to an equivalent automaton M' which has $2(1 + \frac{r-1}{2})$ 1-reversal counters where each counter starts at zero, and on input w, M' accepts if and only if it halts with all counters zero. To insure this, we can add to M' a dummy counter which is incremented at the beginning and only decremented, i.e., becomes zero when the input is accepted. Suppose M' has k 1-reversal counters, $C_1, ..., C_k$ (one of these is the dummy counter). Note that k is fixed since r is fixed. We modify M' to a nondeterministic automaton M'' with $2k$ monotonic counters: $C_1^+, C_1^-, ..., C_k^+, C_k^-$, where C_i^+ and C_i^- are associated with counter C_i. M'' on input w simulates the computation of M', first using counter C_i^+ to simulate C_i when the latter is in a nondecreasing mode. When counter C_i reverses (thus entering the nonincreasing mode), M'' continues the simulation but using counter C_i^-: incrementing this counter when M' decrements C_i. At some time during the computation (which may be different for each i), M' guesses that C_i has reached the zero value. From that point on, M'' will no longer use counters C_i^+ and C_i^-, but continues the simulation. When M'' has guessed that $C_i^+ = C_i^-$ for all i's (note that for sure the two counters corresponding to the dummy counter are equal), it halts. Clearly, w is accepted by M' if and only if M'' on w can reach a configuration with $C_i^+ = C_i^-$ for $i = 1, ..., k$ (this is relation E_e). As noted earlier, M'' on input $a_1^{i_1}...a_n^{i_n}$ can be interpreted as a machine M'''' with an unrestricted counter U and n parameterized constants $A_1, ..., A_n$ whose values are set to $(1+i_1), (1+i_1+i_2), ..., (1+i_1+i_2+...+i_n)$, respectively. The movement of the input head of M'' is simulated by the unrestricted counter U.

For Part 2, given M with monotonic counters $C_1, ..., C_k$, we construct a two-way nondeterministic finite automaton with reversal-bounded counters over a unary input (with endmarkers) M'. M' simulates M faithfully, with the input head of M' simulating the unrestricted counter U of M and k counters simulating the C_i's. At some point, M' guesses that the values of the monotonic counters satisfy the relation E. M' then uses another set of counters to verify that this is the case, and accepts. The result follows from the decidability of the emptiness problem for two-way nondeterministic finite automata augmented with reversal-bounded counters over a unary alphabet[18], and the fact that a Presburger relation on $C_1, ..., C_k$ can be verified by additional reversal-bounded counters.

For Part 3, since M is deterministic and always halts, M' (i.e., the M' in the proof of Part 1) is finite-crossing on any input. The result then follows from the decidability of the emptiness problem for finite-crossing finite automata with reversal-bounded counters (Theorem 3), and the fact that Presburger relations can be verified by reversal-bounded counters.   ∎

In Theorem 4, the relations E and E_e are defined over monotonic counters $C_1, ..., C_k$. In fact, the theorem still holds when the relations are defined over $C_1, ..., C_k, U$, and $A_1, ..., A_n$. This is because $n+1$ additional monotonic counters can be added to "store" the values for U and $A_1, ..., A_n$ at the time when the test for E is performed.

Consider the following restriction on the counters: $C_1, ..., C_k$ are *ordered* in that all the increments of C_i are followed by those of C_j whenever $1 \leq i < j \leq k$. Then it can be shown, using Theorem 2(b) directly, that it is decidable whether M satisfies E.

Open Question: In Part 3, if we do not assume that M always halts, we do not know if the problem is still decidable, since the machine is no longer finite-crossing and, in fact, the set of tuples of reachable values of the monotonic counters is not semilinear in general. For example, consider a deterministic machine M with a two-way unary input and three monotonic counters C_1, C_2, C_3. On input x of length n, M initially stores n in C_1. Then M makes (left-to-right and right-to-left) sweeps of the input, adding 1 to C_2 and n to C_3 after every sweep. M iterates this process without halting. Let $Q = \{(i, j, k) \mid$ there is an instance in the computation where the values of the 3 counters are $i, j, k\}$. Thus Q is the set of all "reachable" counter values. Then Q is not semilinear; otherwise, since semilinear sets are closed under intersection, Q intersected with the semilinear set $\{(n, n, m) \mid n, m$ in $\mathbf{N}\}$ will yield $\{(n, n, n^2) \mid n$ in $\mathbf{N}\}$, which is not semilinear.

4   Generalized Counter Timed Systems

Timed automata [1] are a standard model for studying real-time systems. A timed automaton can be considered as a finite automaton augmented with a finite number of clocks. In this paper, we consider clocks taking values in $\mathbf{N}$ (i.e., clocks are discrete). In a timed automaton, a clock can progress (i.e., $x := x + 1$) or reset ($x := 0$). However, one clock progresses iff all the clock progresses. In addition, clocks can participate tests in the form of comparisons between the difference of two clocks against a constant (e.g., $x - y > 5$).

The result in Part 2 of Theorem 4 can be further generalized. In [3], pushdown timed systems (these are timed automata equipped with a pushdown stack) with "observation" counters were studied. The purpose of the counters is to record information about the evolution of the system and to reason about certain properties (e.g., number of occurrences of certain events in some computation). The counters do not participate in the dynamic of the system, i.e., they are never tested by the system. A transition specifies for each observation counter an integral value (positive, negative, zero) to be added to the counter. Of interest are the values of the counters when the system reaches a specified configuration. Here we only consider the case when the clocks of the system are discrete. It follows from the results in [3] (see also [9]) that these systems have decidable safety properties, and the tuples of "reachable" values of observation counters is Presburger.

A special case is when the pushdown stack is replaced by an unrestricted counter U. Call this a counter timed system. Then the results in [3] obviously hold. Now we generalize the counter timed system in the following way:

1. Instead of observation counters, the system has a finite number of reversal-bounded counters, and these counters participate in the dynamic of the system (i.e., they can be tested by the system).
2. The system is parameterized in the sense that there is a parameterized constant A, and during the computation, the system can check if the unrestricted counter U is equal to A (in addition to being able to test if U is zero).
3. U cannot exceed A.

Note that an observation counter can be simulated by two reversal-bounded counters: one counter keeps track of the increases and another counter keeps track of the decreases. When the system guesses that it has reached the target configuration, the difference of the counter values can be computed in one of the counters, and the sign of the difference can be specified in the other counter, which is set to 0 for negative and 1 for positive. Call a system S defined above a *generalized counter timed system*. Such an S can model part of a memory management system as follows:

1. S dynamically allocates/retrieves memory to/from k processes. The total amount of memory available is equal to A (the parameterized constant).
2. S uses the unrestricted counter U to keep track of the available memory, decrementing (resp. incrementing) U when a process requests (releases) memory. Note that S can check if U is zero or equal to A.
3. Among the reversal-bounded counters, we might use $2k$ monotonic counters for the following purpose: Two counters P_i^+ and P_i^- are associated with process i. During the computation, P_i^+ (resp. P_i^-) is incremented if process i requests (resp. returns) memory from (resp. to) the system.
4. Other reversal-bounded counters can be used for other purposes (e.g., each process can use a monotonic counter to keep track of the "cost" of requesting memory; assume there is a unit charge per memory request).
5. As in a timed automaton, the (discrete) clocks are used to enforce timing constraints.

Questions concerning the relationships among the parameterized constant, unrestricted counter, reversal-bounded counters, and clocks can be asked, such as: Is it always the case that $U = A - (P_1^+ - P_1^-) + ... + (P_k^+ - P_k^-)$? Is it always the case that the total cost of process i is less than the total cost of process j? Is it always the case that S is fair (e.g., $3 \cdot |P_i^+ - P_j^+| < A$ for each i, j)? etc.

   Clearly, the set of tuples of reachable counter values is not semilinear (even when there are no clocks). However, we can still show that safety is decidable. Let S be a generalized counter timed system. A configuration α of S is a tuple (a, q, X, u, R), where a is the value of the parameterized constant A, q is the state, X is the array of clock values, u is the value of the unrestricted counter with $u \leq a$, and R is the array of values of the reversal-bounded counters.

Theorem 5. *It is decidable to determine, given a generalized timed counter system S and two Presburger sets I and B of configurations of S, whether there is a computation of S that starts from some configuration in I and reaches some configuration in B. Thus, it it decidable to determine whether S can go from an initial configuration in I to an "unsafe" configuration in B.*

Proof. We can view the clocks in S as counters, which we shall refer to as clock-counters. Now reversal-bounded counters can only perform *standard tests* (comparing a counter against 0) and *standard assignments* (increment or decrement a counter by 1, or simply nochange). But clock-counters in S have nonstandard tests and nonstandard assignments. This is because a clock constraint allows comparison between two clocks like $x_2 - x_1 > 5$. Note that using only standard tests, we can compare the difference of two clock-counter values against an integer like 5 by computing $x_2 - x_1$ in another counter. But each time this computation is done, it will cause at least one counter reversal, and the number of such tests during a computation can be unbounded. The clock progress $x := x + 1$ is standard, but the clock reset $x := 0$ is not. Since there is no bound on the number of clock resets, clock-counters may not be reversal-bounded (each reset causes a counter reversal).

We first prove an intermediate result. Denote by 2UNCM a 2NCM(c), for some c, over a unary alphabet. Thus, a 2UNCM is a nondeterministic two-way finite automaton over a unary alphabet augmented with finitely many reversal bounded counters. Define a semi-2UNCM as a 2UNCM which, in addition to reversal-bounded counters, has clock-counters that use nonstandard tests and nonstandard assignments as described in the previous paragraph.

Suppose we are given S and Presburger sets I and B. Let M_I and M_B be one-way nondeterministic finite automata with reversal-bounded counters accepting I and B, respectively. (Since I and B are Presburger, these machines can be constructed [17].) We construct a semi-2UNCM M which operates as follows. The input to M is a (which is a concrete instance of the parameterized constant A) with endmarkers. M starts by guessing and storing in its counters the components of the tuple (q, X, u, R), and checks that (a, q, X, u, R) is in I using M_I. M then simulates S starting in configuration $\alpha = (a, q, X, u, R)$. At some point during the simulation, M guesses that it has reached a configuration $\beta = (a, q', X', u', R')$ in B. M' then simulates M_B and accepts if β is in B.

The semi-2UNCM M can now be converted to an 2UNCM M' by simulating the clocks in M by reversal-bounded counters, using the techniques in [9]: First we construct a semi-2UNCM M_1 from M, where clock-counter comparisons are replaced by "finite table look-up", and therefore, nonstandard tests are not present in M_1. Then we eliminate the nonstandard assignments of the form $x := 0$ (clock resets) in M_1 and obtain the desired 2UNCM M'. The details can be found in [9]. The result now follows since emptiness is decidable for 2UNCM's from Theorem 2 (c).   ∎

The safety problem can be generalized as follows (called the multi-safety problem).

- **Given:** A system S and Presburger sets $I, B_1, ..., B_k$ of configurations of S, $k > 0$.
- **Question:** Is there a computation of S that starts from some configuration α in I and goes through some configurations $\alpha_1, ..., \alpha_k$, where α_i is in B_i for $1 \leq i \leq k$? (Note that that $\alpha_1, ..., \alpha_k$ can occur at different times.)

The theorem above easily generalizes to:

Theorem 6. *The multi-safety problem for generalized counter timed system is decidable.*

If S is a generalized counter timed system, define $Reach(S)$ to be the binary reachability set $Reach(S) = \{(\alpha, \beta) \mid$ starting in configuration α, S can reach configuration β}. As we have seen, the binary reachability $Reach(S)$ is not Presburger in general. However, we can give a nice characterization for it.

A 2UNCM is a nondeterministic two-way finite automaton over a unary alphabet augmented with reversal bounded counters. Define a k-2UNCM as a 2UNCM which, in addition to the two-way unary input, is provided with k distinguished counters, called input counters. (Note that the machine has other reversal-bounded counters aside from the input counters.) A tuple $(a, i_1, ..., i_k)$ in $\mathbf{N}^{k+1}$ is accepted by a k-2UNCM M if M, when started with a on the two-way input (i.e., a is the length of the unary input) and $i_1, ..., i_k$ in the k input counters, eventually enters an accepting state. Note that the input counters can be tested against zero and decremented during the computation, but cannot be incremented. The set accepted by M is the set of all $(k + 1)$-tuples accepted by M. A multi-2UNCM is a k-2UNCM for some k. The proof of the following lemma is straightforward from Theorem 2(c).

Lemma 1. *The emptiness problem for multi-2UNCM is decidable.*

Using the above lemma and the ideas in the proof of Theorem 5, we can show the following characterization for the binary reachability $Reach(S)$:

Theorem 7. *If S is a generalized counter timed system, we can effectively construct a multi-2UNCM M accepting $Reach(S)$.*

5 Synchronized Monotonic Counters

Let $A_1, ..., A_n$ be nonnegative integer parameterized constants, and $\mathbf{X} = \{X_1, ..., X_k\}$ be a set of nonnegative integer variables. The variables in $\mathbf{X}$ are incremented synchronously. We use $\mathbf{X} := \mathbf{X} + \delta$ to indicate $X_1 := X_1 + \delta; ... X_k := X_k + \delta$. Consider the following instruction set:

$\quad s : \mathbf{X} := \mathbf{X} + 0$,
$\quad s : \mathbf{X} := \mathbf{X} + 1$,
$\quad s : \mathbf{X} := \mathbf{X} + A_i$, for some $1 \leq i \leq n$,
$\quad s :$ if T then goto p else goto q,
and a nondeterministic choice,
$\quad s :$ goto p or q.
where s is an instruction label and test T is a Presburger formula over $A_1, ..., A_n$ and the variables in $\mathbf{X}$. A *synchronized counter program* P is a sequence of instructions drawn from the instruction set. A configuration of a program is a tuple consisting of an instruction label, values of $A_1, ..., A_n$, and values of variables in $\mathbf{X}$. Notice that, in P, there could be many different tests T's and in general P is nondeterministic.

Theorem 8. *It is decidable to determine, given a synchronized counter program P and two Presburger sets I and B of configurations of P, whether there is a computation of P that starts from some configuration in I and reaches some configuration in B. Thus, it it decidable to determine whether P can go from an initial configuration in I to an "unsafe" configuration in B.*

The proof of Theorem 8 uses Theorem 2 (b). Synchronized counter programs have a close relationship with timed automata (with discrete clocks). Though the standard model of timed automata is quite useful, more powerful clock tests are needed in order to model more complicated systems. In [10], we studied a class of generalized timed automata that allows Presburger clock tests. More precisely, a generalized timed automaton $\mathcal{A}$ is a synchronized counter program augmented with *reset instructions*:

$s : Reset(i),$

where the monotonic variable (i.e., a clock) X_i is reset to 0 and all the other monotonic variables do not change their values. $\mathcal{A}$ is *deterministic* if $\mathcal{A}$ does not contain nondeterministic choice instructions. Since generalized timed automata can simulate two-counter machines [1], though there are practical needs to use generalized timed automata to specify complex real-time systems, the "Turing computing" power of the model prevents automatic verification of simple properties such as reachability. Therefore, decidable approximation techniques are of great interest, since these techniques would provide a user some degree of confidence in analyzing and debugging complex real-time specifications. In contrast to the most direct approximation techniques [2,12,11,13] that bound the number of transitions to a fixed number, the approximation techniques presented in [10] restrict the clock behaviors but do not necessarily bound the number of transition iterations to be finite.

There are three approximation techniques in [10]. The r-reset-bounded approximation bounds the number of clock resets by a given positive integer r for each clock. The n-bounded approximation requires that, for each clock, the clock value is less than a given positive integer n every time the clock resets (but after the last reset, the clock can go unbounded.). Combining these two, the $\langle n, r \rangle$-crossing-bounded approximation requires that, for each clock, there are at most r times that the clock resets after its value is greater or equal to n. It was shown in [10] that, when $\mathcal{A}$ is deterministic, Theorem 8 holds for $\mathcal{A}$ under any one of the three approximations. The case when $\mathcal{A}$ is nondeterministic was left open in [10]. The following result settles the open case.

Theorem 9. *It is decidable to determine, given a generalized timed automaton $\mathcal{A}$ under any one of the three approximations, and two Presburger sets I and B of configurations of $\mathcal{A}$, whether there is a computation of $\mathcal{A}$ that starts from some configuration in I and reaches some configuration in B.*

Proof. We only consider the r-reset-bounded approximation; the other two approximations are similar (see [10]). In an r-reset-bounded execution of $\mathcal{A}$, each clock is reset for at most r times. Therefore, the execution can be partitioned into a concatenation of at most $r \times |\boldsymbol{X}|$ phases such that, in each phase, there is no clock reset, and any two consecutive phases are connected with a clock reset instruction. Notice that a phase can be modeled as an execution of a synchronized counter program (which does not contain resets). The theorem follows by generalizing Theorem 8 into a "concatenation" of the program for each phase. ∎

6   Conclusion

We introduced a model of a two-way nondeterministic finite automaton augmented with monotonic counters operating on inputs over a bounded language for investigating the verification properties of infinite-state systems. We then proved positive and negative results on the safety verification properties for different variations and generalizations of the model (e.g., augmenting the model with reversal-bounded counters, discrete clocks, etc.). In particular, we settled an open problem in [10].

References

1. R. Alur and D. L. Dill. A theory of timed automata. *Theoretical Computer Science*, 126(2):183–235, April 1994.
2. R. Alur, T. A. Henzinger, and M. Y. Vardi. Parametric real-time reasoning. In *Proceedings of the Twenty-Fifth Annual ACM Symposium on the Theory of Computing*, pages 592–601, San Diego, California, 16–18 May 1993.
3. A. Bouajjani, R. Echahed, and R. Robbana. On the automatic verification of systems with continuous variables and unbounded discrete data structures. In *Hybrid Systems II*, volume 999 of *Lecture Notes in Computer Science*. Springer-Verlag, 1995.
4. A. Bouajjani, J. Esparza, and O. Maler. Reachability analysis of pushdown automata: application to model-checking. In *Concurrency (CONCUR 1997)*, volume 1243 of *Lecture Notes in Computer Science*, pages 135–150. Springer-Verlag, 1997.
5. E. M. Clarke and E. A. Emerson. Design and synthesis of synchronization skeletons using branching time temporal logic. In *Workshop of Logic of Programs*, volume 131 of *Lecture Notes in Computer Science*. Springer, 1981.
6. E. M. Clarke, E. A. Emerson, and A. P. Sistla. Automatic verification of finite-state concurrent systems using temporal logic specifications. *ACM Transactions on Programming Languages and Systems*, 8(2):244–263, April 1986.
7. Z. Dang. Phd. dissertation. *Department of Computer Science, University of California at Santa Barbara*, 2000.
8. Z. Dang, O. Ibarra, and Z. Sun. On the emptiness problems for two-way nondeterministic finite automata with one reversal-bounded counter. *Submitted*, 2002.
9. Zhe Dang, O. H. Ibarra, T. Bultan, R. A. Kemmerer, and J. Su. Binary reachability analysis of discrete pushdown timed automata. In *Proceedings of the International Conference on Computer Aided Verification (CAV'00)*, volume 1855 of *Lecture Notes in Computer Science*, pages 69–84. Springer, 2000.
10. Zhe Dang, O. H. Ibarra, and R. A. Kemmerer. Decidable Approximations on Generalized and Parameterized Discrete Timed Automata. In *Proceedings of the 7th Annual International Computing and Combinatorics Conference (COCOON'01)*, volume 2108 of *Lecture Notes in Computer Science*, pages 529–539. Springer, 2001.
11. Zhe Dang and R. A. Kemmerer. A symbolic model-checker for testing ASTRAL real-time specifications. In *Proceedings of the Sixth International Conference on Real-time Computing Systems and Applications*, pages 131–142. IEEE Computer Society Press, 1999.
12. Zhe Dang and R. A. Kemmerer. Using the ASTRAL Model Checker to Analyze Mobile IP. In *Proceedings of the 1999 International Conference on Software Engineering (ICSE'99)*, pages 132–141. IEEE Computer Society Press / ACM Press, 1999.

13. Zhe Dang and R. A. Kemmerer. Three approximation techniques for ASTRAL symbolic model checking of infinite state real-time systems. In *Proceedings of the 2000 International Conference on Software Engineering (ICSE'00)*, pages 345–354. IEEE Computer Society Press, 2000.

14. S. Ginsburg and E. Spanier. Semigroups, presburger formulas, and languages. *Pacific J. of Mathematics*, 16:285–296, 1966.

15. E. M. Gurari and O. H. Ibarra. The complexity of decision problems for finite-turn multicounter machines. *Journal of Computer and System Sciences*, 22:220–229, 1981.

16. G. J. Holzmann. The model checker SPIN. *IEEE Transactions on Software Engineering*, 23(5):279–295, May 1997. Special Issue: Formal Methods in Software Practice.

17. O. H. Ibarra. Reversal-bounded multicounter machines and their decision problems. *Journal of the ACM*, 25(1):116–133, January 1978.

18. O. H. Ibarra, T. Jiang, N. Tran, and H. Wang. New decidability results concerning two-way counter machines. *SIAM J. Comput.*, 24:123–137, 1995.

19. O. H. Ibarra, J. Su, Zhe Dang, T. Bultan, and R. A. Kemmerer. Counter machines: decidable properties and applications to verification problems. In *Proceedings of the 25th International Symposium on Mathematical Foundations of Computer Science (MFCS 2000)*, volume 1893 of *Lecture Notes in Computer Science*, pages 426–435. Springer-Verlag, 2000.

20. K.L. McMillan. *Symbolic Model Checking*. Kluwer Academic Publishers, Norwell Massachusetts, 1993.

21. M. Minsky. Recursive unsolvability of Post's problem of Tag and other topics in the theory of Turing machines. *Ann. of Math.*, 74:437–455, 1961.

22. A. P. Sistla and E. M. Clarke. Complexity of propositional temporal logics. *Journal of ACM*, 32(3):733–749, 1983.

23. M. Y. Vardi and P. Wolper. An automata-theoretic approach to automatic program verification (preliminary report). In *Proceedings 1st Annual IEEE Symp. on Logic in Computer Science, LICS'86, Cambridge, MA, USA, 16–18 June 1986*, pages 332–344, Washington, DC, 1986. IEEE Computer Society Press.

An Infinite Prime Sequence
Can Be Generated in Real-Time by a 1-Bit
Inter-cell Communication Cellular Automaton

Hiroshi Umeo[1] and Naoki Kamikawa[2]

[1] Osaka Electro-Communication Univ.,
Neyagawa-shi, Hastu-cho, 18-8,Osaka, 572-8530, Japan
umeo@umeolab.osakac.ac.jp
[2] Noritsu Koki Co. Ltd.,
Umehara,Wakayama-shi, 640-8550, Japan
naoki@umeolab.osakac.ac.jp

Abstract. It is shown that an infinite prime sequence can be generated in real-time by a cellular automaton having 1-bit inter-cell communications.

1 Introduction

Cellular automata (CA) are considered to be a good model of complex systems in which an infinite one-dimensional array of finite state machines (cells) updates itself in a synchronous manner according to a uniform local rule. In the long history of the study of CA, generally speaking, the number of internal states of each cell is finite and the local state transition rule is defined in a such way that the state of each cell depends on the previous states of itself and its neighboring cells. Thus, in the finite state description of CA, the number of communication bits exchanged in one step between neighboring cells is assumed to be $O(1)$ bits. However, such inter-cell bit-information is hidden in the definition of the conventional automata-theoretic finite state description.

In the present paper, we focus on the inter-cell communication bits and introduce a new class of cellular automata, CA_{1-bit}, for which inter-cell communication is restricted to 1-bit, and then consider an infinite prime sequence generation problem on the 1-bit model. We hereinafter refer to the model as 1-bit CA. The number of internal states of CA_{1-bit} is assumed to be finite in the usual sense. The next state of each cell is determined by the present state of the cell and two binary 1-bit inputs from its left- and right-neighbor cells. Thus, the 1-bit CA can be thought of as being one of the simplest CAs to have a low computational complexity. On the 1-bit CA we consider a sequence generation problem that has been studied extensively on the conventional CA model and propose a real-time prime generation algorithm together with its implementation on a computer.

Although several studies have examined conventional cellular automata [1-4, 7-10], however, few studies have focused on the amount of bit-information exchanged in inter-cell communications. Mazoyer [5, 6] first studied this model under the name of CAs with channels and proposed a time-optimum firing squad

M. Ito and M. Toyama (Eds.): DLT 2002, LNCS 2450, pp. 339–348, 2003.

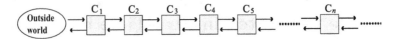

Fig. 1. A one-dimensional cellular automaton having 1-bit inter-cell communication.

synchronization algorithm in which only one-bit information is exchanged. Umeo [11, 13] and Umeo and Inada [12] have studied algorithmic design techniques for sequence generation problems on CA_{1-bit}. Umeo and Kamikawa[14] showed that the prime sequence can be generated in twice real-time on CA_{1-bit}. Worsch [15] established a computational hierarchy between one-way 1-bit CAs. In addition, Arisawa [1], Fisher [3], Korec [4] and Mazoyer and Terrier [7] have considered the sequence generation problem on the conventional cellular automata model. The result of the present paper is an improvement of the model of Umeo and Kamikawa [14] and gives an optimum-time prime generation algorithm on CA_{1-bit} using fewer states and transition rules than the previous design.

First, in Section 2, we introduce a cellular automaton having 1-bit inter-cell communication and define the sequence generation problem on CA_{1-bit}. In Section 3, we give a real-time prime generation algorithm on CA_{1-bit}. The algorithm is based on the sieve of Eratosthenes, and its implementation will be made on a CA_{1-bit} using 34 internal states and 71 transition rules.

2 Cellular Automaton Having 1-Bit Inter-cell Communication

A one-dimensional 1-bit inter-cell communication cellular automaton consists of an infinite array of identical finite state automata, each located at a positive integer point (See Fig. 1). Each automaton is referred to as a cell. A cell at point i is denoted by C_i, where $i \geq 1$. Each C_i, except for C_1, is connected to its left- and right-neighbor cells via a left or right one-way communication link. These communication links are indicated by right- and left-pointing arrows in Fig. 1, respectively. Each one-way communication link can transmit only one bit at each step in each direction. One distinguished leftmost cell C_1, the communication cell, is connected to the outside world. A cellular automaton with 1-bit inter-cell communication (abbreviated by CA_{1-bit}) consists of an infinite array of finite state automaton $A = (Q, \delta, F)$, where

1. Q is a finite set of internal states.
2. δ is a function, defining the next state of any cell and its binary outputs to its left- and right-neighbor cells, such that $\delta\colon Q \times \{0,1\} \times \{0,1\} \to Q \times \{0,1\} \times \{0,1\}$, where $\delta(p, x, y) = (q, x', y')$, $p, q \in Q$, $x, x', y, y' \in \{0,1\}$, has the following meaning. We assume that at step t the cell C_i is in state p and is receiving binary inputs x and y from its left and right communication links, respectively. Then, at the next step, $t+1$, C_i assumes state q and outputs x' and y' to its left and right communication links, respectively. Note that binary inputs to C_i at step t are also outputs of C_{i-1} and C_{i+1} at step t. A quiescent state $q \in Q$ has a property such that $\delta(q, 0, 0) = (q, 0, 0)$.

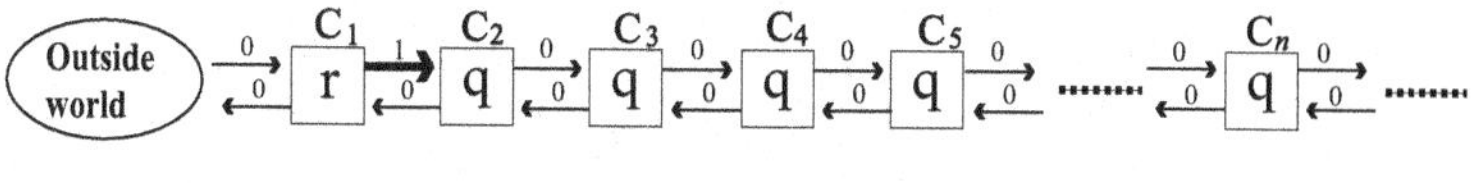

Fig. 2. Initial configuration of CA_{1-bit}.

3. $F(\subseteq Q)$ is a special subset of Q. The set F is used to specify a designated state of C_1 in the definition of sequence generation.

Thus, the CA_{1-bit} is a special subclass of *normal* (i.e., *conventional*) cellular automata. Let N be any normal cellular automaton having a set of states Q and a transition function $\delta : Q^3 \rightarrow Q$. The state of each cell on N depends on the previous states of the cell and its nearest neighbor cells. This means that the total information exchanged per step between neighboring cells consists of $O(1)$bits. By encoding each state in Q with a binary sequence of length $\lceil \log_2 |Q| \rceil$, sending the sequences sequentially bit by bit in each direction via each one-way communication link, receiving the sequences bit-by-bit again, and then decoding the sequences into their corresponding states in Q, the CA_{1-bit} can simulate one step of N in $\lceil \log_2 |Q| \rceil$ steps. This observation yields the following computational relation between the normal CA and CA_{1-bit}.

[Lemma 1] Let N be any *normal* cellular automaton having time complexity $T(n)$. Then, there exists a CA_{1-bit} which can simulate N in $kT(n)$ steps, where k is a positive constant integer such that $k = \lceil \log_2 |Q| \rceil$ and Q is the set of internal states of N.

We now define the sequence generation problem on CA_{1-bit}. Let M be a CA_{1-bit}, and let $\{t_n | n = 1, 2, 3, ...\}$ be an infinite monotonically increasing positive integer sequence defined for natural numbers, such that $t_n \geq n$ for any $n \geq 1$. We then have a semi-infinite array of cells, as shown in Fig. 1, and all cells, except for C_1, are in the quiescent state at time $t = 0$. The communication cell C_1 assumes a special state r in Q and outputs 1 to its right communication link at time $t = 0$ for initiation of the sequence generator. The initial configuration of M is shown in Fig. 2. We say that M generates a sequence $\{t_n | n = 1, 2, 3...\}$ in k *linear-time* if and only if the leftmost end cell of M falls into a special state in $F(\subseteq Q)$ and outputs 1 via its leftmost communication link at time $t = kt_n$, where k is a positive integer. We call M a *real-time* generator when $k = 1$.

3   Real-Time Generation of Prime Sequence

Arisawa [1], Fischer [3] and Korec [4] studied real-time generation of a class of natural numbers on the conventional cellular automata, where $O(1)$ bits of information were allowed to be exchanged at one step between neighboring cells. Fisher [3] showed that prime sequences could be generated in real-time on the cellular automata with 11 states for C_1 and 37 states for $C_i (i \geq 2)$. Arisawa[1] also developed a real-time prime generator and decreased the number of states of each cell to 22. Korec [4] reported a real-time prime generator having 11 states on the same model.

In this section, we present a real-time prime generation algorithm on CA_{1-bit}. The algorithm is implemented on a CA_{1-bit} using 34 internal states and 71 transition rules. Our prime generation algorithm is based on the sieve of Eratosthenes. Imagine a list of all integers greater than 2. The first member, 2, becomes a prime and every second member of the list is crossed out. Then, the next member of the remainder of the list, 3, is a prime and every third member is crossed out. In Eratosthenes' sieve, the procedure continues with 5, 7, 11, and so on. In our procedure, given below, for any odd integer $k \geq 2$, every $2k$th member of the list beginning with k^2 will be crossed out, since the kth members less than k^2 (that is, $\{i \cdot k \mid 2 \leq i \leq k-1\}$) and $2k$th members beginning with $k^2 + k$ (that is, $\{(k+2i-1) \cdot k \mid i = 1, 2, 3, ..\}$) should have been crossed out in the previous stages. Those integers never being crossed out are the primes. Figure 3 is a time-space diagram that shows a real-time detection of odd multiples of three, five and seven. In our detection, we use two 1-bit signals a- and b-waves, which will be described later, and pre-set partitions in which these two 1-bit signals bounce around.

We now outline the algorithm. Each cross-out operation is performed by C_1. We assume that the cellular space is initially divided by the partitions such that a special mark "w" is printed on cell C_k, where $k = i^2$ and $i(\geq 1)$ is any integer. The partitions will be used to generate reciprocating signals for the detection of odd multiples of, for example, three, five, and seven. We denote a subcellular space sandwiched by C_k and C_ℓ as S_i, where $k = i^2$, $\ell = (i+1)^2$, and call the subspace ith partition. Note that S_i contains $(2i+2)$ cells, including both ends. The way to set up the partitions in terms of 1-bit communication will be described in Lemma 3.

[**Lemma 2**] The array given above can generate the ith prime at time $t = i$.

(**Proof**) Consider a unit speed (1-cell/1-step) signal that reciprocates between the left and right ends of S_i, which is shown as the zigzag movement in Fig. 4. This signal is referred to as the *a-wave*. At every reciprocation, the a-wave generates a *b-wave* on the left end of S_i. The b-wave continues to move to C_1 at unit speed to the left through S_{i-1}, S_{i-2}, ..., S_2, and S_1. Both a- and b-waves generated by S_i are responsible for notifying C_1 of odd multiples of the odd integer $(2i+1)$ such that $(2j+1)(2i+1), j \geq i$. In addition, the a-wave, on the second trip to the right end of S_i, initiates a new a-wave for S_{i+1}.

We assume that the initial a-wave on S_i is generated on the left end of S_i at time $t = 3i^2$. Then, as shown in Fig. 4, the b-wave reaches C_1 at step $t = (2i+1)^2 + 2j \cdot (2i+1)$, where $j = 0, 1, 2, .., .$ Moreover, the initiation of the first a-wave in S_{i+1} is successful at step $t = 3(i+1)^2$. We observe from Fig. 5 that the following signals are generated at the correct time. At time $t = 3$, an a-wave starts toward the right from the cell C_1. At time $t = 6$, the a-wave arrives at the right end cell of the partition S_1, and then is reflected toward the left and reaches C_1 at $t = 9$. The a-wave again proceeds toward the right at unit speed and reaches the right end of S_1. The first a-wave for S_2 is generated here. By mathematical induction, we see that the first a-wave for S_i can be generated on the left end of S_i at time $t = 3i^2$ for any $i \geq 1$. In this way, the b-wave generated

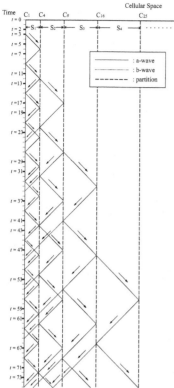

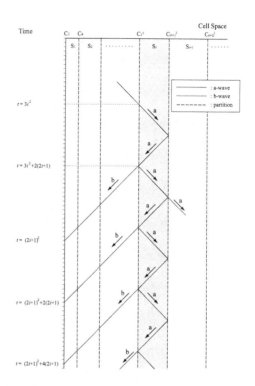

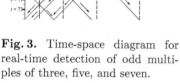

Fig. 3. Time-space diagram for real-time detection of odd multiples of three, five, and seven.

Fig. 4. Time-space diagram for real-time detection of odd multiples of $2i + 1$ greater than $(2i + 1)^2$.

at the left end of S_i notifies C_1 of odd multiples of $(2i + 1)$ that are greater than $(2i + 1)^2$. Multiples of $(2i + 1)$ that are less than $(2i + 1)^2$ have been detected in the previous stages by the b-wave generated at the left end of $S_1, S_2, ..., S_{i-1}$ (See Fig. 5).

Whenever a left-traveling a-wave generated on S_i and the left-traveling b-wave generated on $S_j (j \geq i + 1)$ start simultaneously at the right end of S_i, they are merged into one a-wave. Otherwise, the b-wave meets a reflected a-wave at S_i. Two kinds of unit speed left-traveling 1-bit signals exist in $S_i (i \geq 1)$, that is, an a-wave that is reciprocating on S_i and a b-wave that is generated on $S_j (j \geq i + 1)$. These two left-traveling 1-bit signals must be distinguished, since the latter does not produce a reciprocating a-wave. In order to avoid the reflection of the b-wave at the left end of each partition, we introduce a new *h-wave*, as shown in Fig. 5. Whenever a right-traveling a-wave and a left-traveling b-wave meet on a cell in S_i, the h-wave is newly generated on the cell in which they meet at the next step and one step later, the h-wave begins to follow the

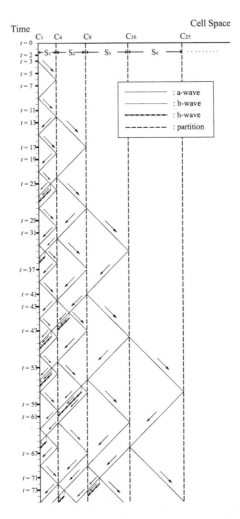

Fig. 5. An h-wave that inhibits the reflection of 1-bit b-signal at the partition.

left-traveling b-wave at unit speed. The h-wave stops the reflection of the b-wave at the left end of S_i, and then both waves disappear. A left-traveling a-wave and b-wave generated on $S_j (j \geq i+1)$ always move with at least one cell interleaved between them. This enables the h-wave to be generated and transmitted toward the left. The left end cell C_1 has a counter that operates in modulo 2 and checks the parity of each step in order to detect every multiple of two. C_1 outputs a 1-bit signal to its left link if and only if it has not received any 1-bit signal from its right link at its previous step t. Then, t is exactly prime.

In this way, the initially partitioned array given above can generate the prime sequence in real-time. In the next lemma, we show that the partition in the cellular space can be set up in time. ■

[**Lemma 3**] For any $i(\geq 3)$, the partition S_i can be set up in time. Precisely, the right end cell C_k of the ith partition S_i, where $k = (i + 1)^2$, can be marked at step $t = 3i^2 + 2i + 3$.

(**Proof**) For the purpose of setting up the partitions in the cellular space in time, we introduce seven new waves: *c-wave, d-wave, e-wave, f-wave, g-wave, u-wave* and *v-wave*. The direction in which these waves propagate and their speeds are as follows:

Wave	Direction	Speed
c-wave	right	1/2
d-wave	right	1/1
e-wave	right	1/3
f-wave	left	1/1
g-wave	right	1/1
u-wave	stays at a cell for four steps	0
v-wave	left	1/1

The u-wave is a kind of partition that appears on a cell for only four steps. Setting up the partitions and the generation of a- and b-waves are performed in parallel on the array. We make a small modification to the a-wave. The first reciprocation of the a-wave in each $S_i (i \geq 3)$ is replaced by the c-, d-, e-, f-, u- and v-waves. In Fig. 6, we present a time-space diagram that shows the generation, propagation and interactions of these waves.

We assume that:

- A_1: The c- and d-waves in S_{i-1} arrive simultaneously at the left end cell C_k of S_i, where $k = i^2$, and prints "w" as a partition marking on the cell at time $t = 3i^2 - 4i + 4$.
- A_2: The g-wave in S_{i-1} hits the left end of S_i at time $t = 3i^2 - 2i$ and generates the c-wave in S_i.
- A_3: The a-wave in S_{i-1} hits the left end of S_i and generates the d- and e-waves in S_i at time $t = 3i^2 + 2$.

Then, the following statements can be obtained from the time-space diagram shown in Fig. 6.

1. The c- and d-waves in S_i arrive at the right end cell of S_i, C_k, where $k = (i + 1)^2$, and the marker "w" is printed on the cell C_k at time $t = 3i^2 + 2i + 3$.
2. The c-wave being propagated generates at every two steps a left-traveling tentacle-like wave that disappears one step after its appearance. This is referred to as the *c'-wave*. At time $t = 3i^2 + 2i$, the c'-wave and the d-wave meet on the cell C_k, where $k = i^2 + 2i - 2$.
3. The left-traveling f-wave, initiated by the intersection of the d- and c'-waves, meets the e-wave on C_ℓ, $\ell = i^2 + i - 1$ at step $t = 3i^2 + 3i - 1$. The g- and u-waves are generated simultaneously on the cell in which the f- and e-waves meet.

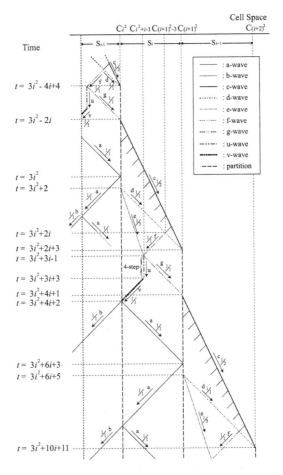

Fig. 6. Time-space diagram for real-time generation of partition S_i.

4. The u-wave remains at C_ℓ, $\ell = i^2 + i - 1$ for only four steps, and then generates a v-wave at step $t = 3i^2 + 3i + 3$.

5. The v-wave hits the left end of S_i and generates an a-wave in S_i and a b-wave in S_{i-1}. The b- and a-waves for the 2nd, 3rd, ... reciprocations in S_i are generated at the same time, as shown in Fig. 4.

Thus, the partition setting for the right end of $S_i (i \geq 3)$ is made inductively. The first two markings on cells S_1 and S_2 at times $t = 7$ and 18, respectively, and the generation of c-, d- and e-waves at the left end of S_2 at steps 8 and 14 are realized in terms of finite state descriptions. Thus, we can set up entire partitions inductively in time.■

In addition to Lemma 3, the generation of a- and b-waves and a number of additional signals in S_1 and S_2, as shown in Fig. 7, are also implemented in terms of finite state descriptions. Figure 7 is our final time-space diagram for

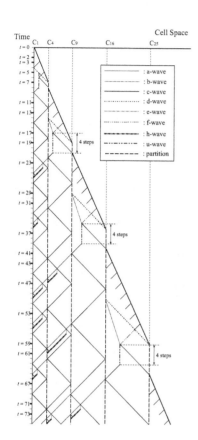

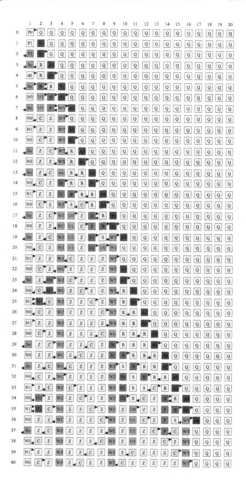

Fig. 7. Time-space diagram for real-time prime generation.

Fig. 8. A configuration of real-time generation on 1-bit CA using 34 states.

the real-time prime generation algorithm. We have implemented the algorithm on a computer. Each cell has 34 internal states and 71 transition rules. We have tested the validity of the rule set from $t = 0$ to $t = 20000$ steps. In Fig. 8, we show a number of snapshots of the configuration from $t = 0$ to 40. Thus, based on Lemmas 2 and 3, we obtain the following theorem.

[**Theorem 4**] A prime sequence can be generated by a CA_{1-bit} in real-time.

4 Conclusion

We have introduced a special class of cellular automata having 1-bit inter-cell communications and have proposed a real-time prime generation algorithm on CA_{1-bit}. The algorithm has been implemented on a CA_{1-bit} using 34 internal states and 71 transition rules. We have considered only prime sequence generation problems on a one-dimensional CA_{1-bit}. However, language recognition problems can be similarly treated within the same framework as in Dyer [2] and Smith [8]. The 1-bit CA is confirmed to be an interesting computational subclass of CAs that merits further study.

References

1. M. Arisawa: On the generation of integer series by the one-dimensional iterative arrays of finite state machines (in Japanese). *The Trans. of IECE*, 71/8 Vol.54-C, No.8, pp.759-766, (1971).

2. C. R. Dyer: One-Way Bounded Cellular Automata. *Information and Control*, Vol.44, pp.261-281, (1980).

3. P. C. Fischer: Generation of primes by a one-dimensional real-time iterative array. *J. of ACM*, Vol.12, No.3, pp.388-394, (1965).

4. I. Korec: Real-time generation of primes by a one-dimensional cellular automaton with 11 states. *Proc. of 22nd Intern. Symp. on MFCS '97, Lecture Notes in Computer Science*, 1295, pp.358-367, (1997).

5. J. Mazoyer: A minimal time solution to the firing squad synchronization problem with only one bit of information exchanged. Technical report of Ecole Normale Superieure de Lyon, No.89-03, pp.51, April, (1989).

6. J. Mazoyer: On optimal solutions to the firing squad synchronization problem. *Theoret. Comput. Sci.*, 168, pp.367-404, (1996).

7. J. Mazoyer and V. Terrier: Signals in one-dimensional cellular automata. *Theoretical Computer Science*, 217, pp.53-80, (1999).

8. A. R. Smith: Real-time language recognition by one-dimensional cellular automata. *J. of Computer and System Sciences*, 6, pp.233-253, (1972).

9. V. Terrier: On real-time one-way cellular array. *Theoretical Computer Science*, 141, pp.331-335, (1995).

10. V. Terrier: Language not recognizable in real time by one-way cellular automata. *Theoretical Computer Science*, 156, pp.281-287, (1996).

11. H. Umeo: Cellular Algorithms with 1-Bit Inter-Cell Communications. *Proc. of MFCS'98 Satellite Workshop on Cellular Automata* (Eds. T. Worsch and R. Vollmar), pp.93-104, Interner Bericht 19/98, University of Karlsruhe, (1998).

12. H. Umeo and M. Inada: A Design of Cellular Algorithms for 1-Bit Inter-Cell Communications. *Proc. of the Fifth International Workshop on Parallel Image Analysis, IWPIA'97*, pp.51-62, (1997).

13. H. Umeo: Linear-time recognition of connectivity of binary images on 1-bit inter-cell communication cellular automaton. *Parallel Computing*, 27, pp.587-599, (2001).

14. H. Umeo and N. Kamikawa: A design of real-time non-regular sequence generation algorithms and their implementations on cellular automata with 1-bit inter-cell communications. *Fundamenta Informaticae*, 52, No.1-3, pp.255-273, (2002).

15. T. Worsch: Linear Time Language Recognition on Cellular Automata with Restricted Communication, *Proc. of LATIN 2000:Theoretical Informatics* (Eds. G. H. Gonnet, D. Panario and A. Viola), LNCS 1776, pp.417-426, (2000).

On the Structure of Graphic DLI-Sets

László Kászonyi*

Department of Mathematics, Berzsenyi College,
H-9700 Szombathely, Hungary

Abstract. The **DLI**-sets were defined in connection with the problem of deciding the context-freeness of the languages $Q_n = Q \cap (ab^*)^n$ [14]. It is conjectured in [3] that Q_n is context-free for all positive numbers n. This conjecture was verified for the cases where $n = p_1^{f_1} \cdots p_k^{f_k}$ with distinct prime numbers $p_1, ..., p_k$ for which

$$\sum_{i=1}^{k} 1/p_i < 4/5. \tag{1}$$

holds. (See.: [18]). To find a proof for the general case it would be useful to find special **DLI** sets other then the boxes used in the proof of the former result. In this paper a sufficient condition is given for **DLI** sets to be stratified semilinear.

1   Introduction

Stratified linear sets play a central role in the theory of bounded context-free languages, developed by Ginsburg [4]. In general, a language $L \subseteq \Sigma^*$, where Σ is a fixed finite alphabet having at least two letters, is a *bounded language* if and only if there exist non-empty words $w_0, \dots, = w_{m-1}$ such that $L \subseteq w_0^* \dots w_{m-1}^*$. Note that for a word $w \in \Sigma^*$ we use w^* as a short-hand notation for $\{w\}^*$. A necessary and sufficient condition for a bounded language to be context-free was given by Ginsburg:

Theorem 1 (Ginsburg [4]). *Let L be a bounded language over the alphabet Σ. Language L is context-free iff set*

$$E(L) = \{\, (e_0, \dots, e_{m-1}) \in \mathbb{N}^m \mid w_0^{e_0} \dots w_{m-1}^{e_{m-1}} \in L \,\}, \tag{2}$$

where words $w_0, \dots, w_{m-1}$ are the corresponding words of L, is a finite union of stratified linear sets.

One interesting subclass of the bounded languages is the class of **DLI**-languages whose exponent set given in (2) is a **DLI**-set, i.e., a vector set **D**efined by **L**inear **I**nequalities:

* This research was supported by the Hungarian-German scientific- technological research project No D 39/2000 in the scope of the treaty contracted by the Hungarian Ministry of Education and his German contractual partner BMBF.

M. Ito and M. Toyama (Eds.): DLT 2002, LNCS 2450, pp. 349–356, 2003.

Definition 2. *The set*

$$E(\Theta, \delta, \epsilon) = \bigcap_{I \in \Theta} \{ (e_0, \dots, e_{m-1}) \in \mathbb{N}^m \mid \epsilon(I) \sum_{i \in I} \delta_i e_i \geq 0 \} \tag{3}$$

is a **DLI**-*set where*

(1) Θ is a system of index-sets, (i.e., of subsets of $\underline{m} = \{0, \dots, m-1\}$). Θ is considered as a multi-set i.e., elements of Θ may have multiplicity greater than one.

(2) $\delta = (\delta_0, \dots, \delta_{m-1})$ is a fixed vector of signs i.e., for $i = 0, \dots m-1$ $\delta_i \in \{-1, 0, 1\}$.

(3) ϵ is a function from Θ into the set $\{-1, 1\}$.

The following "Flip-Flop-Theorem" gives a necessary and sufficient condition for a **DLI**-set to be stratified semilinear. (Kászonyi [12])

Theorem 3 (Flip-Flop Theorem). *Let the set E be a* **DLI**-*set with respect to the sign-vector $\delta = (\delta_0, \dots, \delta_{m-1})$, index-set-system Θ, and function ϵ:*

$$E = E(\Theta, \delta, \epsilon) = \bigcap_{I \in \Theta} \{ (e_0, \dots, e_{m-1}) \in \mathbb{N}^m \mid \epsilon(I) \sum_{i \in I} \delta_i e_i \geq 0 \} \tag{4}$$

E is stratified semilinear if and only if for every $e \in E$ there exists a hypergraph H, having the following properties:

(i) The vertices of H are the vertices of a convex m-polygon, indexed by the elements of a cyclically ordered set $\underline{m}$ according to their cyclical order.

(ii) The edges of H are one- or two-element subsets of the vertex-set $\mathbf{V}(H)$ of H.

(iii) If $\{i, j\}$ is a two-element edge of H, then the signs associated to the endpoints i and j are opposite, i.e., $= \delta_i = -\delta_j \neq 0$.

(iv) The edge f is forbidden if there exists an index-set $I \in \Theta$ such that $f \cap I = \{i\}$ and $\epsilon(I) = -\delta_i$. Hypergraph H doesn't contain forbidden edges.

(v) The edges of H are non-crossing.

(vi) The degreee of each vertex i is e_i.

The Flip-Flop Theorem may be applied in the same way as the so called pumping lemmata. Sometimes the problem is to prove the context-freeness of a given **DLI**-language L. In such cases we can try to construct a context-free grammar G generating L. However if G is of high-level complexity, then such characteristic properties of context-free **DLI**-languages are useful which are given in terms of Θ, δ and ϵ. Such properties are given in the following two lemmata.

Lemma 4 (Flip-Flop Lemma, Kászonyi = ([12])). *For $i = 0, \dots, m-1$ let the δ_i be "signs," i.e., let $\delta_i \in \{-1, 0, 1\}$, and consider the set*

$$E = \{ (e_0, \dots, e_{m-1}) \in \mathbb{N}^m \mid \delta_0 e_0 + \dots + \delta_{m-1} e_{m-1} \neq 0 \}.$$

Then E is a stratified semilinear set.

Definition 5. *The set-system Θ of subsets of $\underline{m}$ is a tree-system over $\underline{m}$ if*

(i) $\underline{m} \in \Theta$ and
(ii) for any two elements I and J of Θ either $I \cap J = \emptyset$ or one of the relations $I \subseteq J$ or $J \subseteq I$ holds.

In other words, the sets of Θ can be arranged in a tree like manner.

The generalized form of Lemma 4 is given by Kászonyi and Holzer [11]:

Lemma 6 (Generalized Flip-Flop lemma). *For $i = 0, \ldots, m - 1$ let the δ_i be "signs," i.e., let $\delta_i \in \{-1, 0, 1\}$, and let Θ be a tree-system of intervals over $\underline{m}$. Define the vector-set $E(\Theta)$ as follows:*

$$E(\Theta) = \{ (e_0, \ldots, e_{m-1}) \in \mathbb{N}^m \mid \bigwedge_{I \in \Theta} (\sum_{i \in I} \delta_i e_i \neq 0) \} \tag{5}$$

$$= \bigcap_{I \in \Theta} \{ (e_0, \ldots, e_{m-1}) \in \mathbb{N}^m \mid \sum_{i \in I} \delta_i e_i \neq 0 \}.$$

Then $E(\Theta)$ is a stratified semilinear set.

2   Main Results

Let us recall that the Flip-Flop Theorem characterizes stratified semilinear **DLI**-sets by associating hypergraphs possessing properties $(i) - (vi)$ with vectors of the **DLI**-sets.

A weakened form of stratified semilinearity is fomulated in the following definition:

Definition 7. *A **DLI**-set E is called* graphic *if for every $e \in E$ there exists a hypergraph G possessing properties $(i) - (iv)$ and (vi). (I.e., G can have crossing edges.)*

In order to give a description of graphic **DLI**-sets we define the forest systems of index sets.

Definition 8. *The set-system Θ of subsets of $\underline{m}$ is a forest system over $\underline{m}$ if*

(FS) for any two elements I and J of Θ either $I \cap J = \emptyset$ or one of the relations $I \subseteq J$ or $J \subseteq I$ holds.

It is easy to see that

Proposition 9. *Every forest system is a union of pair-wise disjoint tree systems.*

In the proof of Theorem 11 we make use of the operation called *reduction* defined as follows.

Definition 10. *Let* $E = E(\Theta, \delta, \epsilon)$ *be an arbitrary* **DLI**-*set and consider a nonempty subset* X *of* $\underline{m}$. *The* **DLI**-*set* $E|X = E(\Theta', \delta', \epsilon')$ *defined over* X *is given by*

$$\Theta' = \{I \in \Theta \mid I \subseteq X\} \tag{6}$$
$$\delta'_i = \delta_i, \; if \; i \in X \tag{7}$$
$$\epsilon'(I) = \epsilon(I), \; if \; I \in \Theta' \tag{8}$$

We will say that $E|X$ *is the* reduction *of* E *to* X. *The natural number* $|X|$ *is the* dimension *of* $E|X$.

The following theorem gives a sufficient condition in terms of Θ, δ and ϵ for **DLI**-sets to be graphic.

Theorem 11. *Let* $E = E(\Theta, \delta, \epsilon)$ *be a* **DLI**-*set such that* Θ *is a forest system. Then* E *is graphic.*

Proof. Let $e \in E$. We have to prove that there exists a hypergraph G possessing properties $(i) - (iv)$ and (vi). By Proposition 9 it is enough to prove the existence of G in such cases when Θ is a tree system.

Case 1. $\Theta = \{\underline{m}\}$. Let $= e = (e_0, ..., e_{m-1})$ and

$$Supp(e) = \{i \mid e_i > 0\} \tag{9}$$

Assume first that $\delta_i = \delta_j$ whenever $i, j \in Supp(e)$. Let $\mathbf{V}(G) = \underline{m}$ and $\mathbf{E}(G)$ the multiset consisting of one element sets $\{i\} \subseteq \underline{m}$ where the multiplicity of $\{i\}$ is e_i. (Here $\mathbf{V}(G)$ and $\mathbf{E}(G)$ denote the point and edge set of hypergraph G respectively.) We show that G satisfies property (iv). If $\delta_i = -\epsilon(\underline{m})$ then $\epsilon(\underline{m}) \leq 0$ implies $e_i = 0$ for $i = 0, ..., m - 1$, i.e., $E = \{0\}$ and (iv) is trivially true. If $\delta_i = \epsilon(\underline{m})$ holds for $i = 0, ..., m - 1$ then $\{i\} \cap \underline{m} = \{i\}$ and (iv) is satisfied. If there exist indices i, j such that $\delta_i = 1 = -\delta_j$ then the proof is by induction on the number $k(e) = \sum_{i=0}^{m-1} e_i$. For $k(e) = 0$ *Case 1* holds. Assume that $k(e) = K > 0$ and for any $e' \in E$ $k(e') < K$ there is a hypergraph G' possessing properties $(i) - (iv)$ and (vi). Actually, if $e' = (e'_0, ..., e'_{m-1})$ where

$$e'_i = e_i - 1,$$
$$e'_j = e_j - 1$$
$$e'_k = e_k \qquad (k \neq i, j) \tag{10}$$

then $k(e') = k(e) - 2 < K$ which assures the existence of G'. Let G be the hypergraph which we get from G' by adding a new edge $\{i, j\}$ to it. It is easy to check that G possesses properties $(i) - (iv)$ and (vi).

Case 2. $\Theta \neq \{\underline{m}\}$. We have to prove that there is a hypergraph G with properties $(i) - (iv)$ and (vi). The proof is by induction on the "dimension" m of the **DLI**-set E. If $m = 0$ then the statement of the theorem is trivially true. Assume that it is true for $m < M$ $(M \geq 1)$. We will show that it holds for $m = M$ as well.

Let $\{J_1, ..., J_\nu\}$ be the class of elements of Θ such that for $k = 1, ..., \nu$ $J_k \neq \underline{m}$ and for any $I \in \Theta$ $J_k \subseteq I \subseteq \underline{m}$ imply either $I = J_k$ or $I = \underline{m}$. Θ is a tree-system, hence $J_k \cap J_l = \emptyset$ holds for $k \neq l$. For $k = 1, ..., \nu$ let $E_k = E|J_k$ where $E|J_k$ is the reduction of E to the set J_k. All of the sets E_k are of tree structure and of dimension smaller then M therefore by our hypothesis, there exist hypergraphs G_k possessing properties $(i) - (iv)$ and (vi) with respect to E_k and $e_k = e_{J_k}$ (Here e_{J_k} is the restriction of vector e to set J_k. It is easy to see, that $e_{J_k} \in E_k$ holds.)

Let $\mathbf{E}^1(G_k)$ be the multiset of one-point edges of G_k. It is easy to see that

$$S_{J_k} = \epsilon(J_k) \sum_{j \in J_k} \delta_j e_j = |\mathbf{E}^1(G_k)| \quad (k = 1, ..., \nu) \tag{11}$$

Without loss of generality we may assume that $\epsilon(\underline{m}) = 1$. Graph G_k has no forbidden edges by (iv) therefore $\delta_i = \epsilon(J_k)$ holds for $i \in \mathbf{E}^1(G_k)$. Let $J = \underline{m} \setminus \cup_{k=1}^\nu J_k$ and $\mathbf{E}_j^1$ be the multiset consisting of e_j copies of the one-point edge $\{j\}$, and define the multisets $\mathbf{E}_+^1$ and $\mathbf{E}_-^1$ as follows:

$$\mathbf{E}_+^1 = \cup_{\epsilon(J_k)=+1}\mathbf{E}^1(G_k) \cup (\cup\{\mathbf{E}_j^1 \mid j \in J, \; \delta_j = +1\}) \tag{12}$$

$$\mathbf{E}_-^1 = \cup_{\epsilon(J_k)=-1}\mathbf{E}^1(G_k) \cup (\cup\{\mathbf{E}_j^1 \mid j \in J, \; \delta_j = -1\}) \tag{13}$$

Let us number the elements of $\mathbf{E}_+^1$ and $\mathbf{E}_-^1$ such that the distinct copies of any edge $\{i\}$ are numbered by distinct indices:

$$\mathbf{E}_+^1 = \{e_1^+, ..., e_\alpha^+\}$$
$$\mathbf{E}_-^1 = \{e_1^-, ..., e_\beta^-\}.$$

Here holds

$$0 \leq S_{\underline{m}} = \sum_{i \in \underline{m}} \delta_i e_i = \alpha - \beta. \tag{14}$$

Let

$$\mathbf{E}^2 = \{e_1^+ \cup e_1^-, ..., e_\beta^+ \cup e_\beta^-\},$$

and consider the hypergraph G given by

$$\mathbf{V}(G) = \underline{m},$$

$$\mathbf{E}(G) = ((\cup_{k=1}^\nu \mathbf{E}(G_k) \cup \cup_{j \in J} = \mathbf{E}_j^1) \setminus (\{e_1^+, ..., e_\beta^+\} \cup \mathbf{E}_-^1)) \cup \mathbf{E}^2.$$

It is easy to show that G posesses properties $(i) - (iv)$ and (vi) with respect to E and e.   □

We conjecture that the existence of a forest system of index sets is a necessary condition for E to be graphic as well.

Conjecture 1. Let $E = E(\Theta, \delta, \epsilon)$ be a graphic **DLI**-set. Then there is a forest system Θ' such that

$$E = E(\Theta', \delta, \epsilon) \tag{15}$$

3   Applications

The most important applications of Lemma 4 and Lemma 6 take place in the theory of primitive words. A *primitive word* over the alphabet Σ is a non-empty word not of the form w^r for any word $w \in \Sigma^+$ and integer $r \geq 2$.

During the last few years, the theory of primitive words, receives major interest in the theory of automata and formal languages (see e.g., [3,2,5,17,19,6,7,8,9]). One = still unsolved problem in that area is the question whether the set of all primitive words (over alphabet Σ), which is usually denoted by Q, is context-free or not. The simplest idea to show that Q is not context-free would be to use one of the pumping lemmata for context-free languages. This approach fails, because Q has some context-free like properties (Dömösi *et al.* [3]). Examples of such properties are given in the following theorems whose proofs heavily rely on the Flip-Flop lemma.

Theorem 12. *Let $a, b \in \Sigma$, with $a \neq b$, $n = p^s q^t r^v$, where p, q and r are pairwise different prime numbers, and $s, t, v \geq 0$. Let further $= L = (ab^*)^n$. Then $Q \cap L$ is a context-free language.*

Theorem 13. *Let $n = p_1^{f_1} \cdots p_k^{f_k}$ where $p_1, \ldots, p_k$ are distinct prime numbers and $f_1, \ldots, f_k$ are positive integers. Assume that*

$$\sum_{i=1}^{k} 1/p_i < 4/5. \tag{16}$$

Then the language $Q \cap (ab^*)^n$ is context-free.

In the proof of Theorem 13 special **DLI**-sets are used, i.e., **DLI**-sets $E = E(\Theta, \delta, \epsilon)$ where the index set Θ consists of so called noncrossing boxes. Let $n = p_1^{f_1} \ldots p_k^{f_k}$ where $p_1, \ldots, p_k$ are pairwise distinct prime numbers and $f_1, \ldots, f_k$ are positive integers. Let $\pi = \{q_1, \ldots, q_r\}$ be a nonempty subset of $\{p_1, \ldots, p_k\}$. We define a π-*box* by:

$$B = B(\xi; \pi) = \{\xi + \rho | \rho = \sum_{i=1}^{r} \rho_i n/q_i, \rho_i \in \{0, 1\}\}$$

For a vector $e = (e_0, \ldots, e_{n-1}) \in \mathbb{N}^n$ and for the box B, the = corresponding *difference* is defined by:

$$\Delta_e(B) = \Delta_e(\xi; \pi) = \sum_{\xi + \rho \in B} (-1)^{\rho_1 + \ldots + \rho_r} e_{\xi + \rho}$$

In other words, a difference defined for a vector e and box B is a signed sum of such components of e whose indices belong to B, and if the index-pair (i, j) is an "edge" of the box B then the corresponding members e_i and e_j of the sum have opposite signs.

For a partition $\{\pi_1, ... \pi_v\}$ of $\{p_1, ..., p_k\}$ let

$$E(\pi_1, ..., \pi_v) = \cup\{e \in \mathbb{N}_+^n \mid \Delta_e(\xi_1, \pi_1) \neq 0, ..., \Delta_e(\xi_v, \pi_v) \neq 0,$$
$$B(\xi_1, \pi_1), ..., B(\xi_v, \pi_v) \ \ are \ \ p.n.c. \ \ boxes, \} \tag{17}$$

where *p.n.c.* is the abbreviation of "pairwise noncrossing". It is proved in [17] that

$$E(Q \cap (ab^*)^n) = \cup\{E(\pi_1, ..., \pi_v) \mid \{\pi_1, ..., \pi_v\} \in = \mathbf{\Pi}\}, \tag{18}$$

where $\mathbf{\Pi}$ is the set of all partitions of the set $\{p_1, ..., p_k\}$. The restriction (16) assures that we can find suitable systems of *p.n.c.* boxes in (17). It is conjectured in [3] that $Q \cap (ab^*)^n$ is context-free for all natural numbers n. In order to find a proof of this conjecture we are looking for **DLI**-sets with index set system Θ consisting of components "thinner" than that of some "thick" boxes. Partial results are given in [13].

References

1. J. Dassow and Gh. Păun. *Regulated Rewriting in Formal Language Theory*, volume 18 of *EATCS Monographs in Theoretical Computer Science*. Springer, Berlin, 1989.

2. P. Dömösi, S. Horváth, M. Ito, L. Kászonyi, and M. Katsura. Some combinatorial properties of words, and the = Chomsky-hierarchy. In *Proceedings of the 2nd International Colloquium on = Words, Languages and Combinatorics*, Kyoto, Japan, 1992. World Scientific, Singapore.

3. P. Dömösi, S. Horváth, M. Ito, L. Kászonyi, and M. Katsura. Formal languages consisting of primitive words. In: Z. Ésik, editor, *Proceedings of the 9th = International Conference on Fundamentals of Computation Theory*, number 710 in LNCS, = pages 194–203, Szeged, Hungary, 1993. Springer.

4. S. Ginsburg. *The Mathematical Theory of Context-Free Languages*. McGraw-Hill, New York, 1966.

5. M. Ito and M. Katsura. Context-free languages consisting of non-primitive words. *International Journal on Computer Mathematics,* = 40:157–167, 1991.

6. H. Petersen. The Ambiguity of Primitive Words, STACS'94

7. Gh. Păun. Morphisms and Primitivity. *EATCS Bulletin, Technical Contributions,* 1997 85–88.

8. A. Salomaa. From Parikh vectors to GO territories. *EATCS Bulletin, Formal Language Column* 1995 89–95.

9. J. Berstel and L. Boasson. The set of Lyndon words is not context-free *EATCS Bulletin, Technical Conributions,* 63 1997 = 139–140.

10. L. Kászonyi. On a class of stratified linear sets. *Pure Mathematics and Applications, Ser. A,* = 6(2):203–210, 1995.

11. L. Kászonyi and M. Holzer. A generalization of the flip-flop lemma *AFL '96, Salgótarján* (1996) *Publ. Math. Debrecen* 54 (1999) suppl., 203-210.

12. L. Kászonyi. On bounded context-free languages In: T. Imaoka and C. Nehaniv, editors, *Proceedings of the First Symposium on Algebra, Languages and Computation* University of Aizu, 1997. Japan

13. L. Kászonyi. On **DLI**-sets of Katsura type. *AFL'2002, Debrecen* , submitted.

14. L. Kászonyi. A pumping lemma for **DLI**-languages *Discrete Mathematics* 258 (2002) 105-122.

15. L. Kászonyi. How to generate binary codes using context-free grammars. In: M. Ito, G. Paun, Sheng Yu, editors, *Words, Semigroups, Transductions, Festschrift in Honor of Gabriel Thierrin* World Scientific, 289-301, 2001.

16. L. Kászonyi and M. Katsura. On an algorithm concerning the languages $Q \cap (ab^*)^n$. *PU.M.A (Pure and Applied Mathematics.)* 10(3):313–322, 1999.

17. L. Kászonyi and M. Katsura. On the context-freeness of a class of primitive words. *Publ. Math. Debrecen* 51 (1997) 1-11.

18. L. Kászonyi and M. Katsura. Some new results on the context-freeness of languages $Q \cap (ab^*)^n$. In: *Proc. of the 8th Conf. Automata and Formal Languages (AFL'96)*, Salgótarján, Hungary, 29 Juli- 2 August. *Publ. Math. Debrecen*, 54 885-894, 1999.

19. A. Mateescu, Gh. Păun, G. Rozenberg, and A. Salomaa. Parikh prime words and GO-like territories. *Journal of of Universal Computer Science*, = 1(12):790–810, 1995.

Finite Completion of Comma-Free Codes. Part I

Huong Lam Nguyen

Hanoi Institute of Mathematics,
P.O.Box 631, Bo Ho, 10 000 Hanoi, Vietnam

1 Introduction

In this paper we consider (finite) comma-free codes and their completion problem. We sketch a few lines on their origin, history and development. Comma-free codes were defined rigorously in 1958, as mathematical objects, in Golomb, Gordon and Welch [4], although a rudimentary notion had been suggested by Crick, Griffith and Orgel earlier in 1957 [2], in connection with the famous discovery of DNA structure; for more details on the biological origin of the problem, see [6], or [1], annotation to Chap. VII.

During the late 1950s and the 1960s there had been quite extensive investigation on comma-free codes of constant length (block codes), free from biological considerations. Then the main line of study was concerned with the maximal size of comma-free codes of a given length and on alphabets with a fixed number of letters. The most impressive achievement was a proof by Eastman [3] of a conjecture of Golomb, Gordon and Welch [5] on the maximal number of words in a comma-free code on k-letter alphabet, k odd; see [1, Chap. VII] for alternative proofs.

The research on variable-length comma-free codes was initiated a bit later, in 1969 [10], and it still goes on as of 1998 [4]. In the later paper some language-theoretic aspects of maximal comma-free codes are discussed.

Several problems often encountered in theory of codes are of the following kind: given code belongs to some special family of codes, is it possible to complete it within this family? For instance, it is trivial to see that every finite prefix code every can be embedded into a finite maximal prefix code. It also represents little difficulty to prove that every finite infix code can be completed to a finite maximal infix code [8]. The case of general codes is more difficult. It was proved by Restivo [9] that there exist finite codes that are not included in any finite maximal codes. More on the negative side we can mention the case of bifix codes [1, Chap. III].

In this paper, we manage to prove that every finite comma-free code has a finite completion, or in other words, is included in a finite maximal comma-free code. Thus we give the affirmative answer to the completion problem for another class of codes.

Our solution is complex, so we divide it into two parts. First, we attempt to complete every finite comma-free code to a (finite) comma-free code of a special kind, which we call *canonical* comma-free codes. Canonical comma-free codes are defined by a neat condition which warrants that, if not maximal, they

M. Ito and M. Toyama (Eds.): DLT 2002, LNCS 2450, pp. 357–368, 2003.
© Springer-Verlag Berlin Heidelberg 2003

can be made into maximal comma-free codes by adding "not very" long words. This part is the content of present paper. Next, we prove, by a concrete and explicit construction, that every canonical comma-free code is completed to a finite maximal comma-free code. This final part we hope to publish in another paper.

The present paper is structured as follows: Following this introduction, Sect. 2 introduces standard notation and basic notions. In Sect. 3, canonical words, the "stones" with which a completion is built, are introduced and studied. Section 4, the most heavy, describes the construction, a repeated process, which is entitled to produce desired finite canonical comma-free codes. Of course, the number of steps to be repeated should be determined by the code at the starting point. This is proved by means of the notions of index and type of right borders. This section prepares the ground for the definition of canonical comma-free codes and the completing procedure itself in the last Sect. 5.

2   Notions and Notation

Our terminology is standard. Let A be a finite alphabet. Then A^* denotes the set of words on A, including the empty word 1, and as usual $A^+ = A^* - \{1\}$. For subsets X, X' of A^*, we denote $XX' = \{xx' : x \in X, x' \in X'\}$ and $\max X = \max \{|x| : x \in X\}$ and

$$X^0 = \{1\}$$
$$X^{i+1} = X^i X, \quad i = 0, 1, 2, \ldots.$$

We define the principal objects of this paper [11].

Definition 2.1. A set $X \subseteq A^+$ is a comma-free code if $X^2 \cap A^+ X A^+ = \emptyset$.

A comma-free code is *maximal* if it is not included in any other comma-free codes. A completion of a comma-free code is a maximal comma-free code containing it. Every comma-free always has completions, in view of Zorn's lemma. In this papers we deal exclusively with finite comma-free codes and finite completions.

Example 2.2. [7]. Let $A^2 = X + Y$ be a partition of the set of words of length 2 and let $k > 0$. Then

$$C = \{a_1 \ldots a_k b_1 \ldots b_k : a_1 b_1 \in X, a_i b_i \in Y, i > 1\}$$

is a comma-free code.

Comma-free codes are closely connected to the notion of overlapping. We say that the two words u and v *overlap* if

$$u = tw, \quad v = ws$$

for some non-empty words $s, t \in A^+$ and $w \in A^+$, or equivalently,

$$us = tv$$

for some non-empty words s, t such that $|t| < |u|$ and $|s| < |v|$. We call the equalities above *overlapping equalities* for a pair of words u, v and we call s a *right border* and t a *left border* of the pair of overlapping words u, v. A right (left) border of a subset X is a right (left, resp.) border of any pair of overlapping words in X. We denote the sets of left borders and right borders of X by $L(X)$ and $R(X)$, respectively.

Throughout this paper, comma-free codes are assumed not to be subsets of the alphabet because for those comma-free codes the completion problem is trivial: the whole alphabet!

3   Canonical Words

In this section we introduce the concept of canonical words and strictly canonical words and study them in detail.

3.1   Canonical Words

Let X be a finite comma-free code, t be a positive integer.

Definition 3.1. The word w is said to be a left (right) X-canonical word of tag t, if every right (left, resp.) factor of w of length longer or equal to t has a left (right, resp.) factor in $R(X)$ (in $L(X)$, resp.).

Clearly, a left X-canonical word w of tag t has the form

$$w = us, \quad u \in A^*, s \in A^+, |s| = t$$

and for every right factor u', including 1, of u

$$u's \in R(X)A^*.$$

Similarly, a right X-canonical word w of tag t has the form

$$w = su, \quad u \in A^*, s \in A^+, |s| = t$$

and for every left factor u', including 1, of u

$$su' \in A^*L(X).$$

We often use the concept of canonical word under the following variation. Let m be a non-negative integer.

Definition 3.2. The word w is said to be a left (right) X, m-canonical word if and only if it is a left (right, resp.) X-canonical word of tag $|w| - m$.

Clearly, if w is a left X, m-canonical word then we can write w in the form

$$w = us, \quad s \in A^+, u \in A^*, |u| = m$$

and for every right factor u' of u

$$u's \in R(X)A^*.$$

Similarly, if w is a right X, m-canonical word then we can write w in the form

$$w = su, \quad s \in A^+, u \in A^*, |u| = m$$

and for every left factor u' of u

$$su' \in L(X)A^*.$$

Here we make no emphasis on the tag.

Definition 3.3. A left (right) X, m-canonical word is called minimal left (right, resp.) X, m-canonical, $m \geq 0$, if it is not a proper left (right, resp.) factor of any other left (right, resp.) X, m-canonical word.

The set of X, m-canonical words is not finite, for any fixed m, but the set of minimal ones is finite because of the following proposition.

Proposition 3.4. If $w = us$ $(w = pu)$, $|u| = m$, is a minimal left (right, resp.) X, m-canonical word then there is a right (left, resp.) factor u' of u such that $u's \in R(X)$ $(pu' \in L(X)$, resp.$)$. As a consequence, $|s| < \max X$ $(|p| < \max X)$.

We present some technical statements which will be in use later.

Proposition 3.5. Let X be a finite comma-free code, Y be a subcode of X and let $a_1 \ldots a_k \ldots a_{k+t}$ $(a_{k+t} \ldots a_k \ldots a_1)$ be a left (right, resp.) Y, k-canonical word (of tag t) for a non-negative integer k, a positive integer t and letters $a_1, \ldots, a_k, \ldots, a_{k+t}$. Then for any integer d such that $d > 0$ and $d \leq k+1$ there exists an integer i such that $i \leq k+t$ and $d \leq i < d + \max Y - 1$ for which $a_1 \ldots a_i$ $(a_i \ldots a_1$, resp.$)$ has no right (left, resp.) factor in $L(X)$ $(R(X)$, resp.$)$ of length less than or equal to d.

Proof. We suppose on the contrary that for every i, $i \leq k+1$ and $d \leq i < d + \max Y - 1$, the word $a_1 \ldots a_i$ has a right factor in $L(X)$ of length less than or equal to d. Set $i(0) = d$. We have

$$a_{i(1)} \ldots a_{i(0)} \in L(X)$$

for some $1 \leq i(1) \leq i(0)$ and $1 + i(1) - i(0) \leq d$. Since $i(1) \leq d \leq k+1$ and

$$|a_{i(1)} \ldots a_k a_{k+1} \ldots a_{k+t}| = t + k + 1 - i(1) \geq t,$$

by definition, $a_{i(1)} \ldots a_k a_{k+1} \ldots a_{k+t}$ has a left factor in $R(Y)$, that is

$$a_{i(1)} \ldots a_{i(2)} \in R(Y)$$

for some $i(1) \leq i(2) \leq k+t$. We must have

$$i(0) < i(2)$$

in virtue of the comma-freeness of X. Note that

$$i(2) = i(1) + i(2) - i(1) \leq i(0) + \max Y - 2 = d + \max Y - 2.$$

Now we repeat the argument with $i(2)$ playing the role of $i(0)$ and so on; the assumption by contradiction allows us to obtain the chain

$$1 \leq \ldots < i(3) < i(1) \leq i(0) < i(2) < \ldots \leq k + t$$

of integers. But this is impossible, as k and t are fixed.

3.2   Strictly Canonical Words

Let X be a finite comma-free code and $m \geq \max X$.

Definition 3.6. A word of the form aus (pvb) is said to be left (right, resp.) strictly X, m-canonical if it satisfies the following conditions:
1. $a \in A$, $|v| = m$, $s \in A^+$ $(b \in A, |v| = m, p \in A^+,$ resp.$)$.
2. us (pv) is a left (right, resp.) X, m-canonical word.
3. $aus \notin R(X)A^*$ $(pvb \notin A^*L(X),$ resp.$)$.
4. $aus \notin A^*XA^*$ $(pvb \notin A^*XA^*,$ resp.$)$.

It is straightforward to see from the definition that all left, as well as all right, strictly X, m-canonical words are not factors of X^2.

Definition 3.7. A left (right) strictly X, m-canonical word is called minimal left (right, resp.) X, m-canonical if it does not contain properly any left (right, resp.) strictly X, m-canonical words as left (right, resp.) factors.

We denote the sets of minimal left and minimal right strictly X, m-canonical words by $T(X, m)$ and $U(X, m)$, respectively. The characterization of the minimal strictly canonical words is similar to the case of minimal canonical words.

Proposition 3.8. *If aus (pvb) is a minimal left (right, resp.) strictly X, m-canonical word then us $(pv,$ resp.$)$ is a minimal left (right, resp.) X, m-canonical word. Consequently, $|s| < \max X \leq m$ $(|p| < \max X \leq m,$ resp.$)$ and $T(X, m)$ $(U(X, m),$ resp.$)$ is finite.*

Let $E(X)$ denote the set of all words that avoid X (i.e. that have no factors in X) and that have no left factors in $R(X)$ and no right factors in $L(X)$. In symbols

$$E(X) = A^* - R(X)A^* - A^*L(X) - A^*XA^*.$$

We have the following criterion

Theorem 3.9. *The set $T(X, m)$ $(U(X, m))$ of minimal left (right, resp.) strictly X, m-canonical words is not empty if and only if $E(X)$ contains a left (right, resp.) X, m-canonical factor. Specifically speaking, if a word of $E(X)$ contains a left (right, resp.) X, m-canonical factor, it contains a word in $T(X, m)$ $(U(X, m),$ resp.$)$.*

Proof. The "if" part is evident by the remark preceding the theorem. For the opposite direction, the "only if" part, let suppose that $E(X)$ contains a word e which has a left X, m-canonical factor us, i.e.

$$e = zusy$$

for some $z, y \in A^*$. First, $z \in A^+$ because $e \notin R(X)A^*$. Further, we can assume that z is chosen as of minimal length among all possible factorizations like one above, i.e. us is chosen as the left-most occurrence of any left X, m-canonical factors. This means that if we write $z = z'a$, for $a \in A$ and $z' \in A^*$, we have $aus \notin R(X)A^*$ showing that aus is a left strictly X, m-canonical which, indeed, has a minimal left strictly X, m-canonical left factor. This proves the "only if" direction and the theorem follows.

Now we reveal one of the main purposes which the concept of (strictly) canonical word is introduced for.

Theorem 3.10. *The sets $X + T(X, m)$ and $X + U(X, m)$ both are comma-free codes for all $m \geq \max X$.*

The proof consists in checking the impossibility of the seven cases of "incidence" of non-comma-freeness. It is straightforward and routine. We further state an important property of $T(X, m)$ and $U(X, m)$.

Proposition 3.11. *Let $Y = X + T(X, m)$ ($Y = X + U(X, m)$). Then every right (left, resp.) border in $R(Y)$ ($L(Y)$, resp.) has a (non-empty) left (right, resp.) factor which is a right (left, resp.) in $R(X)$ ($L(X)$, resp.).*

We derive an immediate consequence, which shows that, when applied the constructions T and U bring no new left and right canonical words, resp. (They could apparently only reduce the sets of strictly canonical words.)

Corollary 3.12. *Let $Y = X + T(X, m)$ ($Y = X + U(X, m)$). Then every left (right, resp.) Y, m-canonical word is also a left (right, resp.) X, m-canonical word.*

The following theorem, which is also an easy corollary of Proposition 3.11, describes a basic property of the constructions T and U.

Theorem 3.13. *Let $Y = X + T(X, m)$ ($Y = X + U(X, m)$). Then $E(Y)$ does not contain left (right, resp.) Y, m-canonical factors.*

Proof. Suppose that a certain word e of $E(Y)$ contains a left Y, m-canonical factor. By the preceding corollary this factor is also a left X, m-canonical word. Since $e \in E(Y) \subseteq E(X)$, by Theorem 3.9, e contains a minimal left strictly X, m-canonical factor, i.e. a factor in $T(X, m) \subseteq Y$. But this is a contradiction, as $E(Y)$ avoids Y, which proves the theorem.

4   Constructions T and U and Their Right Borders

We begin with the main construction which is an iterated process.

4.1   Constructions T and U

Let X_0 be a finite comma-free code, k and λ positive integers, λ is sufficiently large. We define the sequence of finite comma-free codes

$$X_0, Y_1, \ldots, X_{2k}, Y_{2k+1}$$

as follows.

$$Y_1 = X_0 + U(X_0, \lambda n_0)$$
$$X_2 = Y_1 + T(Y_1, n_1)$$
$$\cdots\cdots\cdots$$
$$X_{2k} = Y_{2k-1} + T(Y_{2k-1}, n_{2k-1})$$
$$Y_{2k+1} = X_{2k} + U(X_{2k}, \lambda n_{2k})$$

where $n_0 = \max X_0$, $n_1 = \max Y_1$, ..., $n_{2k-1} = \max Y_{2k-1}$, $n_{2k} = \max X_{2k}$. We get then the ascending chain

$$X_0 \subseteq Y_1 \subseteq \cdots \subseteq X_{2k} \subseteq Y_{2k+1}.$$

We highlight the following inequalities, and what is more important in the sequel,

$$n_{2i} > n_{2i-1} = \max U(X_{2i-2}, \lambda n_{2i-2})$$

for every $i = 1, 2, \ldots k$.

We would like to remark that λ is arbitrarily large but fixed integer throughout the construction. The question how large λ is depends on the later applications and we shall specify it there.

4.2   Right Borders and Simple Right Borders: Index and Type

In this subsection we investigate right borders in detail, more exactly, simple right borders, which are defined as follows.

Definition 4.1. A right border s of Y_{2k+1} is called simple right border if it does not contain any other right border of Y_{2k+1} as a proper left factor, i.e., $s \in R(Y_{2k+1})$ but $s \notin R(Y_{2k+1})A^+$.

We distinguish the right borders by means of their index, defined as follows.

Definition 4.2. Index of a right border s is the least integer i, such that s is a right border of the overlapping equality $y_2s = ly_1$, where $|l| < |y_2|$, $y_1, y_2 \in Y_{2k+1}$ and $y_1 \in T_i$ or $y_1 \in U_i$ respective to i odd or even, or -1 if $y_1 \in X_0$.

The key moment of this section is to define the concept of type. Type of a simple right border s is, in a way, minimal of the indices of some simple right borders that are right factors of s. Once defined, the type $t(s)$ of s satisfies:

1. $-1 \le t(s) \le k$;
2. If $t(s) = -1$ then s is a right factor of X_0; in particular $|s| < \max X_0$.
3. If $t(s) \ge 0$ then s is a right factor of $U_{2t(s)}$; in particular $|s| < \max U_{2t(s)}$ and $|s| > \min\left(\lambda n_{2t(s)}, (\lambda - 1)n_{2t(s)}, (\lambda - 2(k-1))n_{2t(s)}\right) > (\lambda - 2k + 1)n_{2t(s)}$.

These points are embodied in the following theorem.

Theorem 4.3. *For a simple right border s of Y_{2k+1}, if $t(s) = -1$ then s is a right factor of X_0, in particular $|s| < \max X_0$. If $t(s) \geq 0$ then s is a right factor of $U_{2t(s)}$ and $(\lambda - 2k + 1)n_{2t(s)} < |s| < \max U_{2t(s)}$, moreover, s is a right strictly $X_{2t(s)}, m_{2t(s)}$-canonical word with tag less than $n_{2t(s)}$ and $m_{2t(s)} > (\lambda - 2k)n_{2t(s)}$.*

Proof. We have to show only the last claim. Suppose that $t(s) \geq 0$. Since s is a right factor of some word of $U_{2t(s)}$, which is a (minimal) right strictly $X_{2t(s)}, \lambda n_{2t(s)}$-canonical word of tag less than $n_{2t(s)}$ then s is a right strictly $X_{2t(s)}, m_{2t(s)}$-canonical word of the same tag and for some $m_{2t(s)}$. From the fact that $|s| > (\lambda - 2k + 1)n_{2t(s)}$, we see at once that

$$m_{2t(s)} > (\lambda - 2k)n_{2t(s)}$$

which concludes the proof.

5 Canonical Comma-Free Codes

In this section, we introduce the concept of canonical comma-free code and we prove that every finite comma-free code is included in a finite canonical comma-free code.

5.1 Basic Theorems. Canonical Comma-Free Code

Let $\lambda \geq 2k + 2$ and $g = (\lambda + 1)n_0 + (\lambda + 1)n_2 + \cdots + (\lambda + 1)n_{2k} + n_{2k-1}$. We have the following instrumental result.

Theorem 5.1. *If $E(Y_{2k+1})$ has a left Y_{2k+1}, g-canonical factor then there exists a left border $s \in R(Y_{2k+1})$ such that $s \in S(X_0)$ but $s \notin R(X_{2k})$.*

Proof. Let

$$a_1 \ldots a_g a_{g+1} \ldots a_{g+t},$$

where $a_1, \ldots, a_{g+t}$ are letters, be a left Y_{2k+1}, g-canonical factor of $E(Y_{2k+1})$ (of tag t). We prove the following
Claim k. For every positive integer i,

$$i \leq g - (\lambda + 1)n_{2k}$$

the word

$$a_i \ldots a_g a_{g+1} \ldots a_{g+t}$$

has no left factors which are right borders of type k.

In order to prove this, suppose by contradiction that $a_i \ldots a_g a_{g+1} \ldots a_{g+t}$ has a left factor

$$s = a_i \ldots a_h a_{h+1}$$

which is a simple right border of type k. By Theorem 4.3, $|s| < \max U_{2k}$, which is less than $(\lambda + 1)n_{2k}$, hence $h + 1 \leq g$ as $g - i \geq (\lambda + 1)n_{2k}$, and s is a right strictly

U_{2k}, m_{2k}-canonical word of tag $< n_{2k}$ and with $m_{2k} \geq (\lambda - 2k)n_{2k} \geq 2n_{2k}$. To be more precise, we write

$$s = a_i \ldots a_f a_{f+1} \ldots a_h a_{h+1},$$

where $|a_i \ldots a_f| < n_{2k}$, $h - f = m_{2k}$ and

$$a_i \ldots a_f a_{f+1} \ldots a_h$$

is a right U_{2k}-canonical word of tag f. Put $d = h - n_{2k}$, $e = h - 2n_{2k}$. Certainly,

$$i \leq f \leq e < d < h < g.$$

We show that for every integer j in the range

$$e < j \leq d$$

the word

$$a_j \ldots a_g a_{g+1} \ldots a_{g+t}$$

has no left factor which is a simple right border of Y_{2k+1} of type k.
    In fact, if it has such a right border

$$a_j \ldots a_{j+h'+1}$$

as a left factor, which we present, as a right strictly Y_{2k+1}, m_{2k}-canonical word (note that $U_{2k} \subseteq Y_{2k+1}$), as follows:

$$a_j \ldots a_{j+f'} a_{j+f'+1} \ldots a_{j+h'-1} a_{j+h'}$$

where

$$n_{2k} > f' \geq 0, \quad h' - f' = m_{2k}.$$

Now observe that

$$j + f' < j + n_{2k} \leq d + n_{2k} = h$$

and

$$h = e + 2n_{2k} < j + 2n_{2k} < j + f' + 2n_{2k} \leq j + f' + m_{2k} = j + h',$$

or

$$h + 1 \leq j + h' - 1.$$

These two inequalities show that, with respect to the right Y_{2k+1}, m_{2k}-canonical word $a_j \ldots a_{j+h'+1}$ of tag f', the word $a_j \ldots a_{h+1}$, and hence $a_i \ldots a_h a_{h+1}$ has a right factor which is in $L(Y_{2k+1})$. But it is impossible, because $a_i \ldots a_h a_{h+1}$ is strictly right Y_{2k+1}-canonical word. Therefore for every j, $e < j \leq d$,

$$a_j \ldots a_g a_{g+1} \ldots a_{g+t}$$

must have a left a factor which is a right border of type less than k and which is of length less than $\max U_{2(k-1)} = n_{2k-1} < n_{2k}$. But this contradicts Proposition

3.5 with respect to the right X_{2k}-canonical word s (herewith, $X_{2k} = Y$) which says that there should be an integer j such that

$$e = d - n_{2k-1} = d - \max U_{2(k-1)} < j \le d,$$

for which $a_j \dots a_g a_{g+1} \dots a_{g+t}$ has no such a factor! This total contradiction proves Claim k.

Thus, for every $i \le g - (\lambda+1)n_{2k}$, $a_i \dots a_g a_{g+1} \dots a_{g+t}$ has left factors that are simple right borders only of type less than k. By an argument entirely identical to the one above, we can further establish for every $l \le 0$

Claim l. For every $i \le g - (\lambda + 1)n_{2k} - \dots - (\lambda + 1)n_{2l}$ the word

$$a_i \dots a_g a_{g+1} \dots a_{g+t}$$

has no left factors which are simple right borders of type greater or equal to l.

When $l = 0$, we finally get that for every integer i satisfying

$$i \le g - (\lambda+1)n_{2k} - \dots - (\lambda+1)n_0 = n_{2k-1}$$

the word

$$a_i \dots a_g a_{g+1} \dots a_{g+t}$$

has only right simple borders of Y_{2k+1} of type -1 as left factors.

Now that $E(Y_{2k+1}) \subseteq E(X_{2k})$ and $E(X_{2k})$ has no factors that are left X_{2k}, n_{2k-1}-canonical words, it immediately follow that there exist a (simple) right border of Y_{2k+1} of type -1, meaning that it is a right factor of X_0, and it is not a right border of X_{2k}. This achieves the proof of the theorem.

Let $k = |S(X)|+1$ and put $g_i = (\lambda+1)n_0+(\lambda+1)n_2+\dots+(\lambda+1)n_{2i}+n_{2i-1}$ for $i = 1, \dots k$.

Theorem 5.2. *There exists an integer m, $1 \le m \le k$ such that $E(Y_{2m+1})$ does not contain right Y_{2m+1}, g_m-canonical factors. Hence Y_{2m+1} is a g_m-canonical comma-free code.*

Proof. If the theorem fails, then for every $i = 1, 2, \dots k$, $E(Y_{2i+1})$ contains a left Y_{2i+1}, g_i-canonical factor. Since $\lambda \ge 2k + 2 \ge 2i + 2$, applying Theorem 5.1 for each i, we get a right border $s_i \in S(X_0)$ such that $s_i \in R(Y_{2i+1})$ but $s_i \notin R(X_{2i})$. However, $R(Y_{2i-1}) \subseteq R(X_{2i})$ implies that $s_i \in R(Y_{2i+1})$ but $s_i \notin R(Y_{2i-1})$. All together, it means all $s_1, \dots, s_k$ are distinct, hence $k \le |S(X_0)|$. But this is a contradiction, as $k = |S(X_0)| + 1$. Next, note that by Theorem 3.13, $E(Y_{2m+1})$ has no left $Y_{2m+1}, \lambda n_{2m}$-canonical factors, hence Y_{2m+1}, g_m-canonical factors, as $g_m > \lambda n_{2m}$, which proves the theorem.

The Theorem 5.2 motivates the central notion of this paper. Let N be a positive integer.

Definition 5.3. A comma-free code X is said to be N-canonical if $E(X)$ contains no left X, N-canonical factor and no right X, N-canonical factors. A comma-free code is said to be canonical if it is N-canonical for some positive integer N.

Evidently, this definition is equivalent to the next, which is more explicit,

Definition 5.4. A comma-free code X is called N-canonical if for an arbitrary word $w \in E(X)$ and an arbitrary factorization $w = xuy$ with $x, y, u \in A^*$ and $|u| \geq N$, there exist factorizations $u = pp' = ss'$ such that $xp \in E(X)$ and $s'y \in E(X)$, or just the same, $xp \notin A^* L(X)$ and $s'y \notin R(X)A^*$.

5.2 Completing to Canonical Comma-Free Codes

Now we come to the culminating point of this section, the completion theorem.

Theorem 5.5. *Every finite comma-free code X can be completed to a finite N-canonical comma-free code Y with*

$$\max Y \leq 2^k (2k+2)^{k+1} \max X$$

and

$$N = 4^{k+1} \frac{(2k+3)^2}{4k+5} (k+1)^{k+1} + 2^{2k-1}(k+1)^k$$

where $k = |S(X)| + 1$.

Proof. Set $X_0 = X$ and $\lambda = 2k + 2$. The first claim is Theorem 5.2 and the numerical claims follow from an appropriate estimate for g_m.

6   Concluding Remarks

As said before, Theorem 5.2 represents the first stage of the problem of finite completion of comma-free code by proving that there always exists a finite N-canonical containing a given finite comma-free code X. We can figure out that the procedure is actually an effective one, because it involves only finite or regular languages with checking the emptiness problem for them.

Besides, we would like to remark that the estimates do not tend to be best possible; rather, we present them in a more or less quantitative manner to give flavor of the method.

We hope to publish the remaining part of the solution, namely, the proof that every finite N-canonical comma-free code, for all $N > 0$, has finite completions, in another paper. A sketch of it will appear in the Proceedings of the RIMS Symposium on Algebraic Systems, Formal Languages and Conventional and Unconventional Computation Theory , Kyoto, September 24–26, 2002.

References

1. Berstel, J., Perrin, D.: "Theory of Codes", Academic Press, Orlando, 1985
2. F. H. C. Crick, J. S. Griffith, L. E. Orgel, *Codes without Commas*, Proc. Nat. Acad. Sci. USA, **43**(1957), 416–421.

3. Eastman, W. L.: On the Construction of Comma-free Codes. IEEE Trans. Information Theory **IT-11** (1965) 263–267
4. Fan, C. M., Shyr, H. J.: Some Properties of Maximal Comma-free Codes. Tamkang Journal of Mathematics**29** (1998) 121–135
5. Golomb, S. W., Gordon, B., Welch, L. R.: Comma-free Codes. Canad. J. Math. **10** (1958) 202–209
6. Golomb, S. W., Welch, L. R., Delbrück, M: Construction and Properties of Comma-free Codes. Biol. Medd. Dan. Vid. Selsk. **23** (1958) 3–34
7. Ito, M., Jürgensen, H., Shyr, H. J., Thierrin, G.: Outfix and Infix Codes and Related Classes of Languages. Journal of Computer and System Sciences **43** (1991) 484–508.
8. Jiggs, B. H.: Recent Results in Comma-free Codes. Canad. J. Math. **15** (1963) 178–187
9. Restivo, A.: On Codes Having No Finite Completions. Discreet Mathematics **17** (1977) 306-316
10. Scholtz, R. A.: Maximal and Variable Word-length Comma-free Codes. IEEE Trans. Information Theory **IT-15** (1969) 555–559
11. Shyr H. J.: "Free Monoids and Languages", Lecture Notes, Hon Min Book Company, Taichung, 1991

On a Family of Codes
with Bounded Deciphering Delay

Long Van Do[1] and Igor Litovsky[2]

[1] Institute of Mathematics, P.O. Box 631 Bo Ho, 10000 Hanoi, Vietnam
dlvan@thevinh.ncst.ac.vn
[2] ESSI, 930 route des Colles, BP 145, 06903 Sophia Antipolis, France
lito@essi.fr

Abstract. A special kind of codes with bounded deciphering delay, called k-comma-free codes, is considered. The advantage in using these codes is that the decoding can begin "anywhere" in a coded message. This means that a coded message can be deciphered even when it might be lost partially. Related families of codes with bounded deciphering delay are also considered. Criteria to test these codes are established in a unified way.

1 Preliminary

Prefix codes are among the simplest codes, but most important problems in the theory of codes may arise for prefix codes. Codes with bounded deciphering delay (w.b.d.d.) naturally generalize prefix codes in the sense that the decoding of a coded message can begin after a bounded delay (0 delay for prefix codes) without waiting until the end of the message. Sometimes in practice, by technical reasons, an obtained coded message can be lost partially. This prevents the decoding of the rest of the message. We consider in this paper a special kind of codes w.b.d.d., called k-comma-free codes, for which the decoding can begin "anywhere" in a coded message, and therefore the mentioned above difficulty may be overcome.

Let A be a finite *alphabet* and A^* the set of all the *words* over A including the *empty word* ϵ. We denote by $|u|$ the *length* of a word $u \in A^*$. Any subset $X \subseteq A^*$ is a *language* over A. For any languages X, Y we use the following classical operations: the *union* $X \cup Y$, the *intersection* $X \cap Y$, the *difference* $X - Y$, the *concatenation* XY, the *star* X^*, the *plus* $X^+ = X^* - \{\epsilon\}$, the *left* and *right quotients* which are respectively

$$X^{-1}Y = \{u \in A^* |\ \exists x \in X, \exists y \in Y : xu = y\}$$

$$XY^{-1} = \{u \in A^* |\ \exists x \in X, \exists y \in Y : x = uy\}$$

A language $X \subseteq A^+$ is a *code* if $\forall n, m \geq 0, \forall x_1, x_2, \ldots, x_n, y_1, y_2, \ldots, y_m \in X$,

$$x_1 x_2 \ldots x_n = y_1 y_2 \ldots y_m \ \Rightarrow \ n = m \text{ and } x_i = y_i \text{ for all } i$$

More intuitively, it is equivalent to say that any message w encoded by replacing every letter b in w by the corresponding code-word $c(b)$ in X must be deciphered

M. Ito and M. Toyama (Eds.): DLT 2002, LNCS 2450, pp. 369–380, 2003.

in a unique way. The verifying of whether a given language X is a code is not always easy. A procedure for this, known as the Sardinas-Patterson criterion, works as follows. One associates with X a sequence of sets $U_i(X)$ (U_i, for short), $i \geq 0$, where

$$U_0 = X^{-1}X - \{\epsilon\},$$
$$U_{i+1} = X^{-1}U_i \cup U_i^{-1}X, \forall i \geq 0.$$

Theorem 1 ([6, 1]). *For any non-empty subset X of A^+, X is not a code if and only if $\epsilon \in U_i(X)$ for some i.*

Let $u, v \in A^*$. We say that u is a *factor* of v if $v = xuy$ for some $x, y \in A^*$. If $xy \neq \epsilon$ then u is a *proper factor* of v. If $x = \epsilon$ then u is a *prefix* of v, and if moreover $y \neq \epsilon$ then u is a *proper prefix* of v. *Suffixes* and *proper suffixes* are defined in the same way. A subset X of A^+ is a *prefix code* (*suffix code*) if no word in X is a proper prefix (proper suffix, resp.) of another word in X. We call X an *infix code* if no word in X is a proper factor of another word in X. For prefix codes, the deciphering of a coded message can be done instantaneously: reading the message from left to right starting from the *begining* of the message, the code-words found consecutively are exactly the code-words which have been used to encode the original message.

Codes with deciphering delay k, k is a non-negative integer, generalize prefix codes in the following sense: reading a coded message from left to right starting from the begining of the message, every time $k+1$ consecutive code-words can be found one may confirm that the first one is exactly the code-word used to encode the original message. More precisely, a non-empty subset X of A^+ is a *code with deciphering k* if $\forall x, y \in X$:

$$xX^k A^* \cap yX^* \neq \phi \Rightarrow x = y. \tag{1}$$

The smallest k satisfying the above property is the *delay* of the code X. We denote by $\mathcal{D}_k$ the family of all codes with deciphering delay k. If X is a code with deciphering delay k for some k, we say that X is a *code with finite deciphering delay* [1] or *with bounded deciphering delay* [2]. Otherwise, X has an *infinite deciphering delay*. The family of codes w.b.d.d. is denoted by $\mathcal{D}$, $\mathcal{D} = \bigcup_{k \geq 0} \mathcal{D}_k$. It is easy to see that for every $k \geq 0$, the set $\{a, a^k b\}$ is a code with deciphering delay k (see Example 2) whereas the code $\{a, ab, b^2\}$ has an infinite deciphering delay.

The paper is organized as follows. In Section 2 the definition of k-comma-free codes is introduced (Definition 1). It appeared that the family of these codes is strictly included in the family of codes w.b.d.d. (Proposition 1). A characterization of such codes is given (Theorem 2). A more detailed computation lead to a "finitelization" of this characterization (Theorem 3) which allows in particular to show that the 0-comma-free codes coincides with the comma-free codes (Corollary 1). Section 3 is devotd to prove a basic lemma (Lemma 3) by means of which criteria for testing the families of codes under consideration can be established in a unified way. In Section 4 a test for k-comma-free codes is stated

(Theorem 4). As an application, it is shown that there exists an infinite hierarchy after k of k-comma-free codes (Proposition 3). Section 5 considers a family of codes larger than that of k-comma-free codes, named k-cohesive codes. A test for these codes is formulated (Theorem 5). Relationship between k-cohesive codes, k-comma-free codes and codes with deciphering delay k is pointed out (Propositions 5.7, 5.8, 5.9).

2 k-Comma-Free Codes

Definition 1. *Let X be a non-empty subset of A^+, and k a non-negative integer. We call X a k-comma-free code if $\forall l \geq 0, \forall x_0, \ldots, x_k, y_0, \ldots, y_l \in X, \forall u, v \in A^*$:*

$$u x_0 \ldots x_k v = y_0 \ldots y_l \Rightarrow \exists m, 0 \leq m \leq l, : x_0 = y_m, u = y_0 \ldots y_{m-1} \qquad (2)$$

The family of all k-comma-free codes is denoted by $\mathcal{C}_k$, and $\mathcal{C} = \bigcup_{k \geq 0} \mathcal{C}_k$.

Any k-comma-free code is a code with bounded deciphering delay. Namely,

Proposition 1. *For any $k \geq 0$, $\mathcal{C}_k \subset \mathcal{D}_k$ and $\mathcal{C} \subset \mathcal{D}$.*

Proof. The inclusion $\mathcal{C}_k \subseteq \mathcal{D}_k$ is immediate from Definition 1 with $u = \epsilon$. Hence $\mathcal{C} \subseteq \mathcal{D}$. The language $X = \{ab, ba\}$ over the alphabet $A = \{a, b\}$, being prefix, is in $\mathcal{D}_k$ for all $k \geq 0$, hence $X \in \mathcal{D}$. On the other hand, the evident equality $a(ba)^{k+1}b = (ab)^{k+2}$ shows that $X \notin \mathcal{C}_k$ for all k, and therefore $X \notin \mathcal{C}$. It follows the strictness of the inclusions $\square$

The following proposition gives an usefull criterion to distinguish k-comma-free codes among codes w.b.d.d.

Proposition 2. *Let $X \in \mathcal{D}_k$. Then $X \notin \mathcal{C}_k$ if and only if*

$$\exists x, y \in X, \exists z \in P(y) - \{\epsilon\} : zx X^k A^* \cap y X^* \neq \phi \qquad (3)$$

where $P(y)$ is the set of all proper prefixes of y.

Proof. ($\Leftarrow$). Suppose (3) holds. Then $\exists x_0 = x, x_1, \ldots, x_k, y_0 = y, y_1, \ldots, y_l \in X$, $\exists v \in A^*$ such that $z x_0 x_1 \ldots x_k v = y_0 y_1 \ldots y_l$. Since z is a non-empty proper prefix of y_0, there do not exist any $m, 0 \leq m \leq l$ such that $z = y_0 \ldots y_{m-1}$. Therefore $X \notin \mathcal{C}_k$
($\Rightarrow$). Suppose $X \notin \mathcal{C}_k$. By Definition 1, $\exists l \geq 0, \exists x_0, x_1, \ldots, x_k, y_0, y_1, \ldots, y_l \in X, \exists u, v \in A^*$ such that $u x_0 x_1 \ldots x_k v = y_0 y_1 \ldots y_l$ and there do not exist any $m, 0 \leq m \leq l$ such that $x_0 = y_m$ and $u = y_0 \ldots y_{m-1}$. If $u = \epsilon$ then, as $X \in \mathcal{D}_k$, we have $x_0 = y_0$, a contradiction. So $u \neq \epsilon$. If $|u| < |y_0|$ then (3) holds. If $|u| = |y_0|$ then, again because $X \in \mathcal{D}_k$, we have $x_0 = y_1$, a contradiction. Consider now the case $|u| > |y_0|$. We have then $l \geq 1$ and $\exists m, 1 \leq m \leq l$ such that $|y_0 y_1 \ldots y_{m-1}| < |u| \leq |y_0 y_1 \ldots y_m|$, which means that there exist $z \in A^*$ and $t \in A^+$ such that $u = y_0 y_1 \ldots y_{m-1} z$, $zt = y_m$, $z x_0 x_1 \ldots x_k v = y_m \ldots y_l$. If $z = \epsilon$ then $u = y_0 \ldots y_{m-1}$ and $x_0 x_1 \ldots x_k v = y_m \ldots y_l$. Again because $X \in \mathcal{D}_k$ it follows $x = y_m$, a contradiction. Thus z must be a non-empty proper prefix of y_m and $z x_0 x_1 \ldots x_k v = y_m \ldots y_l$, i.e. the condition (3) holds $\square$

Example 1. Let $X = \{aab, abab, (abab)^k b\}$. It is easy to see that $X \in \mathcal{D}_k$. Choosing $x_0 = x_1 = \ldots = x_k = abab$, $y_0 = aab$, $y_1 = abab$, $y_2 = (abab)^k b$, $z = a \in P(y_0) - \{\epsilon\}$ and $u = abb$ we have

$$zx_0 x_1 \ldots x_k u = a(abab)^{k+1} abb = aab.abab.(abab)^k b = y_0 y_1 y_2.$$

Thus, by Proposition 2, $X \notin \mathcal{C}_k$.

Proposition 2 allows us to prove a characterization of k-comma-free codes.

Theorem 2. *Let $\phi \neq X \subseteq A^+$. Then X is in $\mathcal{C}_k$ if and only if $\forall x, y \in X$, $\forall z \in P(y)$:*

$$zxX^k A^* \cap yX^* \neq \phi \Rightarrow z = \epsilon \& x = y. \tag{4}$$

Proof. ($\Rightarrow$). Suppose the contrary that $X \in \mathcal{C}_k$ but (4) does not hold. Then $\exists x, y \in X$, $\exists z \in P(y)$ such that

$$(zxX^k A^* \cap yX^* \neq \phi) \& ((z \neq \epsilon) \vee (z = \epsilon \& x \neq y)),$$

which is equivalent to

$$((zxX^k A^* \cap yX^* \neq \phi) \& (z \neq \epsilon)) \vee ((zxX^k A^* \cap yX^* \neq \phi) \& ((z = \epsilon) \& (x \neq y))).$$

This in its turn, because $X \in \mathcal{D}_k$, is equivalent to the fact that

$$\exists x, y \in X, \exists z \in P(y) : (zxX^k A^* \cap yX^* \neq \phi) \& (z \neq \epsilon),$$

or equivalently

$$\exists x, y \in X, \exists z \in P(y) - \{\epsilon\} : zxX^k A^* \cap yX^* \neq \phi,$$

which implies, by Proposition 2, $X \notin \mathcal{C}_k$, a contradiction.

($\Leftarrow$). Suppose (4) holds true. With $z = \epsilon$ (4) becomes $\forall x, y \in X : xX^k A^* \cap yX^* \Rightarrow x = y$, i.e. $X \in \mathcal{D}_k$. If $X \notin \mathcal{C}_k$ then, by Proposition 2, we have

$$\exists x, y \in X, \exists z \in P(y) - \{\epsilon\} : zxX^k A^* \cap yX^* \neq \phi,$$

or equivalently

$$\exists x, y \in X, \exists z \in P(y) : (zxX^k A^* \cap yX^* \neq \phi) \& (z \neq \epsilon),$$

which contradicts (4). Thus $X \in \mathcal{C}_k$ □

In fact, the left side of the implication (4) in Theorem 2 can be replaced by a weaker condition, namely we may replace X^* by X^m for some m depending on k. For this we need two more lemmas.

Lemma 1. *Let $X \in \mathcal{C}_k$. Then, $\forall m \geq 0, \forall x_1, x_2, \ldots x_m, y \in X, \forall z, v \in A^*$:*

$$zx_1 x_2 \ldots x_m v = y \& zv \neq \epsilon \Rightarrow m \leq k.$$

Proof. Suppose the contrary that $m > k$. By putting $w = x_{k+2} \ldots x_m v$ we have $zx_1 x_2 \ldots x_{k+1} w = y$. If $z \neq \epsilon$ then the condition (4) in Theorem 2 does not hold, therefore $X \notin \mathcal{C}_k$, a contradiction. If $z = \epsilon$ then $v \neq \epsilon$ and hence $w \neq \epsilon$. We obtain then $x_1 \ldots x_{k+1} w = y$ with $x_1 \neq y$ which implies $X \notin \mathcal{D}_k$, again a contradiction. So $m \leq k$ □

Using Lemma 1, by induction on n, it is not difficult to prove

Lemma 2. *Let $X \in \mathcal{C}_k$. Then, $\forall m, n \geq 0$, $\forall x, x_1, \ldots, x_m, y_0, y_1, \ldots, y_n \in X$, $\forall z \in P(x)$:*

$$x x_1 \ldots x_m \preceq z y_0 y_1 \ldots y_n \Rightarrow m \leq (n+2)(k+1) - 2.$$

where $\preceq$ denotes the prefix relation on words.

Theorem 3. *Given $\phi \neq X \subseteq A^+$. Then $X \in \mathcal{C}_k$ if and only if $\forall x, y \in X$, $\forall z \in P(y)$, $\forall m \leq (k+1)(k+2) - 1$:*

$$zxX^k A^* \cap yX^m \neq \phi \Rightarrow z = \epsilon \& x = y. \tag{5}$$

Proof. The "only if" part is immediate from Theorem 2. For the "if" part it suffices to show that if (5) holds true then so does (4). Indeed, let us have $zxX^k A^* \cap yX^* \neq \phi$ for $x, y \in X, z \in P(y)$. Then there is an $m \geq 0$ such that $zxX^k A^* \cap yX^m \neq \phi$. We may assume that m is the smallest integer satisfying this condition. More concretely, there are $x_1, \ldots, x_k, y_1, \ldots, y_m \in X$ such that $zxx_1 \ldots x_k u = yy_1 \ldots y_m$ for some $u \in A^*$. he smallestness of m implies $|u| < |y_m|$. If $m \leq (k+1)(k+2) - 1$ then by (5) it follows $z = \epsilon$ and $x = y$, i.e. (4) holds true. Suppose now $m > (k+1)(k+2) - 1$. If $u = \epsilon$ then $yy_1 \ldots y_m \preceq zxx_1 \ldots x_k$ and $z \in P(y)$. By Lemma 2 we have $m \leq (k+1)(k+2) - 2 < (k+1)(k+2) - 1$, a contradiction. Suppose that $u \neq \epsilon$. We distinguish two cases:
Case 1: $|zx| > |yy_1 \ldots y_{m-1}|$. Then $yy_1 \ldots y_{m-1} \preceq zx$ with $z \in P(y)$. By Lemma 2, $m - 1 \leq 2(k+1) - 2$. Hence $m \leq 2(k+1) - 1 \leq (k+1)(k+2) - 1$, a contradiction.
Case 2: $|zx| \leq |yy_1 \ldots y_{m-1}|$. Then we must have $k \geq 1$ and $\exists l, 1 \leq l \leq k$, such that

$$|zxx_1 \ldots x_{l-1}| \leq |yy_1 \ldots y_{m-1}| < |zxx_1 \ldots x_l|.$$

It follows that there are $v, w \in A^*$ with $w \neq \epsilon$ such that $zxx_1 \ldots x_{l-1} v = yy_1 \ldots y_{m-1}$, $x_l = vw$, $wx_{l+1} \ldots x_k u = y_m$. Thus $yy_1 \ldots y_{m-1} \preceq zxx_1 \ldots x_l$ with $z \in P(y)$. By Lemma 2 we have $m - 1 \leq (k+1)(l+2) - 2 \leq (k+1)(k+2) - 2$, hence $m \leq (k+1)(k+2) - 1$, a contradiction. This completes the proof of the theorem □

A code X over an alphabet A is called *uniformly synchronous* if there is an integer $s \geq 0$ such that

$$\forall x \in X^s, \forall u, v \in A^* : uxv \in X^* \Rightarrow ux, xv \in X^*$$

The smallest integer s satisfying the above condition is called the *synchronization delay* of X, denoted by $\sigma(X)$. The notions of synchronization delay and of uniformly synchronous code have been introduced in [3] (see also [1]). A code $X \subseteq A^+$ is called *comma-free* if X is bifix and $\sigma(X) = 1$. Comma-free codes were introduced in [4]. Different characterizations of comma-free codes have been given (see [1,7]). One of them is: $X \subseteq A^+$ is a comma-free code if and only if $A^+ X A^+ \cap X^2 = \phi$ (see [7]). It appears that the 0-comma-free codes are exactly the comma-free codes. This explains the terminology of k-comma-free code.

Corollary 1. *The family C_0 of 0-comma-free codes coincides with the family of comma-free codes.*

Proof. Let $X \subseteq A^+$ be a 0-comma-free code. By Definition 1, $\forall x, y, y' \in X, \forall u, v \in A^*$:

(a) $uxv = y \Rightarrow x = y \& u = \epsilon$,
(b) $uxv = yy' \Rightarrow (x = y \& u = \epsilon) \vee (x = y' \& u = y)$.

It is easy to see that these conditions are equivalent to the following conditions (a') and (b'):

(a') X is infix,
(b') $uxv = yy' \Rightarrow u = \epsilon \vee v = \epsilon$.

which in turn is equivalent to the condition $A^+ X A^+ \cap X^2 = \phi$. Thus X is a comma-free code.

Conversely, let X be a comma-free code. To prove that X is a 0-comma-free code, by Theorem 3, it suffices to verify (4) for $m \leq 1$, i.e. to justify (a) and (b). But, as seen above, these conditions are equivalent to the fact that X is a comma-free code ☐

Remark 1. Clearly, every k-comma-free code is a uniformly synchronous code having a synchronization delay not greater than $k + 1$. The code $X = \{a, aba\}$ however, as easilly verified, is a code with synchronization delay $\sigma(X) = 1$, but it is not a k-comma-free code for any $k \geq 0$ (see Example 3).

3   Basic Lemmas

In this section we formulate two lemmas needed to establish in a uniform way criteria for testing k-comma-free codes as well as related codes under consideration.

Definition 2. *Given $\phi \neq X, Y \subseteq A^*$. We define a sequence of sets $W_i(X, Y)$, $i \geq 0$, in the following way:*

$W_0(X, Y) = Y$,
$W_{i+1}(X, Y) = X^{-1} W_i \cup W_i^{-1} X$ *for all $i \geq 0$.*

For instance, with $X = \{a, ba\}$, $Y = \{ab, baba^2\}$ we have $W_0 = Y = \{ab, baba^2\}$, $W_1 = \{b, ba^2\}$, $W_2 = \{a\}$, $W_3 = \{\epsilon\}$, ...

The following properties of the sets $W_i(X, Y)$ can be easily verified.

Remark 2. (i) If X and Y are regular then the number of the different sets $W_i(X, Y)$ is finite.

(ii) If $\epsilon \in W_i(X, Y)$ for some i then $W_j(X, Y) \neq \phi$ for all $j \geq 0$.

(iii) If $W_i(X, Y) = \phi$ for some i then $W_j(X, Y) = \phi$ for all $j \geq i$.

By induction on h we can prove the following lemma which is fundamental in establishing criteria to test codes under consideration.

Lemma 3. *Let* $\phi \neq X, Y \subseteq A^*$. *Then* $\forall n \geq 1, \forall h \in \{0, \ldots, n\} : W_n(X, Y) \neq \phi$ *if and only if* $\exists u \in W_h(X, Y), \exists i, j \geq 0, \exists v, w \in A^*$ *such that*

(i) $uX^i v \cap X^j w \neq \phi$, *i.e*

$\exists x_1, x_2, \ldots, x_i, y_1, y_2, \ldots, y_j \in X : ux_1 x_2 \ldots x_i v = y_1 y_2 \ldots y_j w;$

(ii) $i + j + h = n;$

(iii) $v = \epsilon \lor w = \epsilon;$

(iv) $i \neq 0 \Rightarrow |w| \leq |x_i|;$

(v) $j \neq 0 \Rightarrow |v| \leq |y_j|.$

Choosing $Y = X^{-1}X - \{\epsilon\}$ we have $W_i(X, Y) = U_i(X)$ for all $i \geq 0$. Using Lemma 3 it is not difficult to prove

Lemma 4. *For any* $X, \phi \neq X \subseteq A^+$, *and for any* $k \geq 0$,

(i) $U_k(X) = \phi \Rightarrow X \in \mathcal{D}_k,$

(ii) $X \in \mathcal{D}_k \Rightarrow U_{2k+1}(X) = \phi.$

Example 2. Consider the language $X = \{a, a^{k+1}b\}$. On one hand we have $U_0 = \{a^k b\}$, $U_1 = \{a^{k-1}b\}, \ldots, U_k = \{b\}, U_{k+1} = \phi$. Hence, by Lemma 4(i), $X \in \mathcal{D}_{k+1}$. On the other hand the equality $a^{k+1}b = a.a^k.b$ implies $aX^k A^* \cap a^{k+1}bX^* \neq \phi$, i.e. $X \notin \mathcal{D}_k$.

Remark 3. The converses of (i) and (ii) in Lemma 4 are not true. Indeed, consider $X = \{ab, abb, baa, aabb\}$. It is easy to check that $X \in \mathcal{D}_2 - \mathcal{D}_1$. On the other hand we have:

$$U_0 = \{b\}, U_1 = \{aa\}, U_2 = \{bb\}, U_3 = \phi.$$

Thus, $X \in \mathcal{D}_2$ and $U_2(X) \neq \phi$ show that the converse of (i) is not true. The converse of (ii) is not true either because $U_{2*1+1}(X) = U_3(X) = \phi$ and $X \notin \mathcal{D}_1$.

From Lemma 3 it follows immediately

Corollary 2 ([1], p.138). *For any* X, $\phi \neq X \subseteq A^+$,

$$X \in \mathcal{D} \Leftrightarrow \exists k \geq 1 : U_k(X) = \phi \Leftrightarrow \exists n \geq 1, \forall k \geq n : U_k(X) = \phi.$$

Remark 4. If X is regular then so is $Y = X^{-1}X - \{\epsilon\}$. Thus, by virtue of Remark 2(i), Corollary 2 gives an algorithm to decide, for any regular language X, whether X is a code w.b.d.d. or not

4   Test for k-Comma-Free Codes

Definition 3. *Given $\phi \neq X \subseteq A^*$. We define a sequence of sets $V_i(X)$ as follows:*
$$P = P(X) - \{\epsilon\},$$
$$Y = P^{-1}X - \{\epsilon\},$$
$$V_i(X) = W_i(X, Y) \text{ for all } i \geq 0 .$$

Theorem 4. *Let $\phi \neq X \subseteq A^+$ and $k \geq 0$ be an integer. Then we have*
(i) $X \in \mathcal{D}_k$ & $V_{k+1}(X) = \phi \Rightarrow X \in \mathcal{C}_k$.
(ii) $X \in \mathcal{C}_k \Rightarrow V_{(k+1)(k+3)}(X) = \phi$.

Proof. (i) Let $X \in \mathcal{D}_k$ and $V_{k+1}(X) = \phi$. Suppose $X \notin \mathcal{C}_k$. Then, by Proposition 2, $\exists l \geq 0$, $\exists x, x_1, \dots, x_k, y, y_1, \dots, y_l \in X$, $\exists z \in P(y) - \{\epsilon\}$, $\exists v \in A^*$ such that $zxx_1 \dots x_k v = yy_1 \dots y_l$. Then there is $u \in A^+$ such that $y = zu$. Hence $u \in P^{-1}X - \{\epsilon\} = V_0(X)$ and

$$uy_1 \dots y_l = xx_1 \dots x_k v.$$

If either $v = \epsilon$ or $v \neq \epsilon$ and $l = 0$ then, by Lemma 3, $V_{l+k+1}(X)$ and therefore $V_{k+1}(X)$ is not empty, a contradiction. Let now $v \neq \epsilon$ and $l \neq 0$. We distinguish two cases:
Case 1: $|u| < |xx_1 \dots x_k|$. There must be $m, 1 \leq m \leq l$, such that

$$|uy_1 \dots y_{m-1}| \leq |xx_1 \dots x_k| < |uy_1 \dots y_m|.$$

Then there are $z \in A^*$ and $w \in A^+$ such that

$$uy_1 \dots y_m = xx_1 \dots x_k w, y_m = zw, y_m \dots y_l = zv.$$

By Lemma 3, $V_{m+k+1}(X)$ and therefore $V_{k+1}(X)$ is not empty, a contradiction.
Case 2: $|u| \geq |xx_1 \dots x_k|$. Then there is $w \in A^*$ such that $u = xx_1 \dots x_k w$. Again by Lemma 3 $V_{k+1}(X) \neq \phi$, a contradiction. Thus the assertion (i) of the lemma is true.

(ii) Suppose $X \in \mathcal{C}_k$ but $V_{(k+1)(k+3)}(X) \neq \phi$. By Lemma 3 $\exists u \in V_0(X)$, $\exists i, j \geq 0$, $\exists x_1, \dots, x_i, y_1, \dots, y_j \in X$, $\exists v, w \in A^*$ such that all the conditions (i)-(v) in Lemma 3 hold true. In particular we have

$$ux_1 \dots x_i v = y_1 \dots y_j w , \quad i + j = (k+1)(k+3).$$

Since $V_0(X) = P^{-1}X - \{\epsilon\}$, there are $z \in P(X) - \{\epsilon\}$ and $x \in X$ such that $x = zu$. Thereby

$$zy_1 \dots y_j w = xx_1 \dots x_i v.$$

Case 1: $j \geq k + 2$. By the condition (iii) in Lemma 3, either $v = \epsilon$ or $w = \epsilon$. If $v = \epsilon$ then $zy_1 \dots y_j w = xx_1 \dots x_i$ with $z \in P(x) - \{\epsilon\}$. By Proposition 2, $X \notin \mathcal{C}_k$ which contradicts the hypothesis. If $w = \epsilon$ then $zy_1 \dots y_j = xx_1 \dots x_i v$. By the condition (v) in Lemma 3, $|v| \leq |y_j|$ and therefore $y_j = v'v$ for some $v' \in A^*$. It follows $zy_1 \ldots y_{j-1}v' = xx_1 \ldots x_i$ with $z \in P(x) - \{\epsilon\}$ and $j - 1 \geq k + 1$. Again by Proposition 2, $X \notin C_k$, a contradiction.

Case 2: $j < k+2$. From $i+j = (k+1)(k+3)$ it follows $i > (k+1)(k+2) - 1$. If $v = \epsilon$ then we have $zy_1 \ldots y_j w = xx_1 \ldots x_i$. By the condition (iv) in Lemma 3, $|w| \leq |x_i|$. Therefore $xx_1 \ldots x_{i-1} \preceq zy_1 \ldots y_j$ with $z \in P(x) - \{\epsilon\}$. From Lemma 2 it follows $i-1 \leq (j+1)(k+1) - 2 \leq (k+2)(k+1) - 2$. Hence $i \leq (k+1)(k+2) - 1$, a contradiction. If $w = \epsilon$ then we have $zy_1 \ldots y_j = xx_1 \ldots x_i v$. Since $|z| < |x|$, $j \geq 1$. We distinguish two subcases according as $j = 1$ or $j \geq 2$:

Subcase 1: $j = 1$. Then $xx_1 \ldots x_i \preceq zy_1$ with $z \in P(x) - \{\epsilon\}$. By Lemma 2, $i \leq 2(k+1) - 2 \leq (k+1)(k+2) - 2$. Hence $i < (k+1)(k+2) - 1$, a contradiction.

Subcase 2: $j \geq 2$. By the condition (v) in Lemma 3, $|v| \leq |y_j|$, say $y_j = v'v$ for some $v' \in A^*$. We have therefore

$$zy_1 \ldots y_{j-1}v' = xx_1 \ldots x_i.$$

If $v' = \epsilon$ then $xx_1 \ldots x_i \preceq zy_1 \ldots y_{j-1}$ with $z \in P(x) - \{\epsilon\}$. By Lemma 2, $i \leq ((j-2)+2)(k+1) - 2 < (k+2)(k+1) - 2 < (k+1)(k+2) - 1$, a contradiction. Suppose now $v' \neq \epsilon$. There exists then an integer l, $0 \leq l \leq i$, such that

$$|x_0 \ldots x_{l-1}| \leq |zy_1 \ldots y_{j-1}| < |x_0 \ldots x_l|,$$

where x_0 stands for x. Then there are $v_1 \in A^*$ and $v_2 \in A^+$ such that

$$x_0 \ldots x_{l-1}v_1 = zy_1 \ldots y_{j-1}, \quad x_l = v_1v_2, \quad v' = v_2x_{l+1} \ldots x_i.$$

The last equality implies $v_2x_{l+1} \ldots x_i v = v'v = y_j$ with $v_2v \neq \epsilon$. By Lemma 1, it follows $i - (l+1) + 1 \leq k$, that is $i \leq l + k$. On the other hand, the first equality above means that $x_0 \ldots xl - 1 \preceq zy_1 \ldots y_{j-1}$ with $z \in P(x_0) - \{\epsilon\}$. By Lemma 2 we have $l - 1 \leq j(k+1) - 2 \leq (k+1)(k+1) - 2$, i.e. $l \leq (k+1)(k+1) - 1$. Putting all together we get

$$i \leq l + k \leq (k+1)(k+1) - 1 + k = (k+1)(k+2) - 2 < (k+1)(k+2) - 1,$$

a contradiction. This completes the proof of (ii) of the theorem and therefore the theorem itself   □

Example 3. Let $X = \{a, aba\}$. We have $P = P(X) - \{\epsilon\} = \{a, ab\}$, $V_0(X) = P^{-1}X - \{\epsilon\} = \{ba, a\}$, $V_1(X) = \{\epsilon, ba\}$. Hence $V_k(X) \neq \phi$ for all $k \geq 0$. By Theorem 4(ii) $X \notin C_k$ for all $k \geq 0$.

As another application of Theorem 4 we obtain

Proposition 3. *For all $k \geq 0$, $C_k \subset C_{k+1}$.*

Proof. The inclusion $C_k \subseteq C_{k+1}$ is evident. Consider again the code $X = \{a, a^{k+1}b\}$. As seen in Example 2, $X \in \mathcal{D}_{k+1} - \mathcal{D}_k$. Therefore $X \notin C_k$. On the other hand, an easy computation gives: $P = P(X) - \{\epsilon\} = \{a, a^2, \ldots, a^{k+1}\}$, $V_0 = P^{-1}X - \{\epsilon\} = \{a^kb, a^{k-1}b, \ldots, b\}$, $V_1 = X^{-1}V_0 \cup V_0^{-1}X = \{a^{k-1}b, a^{k-2}b, \ldots, b\}$, $\ldots$, $V_k = \{b\}$, $V_{k+1} = \phi$, $V_{k+2} = \phi$. Thus, by Theorem 4(i), $X \in C_{k+1}$. This proves the strictness of the inclusion   □

Put $\overline{V}_i(X) = U_i(X) \cup V_i(X)$. Combining Lemma 4 and Theorem 4 we obtain immediately

Corollary 3. *Let* $\phi \neq X \subseteq A^+$. *Then we have*
  (i) $\overline{V}_k(X) = \phi \Rightarrow X \in \mathcal{C}_k$.
  (ii) $X \in \mathcal{C}_k \Rightarrow \overline{V}_{(k+1)(k+3)}(X) = \phi$.

Corollary 3 in its turn gives directly

Corollary 4. *For any* X, $\phi \neq X \subseteq A^+$, *we have*

$$X \in \mathcal{C} \Leftrightarrow \exists k \geq 1 : \overline{V}_k(X) = \phi \Leftrightarrow \exists n \geq 1, \forall k \geq n : \overline{V}_k(X) = \phi.$$

Remark 5. Because $Y = P^{-1}X - \{\epsilon\}$ is regular if so is X, Corollary 4 provides us with an algorithm to decide whether a given regular language X is a code in $\mathcal{C}$ or not.

5   k-Cohesive Codes

Motivated by the notion of *cohesive-prefix code* and relationship between such codes and infix codes [5], we introduce in this section the so called k-cohesive codes and consider relations between them and k-comma-free codes.

Definition 4. *Let* $\phi \neq X \subseteq A^+$ *and* $k \geq 0$. *We say that* X *is a* k-cohesive code *if* $\forall x, y \in X$, $\forall z \in P(y)$:

$$zxX^k A^* \cap yX^* \neq \phi \Rightarrow zX \subseteq X \ \& \ zx = y.$$

The family of these codes is denoted by $\mathcal{H}_k$ and we put $\mathcal{H} = \bigcup_{k \geq 0} \mathcal{H}_k$.

Definition 5. *With every language* X, $\phi \neq X \subseteq A^+$, *we associate a sequence of sets* $T_i(X), i \geq 0$, *defined by*
  $P = P(X) - \{\epsilon\}$,
  $Q = \{p \in P : pX \not\subseteq X\}$,
  $Y = Q^{-1}X - \{\epsilon\}$,
  $T_i(X) = W_i(X, Y)$ *for all* $i \geq 0$.

Based on Lemma 3 and on the evident fact that for any $X \in \mathcal{D}_k$, X is not in $\mathcal{H}_k$ if and only if $\exists l \geq 0$, $\exists x, y \in X$, $\exists z \in P(y)$ such that $zxX^k A^* \cap yX^* \neq \phi$ and $zX \not\subseteq X$, one can get the following result whose proof is similar to that of Theorem 4 and therefore omited.

Theorem 5. *For any language* X, $\phi \neq X \subseteq A^+$, *we have*
  (i) $X \in \mathcal{D}_k \ \& \ T_{k+1}(X) = \phi \Rightarrow X \in \mathcal{H}_k$.
  (ii) $X \in \mathcal{H}_k \Rightarrow T_{(k+1)(k+3)}(X) = \phi$.

Put $\overline{T}_k(X) = T_k(X) \cup U_k(X)$. As an immediate consequence of Lemma 4 and Theorem 5 we obtain

Corollary 5. *For any* X, $\phi \neq X \subseteq A^+$, *the following implications hold*
(i) $\overline{T}_k(X) = \phi \Rightarrow X \in \mathcal{H}_k$.
(ii) $X \in \mathcal{H}_k \Rightarrow \overline{T}_{(k+1)(k+3)}(X) = \phi$.

Corollary 5 in its turn gives directly

Corollary 6. *For any* X, $\phi \neq X \subseteq A^+$,

$$X \in \mathcal{H} \Leftrightarrow \exists k \geq 1 : \overline{T}_k(X) = \phi \Leftrightarrow \exists n \geq 1, \forall k \geq n : \overline{T}_k(X) = \phi.$$

Remark 6. It is not difficult to see that if X is regular then so is $Y = Q^{-1}X - \{\epsilon\}$. Thus, Corollary 6 provides us with an algorithm to decide, for any regular language X, whether X is a code in $\mathcal{H}$ or not.

Theorem 5 can be used to make clear relations between families of codes under consideration. Namely we have

Proposition 4. *For any* $k \geq 0$,
(i) $\mathcal{H}_k \subset \mathcal{H}_{k+1}$.
(ii) $\mathcal{C}_k \subset \mathcal{H}_k \subset \mathcal{D}_k$.
(iii) $\mathcal{C} \subset \mathcal{H} \subset \mathcal{D}$.

Proof. All the inclusions $\mathcal{H}_k \subseteq \mathcal{H}_{k+1}$, $\mathcal{C}_k \subseteq \mathcal{H}_k \subseteq \mathcal{D}_k$ and $\mathcal{C} \subseteq \mathcal{H} \subseteq \mathcal{D}$ are immediate from the corresponding definitions. We now show that they are all strict.

(i) Consider the language $X = a^*(ab) \cup a^*(ab)^{k+1}b$. It is easy to see that $U_{k+1}(X) = \phi$. So, by Lemma 4, $X \in \mathcal{D}_{k+1}$. Let's compute
$P = P(X) - \{\epsilon\} = a^+ \cup a^*(ab) \cup a^*(ab)^2 \cup \ldots \cup a^*(ab)^k$,
$Q = \{p \in P : pX \not\subseteq X\} = a^*(ab) \cup a^*(ab)^2 \cup \ldots \cup a^*(ab)^k$,
$T_0(X) = Q^{-1}X - \{\epsilon\} = \{(ab)^kb, \ldots, (ab)b\} \cup \{b(ab)^{k-1}b, \ldots, bb\}$,
$T_1(X) = X^{-1}T_0 \cup T_0^{-1} = \{(ab)^{k-1}b, \ldots, (ab)b, b\}$,
$\ldots$
$T_k(X) = \{b\}$,
$T_{k+1}(X) = \phi$,
which implies $T_{k+2}(X) = \phi$. By Theorem 5(i), $X \in \mathcal{H}_{k+1}$. On the other hand the equality $ab.(ab)^k.b = (ab)^{k+1}b$ shows that $X \notin \mathcal{D}_k$ and therefore $X \notin \mathcal{H}_k$. This proves the strictness of the inclusion $\mathcal{H}_k \subseteq \mathcal{H}_{k+1}$.

(ii)-(iii) Consider the language $X = \{ab, aab, (ab)^kb\}$. It is easy to see that $U_k(X) = \phi$. By Lemma 4, $X \in \mathcal{D}_k$. Hence $X \in \mathcal{D}$. On the other hand we have
$P = P(X) - \{\epsilon\} = \{a, a^2, ab, (ab)a, (ab)^2, (ab)^2a, \ldots, (ab)^{k-1}, (ab)^{k-1}a\}$,
$Q = \{p \in P : pX \not\subseteq X\} = P$,
$T_0(X) = Q^{-1}X - \{\epsilon\} = \{b, ab, b(ab)^{k-1}b, (ab)^{k-1}b, b(ab)^{k-2}b, \ldots, abb, bb\}$.
Therefore $\epsilon \in T_1(X)$. Hence $T_i(X) \neq \phi$ for all $i \geq 0$. By Theorem 5(ii), $X \notin \mathcal{H}_i$ for all $i \geq 0$. Hence $X \notin \mathcal{H}$. Thus the inclusions $\mathcal{H}_k \subseteq \mathcal{D}_k$ and $\mathcal{H} \subseteq \mathcal{D}$ are strict. Let now consider the langue $X = a^*(ab)$. As seen above, $X \in \mathcal{H}_0$ and therefore $X \in \mathcal{H}$. The equality $a.ab.(ab)^k = aab.(ab)^k$ shows that $X \notin \mathcal{C}_k$ for all $k \geq 0$. Hence $X \notin \mathcal{C}$. Thus the inclusions $\mathcal{C}_k \subseteq \mathcal{H}_k$ and $\mathcal{C} \subset \mathcal{H}$ are strict also. The proof is completed □

Relationship between k-Comma-free codes and k-cohesive codes is given in the following two propositions whose proofs are easy.

Proposition 5. *Let $\phi \neq X \subseteq A^+$. For any $k \geq 0$, X is a k-comma-free code if and only if every finite subset of X is a k-cohesive code.*

Thus, in the framework of finite languages, k-comma-free codes coincide with k-cohesive codes.

Proposition 6. *Let $\phi \neq X \subseteq A^+$ and $k \geq 0$. For each code $X \in \mathcal{H}_k$ there exists a code $X' \in \mathcal{C}_k$ such that $X = R^* X'$, where R is either a suffix code or $R = \{\epsilon\}$.*

Example 4. Consider again $X = a^* ab \cup a^* (ab)^k b$. As seen in the proof of Proposition 4, X is a k-cohesive code. Obviously, $X = R^* Z$ with $Z = \{ab, (ab)^k b\}$ and $R = \{a\}$, a suffix code. Now we check that Z is in $\mathcal{C}_k$. Obviously, $Z \in \mathcal{D}_k$. Let us compute

$$P = P(Z) - \{\epsilon\} = \{a, ab, aba, (ab)^2, \ldots, (ab)^{k-1}a, (ab)^k\},$$
$$V_0(Z) = P^{-1}Z - \{\epsilon\} =$$
$$= \{(ab)^{k-1}b, (ab)^{k-2}b, \ldots, (ab)b, b\} \cup \{b, b(ab)^{k-1}b, b(ab)^{k-2}b, \ldots, bb\},$$
$$V_1(Z) = Z^{-1}V_0 \cup V_0^{-1}Z = \{(ab)^{k-2}b, \ldots, b\},$$
$$\ldots$$
$$V_{k-1}(Z) = \{b\},$$
$$V_k(Z) = \phi,$$
$$V_{k+1}(Z) = \phi.$$

Thus, by Theorem 4(i), X is a k-comma-free code.

References

1. Berstel, J. and Perrin, D.: Theory of Codes. Academic Press, New York, London, 1985.
2. Devolder, J., Latteux, M., Litovsky, I., Staiger, L.: Codes and infinite words. Acta Cybern. **11** (1994) 241-256.
3. Golomb S. W., Gordon B.: Codes with bounded synchronization delay. Inform. and Contr. **8** (1965) 355-372.
4. Golomb, S. W., Gordon B., Welch L. R.: Comma-free codes. Canad. J. Math. **10** (1958) 202-209.
5. Ito, M., Thierrin, G.: Congruences, infix and cohesive prefix codes. Theoret. Comput. Sci. **136** (1984) 471-485.
6. Sardinas A. A., Patterson C. W.: A necessary and sufficient condition for the unique decomposition of coded messages. IRE Internat. Conv. Rec. **8** (1953) 104-108.
7. Shyr, H. J.: Free Monoids and Languages. Lecture Notes, Hon Min Book Company, Taichung, 1991.

Abstract Families of Graphs

Tanguy Urvoy

[1] IRISA, Rennes,
Tanguy.Urvoy@irisa.fr,
http://www.irisa.fr/galion
[2] Campus Universitaire de Beaulieu,
Avenue du Général Leclerc,
35042 Rennes - France

Abstract. A natural way to describe a family of languages is to use rational transformations from a generator. From these transformations, Ginsburg and Greibach have defined the Abstract Family of Languages (AFL). Infinite graphs (also called infinite automata) are natural tools to study languages. In this paper, we study families of infinite graphs that are described from generators by transformations preserving the decidability of monadic second order logic. We define the Abstract Family of Graphs (AFG). We show that traces of AFG are rational cones and traces of AFG that admit a rationally colored generator are AFL. We generalize some properties of prefix recognizable graphs to AFG. We apply these tools and the notion of geometrical complexity to study sub-families of prefix recognizable graphs.

A large number of families of formal languages arising from either computer science or linguistics were investigated for their properties. These families first described with sophisticated acceptors, were also described by combinatorial or algebraic properties. A well known example is the Chomsky-Schützenberger characterization of context-free languages as rational transductions from the Dyck language [6]. A lot of language families were found to be closed by rational transductions and some other operations like union or concatenation. From these closure operations, Ginsburg and Greibach have derived the Abstract Family of Languages (AFL) [10]. In the same paper, they defined the Abstract Family of Acceptors (AFA). They showed the direct relation between these language and machine families. The notion of AFL was intensively studied and refined. Many structures defined by closure operations like rational cones, or cylinders were also considered (see [2] or [1] for a survey). The notion of AFA fell into disuse.

The development of model checking arose a new interest for behavioral properties of machines. A main task of model checking is to decide if a given machine modelized by a (finite or infinite) graph satisfies or not a given logical sentence. This problem is undecidable in general but Muller and Shupp showed that monadic second order (MSO) logic, is decidable for the transition graphs of pushdown automata [14]. This result has been extended to HR-equational graphs by Courcelle [7] and to prefix recognizable graphs by Caucal [4]. Different characterizations of these graph families are summarized in [3].

M. Ito and M. Toyama (Eds.): DLT 2002, LNCS 2450, pp. 381–392, 2003.

Verification of logical properties is not the only interest of graph families: they also provide a natural tool for the study of formal languages: a trace of a graph is the language of path labels from and to finite vertex sets. A graph hierarchy is constructed in parallel of the Chomsky hierarchy: the traces of finite graphs are the rational languages, the traces of prefix recognizable graphs are the context-free languages [4], the traces of rational and synchronized rational graphs are the context-sensitive languages [13], [15] and the traces of Turing graphs are the recursively enumerable languages [5]. A survey about this hierarchy is done in [16].

In [5], the family of prefix recognizable graphs is generated from the Dyck graph Δ_2 (see Figure 2) with two closure operations: rational coloring and inverse rational substitution. We study these operators in a more general context: given a graph Γ, we consider the family of graphs generated from a rational coloring of Γ by inverse rational substitution and product with a finite graph. We call Abstract Family of Graphs (AFG) a graph family which is closed by inverse substitution and product with a finite graph. We show that traces of AFG are rational cones. We also show that traces of AFG that admit a rationally colored generator (we call them *principal*) are AFL. We establish that geometrical complexity of graphs is preserved by AFG transformations and we use these tools to study sub-families of prefix recognizable graphs like linear or polynomial graphs.

This paper is divided into three sections. In Section 1 some facts about language families are reviewed. In Section 2 we define generic graph families and we study their properties. In section 3 we apply the concepts of Section 2 to sub-families of prefix recognizable graphs. We also define the geometrical complexity of a graph to exhibit a strictly increasing chain of AFG.

1   Language Families

The reader is supposed to be familiar with the basic notions of formal languages described in [2]. We recall here briefly some notions about language families; see [1], [2] or [9] for a complete reference.

1.1   Language Transformations

A monoid *morphism* f from X^* to Y^* is a mapping which associates a word of Y^* to each letter of X. It is extended to words by: $f(\varepsilon) = \varepsilon$ and $f(au) = f(a)f(u) \ \forall a \in X, u \in X^*$. When $f^{-1}(\varepsilon) = \{\varepsilon\}$, f is called *non-erasing*; when $f(X) \subseteq Y \cup \{\varepsilon\}$, f is called *alphabetic*. The *projection* or *erasing* $p_a : X^* \longrightarrow Y^*$ is the alphabetic morphism defined by $p_a(a) = \varepsilon$ and $p_a(b) = b \ \forall b \neq a$. The *copy* $c : X^* \longrightarrow Y^*$ is an alphabetic morphism defined by an injection from X to Y. A *substitution* of domain X is a mapping which associates a language to each letter of X. A subtitution is finite (resp. rational) when the associated language is finite (resp. rational). Substitutions are extended to words by morphism.

We use the notation $Fin(M)$ to design the finite subsets of a set M. If M is a monoid, we use the notation $Rat(M)$ to design its rational subsets. A *rational transduction* (or simply *transduction*) between two monoids X^* and Y^* is a relation $T \in Rat(X^* \times Y^*)$; it is *non-erasing* if for any $y \in Y^*$, $T^{-1}(y) \in Fin(X^*)$. A non-erasing transduction is the composition of an inverse morphism, an intersection with a rational set and a non-erasing morphism. A transduction is the composition of a non-erasing transduction and a morphism. Inverse rational (resp. finite) substitutions are particular cases of transductions (resp. non-erasing transductions).

1.2   AFL and Rational Cones

A *family of languages* is a nonempty set of languages that is closed under copy. A *cone* is a family of languages that is closed by transduction. A cone C is *principal* of generator Λ if $\Lambda \in C$ and if any L in C is the image of Λ by a transduction. Using the notation of [9] we write $C = \mathcal{C}(\Lambda)$. The relation $L \preceq L'$ when L is the image by a transduction of L' is a way to measure the "complexity" of a language, in that sense the generators of C are its most complex languages. Consider the Dyck language $D_2 = aD_2\bar{a}D_2 + bD_2\bar{b}D_2 + \varepsilon$, which is the language of "well formed" words over two types of parentheses; the Chomsky-Schützenberger theorem [6] gives a characterization of context-free languages as being the principal cone $CF = \mathcal{C}(D_2)$.

types	inverse morphism	rational $\cap$	non-erasing morphism	erasing morphism	$\cup$	L^+
semi-cone	×	×	×			
cone	×	×	×	×		
AFL	×	×	×		×	×
full-AFL	×	×	×	×	×	×

Fig. 1. Different types of language families.

There are different types of language families classified by closure properties (see Fig. 1). All families of the Chomsky hierarchy except context sensitive languages are full-AFL. The family Ocl of one-counter languages is a principal full-AFL. The family Lin of linear languages and the family $Rocl$ of restricted one counter languages are examples of principal cones (see [2]).

2   Graph Families

2.1   Preliminaries on Graphs

Given an arbitrary set V of vertices, and an arbitrary set Σ of symbols, a *graph* G is a subset of $V \times \Sigma \times V$ where $\Sigma_G := \{a \in \Sigma \mid (s, a, t) \in G\}$ is finite and $V_G := \{s \in V \mid \exists t, (s, a, t) \in G \vee (t, a, s) \in G\}$ is countable. The existence of an arc (s, a, t) of *source* s and of *goal* t in G is denoted $s \xrightarrow{a}_{G} t$, or simply $s \xrightarrow{a} t$

when G is understood. We write $s \overset{w}{\Longrightarrow} t$ with a word $w = w_1 \ldots w_n \in \Sigma_G^*$ the existence of a path $s \overset{w_1}{\longrightarrow} s_1 \ldots s_{n-1} \overset{w_n}{\longrightarrow} t$ in G.

When $I \in Fin(V_G)$ and $F \in Fin(V_G)$, the language $L(G, I, F) = \{w \in \Sigma_G^* \mid \exists i \in I, f \in F, i \underset{G}{\overset{w}{\Longrightarrow}} f\}$ is called a *trace* of G. With a special symbol $\tau \in \Sigma_G$, we may also consider the trace with ε-transitions $p_\tau(L(G, I, F))$. A graph G is deterministic if for all $s, t, t' \in V_G$ we have $s \underset{G}{\overset{a}{\longrightarrow}} t \wedge s \underset{G}{\overset{a}{\longrightarrow}} t' \Rightarrow t = t'$. It is *strongly connected* when for any $s, t \in V_G$ there is a word w such that $s \underset{G}{\overset{w}{\Longrightarrow}} t$; it is simply *connected* if $G \cup G^{-1}$ is *strongly connected* ($G^{-1} = \{(t, a, s) \mid s \underset{G}{\overset{a}{\longrightarrow}} t\}$). The distance between two vertices in a connected graph G is the function defined by $d_G(s, t) = min\{|w| \mid s \underset{G \cup G^{-1}}{\overset{w}{\Longrightarrow}} t\}$. The degree of a vertex s is $d_G(s) = |\{(s, a, t) \in G \mid a \in \Sigma_G \wedge t \in V_G\}| + |\{(t, a, s) \in G \mid a \in \Sigma_G \wedge t \in V_G\}|$.

2.2   Graph Transformations

The most general aproach to transform a graph by preserving MSO decidability is to use all MSO-interpretations [8]. Our approach is different since we want to define graph families in relation with language families: we want a reduced set of transformations corresponding to language transformations. Given a graph G, a transformation T is *internal* if $V_{T(G)} \subseteq V_G$.

Internal Transformations

Copy: Given a bijection c between two alphabets, the *copy* by c of a graph G is the graph $c(G) = \{(s, c(a), t) \mid s \underset{G}{\overset{a}{\longrightarrow}} t\}$.

Rational Coloring: Let G be a graph with a distinguished vertex v_0 and let f be a rational substitution. The *coloring* of G from v_0 according to f is the graph $\#_{v_0, f} G = G \cup \{(t, a, t) \mid \exists w \in f(a), v_0 \underset{G}{\overset{w}{\Longrightarrow}} t\}$. When v_0 is understood, we simply write $\#_f G$.

Inverse substitution: Let G be a graph and f be a substitution from an alphabet Σ to 2^{Σ_G}. The *inverse substitution of G according to f* is the graph $f^{-1}(G) = \{(s, a, t) \mid \exists w \in f(a), s \underset{G}{\overset{w}{\Longrightarrow}} t\}$.

ε-Closure: The ε-closure is a particular inverse substitution. Let G be a graph with a special symbol τ for ε-transitions and let $\#$ be a 'transparent' color. Consider the rational substitution ϕ from $\Sigma_G - \{\tau\}$ to $Rat((\Sigma_G + \#)^*)$ defined by $\phi(a) = \#\tau^* a \tau^* \#$ for all $a \in \Sigma_G - \{\tau, \#\}$. We call ε-closure of G the graph $\overline{G} = \phi^{-1}(G)$. Let p_τ be the erasing of τ. If f is the mapping defined by $f(\#) = \{\varepsilon\}$ and if $\varepsilon \notin p_\tau(L(G, I, F))$ then the trace $p_\tau(L(G, I, F))$ with ε-transition is $L(\overline{f^{-1}(G)}, I, F)$.

Other Transformations

Isomorphism: If f is a mapping of domain V_G then $f(G) = \{(f(s), a, f(t)) \mid s \xrightarrow{a}_G t\}$. A *graph isomorphism* from G to H is an injective mapping $f : V_G \longrightarrow V_H$ such that $f(G) = H$. The graphs G and H are isomorphic and we write $G \sim H$ when there is a graph isomorphism between G and H.

Products: To study the intersection between the traces of two graphs, it may be useful to consider the *synchronized product* of these graphs but the *concatenation* and *desynchronized product* are more easy to handle. Given graphs G and H, we define products between G and H as subsets of $(V_G \times V_H) \times \Sigma_G \cup \Sigma_H \times (V_G \times V_H)$. Let v be a vertex of H. The *concatenation* of H to G at $v \in V_H$ is the graph $G \cdot_v H = \{((s, s'), a, (t, t')) \mid (s \xrightarrow{a}_G t \wedge s' = t' = v) \vee (s = t \wedge s' \xrightarrow{a}_H t')\}$. The *desynchronized product* of G by H is the graph $G \square H = \{((s, s'), a, (t, t')) \mid (s \xrightarrow{a}_G t \wedge s' = t') \vee (s' \xrightarrow{a}_H t' \wedge s = t)\}$. The *synchronized product* of G by H is the graph $G \times H = \{((s, s'), a, (t, t')) \mid s \xrightarrow{a}_G t \wedge s' \xrightarrow{a}_H t'\}$.

2.3   Abstract Graph Families

An *internal graph family* is a set of graphs closed by copy. A *graph family* is a non-empty set of graphs that is closed by copy and isomorphism. Some operations like union or intersection of graphs may be well defined in an internal family but does not make sense when considered up to isomorphism.

Closure Operators. Each graph transformation considered in Section 2.1 is naturally extended to graph families. Let $\mathcal{F}$ be a graph family, we denote respectively by $\#\mathcal{F}$, $Fin^{-1}(\mathcal{F})$, $Rat^{-1}(\mathcal{F})$ and $\overline{\mathcal{F}}$ the closures by rational coloring, inverse finite substitution, inverse rational substitution and ε-closure of $\mathcal{F}$. We denote by Fin the family of finite graphs. We denote by $\mathcal{F} \cdot Fin$, $\mathcal{F} \square Fin$ and $\mathcal{F} \times Fin$ the smallest family of graphs containing $\mathcal{F}$ that is respectively closed by concatenation, desynchronized product and synchronized product with a finite graph. Most operators do not commute. For instance $\#Fin^{-1}(G) \subseteq Fin^{-1}(\#G)$ but this inclusion is strict for $G = \{(0, a, 1), (1, a, 0)\}$: $\{(0, a, 1)\}$ is in $Fin^{-1}(\#G)$ but not in $\#Fin^{-1}(G)$.

Finite Product Equivalence. We consider a particular family of finite graphs for concatenation and show the equivalence of finite products. Let $n \in \mathbb{N}$ and the alphabet $\Sigma_n = \Sigma_{n,+} \cup \Sigma_{n,-}$ where $\Sigma_{n,+} = \{\sigma_0, \ldots, \sigma_n\}$ and $\Sigma_{n,-} = \{\overline{\sigma_i} \mid \sigma_i \in \Sigma_{n,+}\}$. We extend the bijection $\overline{a}$ to words by the rules $\overline{a \cdot w} = \overline{w} \cdot \overline{a}$ and $\overline{\overline{a}} = a$. The *segment* $\Delta_0^n \subset \mathbb{N} \times \Sigma_n \times \mathbb{N}$ is the graph

$$\Delta_0^n = \{(i-1, \sigma_i, i) \mid 1 \le i \le n\} \cup \{(i, \overline{\sigma_i}, i-1) \mid 1 \le i \le n\} \cup \{(0, \sigma_0, 0), (0, \overline{\sigma_0}, 0)\}$$

This graph is deterministic and strongly connected. For any $0 \leq i \leq j \leq n$, we have: $i \underset{\Delta_0^n}{\overset{\sigma_i...\sigma_j}{\Longrightarrow}} j$ and $j \underset{\Delta_0^n}{\overset{\overline{\sigma_i...\sigma_j}}{\Longrightarrow}} i$. When $\mathcal{F}$ is a graph family, we denote by $\mathcal{F} \cdot_0 \Delta_0^n$ the smallest family of graphs containing $\mathcal{F}$ that is closed under concatenation with Δ_0^n at 0.

Property 1. If $\mathcal{F} = Fin^{-1}(\mathcal{F})$ then these four statements are equivalent:

1. $\mathcal{F} = \mathcal{F} \cdot_0 \Delta_0^n \ \forall n \in \mathbb{N}$;
2. $\mathcal{F} = \mathcal{F} \cdot Fin$;
3. $\mathcal{F} = \mathcal{F} \Box Fin$;
4. $\mathcal{F} = \mathcal{F} \times Fin$.

Abstract Families. As for languages, we define different types of families by closure properties:

	$\mathcal{F} = Fin^{-1}(\mathcal{F})$	$\mathcal{F} = \mathcal{F} \cdot \Delta_0^n$	$\mathcal{F} = \overline{\mathcal{F}}$
AFG	×	×	
full-AFG	×	×	×

The rational coloring ($\#$) does not appear in the definition of AFG because it is too restrictive and because it does not commute. The Rat^{-1} operator neither appears but the following property explain this disparition.

Property 2. If $\mathcal{F} = \mathcal{F} \cdot Fin$ then $\overline{Fin^{-1}(\mathcal{F})} = Rat^{-1}(\mathcal{F})$.

For any set $\mathcal{G}$ of graphs, we denote by $AFG(\mathcal{G})$ (resp. $\overline{AFG}(\mathcal{G})$) the smallest AFG (resp. full-AFG) containing $\mathcal{G}$. A *rationally colored generator* of $\mathcal{F}$ is a strongly connected deterministic graph $\Gamma \in \mathcal{F}$ such that $\mathcal{F} = AFG(\#\Gamma)$ (resp. $\mathcal{F} = \overline{AFG}(\#\Gamma)$ for full-AFG). We call *principal* an AFG that admit such a generator. Principality is a strong property since it gives sense to the union operator.

Lemma 1. *If $\mathcal{F}$ is a principal AFG (resp. full AFG) then*

1. *$\mathcal{F}$ is closed under disjoint union;*
2. *$\mathcal{F}$ is closed under union with finite graphs.*

Traces of AFG. We denote the traces of a family $\mathcal{F}$ by $L(\mathcal{F}) := \{L(G, I, F) \mid G \in \mathcal{F} \wedge I, F \in Fin(V_G)\}$.

Theorem 1. *If $\mathcal{F}$ is an AFG (resp. a full-AFG) then $L(\mathcal{F})$ is a semi-cone (resp. a cone). If $\mathcal{F}$ is a principal AFG (resp. a full-AFG) then $L(\mathcal{F})$ is an AFL (resp. a full-AFL).*

We give a proof sketch. Let $\mathcal{F}$ be an AFG and $L \in L(\mathcal{F})$. There is a graph $G \in \mathcal{F}$, and two sets $I_G, F_G \in Fin(V_G)$ such that $L = L(G, I_G, F_G)$. To show that $L(\mathcal{F})$ is a cone (resp. a semi-cone) we show that $L(\mathcal{F})$ is closed under inverse morphism, morphism (resp. non-erasing morphism) and intersection with rational set.

The intersection with a rational set is a consequence of Property 1. To show that $L(\mathcal{F})$ is an AFL when $\mathcal{F}$ is a principal AFG we show that $L \cup L'$, L^* and $L \cdot L'$ stay in $L(\mathcal{F})$. The proof is similar to the one of Kleene theorem and relies on Lemma 1.

3   Application to Pushdown Graphs

In this section, we consider some properties of *pushdown graphs* and *prefix recognizable graphs* from the AFG point of view. We define the notion of *geometrical complexity* which is preserved by AFG transformations. We then consider subfamilies of prefix recognizable graphs like *one-reversal pushdown graphs* or *polynomial graphs*.

3.1   Pushdown Graphs

A *pushdown automaton* is a finite set of rules $R \subseteq QS \times T \times QS^*$ where Q is a finite set of *states*, S is a (disjoint) *stack* alphabet and T is an alphabet of *terminals*. A *configuration* of the machine is defined by a word $qw \in QS^*$ where q is a state and w is a stack word. Its transition graph restricted to a rational set $L \subseteq QS^*$ of acceptable configurations is the graph

$$G := (RS^*)_{|L} = \{uw \xrightarrow{a} vw \mid u \xrightarrow{a}_{R} v \land uw, vw \in L\} . \tag{1}$$

Any graph isomorphic to such a graph is called a *pushdown graph*. We denote this family PDG.

Prefix Recognizable Graphs: The family of prefix recognizable graphs is a natural extension of pushdown graphs to recognizable sets of rewriting rules. We give here an internal representation, similar to the one of Equation (1). Let X be an alphabet, we denote $U \xrightarrow{a} V := \{(u, a, v) \mid u \in U \land v \in V\}$ the graph of transitions from $U \subseteq X^*$ to $V \subseteq X^*$ labelled by $a \in \Sigma$. A *recognizable graph* is a finite union of graphs $U \xrightarrow{a} V$ where $U, V \in Rat(X^*)$. A *prefix recognizable graph* is a graph of the form

$$(RX^*)_{|L} = \{uw \xrightarrow{a} vw \mid u \xrightarrow{a}_{R} v \land uw, vw \in L\} \tag{2}$$

where R is a recognizable graph and $L \in Rat(X^*)$ is a set of acceptable configurations. Any graph isomorphic to such a graph is called a *prefix recognizable graph*. We denote this family REC.

The Dyck Graph: We define the two parentheses semi-Dyck graph by $\Delta_2 = \{u \xrightarrow{a} au \mid u \in (a+b)^*\} \cup \{u \xrightarrow{b} bu \mid u \in (a+b)^*\} \cup \{au \xrightarrow{\bar{a}} u \mid u \in (a+b)^*\} \cup \{bu \xrightarrow{\bar{b}} u \mid u \in (a+b)^*\}$. A representation of Δ_2 is given in Figure 2. The trace $L(\Delta_2, \{\varepsilon\}, \{\varepsilon\})$ is the Dyck language D_2 presented in Section 1.2.

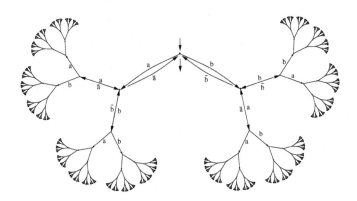

Fig. 2. The two letters Dyck graph Δ_2.

Theorem 2. *[4]*

1. $PDG = Fin^{-1}(\#\Delta_2)$;
2. $REC = Rat^{-1}(\#\Delta_2)$;

As PDG is a full-AFL and from Property 2, we deduce the following result originally due to Knapik.

Property 3. $REC = \overline{PDG}$.

3.2   Other Generators

One-reversal Pushdown Graphs. A *one-reversal pushdown automaton* is a pushdown automaton which first writes informations in the stack and secondly reads these informations. We define the diamond graph of Figure 3 by

$$\diamondsuit_2 = \{(u, a, au) \mid u \in (a+b)^*\} \qquad \cup \{(u, b, bu) \mid u \in (a+b)^*\}$$
$$\cup \{(u, \tau, \#u) \mid u \in (a+b)^*\} \qquad \cup \{(\#au, \bar{a}, \#u) \mid u \in (a+b)^*\}$$
$$\cup \{(\#bu, \bar{b}, \#u) \mid u \in (a+b)^*\} \,.$$

The trace $p_\tau(L(\diamondsuit_2, \varepsilon, \#))$ is the symmetric language $S_2 = aS_2\bar{a} + bS_2\bar{b} + \varepsilon$. The $\diamondsuit_2$ graph is a generator of the *one-reversal pushdown graphs* which traces up to ε define the cone of linear languages Lin. Note that $\diamondsuit_2$ is not strongly connected. This AFG is not principal because Lin is not an AFL.

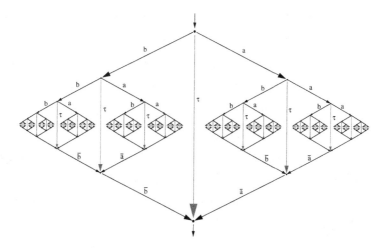

Fig. 3. A fractal representation of the one-reversal pushdow graphs generator: $L\left(\overline{Fin^{-1}(\#\Diamond_2)}\right) = Lin.$

Linear Pushdown Graphs. A *one-counter automaton* is a pushdown automaton with only one stack symbol. As only the length of the stack can be used for computation, it can also be considered as a finite state machine associated with an integer register. The traces of *one-counter automaton* transition graphs define the *Ocl* AFL which generator is the Dyck set with one symbol $D_1 = aD_1\bar{a}D_1 + \varepsilon$. The linear Dyck graph Δ_1 described on Figure 4, is a generator of this graph family.

Fig. 4. The one letter Dyck graph Δ_1.

3.3   Geometrical Complexity

We will show that *geometrical complexity* is a measure preserved by AFG transformations.

General Definition. Let G be a graph of finite degree and let $v \in V_G$. The distance between two vertices in a same connected component of G is the function $d_G(s,t) = min\{|w| \mid s \overset{w}{\underset{G \cup G^{-1}}{\Longrightarrow}} t\}$. Let us consider the set $B_G(v,n) := \{s \in V_G \mid d_G(v,s) \leq n\}$. We call *complexity* of G from v the function defined by $c_{G,v}(n) = |B_G(v,n)|$. Taking two functions $f,g : \mathbb{N} \longrightarrow \mathbb{N}$, g *dominates* f and we write $f \preceq g$ when $\exists \lambda > 0, \mu > 0, n_0, \ \forall n \geq n_0, \ f(n) \leq \lambda g(\mu n)$. We say that f and g are *equivalent* and we write $f \equiv g$ when both $f \preceq g$ and $g \preceq f$.

Property 4. Let G be a connected graph of finite degree. For any $s, t \in V_G$ we have $c_{G,s} \equiv c_{G,t}$.

Proof. Let $\delta = d_G(s, t)$. For any $n \in \mathbb{N}$ we have $B_G(t, n) \subseteq B_G(s, \delta + n)$. For every $n \geq \delta$ we have $c_{G,t}(n) \leq c_{G,s}(2n)$ i.e. $c_{G,t} \preceq c_{G,s}$. By permutation we get $c_{G,s} \preceq c_{G,t}$ therefore $c_{G,s} \equiv c_{G,t}$. □

This property allows us to define a *complexity order* $c_G(n)$ for each connected graph G. This order is a criterion to classify graph families.

Property 5. Let G, H be two connected graphs of finite degree such that H is image by AFG transformations from G then $c_H \preceq c_G$.

Complexity in REC. The concept of *geometrical complexity* was initialy studied by [11]. He first defined the complexity for pushdown graphs and gave an extended definition for HR-equational graphs (which may have infinite degree). Relying on the property that a pushdown graph only admits a finite set of isomorphic connected components [14], we extend the complexity measure to any prefix recognizable graph: Let $G \in REC$, $H \in PDG$ such that $G = \overline{H}$ and $C_1, \ldots, C_n$ be the classes of connected components of H. We define $c_G = \max\{c_{C_i} \mid 1 \leq i \leq n\}$.

Polynomial Pushdown Graphs. A generalisation of one-counter automaton is the notion of *counter-stack automaton*. A counter-stack automaton is a finite automaton associated with a finite stack of positive counters. Each counter may be increased or decreased when on top of the stack. A counter can be added to (resp. removed from) the top stack only when its value is zero.

We define the *alternation languages* A_n by $A_0 = \varepsilon$, $A_{2k+1} = a^* A_{2k}$ and $A_{2k+2} = b^* A_{2k+1}$. The *comb graph* of order n is the graph defined by restriction from the Dyck graph by $\Delta_1^n \sim \{(u, a, v) \in \Delta_2 \mid u, v \in A_n\}$. The family of polynomial pushdown graphs X^n is the AFG of generator Δ_1^n.

Property 6. If $G \in X^n$ then $c_G \preceq k^n$.

This result is a direct consequence of Property 5. We exhibit a strictly increasing chain of AFG in PDG: for any n, $X^{n+1} - X^n \neq \emptyset$.

4   Conclusion

The general model of acceptors used by [10] to define AFA is a finite state machine associated with an auxiliary storage system. The infinite graph formalism is an elegant way to modelize these machines. A language is defined as a trace of an infinite graph. This infinite graph is obtained by simple transformations from a generator which may be considered as an auxiliary storage structure (e.g. Δ_2

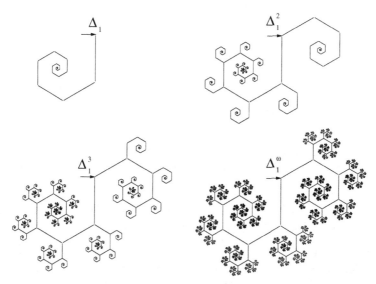

Fig. 5. A fractal representation of polynomial generators: $c_{\Delta_1} \equiv k$, $c_{\Delta_1^2} \equiv k^2$, $c_{\Delta_1^3} \equiv k^3$, $c_{\Delta_1^n} \equiv k^n$.

trace & closure	trace	$\mathcal{F} = Fin^{-1}(\mathcal{F})$	$\mathcal{F} = \mathcal{F} \cdot \Delta_0^n$	$\mathcal{F} = \#\mathcal{F}$	$\mathcal{F} = \overline{\mathcal{F}}$
Turing graphs	RE	yes	yes	yes	yes
Rational graphs [13]	CS	yes	yes	**no**	no
Commutative rational [12]	?	yes	?	?	no
Congruential graphs [17]	?	yes	yes	?	no
Synchronized rational [15]	CS	yes	yes	**no**	no
Prefix recognizable	CF	yes	yes	yes	yes
Prefix	CF	yes	yes	yes	no
One-counter pushdown (ε-closure)	Ocl	yes	yes	yes	yes
One-counter pushdown	Ocl	yes	yes	yes	no
One-reversal pushdown (ε-closure)	Lin	yes	yes	yes	yes
Finite	Rat	yes	yes	yes	yes

Fig. 6. Traces and closure properties for some AFG.

is a stack, Δ_1 is a counter...). The intrinsic complexity of the graph and therefore of its traces are inherited from the generator. Figure 6 summaries traces and closure properties for some AFG. Note that rational and synchronized rational graphs are not closed under rationnal colouring. Some open questions are pointed with a '?'.

Acknowledgement

Many thanks to D. Caucal and T. Colcombet for their advices and remarks.

References

1. J.-M. Auteber and L. Boasson. *Transductions Rationnelles – Application aux Langages Algébriques*. Masson, Paris, 1988.
2. J. Berstel. *Transductions and Context-Free-Languages*. B.G. Teubner, Stuttgart, 1979.
3. A. Blumensath. Prefix-recognizable graphs and monadic second order logic. Technical Report AIB-06-2001, RWTH Aachen, April 2001.
4. D. Caucal. On infinite transition graphs having a decidable monadic theory. In *ICALP 1996*, volume 1099 of *LNCS*, pages 194–205, 1996.
5. D. Caucal. On the transition graphs of Turing machines. In *MCU 2001*, LNCS, pages 177–189, 2001.
6. N. Chomsky and MP. Schützenberger. *The algebraic theory of context-free languages in computer programming and formal systems*. North Holland, 1963.
7. B. Courcelle. The monadic second-order logic of graphs ii: infinite graphs of bounded tree width. *Math. Systems Theory*, 21:187–221, 1989.
8. B. Courcelle. Monadic-second order definable graph transductions : a survey. *TCS*, vol. 126:pp. 53–75, 1994.
9. S. Ginsburg. *Algebraic and Automata-Theoretic Properties of Formal Languages*. North Holland/American Elsevier, 1975.
10. S. Ginsburg and S. Greibach. Abstract families of languages. *Mem. Am. Math. Soc.*, 87, 1969.
11. A. Maignan. Sur la complexité des graphes réguliers. rapport de DEA (IFSIC), Rennes, 1994.
12. C. Morvan. *Les graphes rationnels*. Thèse de doctorat, Université de Rennes 1, Novembre 2001.
13. C. Morvan and C. Stirling. Rational graphs traces context-sensitive languages. In *MFCS 2001*, number 2136 in LNCS, pages 436–446, 2001.
14. D. Muller and P. Schupp. The theory of ends, pushdown automata, and second-order logic. *TCS*, 37:51–75, 1985.
15. C. Rispal. Graphes rationnels synchronisés. Rapport de DEA, Université de Rennes 1, 2001.
16. W. Thomas. A short introduction to infinite automata. In *Proceedings of the 5th international conference Developments in Language Theory*, volume 2295, pages 130–144. LNCS, 2001.
17. T. Urvoy. Regularity of congruential graphs. In *MFCS 2000*, volume 1893 of *LNCS*, pages 680–689, 2000.

Automaton Representation
of Linear Conjunctive Languages

Alexander Okhotin

School of Computing, Queen's University,
Kingston, Ontario, Canada K7L 3N6
okhotin@cs.queensu.ca

Abstract. Triangular trellis automata, also studied under the name of one-way real-time cellular automata, have been known for several decades as a purely abstract model of parallel computers. This paper establishes their computational equivalence to linear conjunctive grammars, which are linear context-free grammars extended with an explicit intersection operation. This equivalence allows to combine the known results on the generative power and closure properties of triangular trellis automata and linear conjunctive grammars and to obtain new previously unexpected results on this language family – for instance, to determine their exact relationship with other comparable families of languages.

1 Introduction

The main properties of linear context-free grammars and the languages they generate have been uncovered already in the early days of formal language theory. One of their most attractive qualities is low computational complexity: the general membership problem is known to be **NLOGSPACE**-complete [7], while every particular language can be recognized in square time and linear space using a well-known algorithm that inductively computes the sets of nonterminals deriving every substring of the input string.

This algorithm is known to have a simple extension for the more general case of linear conjunctive grammars [4], in which all rules are of the form $A \rightarrow u_1 B_1 v_1 \& \ldots \& u_m B_m v_m$ (where $m \geqslant 1$, B_i are nonterminals and u_i, v_i are terminal strings) or $A \rightarrow w$ (where w is a terminal string), and conjunction is interpreted as set-theoretic intersection. These grammars can be effectively transformed to a normal form similar to the context-free linear normal form, for which the sets of nonterminals deriving all substrings of the input string can be computed with the same simplicity as in the context-free case.

The attempts to generalize and refine this kind of computation that preceded the present research led the author to a previously known family of acceptors – the triangular trellis automata [2], also known as one-way real-time cellular automata [1,3]. In the present paper it is first proved that they are at least as powerful as linear conjunctive grammars, and therefore can accept numerous noteworthy languages for which there exists such grammars [6]; somewhat surprisingly, linear conjunctive grammars can in turn simulate these automata,

M. Ito and M. Toyama (Eds.): DLT 2002, LNCS 2450, pp. 393–404, 2003.

which implies complete computational equivalence of the two models, and also gives a way to construct linear conjunctive grammars for some languages, such as the von Dyck language, which were not previously known to be linear conjunctive.

In Section 2 we give a definition of triangular trellis automata, which attempts to unify the "trellis" and "cellular" semantics used in the earlier research; a definition of linear conjunctive grammars is then presented, following [6], and the connections between these grammars and triangular trellis automata are investigated: it is shown how these automata can simulate linear conjunctive grammars and vice versa.

Section 3 addresses the generative power of these automata, showing that, for instance, the language $\{a^n b^{2^n} \mid n \geqslant 1\}$ is accepted by a triangular trellis automaton, while the language $\{a^n b^{2^{n \lfloor \log n \rfloor}} \mid n \geqslant 1\}$ is not; these results give a new insight on the generative power of linear conjunctive grammars and allow to prove that this language family is not closed under quotient with a^*.

In Section 4 the closure properties of this language family are summarized, basing upon the results obtained earlier using triangular trellis automata [8] and conjunctive grammars [6]. Some known closure properties of linear conjunctive languages are thus given new clearer proofs, and some of the open problems raised in [6] are immediately solved using the known results on automata. This also leads to better understanding of the scope of linear conjunctive languages in comparison to other language families, summarized in the conclusion.

2   Automata, Grammars and Their Equivalence

In this section we provide a short overview of linear conjunctive grammars and triangular trellis automata, specifying the exact notation for the latter and consequently proving these two formalisms computationally equivalent to each other.

2.1   Triangular Trellis Automata

Systolic trellis automata [1,2] were introduced in early 1980s as a model of a massively parallel computer with simple identical processors connected in a uniform pattern, as in Figure 1(a); they are used as acceptors of strings loaded from the bottom, and the states in the column shown with a dotted line determine whether the input is accepted.

Triangular trellis automata are a particular case of trellis automata, in which the connections between nodes form a figure of triangular shape shown in Figure 1(b), and acceptance is determined by the topmost element. In this restricted form they are immediately seen to be equivalent to *one-way real-time cellular automata* [3], which in their general form are also used as a model of parallel computation.

Let us define these automata in the following way:

Definition 1. *A triangular trellis automaton is a quintuple $M = (\Sigma, Q, I, \delta, F)$, where Σ is the input alphabet, Q is a finite nonempty set of states (these are*

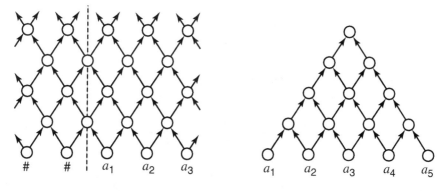

Fig. 1. Trellis automata: (a) of general form; (b) triangular.

states of individual processing units), $I : \Sigma \to Q$ is a homomorphism that sets the initial states (loads values into the bottom processors), $\delta : Q \times Q \to Q$ is the transition function (the function computed by processors) and $F \subseteq Q$ is the set of final states (effective in the top processor).

Let us give two equivalent definitions of the semantics of these automata. According to "trellis" semantics, given a string $a_1 \ldots a_n$ $(a_i \in \Sigma, n \geqslant 1)$, every node corresponds to a certain substring $a_i \ldots a_j$ $(1 \leqslant i \leqslant j \leqslant n)$ of symbols on which its value depends. The value of a bottom node corresponding to one symbol of the input is $I(a_i)$; the value of a successor of two nodes is δ of the values of these ancestors. Denote the value of a node corresponding to $a_i \ldots a_j$ as $\Delta(I(a_i \ldots a_j)) \in Q$; by definition, $\Delta(I(a_i)) = I(a_i)$ and $\Delta(I(a_i \ldots a_j)) = \delta(\Delta(I(a_i \ldots a_{j-1})), \Delta(I(a_{i+1} \ldots a_j)))$. Now define $L(M) = \{w \mid \Delta(I(w)) \in F\}$.

According to "cellular" semantics, an instantaneous description (ID) of an automaton $(\Sigma, Q, I, \delta, F)$ is an arbitrary nonempty string over Q; the initial ID on an input $a_1 \ldots a_n$ $(a_i \in \Sigma, n \geqslant 1)$ is defined as $I(w) = I(a_1) \ldots I(a_n)$; every next ID is constructed out of the previous one by replacing a string of the form $q_1 \ldots q_m$ with the string $\delta(q_1, q_2), \delta(q_2, q_3), \ldots, \delta(q_{m-1}, q_m)$, which is denoted as $\bar{\delta}(q_1 \ldots q_m)$. This is being done until the string of states shrinks to a single state; then the string is accepted if and only if this single state belongs to the set F. Define $L(M) = \{w \mid \bar{\delta}^{|w|-1}(I(w)) \in F\}$.

The equivalence of these two semantics is obvious, as $\Delta(I(w)) = \bar{\delta}^{|w|-1}(I(w))$ for all $w \in \Sigma^+$. The operation of such an automaton on an input $w = a_1 \ldots a_n$ is illustrated in Figure 2, where the value $\Delta(I(a_i \ldots a_j))$ computed by a single processing unit of a trellis automaton is denoted as Δ_{ij}. At the same time, the values in every single horizontal level of the trellis can be read as an instantaneous description of a cellular automaton. The single element of the top row holds the value of the top processor $\Delta(I(w))$ (according to the trellis semantics), or, equivalently, the value of the leftmost cell in the id of a cellular automaton after $|w| - 1$ transitions.

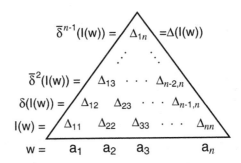

Fig. 2. The computation of an automaton on a given input string.

One evident limitation of these automata is their inability to accept or reject the empty string; however, this is only a technical limitation which does not affect their generative power on longer strings.

2.2 Linear Conjunctive Grammars

Let us start with a definition of conjunctive grammars of the general form [4]:

Definition 2. *A conjunctive grammar is a quadruple* $G = (\Sigma, N, P, S)$, *where* Σ *and* N *are disjoint finite nonempty sets of terminal and nonterminal symbols;* P *is a finite set of grammar rules of the form*

$$A \to \alpha_1 \& \ldots \& \alpha_n \quad (A \in N; \; n \geqslant 1; \; \alpha_1, \ldots, \alpha_n \in (\Sigma \cup N)^*), \qquad (1)$$

where the strings α_i *are distinct and their order is considered insignificant;* $S \in N$ *is a nonterminal designated as the start symbol.*

For every rule of the form (1) and for every i *(*$1 \leqslant i \leqslant n$*),* $A \to \alpha_i$ *is called a conjunct. Let* $conjuncts(P)$ *denote the sets of all conjuncts.*

A conjunctive grammar generates strings by deriving them from the start symbol, generally in the same way as the context-free grammars do. Intermediate strings used in course of a derivation are defined as follows:

Definition 3. *Let* $G = (\Sigma, N, P, S)$ *be a conjunctive grammar. The set of conjunctive formulae* $\mathcal{F} \subset (\Sigma \cup N \cup \{\text{"("}, \text{"\&"}, \text{")"}\})^*$ *is defined inductively: (i)* ϵ *is a formula; (ii) Any symbol from* $\Sigma \cup N$ *is a formula; (iii) If* $\mathcal{A}$ *and* $\mathcal{B}$ *are nonempty formulae, then* $\mathcal{AB}$ *is a formula. (iv) If* $\mathcal{A}_1, \ldots, \mathcal{A}_n$ *(*$n \geqslant 1$*) are formulae, then* $(\mathcal{A}_1 \& \ldots \& \mathcal{A}_n)$ *is a formula.*

There are two types of derivation steps: First, a nonterminal can be rewritten with a body of a rule enclosed in parentheses – $s_1 A s_2 \overset{G}{\Longrightarrow} s_1(\alpha_1 \& \ldots \& \alpha_n)s_2$, if $A \to \alpha_1 \& \ldots \& \alpha_n \in P$ and $s_1 A s_2 \in \mathcal{F}$, where $s_1, s_2 \in (\Sigma \cup N \cup \{\text{"("}, \text{"\&"}, \text{")"}\})^*$. Second, a conjunction of one or more identical terminal strings enclosed in parentheses can be replaced with one such string without the parentheses – $s_1(w \& \ldots \& w)s_2 \overset{G}{\Longrightarrow} s_1 w s_2$.

Definition 4. *Let* $G = (\Sigma, N, P, S)$ *be a conjunctive grammar. The language of a formula is the set of all terminal strings derivable from the formula:* $L_G(\mathcal{A}) = \{w \in \Sigma^* \mid \mathcal{A} \overset{G}{\Longrightarrow} {}^* w\}$. *Define* $L(G) = L_G(S)$.

Let us now restrict general conjunctive grammars to obtain the subclass called *linear conjunctive grammars*:

Definition 5. *A conjunctive grammar* $G = (\Sigma, N, P, S)$ *is said to be linear, if every rule in* P *is of the form* $A \to u_1 B_1 v_1 \& \ldots \& u_m B_m v_m$, *where* $m \geqslant 1$, $u_i, v_i \in \Sigma^*$ *and* $B_i \in N$, *or of the form* $A \to w$ $(w \in \Sigma^*)$.
  G is said to be in the linear normal form, if each rule in P *is of the form* $A \to bB_1 \& \ldots \& bB_m \& C_1 c \& \ldots \& C_n c$, *where* $m + n \geqslant 1$, $B_i, C_j \in N$ *and* $b, c \in \Sigma$, *or of the form* $A \to a$, *where* $A \in N$ *and* $a \in \Sigma$. *If* S *does not appear in right parts of rules, then the rule* $S \to \epsilon$ *is also allowed.*

It has been proved in [4] that every linear conjunctive grammar can be effectively transformed to the linear normal form.

Linear conjunctive grammars are known to be able to generate the classical non-context-free languages $\{a^n b^n c^n \mid n \geqslant 0\}$, $\{a^m b^n c^m d^n \mid m, n \geqslant 0\}$ and $\{wcw \mid w \in \{a, b\}^*\}$ [4], as well as more complex languages, such as the language $\{ba^2 ba^4 b \ldots ba^{2n-2} ba^{2n} b \mid n \geqslant 0\}$ with square growth property and the language of all derivations within a finite string-rewriting system [6]. Despite the increased generative power in comparison with linear context-free grammars, any language generated by a linear conjunctive grammar is still a quadratic-time language. Given a string $a_1 \ldots a_n \in \Sigma^+$ $(n \geqslant 1)$, the $O(n^2)$ recognition algorithm of [4] computes the sets $T_{ij} = \{A \mid A \in N, A \overset{G}{\Longrightarrow} {}^* a_i \ldots a_j\}$ for all i and j $(1 \leqslant i \leqslant j \leqslant n)$. The sets T_{ii} can be computed immediately as $T_{ii} = \{A \mid A \in N, A \to a_i \in P\}$. In order to compute the sets T_{ij} for $i < j$, it suffices to note that

$$T_{ij} = \{A \mid \text{there is a rule } A \to bB_1 \& \ldots \& bB_m \& C_1 c \& \ldots \& C_n c \in P,$$
$$\text{such that } b = a_i, \ c = a_j, \text{ for all } p\,(1 \leqslant p \leqslant m)\, B_p \in T_{i+1,j}, \quad (2)$$
$$\text{and for all } q\,(1 \leqslant q \leqslant n)\, C_q \in T_{i,j-1}\},$$

i.e., T_{ij} depends on the sets $T_{i+1,j}$ and $T_{i,j+1}$, and the symbols a_i and a_j.

2.3 Simulation of Grammars by Automata

Let us show that triangular trellis automata can accept any language generated by a linear conjunctive grammar by computing the same sets T_{ij} as above.

Each state of the automaton will be associated with some subset of N, so that the state $\Delta(I(a_i \ldots a_j))$ "knows" the set T_{ij}. In order to eliminate the direct dependence of every step of computation on the input string, the characters of the input string will also be encoded in the states, so that each processing unit will additionally remember the symbols a_i and a_j in the state $\Delta(I(a_i \ldots a_j))$. The computation of each T_{ii} out of a_i will be done by the function I. The assignment (2) to every T_{ij} $(i < j)$ will be modeled by the function δ.

Formally, let $G = (\Sigma, N, P, S)$ be a linear conjunctive grammar in the linear normal form. We construct the automaton $M = M(G) = (\Sigma, Q, I, \delta, F)$, where $Q = \Sigma \times 2^N \times \Sigma$, $I(a) = (a, \{A \mid A \rightarrow a \in P\}, a)$, $\delta((b, Q, b'), (c', R, c)) = (b, \{A \mid \exists A \rightarrow bB_1 \& \ldots bB_m \& C_1 c \& \ldots \& C_n c : B_i \in R, C_j \in Q\}, c)$ and $F = \{(a, R, b) \mid R \subseteq N,\ S \in X,\ a, b \in \Sigma\}$.

The correctness of our construction is stated in the following lemma, which can be proved by a straightforward induction on the length of the string w.

Lemma 1. *For each string $w \in \Sigma^+$, let $\Delta(I(w)) = (b, R, c)$.*

Then, for each nonterminal $A \in Q$, $A \stackrel{G}{\Longrightarrow}{}^* w$ if and only if $A \in R$.

Theorem 1. *For every linear conjunctive grammar $G = (\Sigma, N, P, S)$ there exists and can be effectively constructed a triangular trellis automaton $M = (\Sigma, Q, I, \delta, F)$, such that $L(M) = L(G) \pmod{\Sigma^+}$.*

2.4   Simulation of Automata by Grammars

Let $M = (\Sigma, Q, I, \delta, F)$ be an automaton. We construct the grammar $G = G(M) = (\Sigma, N_Q \cup \{S\}, P, S)$, where $N_Q = \{A_q \mid q \in Q\}$ and P contains the following rules: (i) $S \rightarrow A_q$ for all $q \in F$; (ii) $A_{I(a)} \rightarrow a$ for all $a \in \Sigma$; (iii) $A_{\delta(q_1, q_2)} \rightarrow bA_{q_2} \& A_{q_1} c$ for all $q_1, q_2 \in Q$ and $b, c \in \Sigma$.

The following lemma states the correctness of the construction. It can be proved by induction on $|w|$.

Lemma 2. *For every $w \in \Sigma^+$ and $q \in Q$, $A_q \stackrel{G}{\Longrightarrow}{}^* w$ iff $\Delta(I(w)) = q$.*

Theorem 2. *For every triangular trellis automaton M there exists and can be effectively constructed a linear conjunctive grammar G, such that $L(G) = L(M)$.*

Together with the earlier Theorem 1, this allows to make the conclusion that a language $L \subseteq \Sigma^+$ is accepted by some triangular trellis automaton if and and only if it is generated by some linear conjunctive grammar.

3   Generative Power

3.1   The Case of a Unary Alphabet

It was proved in [6] that every linear conjunctive language over a unary alphabet is regular. The automaton representation of linear conjunctive languages developed in this paper allows us to give an entirely new proof of the same assertion:

Theorem 3. *For every triangular trellis automaton M over a unary alphabet Σ there exists and can be effectively constructed a DFA accepting $L(M)$.*

Proof. Consider an arbitrary automaton $M = (\{a\}, Q, I, \delta, F)$. It is easy to note that for every string $a^n \in \Sigma^+$ and for every number i $(0 \leqslant i < n)$, $\overline{\delta}^i(I(a^n)) = (q_i)^{n-i}$ for some state $q_i \in Q$ that depends only on i, and that $q_{i+1} = \delta(q_i, q_i)$. Now let us construct a DFA $D = (\{a\}, Q_D, q_0, \delta_D, F_D)$, where $Q_D = Q \cup \{q_0\}$, $\delta_D(q_0, a) = I(a)$, $\delta_D(q, a) = \delta(q, q)$, and $F_D = F$. By a straightforward induction on n it is not hard to show that $\hat{\delta}_D(q_0, a^n) = \Delta(I(a^n))$ for every $n \geqslant 1$. Therefore, $a^n \in L(D)$ iff $\hat{\delta}_D(q_0, a^n) \in F_D$ iff $\Delta(I(a^n)) \in F$ iff $a^n \in L(M)$.

3.2   Keeping Count of the Number of Symbols

In this section we consider the limitations of the generative power of triangular trellis automata (and therefore of linear conjunctive grammars) for the languages of the form

$$\{a^n b^{f(n)} \mid n \geqslant 1\} \quad (f : \mathbb{N} \to \mathbb{N} \text{ is some function}) \tag{3}$$

A language $L \subseteq a^* b^*$ is of the form (3) if for every $n \geqslant 1$ there exists one and only one $k \geqslant 1$, such that $a^n b^k \in L$. The task of recognizing such a language involves counting the symbols. It is known that finite automata *cannot keep count* in the sense that a language of the form (3) can be regular only if $f(n)$ is bounded by some constant. Context-free grammars *can* keep count, for instance, being able to generate the language $\{a^n b^n \mid n \geqslant 1\}$, but, to put it informally, they *count on their fingers*, and thus a language of the form (3) can be context-free only if $f(n)$ is linearly bounded.

We shall now show that for every linear conjunctive language of the form (3) the function f is necessarily bounded by some exponential function. The argument will be in some way similar to the regular and context-free pumping lemmata applied to the languages of the form (3).

Let us consider the computation of an arbitrary triangular trellis automaton $M = (\Sigma, Q, I, \delta, F)$ over the alphabet $\Sigma = \{a, b\}$ on some input string from $a^+ b^+$. Denote $\Delta(I(b^i)) = B_i \in Q$ $(i > 0)$. It is easy to notice that $B_{i+1} = \delta(B_i, B_i)$, i.e. each element of the sequence $\{B_i\}_{i=1}^{\infty}$ depends solely on the previous element. Therefore, the sequence $\{B_i\}_{i=1}^{\infty}$ is periodic with a period no more than $|Q|$. Similarly, denote $\Delta(I(a^l)) = A_l \in Q$.

Now fix an arbitrary number $n \geqslant 1$ and consider the strings of the form $a^n b^k$ for different $k \geqslant 1$. Denote $\alpha_{ij} = \Delta(I(a^{n-j+1} b^i))$ $(1 \leqslant i \leqslant k, 1 \leqslant j \leqslant n)$ and $\alpha_i = \alpha_{i1} \ldots \alpha_{in}$ $(1 \leqslant i \leqslant k)$, and consider the sequence of strings of states $\{\alpha_i B_i\}_{i=1}^{\infty}$. Every B_{i+1} depends on B_i; $\alpha_{i+1,n}$ depends on $\alpha_{i,n}$ and B_{i+1}; $\alpha_{i+1,n-1}$ depends on $\alpha_{i,n-1}$ and $\alpha_{i+1,n}$, etc. Accordingly, every string $\alpha_{i+1} B_{i+1}$ is dependent solely on the previous string $\alpha_i B_i$; this dependence is illustrated in Figure 3(a). Each of the strings $\alpha_i B_i$ is of length $n+1$, and thus there are $|Q|^{n+1}$ distinct strings of this kind. Therefore, the sequence $\{\alpha_i B_i\}_{i=1}^{\infty}$ is guaranteed to start repeating itself after $|Q|^{n+1}$ steps.

Theorem 4. *If a linear conjunctive language $L \subseteq a^+ b^+$ is of the form (3), then there exists a constant $C > 0$, such that $f(n) \leqslant C^n$ for all $n \geqslant 1$.*

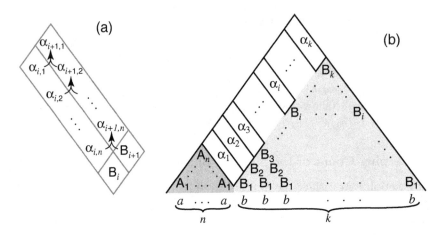

Fig. 3. Computation of an automaton on the string of the form $a^n b^k$.

Proof. Let $f = f(n)$ be a function, such that for every constant $C > 0$ there exists a number $n = n(C) \geqslant 1$, for which $f(n) > C^n$. Now suppose there is an automaton $M = (\Sigma, Q, I, \delta, F)$ that accepts the language $L = \{a^n b^{f(n)} \mid n \geqslant 1\}$. Choose $C = |Q|^2$ and let $n = n(C) \geqslant 1$ be the corresponding number, such that $f(n) > C^n$. Let $k = f(n)$ and consider the computation of M on the string $a^n b^k$. Because the sequence $\{\alpha_t B_t\}_{t=1}^{\infty}$ starts to repeat itself after $|Q|^{n+1}$ steps, there exist numbers i and j ($1 \leqslant i < j \leqslant |Q|^{n+1} + 1$), such that $\alpha_i B_i = \alpha_j B_j$. Since $k = f(n) > C^n = |Q|^{2n} \geqslant |Q|^{n+1} \geqslant i$, it follows that $\alpha_k B_k = \alpha_{k+(j-i)} B_{k+(j-i)}$. On the other hand, since $a^n b^k \in L(M)$, $\alpha_{k,1} \in F$, and therefore $\alpha_{k+(j-i),1} \in F$. This implies that $a^n b^{k+(j-i)} \in L(M)$, which is a contradiction.

Corollary 1. *If the function $f(n)$ is asymptotically greater than C^n for any $C > 0$, then the language $\{a^n b^{f(n)} \mid n \geqslant 1\}$ is not linear conjunctive.*

Corollary 1 provides a certain asymptotical upper bound for f, which at the first glance seems to be not very tight. However, it will follow from the results of the next section that this bound is in fact precise.

3.3   Counting Using Positional Notation

As stated in the previous section, context-free grammars count on their fingers, because using pushdown to count means counting in the unary notation. In this section we show that triangular trellis automata can in fact count using *positional notation*, which allows them to exceed the generative capacity of context-free grammars on the $a^n b^{f(n)}$-languages.

Fix an integer $C \geqslant 2$ and consider the set of states $Q = \{A, B, 0, 1, \dots, C - 1, 0^+\}$, in which 0^+ means "zero with carry". Define the mappings I and δ as follows: $I(a) = A$, $I(b) = B$, $\delta(A, A) = A$, $\delta(B, B) = B$, $\delta(A, B) = 1$, $\delta(A, 1) = 0$,

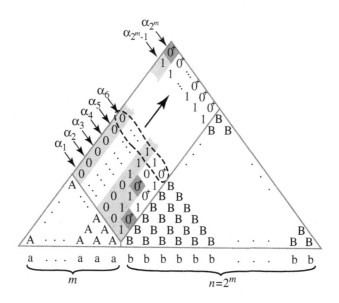

Fig. 4. Triangular trellis automata can count using positional notation.

$\delta(A, 0) = 0$, $\delta(x, B) = \delta(x, 0^+) = x + 1$ for all $x \in \{0, \ldots, C - 2\}$, $\delta(C - 1, B) = \delta(C - 1, 0^+) = 0^+$, $\delta(0^+, B) = 1$, $\delta(x, y) = x$ for all $x, y \in \{0, \ldots, C - 1\}$, and $\delta(0^+, x) = 0$ for all $x \in \{0, \ldots, C - 1\}$.

A sample computation of the automaton for the case $C = 2$ on the input $a^m b^{2^m}$ is given in Figure 4; the shades of gray ought to be ignored for the moment. If, given an arbitrary string $w = a^m b^n$ $(m, n \geqslant 1)$, one uses an arrangement of states into strings similar to that from Section 3.2,

$$\alpha_k = \Delta(a^m b^k)\Delta(a^{n-1} b^k) \ldots \Delta(a^2 b^k)\Delta(a^1 b^k) \quad (1 \leqslant k \leqslant n) \qquad (4)$$

it becomes evident that every α_k holds base C notation of the number k modulo 2^m, if both 0 and 0^+ are interpreted as zero. If the set of accepting states is chosen as $F = \{0^+\}$, then the given automaton accepts the language $\{a^n(b^{C^n})^+ | n \geqslant 1\}$.

We shall now apply this method to construct a more complicated automaton accepting the exponential-growing $a^n b^{f(n)}$-language $\{a^n b^{C^n} | n \geqslant 1\}$. Choose $Q' = \{A, B\} \cup \{0, \ldots, C - 1, 0^+\} \times \{t, f\}$ and define I' and δ' in the same way as I and δ above, but where the second component of a pair *(digit, flag)* determines whether there are *no accepting states among the left predecessors* of an element of the trellis. In Figure 4 the states with true second component are emphasized with gray; dark gray is used for the state $(0^+, t)$, which is set to be the sole accepting state. Once this is done, the automaton will accept every string $a^n b^{2^n}$, but will not accept any $a^n b^{k \cdot 2^n}$ $(k \geqslant 2)$, since the second component of $\{\Delta(I(a^n b^i))\}_{i=1}^{\infty}$ will be reset to false right after $i = 2^n$.

3.4   Matching Parentheses: The von Dyck Language

The *von Dyck language* over the alphabet $\Sigma = \{a_1, \ldots, a_n, b_1, \ldots, b_n\}$ is naturally defined by a context-free grammar with a single nonterminal S and the rules $S \to \epsilon$, $S \to SS$ and $S \to a_i S b_i$ (for all $1 \leqslant i \leqslant n$). This language is a common example of a language which is not linear context-free.

The paper [3] gives a general idea of constructing a triangular trellis automaton for this language. Let us give an actual construction for the case $n = 1$ that more or less follows [3]. We define the set of states as $Q = \{\nearrow, \nwarrow, Acc, \smile\}$, and then $I(a) = \nearrow$, $I(b) = \nwarrow$, $\delta(\nearrow, \nwarrow) = Acc$, $\delta(\nearrow, q) = \nearrow$ (for all $q \neq Acc$), $\delta(q, \nwarrow) = \nwarrow$ (for all $q \neq Acc$), $\delta(q, \searrow) = \nwarrow$, $\delta(Acc, \smile) = \nearrow$, $\delta(\smile, Acc) = \nwarrow$ and the rest of transitions lead to $\smile$. Also let the set of accepting states be $F = \{Acc\}$.

Using the results of the present paper, one can immediately construct a linear conjunctive grammar for the same language that contains as few as five nonterminals. However, this grammar will be anything but human-readable. It remains an open question whether one can construct an intuitively understandable linear conjunctive grammar for the von Dyck language.

4   Closure Properties

4.1   Set-Theoretic Operations

The languages generated by linear conjunctive grammars are trivially closed under union and intersection. Their closure under complement was shown in [6] by a direct construction of a grammar for the complement of the language generated by a given grammar; actually, this closure result came quite unexpected. The use of triangular trellis automata allows to give new proofs of these facts.

Let $M_i = (\Sigma, Q_i, I_i, \delta_i, F_i)$ $(i = 1, 2)$ be the original automata. We take the set of states $Q = Q_1 \times Q_2$, define the mappings I and δ as $I(a) = (I_1(a), I_2(a))$ and $\delta((q_1', q_2'), (q_1'', q_2'')) = (\delta_1(q_1', q_1''), \delta_2(q_2', q_2''))$, and then the automata $M_\cup = (\Sigma, Q, I, \delta, \{(q_1, q_2) \mid q_i \in F_i \text{ for some } i\})$, $M_\cap = (\Sigma, Q, I, \delta, F_1 \times F_2)$, are easily seen to accept $L(M_1) \cup L(M_2)$ and $L(M_1) \cap L(M_2)$ respectively. Turning to the case of the complement, for any given $M = (\Sigma, Q, I, \delta, F)$, the automaton $M' = (\Sigma, Q, I, \delta, Q \setminus F)$ is easily verified to accept $\Sigma^* \setminus L(M)$.

It is interesting to note the similarity between this construction and the classical proofs of the same closure results for DFAs. In particular, the new proof for complement is immediate, and the construction is confined to taking the complement of the set of accepting states.

4.2   Quotient

Linear conjunctive languages were proved to be closed under quotient with finite languages in [6]; due to their known closure under union, it was enough to demonstrate that they are closed under quotient with the language $\{a\}$. On the other hand, linear conjunctive languages are not closed under quotient with regular languages [6].

It is possible to give a simpler automaton-based proof of the former fact, which is omitted due to space considerations. Additionally, the results of the present paper allow to improve the latter nonclosure result as follows:

Theorem 5. *Let Σ be a finite alphabet, such that $|\Sigma| \geqslant 2$, and let $a \in \Sigma$.*

Then the family of linear conjunctive languages over Σ is not closed under left- and right- quotient with a^*.

Proof. It suffices to note that the language $((a^*)^{-1} \cdot \{a^n b^{2^n} \mid n \geqslant 1\}) \cap b^* = \{b^{2^n} \mid n \geqslant 1\}$, is not a linear conjunctive language by Theorem 3, while the language $\{a^n b^{2^n} \mid n \geqslant 1\}$ is linear conjunctive by the results of Section 3.3.

4.3 Concatenation

Some partial positive results on concatenation were obtained in [6]: linear conjunctive languages are closed under left- and right-concatenation with regular languages, under concatenation over disjoint alphabets (if $L_i \subseteq \Sigma_i^*$ ($i = 1, 2$) are linear conjunctive languages and $\Sigma_1 \cap \Sigma_2 = \emptyset$, then $L_1 \cdot L_2$ is linear conjunctive), under marked concatenation (in $L_1, L_2 \subseteq \Sigma^*$ are linear conjunctive, then $L_1 \cdot \{c\} \cdot L_2$ is linear conjuntive for any $c \notin \Sigma$). The question of whether linear conjunctive languages are closed under general concatenation was left open [6].

While these partial positive results would most probably be not very easy to obtain using triangular trellis automata, on the other hand, the problem of the closure under general concatenation, which looks so hard for linear conjunctive grammars, has actually been solved for the family of cellular automata isomorphous to triangular trellis automata. It was proved by V. Terrier in 1995 [8] that while the language $L_{Terrier} = \{a^n b^n\} \cup \{a^n b(a|b)^* a b^n\}$ is a linear context-free language, there exists no one-way real-time cellular automaton to accept the square of this language, which is obviously context-free.

This immediately gives us a negative solution for the problem of the closure of linear conjunctive languages under concatenation proposed in [6]. This also allows to conclude that the families of context-free and linear conjunctive languages are incomparable subsets of the family of languages generated by conjunctive grammars of the general form.

5 Concluding Remarks

It was proved that linear conjunctive grammars [4] have a simple automaton representation by a computing device that has long been known as purely theoretical model. This allows to join the existing results on these formalisms and obtain several new ones. In particular, the positive and negative closure results [8,4,6] help clarifying the relationship between context-free, conjunctive, linear context-free and linear conjunctive languages by showing that all the obvious inclusions are proper (see Figure 5).

The general formal simplicity of triangular trellis automata leaves hope that

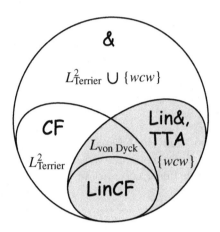

Fig. 5. Relationship between families of languages.

they could eventually be used to solve further open problems concerning linear conjunctive languages, and, as they can be simulated in just $n(n + 1)/2$ elementary table lookups, to be utilized as a low-level computational model.

The relationship between linear conjunctive grammars and triangular trellis automata resembles that between regular expressions and finite automata: while grammars and regular expressions are usually more convenient for human use, the automata are much better suitable for machine implementation. Perhaps in the case of linear conjunctive languages the coexistence of so different representations might become as useful for their study as it proved to be for regular languages.

The author is grateful to Kai Salomaa for many inspiring discussions on the subject of this paper.

References

1. C. Choffrut and K. Culik II, "On real-time cellular automata and trellis automata", *Acta Informatica*, 21 (1984), 393–407.
2. K. Culik II, J. Gruska and A. Salomaa, "Systolic trellis automata" (I, II), *International Journal of Computer Mathematics*, 15 (1984) 195–212; 16 (1984) 3–22.
3. C. Dyer, "One-way bounded cellular automata", *Information and Control*, 44 (1980), 261–281.
4. A. Okhotin, "Conjunctive grammars", *Journal of Automata, Languages and Combinatorics*, 6:4 (2001), 519–535.
5. A. Okhotin, "A recognition and parsing algorithm for arbitrary conjunctive grammars", *Theoretical Computer Science*, to appear.
6. A. Okhotin, "On the closure properties of linear conjunctive languages", *Theoretical Computer Science*, to appear.
7. I. H. Sudborough, "A note on tape-bounded complexity classes and linear context-free languages", *Journal of the ACM*, 22:4 (1975), 499–500.
8. V. Terrier, "On real-time one-way cellular array", *Theoretical Computer Science*, 141 (1995), 331–335.

On-Line Odometers
for Two-Sided Symbolic Dynamical Systems

Christiane Frougny[1,2]

[1] LIAFA
Case 7014, 2 place Jussieu, 75251 Paris Cedex 05, France
Christiane.Frougny@liafa.jussieu.fr
[2] Université Paris 8

Abstract. We consider biinfinite sequences on a finite alphabet of digits which satisfy a constraint of finite type. Such sequences are perturbed by adding a 1 in position 0. The odometer is the function which transforms the initial sequence into an admissible sequence equivalent to the perturbed one. It is shown that the odometer can be realized by an on-line finite automaton when the constraint is linked to numeration in base β, where β is a Pisot number satisfying the equation $\beta^m = t_1\beta^{m-1} + \cdots + t_m$, where $t_1 \geq t_2 \geq \cdots t_m \geq 1$ are integers.

Keywords: Finite automata, symbolic dynamical systems, numeration systems, odometer.

1 Introduction

We present the problem by an example. Consider the set V of biinfinite sequences of 0's and 1's satisfying the constraint that no factor 11 occurs. Denote such a sequence by $a = (a_n)_{n\in\mathbb{Z}} = \cdots a_2 a_1 a_0 . a_{-1} a_{-2} a_{-3} \cdots$. The set V is a symbolic dynamical system of finite type, called the golden ratio shift, see [14]. Now we introduce a small perturbation consisting of adding 1 at coordinate of index 0, and we want to eliminate the non-admissible factors, namely those containing 11 or 2, in order to obtain an "equivalent" sequence $b = (b_n)_{n\in\mathbb{Z}}$ belonging to V. The *odometer*, denoted by τ, is the function that maps the sequence a onto the sequence b. Here we use the fact that $011 \sim 100$ and that $0200 \sim 1001$ in base $\beta = (1 + \sqrt{5})/2$. We give two examples.

Let $a = \cdots 0000101.01010000\cdots$. Then

$$a + 1 = \cdots 0000102.01010000\cdots$$
$$\sim \cdots 0000110.02010000\cdots$$
$$\sim \cdots 0001000.10020000\cdots$$
$$\sim \cdots 0001000.10100100\cdots$$

Thus $\tau(a) = b = \cdots 0001000.10100100\cdots$.

M. Ito and M. Toyama (Eds.): DLT 2002, LNCS 2450, pp. 405–416, 2003.

Let $a' = \cdots 010101010101.010101010101\cdots$. Then

$$a' + 1 = \cdots 010101010102.010101010101 \cdots$$
$$\sim \cdots 010101010110.020101010101 \cdots$$
$$\sim \cdots 010101011000.100201010101 \cdots$$
$$\sim \cdots 010101100000.101002010101 \cdots$$
$$\cdots$$
$$\sim \cdots 000000000000.101010101010 \cdots$$

Thus $\tau(a') = b' = \cdots 000000000000.101010101010\cdots$.

In this paper we show that the odometer can be realized by an on-line algorithm starting from the middle of the input, going simultaneously in both directions, and needing only a finite auxiliary storage memory, that is to say by a bidirectional *on-line finite automaton*, see Section 2.5 for definitions. On-line finite automata are a special kind of sequential finite automata, processing data from left to right, see [16]. In integral base b on the canonical digit set $\{0, \ldots, b-1\}$, addition is not on-line computable, but with a balanced alphabet of signed digits of the form $\{-a, \ldots, a\}$ with $b/2 \le a \le b-1$, using the algorithms of Avizienis [1] and Chow and Robertson [4], addition is computable by an on-line finite automaton.

Now we introduce the problem in full generality. Let β be a non-integer real number > 1. For any $\beta > 1$, by the greedy algorithm of Rényi [18], one can compute a representation in base β of any real number x from the interval $[0, 1]$, called its β-expansion and denoted by $d_\beta(x)$, and where the digits are elements of the canonical digit set $A_\beta = \{0, \ldots, \lfloor \beta \rfloor\}$. In such a representation, not all the digits patterns are allowed (see [17] for instance), and the numeration system defined by β is thus redundant. The β-expansion of 1 plays a special role. If $d_\beta(1) = t_1 \cdots t_m$ is finite (meaning that it ends with infinitely many zeroes), the associated dynamical system is said to be of *finite type*.

The *normalization* is the fonction ν which maps an infinite sequence of digits $(d_k)_{k \ge 1}$ onto $d_\beta(\sum_{k \ge 1} d_k \beta^{-k})$. Recall that a Pisot number (also called a Pisot-Vijayaraghavan number) is an algebraic integer such that all its algebraic conjugates have modulus less than 1. The natural numbers and the golden ratio are Pisot numbers. It is known that if β is a Pisot number, then normalization on any alphabet is computable by a finite automaton [6], and that non-normalized digit set conversion is computable by an on-line finite automaton [8].

Two-sided beta-expansions have been considered in [20] for the golden ratio case, and in full generality in [19]. In [20] two-sided normalization is introduced and studied in relation with the Erdős measure.

The one-sided odometer has been introduced in [12]. Roughly speaking, to a number $\beta > 1$ one can canonically associate a numeration system defined by a sequence of integers U_β, allowing the representation of the natural numbers [2]. The odometer is the left-infinite extension of the successor function in the numeration system U_β. It is shown in [12] that the one-sided odometer associated with β is continuous if and only if the β-expansion of 1 is finite. We have shown in [7] that the successor function in U_β is realizable by a right subsequential finite automaton if and only if $d_\beta(1)$ is finite. From this follows, by a result of [10], that the associated one-sided odometer can be realized by a right on-line finite automaton. If $d_\beta(1) = t_1 \cdots t_m$, one can give a direct construction of a right on-line finite automaton with delay $m - 1$, [9].

Since we are interested in on-line functions, which are continuous by the result just recalled, $d_\beta(1)$ must be finite, so we define the two-sided odometer only in this case. Let β such that $d_\beta(1) = t_1 \cdots t_m$. We consider the set V_β of biinfinite sequences on the alphabet A_β such that no factor greater than or equal in the lexicographic order to $t_1 \cdots t_m$ occurs. This defines a two-sided symbolic dynamical system of finite type. Admissibility of such sequences can be checked on finite factors of length m. These sequences are perturbed by adding a 1 in position 0. The two-sided odometer can be seen as the biinfinite extension of the successor function. It is the function which transforms a sequence a of V_β into a sequence b of V_β equivalent to the sequence $a + 1$ (see below for a formal definition of equivalence).

From the previous results, it is natural to consider numbers β that are Pisot numbers. In this paper we consider numbers β such that $d_\beta(1) = t_1 \cdots t_m$, with $t_1 \geq t_2 \geq \cdots \geq t_m \geq 1$. It is proved in [3] that they are Pisot numbers. These Pisot numbers appear in several contexts, e.g. [11] and [13]. We show that, for these numbers β, the two-sided odometer associated with β is computable by a bidirectional on-line finite automaton with delay m, hence it is a continuous function. This delay is optimum. There exist other Pisot numbers with $d_\beta(1) = t_1 \cdots t_m$ such that the two-sided odometer is computable by an on-line finite automaton, but the delay is larger than m. In fact the major difficulty is the determination of the exact delay in general. For instance, for the Pisot number β such that $d_\beta(1) = 201$, with similar arguments one can show that the two-sided odometer is computable by an on-line finite automaton with delay 4.

The paper is organized as follows. After the definitions, the result is proved, and the two-sided odometer associated with the golden ratio is developed in Example 1.

2   Preliminaries

2.1   Words

An *alphabet* A is a finite set. A finite sequence of elements of A is called a *word*, and the set of words on A is the free monoid A^*. The *length* of a word u is equal to the number of its letters, and is denoted by $|u|$. The *empty word* is denoted by ε. Let v be a word of A^*, denote by v^n the concatenation of n times the word v, and by A^n the set of words on A of length n.

The set of *(right) infinite* words on A is denoted by $A^{\mathbb{N}}$, and by v^ω the infinite concatenation $vvv \cdots$. The set of *left infinite* words on A is denoted by $^{\mathbb{N}}A$, and the sequence $\cdots vvv$ is denoted by $^\omega v$.

One defines a distance ρ on $A^{\mathbb{N}}$ as follows: let $v = (v_i)_{i\in\mathbb{N}}$ and $w = (w_i)_{i\in\mathbb{N}}$ be in $A^{\mathbb{N}}$, $\rho(v,w) = 2^{-r}$ where $r = \min\{i \geq 0 \mid v_i \neq w_i\}$ if $v \neq w$, and $\rho(v,w) = 0$ otherwise. The set $A^{\mathbb{N}}$ equipped with this metric becomes a compact metric space. The topology induced by the metric is the same as the product topology of $A^{\mathbb{N}}$ where A is equipped with the discrete topology.

The set of biinfinite words is denoted by $A^{\mathbb{Z}}$, and $\cdots vvv \cdots$ by $^{\omega}v^{\omega}$. In this paper, a biinfinite word $v \in A^{\mathbb{Z}}$ is denoted by $\cdots v_2 v_1 v_0 \cdot v_{-1} v_{-2} v_{-3} \cdots$ with v_i in A for every i in $\mathbb{Z}$.

One defines a distance ρ on $A^{\mathbb{Z}}$ as follows: let $v = (v_i)_{i\in\mathbb{Z}}$ and $w = (w_i)_{i\in\mathbb{Z}}$ be in $A^{\mathbb{Z}}$, $\rho(v,w) = 2^{-r}$ if $v \neq w$ and $r = \max\{i \mid v_i \cdots v_{-i} = w_i \cdots w_{-i}\}$, and $\rho(v,w) = 0$ if $v = w$. The set $A^{\mathbb{Z}}$ is then a compact metric space. This topology is equal to the product topology.

2.2   On-Line Functions

Let A and B be two alphabets, and let $\varphi : (a_n)_{n\in\mathbb{N}} \mapsto (b_n)_{n\in\mathbb{N}}$ be a function from $A^{\mathbb{N}}$ to $B^{\mathbb{N}}$. Then φ is said to be *on-line computable with delay* δ if there exists an integer δ such that for every $n \geq 0$ there exists a function $\Phi_n : A^{n+\delta} \to B$ such that $b_n = \Phi_n(a_0 \cdots a_{n+\delta-1})$. An on-line computable function with delay δ is 2^{δ}-Lipschitz and is thus uniformously continuous [8].

A function $\varphi : (a_n)_{n\in\mathbb{Z}} \mapsto (b_n)_{n\in\mathbb{Z}}$ from $A^{\mathbb{Z}}$ to $B^{\mathbb{Z}}$ is said to be *on-line computable with delay* δ if there exists an integer δ such that for every $n \geq 0$ there exists a function $\Phi_n : A^{n+\delta} \to B$ and for every $p \geq 1$ a function $\Psi_p : A^{p+\delta} \to B$ such that $b_n = \Phi_n(a_{n+\delta-1} \cdots a_0)$ and $b_{-p} = \Psi_p(a_{-1} \cdots a_{-p-\delta})$.

As for infinite words, we have that

Proposition 1. *Any function $\varphi : A^{\mathbb{Z}} \longrightarrow B^{\mathbb{Z}}$ which is on-line computable with delay δ is 2^{δ}-Lipschitz and is thus uniformously continuous.*

2.3   Beta-Representations

We refer the reader to [15, Chap. 7] for results and proofs on these topics. Let $\beta > 1$ be a real number. A *β-representation* on a digit set D of a number x of $[0,1]$ is an infinite sequence $(d_k)_{k\geq 1}$ of digits d_k in D such that $\sum_{k\geq 1} d_k \beta^{-k} = x$.

Any real number $x \in [0,1]$ can be represented in base β by the following greedy algorithm [18]. Denote by $\lfloor . \rfloor$ and by $\{.\}$ the integral part and the fractional part of a number, respectively. Let $r_0 = x$ and for $k \geq 1$, let $x_k = \lfloor \beta r_{k-1} \rfloor$ and $r_k = \{\beta r_{k-1}\}$.
Thus $x = \sum_{k\geq 1} x_k \beta^{-k}$, where the digits x_k are elements of the *canonical* alphabet $A_{\beta} = \{0, \ldots, \lfloor \beta \rfloor\}$ if $\beta \notin \mathbb{N}$, and $A_{\beta} = \{0, \ldots, \beta-1\}$ otherwise. The sequence $(x_k)_{k\geq 1}$ of $A_{\beta}^{\mathbb{N}}$ is called the *β-expansion* of x, and is denoted by $d_{\beta}(x)$. When β is not an integer, a number x may have several different β-representations on A_{β}: this system is naturally redundant. The β-expansion obtained by the greedy algorithm is the greatest one in the lexicographic order. A β-representation is said to be *finite* if it ends with infinitely many zeroes, and the zeroes are omitted.

The β-expansion of an $x > 1$ is obtained as follows: let $n \geq 0$ such that $\beta^n \leq x < \beta^{n+1}$. The β-expansion of x is $x_n \cdots x_0 \cdot x_{-1} x_{-2} \cdots$ if and only if $x_n \cdots x_0 x_{-1} x_{-2} \cdots$ is the β-expansion of x/β^{n+1}.

The set $A_\beta^{\mathbb{N}}$ is endowed with the lexicographic order, denoted by $\prec$, and the shift σ, defined by $\sigma((x_i)_{i \geq 1}) = (x_{i+1})_{i \geq 1}$. Denote by D_β the set of β-expansions of numbers of $[0, 1[$. It is a shift-invariant subset of $A_\beta^{\mathbb{N}}$. The β-shift S_β is the topological closure of D_β in $A_\beta^{\mathbb{N}}$.

Let us introduce the following notation. Let $d_\beta(1) = (t_i)_{i \geq 1}$ and set $d_\beta^*(1) = d_\beta(1)$ if $d_\beta(1)$ is infinite and $d_\beta^*(1) = (t_1 \cdots t_{m-1}(t_m - 1))^\omega$ if $d_\beta(1) = t_1 \cdots t_{m-1} t_m$ is finite. Accordingly $d_\beta^*(x) = d_\beta(x)$ if the latter is infinite, $d_\beta^*(x) = x_1 \cdots x_{k-1}(x_k - 1)d_\beta^*(1)$ if $d_\beta(x) = x_1 \cdots x_k$ is finite.

By a result from [17], one has that

$$D_\beta = \{s \in A_\beta^{\mathbb{N}} \mid \forall p \geq 0 \; \sigma^p(s) \prec d_\beta^*(1)\}$$

and

$$S_\beta = \{s \in A_\beta^{\mathbb{N}} \mid \forall p \geq 0 \; \sigma^p(s) \preceq d_\beta^*(1)\}.$$

We now give the definition of the two-sided β-shift, denoted by V_β (see [19])

$$V_\beta = \{s \in A_\beta^{\mathbb{Z}} \mid \forall k \in \mathbb{Z} \; (s_n)_{n \leq k} \in S_\beta\}.$$

Denote by L_β the set $F(D_\beta)$ of finite factors of D_β. Clearly $L_\beta = F(S_\beta) = F(V_\beta)$. Set $V_\beta^+ = \{\ldots s_3 s_2 s_1 s_0 \mid s \in V_\beta\}$ and $V_\beta^- = \{s_{-1} s_{-2} s_{-3} \cdots \mid s \in V_\beta\}$.

When $d_\beta(1) = t_1 \cdots t_m$, the β-shift is a system of *finite type*, and to determine if a word belongs to L_β can be checked on factors of length m. The set of maximal words in the lexicographic order is $M_\beta = (t_1 \cdots t_{m-1}(t_m - 1))^* \{\varepsilon, t_1, t_1 t_2, \ldots, t_1 t_2 \cdots t_{m-1}\}$.

The set L_β is recognized by a finite automaton $\mathcal{A}_\beta$ with m states $q_1, \ldots, q_m$ defined as follows. For each i, $1 \leq i < m$, there is an edge labelled t_i from q_i to q_{i+1}. For $1 \leq i \leq m$, there are edges labelled by $0, 1, \ldots, t_i - 1$ from q_i to q_1. Let q_1 be the only initial state, and all states be terminal. The β-shift S_β is thus the set of labels of infinite paths in $\mathcal{A}_\beta$, and the two-sided β-shift V_β is the set of labels of biinfinite paths in $\mathcal{A}_\beta$.

2.4   Two-Sided Odometer Associated with β

Let β such that $d_\beta(1) = t_1 \cdots t_m$. We introduce some notations: let $s = \cdots s_2 s_1 s_0 \cdot s_{-1} s_{-2} \cdots$, set $s + 1 = \cdots s_2 s_1 (s_0 + 1) \cdot s_{-1} s_{-2} \cdots$, and similarly for finite or infinite words. Let $v = v_j v_{j-1} \cdots$, $x = \sum_{i \leq j} v_i \beta^i$, and let $\lambda(v) = d_\beta(x)$ if v does not end with $(t_1 \cdots t_{m-1}(t_m - 1))^\omega$, $\lambda(v) = d_\beta^*(x)$ otherwise.

The *two-sided odometer* associated with β is the function $\tau : V_\beta \to V_\beta$ defined as follows. Let $s = \cdots s_2 s_1 s_0 \cdot s_{-1} s_{-2} \cdots \in V_\beta$. There are two cases.

1. There exists $N_s \geq 0$ such that for every $j \geq N_s$, $s_j \cdots s_0 \notin M_\beta$. This means that the carry does not propagate infinitely far to the left. For $j \geq N_s$ let $\tau(s) = \cdots s_{j+2} s_{j+1} \lambda(s_j \cdots s_1 (s_0 + 1) \cdot s_{-1} s_{-2} \cdots)$.

2. The carry does propagate infinitely far to the left.

If $\cdots s_2 s_1 s_0 = {}^{\omega}(t_1 \cdots t_{m-1}(t_m - 1))$ then $\tau(s) = {}^{\omega}0.s_{-1}s_{-2}\cdots$. If $\cdots s_2 s_1 s_0 = {}^{\omega}(t_1 \cdots t_{m-1}(t_m - 1))t_1 \cdots t_k$, with $1 \leq k \leq m - 1$, then $\tau(s) = {}^{\omega}0 y_k y_{k-1} \cdots$, where $y_{k+1} y_k \cdots = \lambda(0t_1 \cdots t_{k-1}(t_k + 1).s_{-1}s_{-2}\cdots)$.

The one-sided odometer is thus the restriction of the two-sided odometer to the set V_β^+.

If u is a word equal to $u_1 \cdots u_n$, we denote by $\pi_\beta(u) = u_1\beta^{n-1} + \cdots + u_n$ and by $\mu_\beta(u) = u_1\beta^{-1} + \cdots + u_n\beta^{-n}$.

2.5   Automata

We refer the reader to [5]. An *automaton over* A, $\mathcal{A} = (Q, A, E, I, T)$, is a directed graph labelled by elements of A. The set of vertices, traditionally called *states*, is denoted by Q, $I \subset Q$ is the set of *initial* states, $T \subset Q$ is the set of *terminal* states and $E \subset Q \times A \times Q$ is the set of labelled *edges*. If $(p, a, q) \in E$, we note $p \xrightarrow{a} q$. The automaton is *finite* if Q is finite. The automaton $\mathcal{A}$ is *deterministic* if E is the graph of a (partial) function from $Q \times A$ into Q, and if there is a unique initial state. A subset H of A^* is said to be *recognizable by a finite automaton* if there exists a finite automaton $\mathcal{A}$ such that H is equal to the set of labels of paths starting in an initial state and ending in a terminal state. A subset K of $A^{\mathbb{N}}$ is said to be *recognizable by a finite automaton* if there exists a finite automaton $\mathcal{A}$ such that K is equal to the set of labels of infinite paths starting in an initial state and going infinitely often through a terminal state (Büchi acceptance condition, see [5]).

On-line automata are a special kind of 2-tape automata or transducers. Let A and B be two alphabets. An *on-line finite automaton with delay* δ, $\mathcal{A} = (Q, A \times (B \cup \varepsilon), E, \{i_0\})$, is a finite automaton labelled by elements of $A \times (B \cup \varepsilon)$ which is deterministic when projected on the input tape A. Moreover it is composed of a transient part and of a synchronous part (see [16]). The set of states is equal to $Q = Q_t \cup Q_s$, where Q_t is the set of transient states and Q_s is the set of synchronous states. In the transient part, every path of length δ starting in the initial state i_0 is of the form

$$i_0 \xrightarrow{x_1/\varepsilon} i_1 \xrightarrow{x_2/\varepsilon} \cdots \xrightarrow{x_\delta/\varepsilon} i_\delta$$

where $i_0, \ldots, i_{\delta-1}$ are in Q_t, x_j in A, for $1 \leq j \leq \delta$, and every edge arriving in a state of Q_t is as above. In the synchronous part, edges are labelled by elements of $A \times B$. This means that the automaton starts reading words of length $\leq \delta$ outputting nothing, and after δ readings, outputs serially one digit for each input digit. In practice we shall not consider transient states and edges, but we will replace a transient path $i_0 \xrightarrow{x_1/\varepsilon} i_1 \xrightarrow{x_2/\varepsilon} \cdots \xrightarrow{x_\delta/\varepsilon} i_\delta$ by defining $x_1 \cdots x_\delta$ as an *initial state*. Edges will then be all synchronous.

A function $\varphi : A^{\mathbb{N}} \longrightarrow B^{\mathbb{N}}$ is computable by an on-line finite automaton if its graph is recognizable by an on-line finite automaton.

This definition is implicitly the one of a *left* automaton, where words are processed from left to right. An automaton processing words from right to left will be called a *right* automaton. A *bidirectional on-line finite automaton with delay* δ is composed of a left on-line finite automaton with delay δ and of a right on-line finite automaton with delay δ. Such a device starts from the middle, and then goes synchronously in both directions. A function $\varphi : A^{\mathbb{Z}} \longrightarrow B^{\mathbb{Z}}$ is computable by an on-line finite automaton if its graph is recognizable by a bidirectional on-line finite automaton.

3   On-Line Two-Sided Odometer

Our purpose is to show the following result.

Theorem 1. *Let β be a Pisot number such that $d_\beta(1) = t_1 \cdots t_m$ with $t_1 \geq t_2 \geq \cdots t_m \geq 1$. The two-sided odometer associated with β is computable by a bidirectional on-line finite automaton with delay m.*

The proof is composed of three parts. First we define the central initial states, then the right on-line finite automaton $\mathcal{P}$ working on V_β^+ and finally the left on-line finite automaton $\mathcal{N}$ working on V_β^-. States of $\mathcal{P}$ and of $\mathcal{N}$ will be composed of pairs (u, v) where v is the last factor of length $m-1$ of the input having been read before arriving to that state. This allows the checking of the admissibility of the input (see Example 1 below). The input is a word $a = (a_k)_{k \in \mathbb{Z}}$ belonging to V_β and the output is a word $b = (b_k)_{k \in \mathbb{Z}}$ belonging to V_β such that $b = \tau(a)$. Two words u and v are *equivalent*, denoted by $u \sim v$, if v is obtained from u by replacing every factor $0t_1 \cdots t_m$ by the equivalent factor 10^m and $0t_1 \cdots t_{j-1}(t_j + 1)0^m$, for $1 \leq j \leq m-1$, by $10^j(t_1 - t_{j+1}) \cdots (t_{m-j} - t_m)t_{m-j+1} \cdots t_m$. The delay m is taken into account in the length m of the states.

First we give a fundamental lemma, which uses the hypothesis of decrease of the coefficients t_i's.

Lemma 1. *Let $w = 0t_1 \cdots t_{j-1}t_j.t_{j+1} \cdots t_{j+k-1}(t_{j+k} - h)a_{-k-1} \cdots a_{-m-1}$ be in L_β, with $1 \leq j \leq m-1$, $1 \leq k \leq m$, $1 \leq h \leq t_{j+k}$, and let $v = w + 1$. Then $v \sim y = 10^j.b_{-1}u$ with b_{-1} in A_β, $|u| = m$, and $0 \leq \mu_\beta(u) \leq 1 - \beta^{-m}$.*

Proof of Theorem 1. Let $(a_k)_{k \in \mathbb{Z}}$ be fixed in V_β.

1. *Central initial states*
A *central initial state* is a word of the form $c = a_{m-1} \cdots a_1(a_0 + 1).a_{-1} \cdots a_{-m}$. Let $w = a_m ca_{-m-1}$. The general scheme is the following : there are two edges outgoing from c, one going into $\mathcal{P}$ and the other one going into $\mathcal{N}$

$$p_0 \xleftarrow{a_m/b_0} c \xrightarrow{a_{-m-1}/b_{-1}} n_1$$

where $p_0 = (s, t)$ and $n_1 = (u, v)$, where s and u are words of length m such that $sb_0.b_{-1}u \sim w$. The words t and v represent the past of s and u respectively. We say that p_0 is an initial state of $\mathcal{P}$ and n_1 is an initial state of $\mathcal{N}$.

Denote by R_p the set of words of L_β of length p. There are three cases.

1. If w is in L_β, then we set $b_0 = a_0 + 1$ and $b_{-1} = a_{-1}$. Thus $p_0 = (a_m \cdots a_1, a_m \cdots a_2)$ and $n_1 = (a_{-2} \cdots a_{-m-1}, a_{-3} \cdots a_{-m-1})$ are elements of $R_m \times R_{m-1}$.
2. If w contains a factor $t_1 \cdots t_m$, we take the leftmost factor. There exists j, $1 \leq j \leq m$ such that

$$w = a_m \cdots a_j t_1 \cdots t_j . t_{j+1} \cdots t_m a_{-m+j-1} \cdots a_{-m-1}.$$

Thus

$$w \sim a_m \cdots a_{j+1}(a_j + 1) 0^j . 0^{m-j} a_{-m+j-1} \cdots a_{-m-1}.$$

Set $b_0 = 0$ and $p_0 = (a_m \cdots a_{j-1}(a_j + 1)0^{j-1}, a_m \cdots a_j t_1 \cdots t_{j-2})$ which belongs to $R_m \times R_{m-1}$ since $a_m \cdots a_j \preceq t_1 \cdots t_{m-j}(t_{m-j} - 1)$, except if $j = 1$ and $w = t_1 \cdots t_{m-1}(t_m - 1)t_1 . t_2 \cdots t_m a_{-m-1}$, in which case $p_0 = (t_1 \cdots t_m, t_1 \cdots t_{m-1})$. If $j < m$ then $b_{-1} = 0$ and $n_1 = (0^{m-j} a_{-m+j-1} \cdots a_{-m-1}, t_{j+2} \cdots t_m a_{-m+j-1} \cdots a_{-m-1})$ is in $R_m \times R_{m-1}$. If $j = m$ then $b_{-1} = a_{-1}$ and $n_1 = (a_{-2} \cdots a_{-m-1}, a_{-3} \cdots a_{-m-1})$ is in $R_m \times R_{m-1}$.
3. If w contains a factor $\succ t_1 \cdots t_m$, we take the leftmost one, and there exist $1 \leq j \leq m-1$, $1 \leq k \leq m$ and $1 \leq h \leq t_{j+k}$ such that

$$w = a_m \cdots a_j t_1 \cdots t_{j-1}(t_j + 1).t_{j+1} \cdots t_{j+k-1}(t_{j+k} - h)a_{-k-1} \cdots a_{-m-1}.$$

By Lemma 1, $w \sim a_m \cdots a_{j-1}(a_j + 1)0^j . b_{-1} u$ with $\mu_\beta(u) \leq 1 - \beta^{-m}$. Thus we set $b_0 = 0$, and $p_0 = (a_m \cdots a_{j-1}(a_j + 1)0^{j-1}, a_m \cdots a_j t_1 \cdots t_{j-1})$; then $p_0 \in R_m \times R_{m-1}$ except if $j = 1$ and $w = t_1 \cdots t_{m-1}(t_m - 1)t_1 . t_2 \cdots t_m a_{-m-1}$, in which case $p_0 = (t_1 \cdots t_m, t_1 \cdots t_{m-1})$. We set $n_1 = (u, t_{j+2} \cdots t_{j+k-1}(t_{j+k} - h)a_{-k-1} \cdots a_{-m-1})$.

2. Right on-line finite automaton $\mathcal{P}$

There are four classes of states of $\mathcal{P}$:
$C_1 = \{p_k = (a_{m+k} \cdots a_{k+1}, a_{m+k} \cdots a_{k+2}) \in R_m \times R_{m-1} \mid k \geq 0\}$,
$C_2 = \{p_k = (a_{m+k} \cdots a_{j+k-1}(a_{j+k} + 1)0^{j-1}, a_{m+k} \cdots a_{j+k} t_1 \cdots t_{j-2}) \in R_m \times R_{m-1}, 2 \leq j \leq m \mid k \geq 0\}$,
$C_3 = \{p_k = (a_{m+k} \cdots a_{k+2}(a_{k+1} + 1), a_{m+k} \cdots a_{k+2}) \in R_m \times R_{m-1} \mid k \geq 0\}$,
$C_4 = \{(t_1 \cdots t_m, t_1 \cdots t_{m-1})\}$.
The set of initial states is $\{p_0 \in \cup_{1 \leq i \leq 4} C_i\}$.
There are two types of transitions, the reduction edges and the shift edges. First we describe the shift edges.

Let $p_k \in C_1$. There is an edge $p_k \xrightarrow{a_{m+k+1}/a_{k+1}} p_{k+1} \in C_1$.

Let $p_k \in C_2$. There is an edge $p_k \xrightarrow{a_{m+k+1}/0} p_{k+1} = (a_{m+k+1} \cdots a_{j+k}(a_{j+k+1} + 1)0^{j-2}, a_{m+k+1} \cdots a_{j+k+1} t_1 \cdots t_{j-3})$, and p_{k+1} belongs to C_2 if $j \geq 3$. If $j = 2$, p_{k+1} belongs to C_3, or to C_4 if p_k is of the form $p_k = (t_2 \cdots t_m 0, t_2 \cdots t_{m-1})$ and the edge is $p_k \xrightarrow{t_1/0} p_{k+1} = (t_1 \cdots t_m, t_1 \cdots t_{m-1}) \in C_4$.

Let $p_k \in C_3$. There is an edge $p_k \xrightarrow{a_{m+k+1}/a_{k+1}+1} p_{k+1} \in C_1$.

The only reduction edge is the following one $(t_1 \cdots t_m, t_1 \cdots t_{m-1}) \xrightarrow{a_{m+k+1}/0} p_{k+1} = ((a_{m+k+1} + 1)0^{m-1}, a_{m+k+1} t_1 \cdots t_{m-2})$ which belongs to C_2.

Clearly, if $p_k = (u_k, v_k)$ and if $a_{m+k+1}v_k \in R_m$ then there is an edge $p_k \xrightarrow{a_{m+k+1}/b_{k+1}} p_{k+1}$ with $p_{k+1} = (u_{k+1}, v_{k+1})$ and $\pi_\beta(a_{m+k+1}u_k) = \pi_\beta(u_{k+1}b_{k+1})$.

Let $p_0 \xrightarrow{a_{m+1}/b_1} p_1 \xrightarrow{a_{m+2}/b_2} \cdots p_{n-1} \xrightarrow{a_{m+n}/b_n} p_n \cdots$ be an infinite path in $\mathcal{P}$. We have to show that the sequence $(b_i)_{i \geq 0}$ belongs to V_β^+.

First suppose there exists infinitely many k's such that p_k belongs to C_4. Thus the input is of the form ${}^\omega(t_1 \cdots t_{m-1}(t_m - 1))z.a_{-1}\cdots$ with $z = t_1 \cdots t_m$ or $z = t_1 \cdots t_{j-1}(t_j + 1)$. The carry propagates infinitely to the left, and the positive part of the output is ${}^\omega 0$.

Second, there exists a finite number of indices k such that p_k belongs to C_4, and let K be the greatest of them. This means that the carry does not propagate after state p_{K+1}. After that, there are only shift edges, which preserve admissibility.

3. Left on-line finite automaton $\mathcal{N}$

The set of states of $\mathcal{N}$ is taken as $S = \{n_k = (u_k, v_k) \mid k \geq 1, |u_k| = m, 0 \leq \mu_\beta(u_k) < 1, v_k \in R_{m-1}\}$. Note that, when u_k belongs to R_m then $\mu_\beta(u_k) \leq 1 - \beta^{-m}$, so the set of initial states is $I = \{n_1 = (u_1, v_1) \mid |u_1| = m, 0 \leq \mu_\beta(u_1) \leq 1 - \beta^{-m}, v_1 = a_{-3}\cdots a_{-m-1}\}$. The proof that, if β is a Pisot number then the set S is finite, is the same as the one given in Theorem 2 of my paper [8], so I do not give it here. This implies that the component u_k of state n_k is written on a finite alphabet of digits.

Let $k \geq 1$. There is an edge

$$n_k = (u_k, v_k) \xrightarrow{a_{-m-k-1}/b_{-k-1}} n_{k+1} = (u_{k+1}, v_{k+1})$$

if $v_k a_{-m-k-1} \in R_m$, and b_{-k-1} is chosen as follows.
If $\mu_\beta(u_k a_{-m-k-1}) < 1$ take $b_{-k-1} = \lfloor \beta\mu_\beta(u_k a_{-m-k-1}) \rfloor$ and let $y_{k+1} = \{\beta\mu_\beta(u_k a_{-m-k-1})\}$. Hence $b_{-k-1} \in A_\beta$ and $y_{k+1} < 1$. Take u_{k+1} of length m such that $\mu_\beta(u_{k+1}) = y_{k+1}$.
If $\mu_\beta(u_k a_{-m-k-1}) \geq 1$ take $b_{-k-1} = t_1$ and u_{k+1} of length m such that $u_k a_{-m-k-1} \sim t_1 u_{k+1}$. Then $\mu_\beta(u_{k+1}) = \beta\mu_\beta(u_k) + a_{-m-k-1}\beta^{-m} - t_1 < \beta + t_1\beta^{-m} - t_1 \leq 1$.

So the relation $\mu_\beta(u_k a_{-m-k-1}) = \mu_\beta(b_{-k-1}u_{k+1})$ holds for each k, with n_{k+1} in S and b_{-k-1} in A_β.

Let $n_1 \xrightarrow{a_{-m-2}/b_{-2}} n_2 \xrightarrow{a_{-m-3}/b_{-3}} \cdots n_{k-1} \xrightarrow{a_{-m-k}/b_{-k}} n_k \cdots$ be an infinite path in $\mathcal{N}$. We have to show that the sequence $(b_{-k})_{k \geq 1}$ belongs to V_β^-.

1. First note that if $n_k = (u_k, v_k)$ is such that u_k belongs to R_m, then the sequence $b_{-k-1}b_{-k-2}\cdots$ belongs to V_β^-, because edges are shift edges (there is no carry propagating to the right).
2. Suppose that there exists a minimal k such that u_k does not belong to R_m and $\mu_\beta(u_k) \leq 1 - \beta^{-m}$, and the output sequence $b_{-k-1}\cdots b_{-k-m}$ is not admissible. Then it follows that $\mu_\beta(u_k a_{-m-k-1}\cdots a_{-2m-k}) > 1$, which is impossible.
3. Suppose that there exists a minimal k such that u_k does not belong to R_m and $1 - \beta^{-m} < \mu_\beta(u_k) < 1$, and the output sequence $b_{-k-1}\cdots b_{-k-m}$ is not

admissible. Since $\mu_\beta(u_k) > 1 - \beta^{-m}$, the β-expansion of u_k is of the form $t_1 \cdots t_{m-1}(t_m - 1)w$ where w is not 0. Thus there exists $z_1 \cdots z_m$ in R_m such that $\mu_\beta(u_k z_1 \cdots z_m) > 1$. The existence of a minimal j, $1 \leq j \leq m$, such that $\mu_\beta(u_k z_1 \cdots z_j) > 1$ is a contradiction since, in the automaton $\mathcal{N}$, no path starting from n_k with input $z_1 \cdots z_j$ can exist. $\qquad\square$

Example 1. Take $\beta = \frac{1+\sqrt{5}}{2}$ the golden ratio. Then $A_\beta = \{0,1\}$, $d_\beta(1) = 11$ and the factor 11 is forbidden. The on-line finite odometer with delay 2 associated with the golden ratio is shown on Figure 1, Figure 2 and Figure 3 below. A state of the form (u, v) is represented as $\frac{u}{v}$. On Figure 2 and Figure 3 initial states of $\mathcal{P}$ and $\mathcal{N}$ are indicated by an ingoing arrow. The signed digit -1 is denoted by $\bar{1}$.

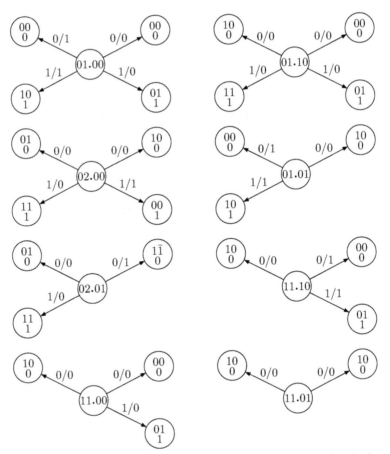

Fig. 1. Central part of the two-sided odometer with delay 2 associated with the golden ratio

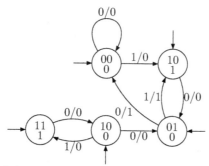

Fig. 2. Right on-line finite automaton $\mathcal{P}$ with delay 2 for the two-sided odometer associated with the golden ratio

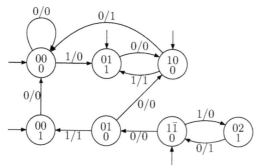

Fig. 3. Left on-line finite automaton $\mathcal{N}$ with delay 2 for the two-sided odometer associated with the golden ratio

References

1. A. Avizienis, Signed-digit number representations for fast parallel arithmetic. *IRE Transactions on electronic computers* **10** (1961), 389–400.
2. A. Bertrand-Mathis, Comment écrire les nombres entiers dans une base qui n'est pas entière. *Acta Math. Acad. Sci. Hungar.* **54** (1989), 237–241.
3. A. Brauer, On algebraic equations with all but one root in the interior of the unit circle. *Math. Nachr.* **4** (1951), 250-257.
4. C.Y. Chow and J.E. Robertson, Logical design of a redundant binary adder. *Proc. 4th Symposium on Computer Arithmetic*, I.E.E.E. Computer Society Press (1978), 109–115.
5. S. Eilenberg, *Automata, Languages and Machines*, vol. A, Academic Press, 1974.
6. Ch. Frougny, Representation of numbers and finite automata. *Math. Systems Theory* **25** (1992), 37–60.
7. Ch. Frougny, On the sequentiality of the successor function. *Inform. and Computation* **139** (1997), 17–38.
8. Ch. Frougny, On-line digit set conversion in real base. *Theor. Comp. Sci.* **292** (2002), 221–235.
9. Ch. Frougny, On-line odometers, manuscript.

10. Ch. Frougny and J. Sakarovitch, Synchronisation déterministe des automates à délai borné. *Theor. Comp. Sci.* **191** (1998), 61–77.

11. Ch. Frougny and B. Solomyak, Finite beta-expansions. *Ergodic Theory & Dynamical Systems* **12** (1992), 713–723.

12. P. Grabner, P. Liardet, R. Tichy, Odometers and systems of numeration. *Acta Arithmetica* **LXXX.2** (1995), 103–123.

13. S. Ito and Y. Sano, Substitutions, atomic surfaces, and periodic beta expansions. *Analytic Number Theory*, C. Jia and K. Matsumoto eds., Developments in Mathematics, Volume 6, 2002.

14. D. Lind and B. Marcus, *An introduction to symbolic dynamics and coding*, Cambridge University Press, 1995.

15. M. Lothaire, *Algebraic Combinatorics on Words*, Cambridge University Press, 2002.

16. J.-M. Muller, Some characterizations of functions computable in on-line arithmetic. *I.E.E.E. Trans. on Computers*, **43** (1994), 752–755.

17. W. Parry, On the β-expansions of real numbers. *Acta Math. Acad. Sci. Hungar.* **11** (1960), 401–416.

18. A. Rényi, Representations for real numbers and their ergodic properties. *Acta Math. Acad. Sci. Hungar.* **8** (1957), 477–493.

19. K. Schmidt, Algebraic coding of expansive group automorphisms and two-sided beta-shifts. *Monatsh. Math.* **129** (2000), 37–61.

20. N. Sidorov and A. Vershik, Ergodic properties of the Erdős measure, the entropy of the golden shift, and related problems. *Monatsh. Math.* **126** (1998), 215–261.

Characteristic Semigroups
of Directable Automata

Tatjana Petković[1,2], Miroslav Ćirić[2], and Stojan Bogdanović[3]

[1] Turku Centre for Computer Science and Department of Mathematics,
University of Turku, FIN-20014 Turku, Finland
tatpet@utu.fi
[2] Faculty of Sciences and Mathematics, University of Niš,
P. O. Box 91, 18000 Niš, Yugoslavia
ciricm@bankerinter.net
[3] Faculty of Economics, University of Niš,
P. O. Box 121, 18000 Niš, Yugoslavia
sbogdan@pmf.pmf.ni.ac.yu

Abstract. In some of their earlier papers the authors established correspondences between varieties of automata and semigroups and then found inconvenient the usual notion of transition semigroup of an automaton for dealing with automata satisfying some irregular identities. This is the motivation for assigning a new kind of semigroups to directable automata, which is done in this paper. The properties of these, so-called characteristic, semigroups are studied and semigroups that can be characteristic semigroups of directable automata are characterized. Finally, a correspondence between irregular varieties of automata and certain varieties of semigroups is given.

1 Introduction

Directable automata, known also as synchronizable, cofinal and reset automata, are a significant type of automata that has been a subject of interest of many eminent authors since 1964, when these automata were introduced in a paper by Černý [5]. Some of their special types were investigated even several years earlier. As automata that correspond to definite languages, definite automata were studied in 1956 in [10] and in 1963 in the paper [11], while nilpotent automata were investigated in 1962 in [16] (see also the book [7]). Various other specializations and generalizations of directable automata, such as trap-directable, trapped, generalized directable, locally directable, uniformly locally directable, have appeared recently in the papers [13], [3] and [12]. Transition semigroups of directable automata were investigated in [6] in 1978, and recently in [13], [14] and [15].

It is well-known that structures of automata and their transition semigroups are closely related. But, since the notion of transition semigroup of an automaton treats only regular identities satisfied on it, they do not contain enough information about automata satisfying irregular identities. In this paper we propose

M. Ito and M. Toyama (Eds.): DLT 2002, LNCS 2450, pp. 417–427, 2003.

a new concept for characteristic semigroups of irregular automata, i.e., of directable automata, and investigate its properties. Also, we describe semigroups that can be characteristic semigroups of some directable automata. Especially, semigroups corresponding to automata belonging to some important subclasses of the class of directable automata are considered. In [14], [15] the authors gave two kinds of correspondences between varieties of automata and semigroups. As a tool there they used transition semigroups of automata, and hence the obtained results covered only regular varieties of automata. Here, using characteristic semigroups, the correspondence between irregular varieties of automata and varieties of semigroups is established.

First we give some definitions concerning semigroups. An element z of a semigroup S is a *right (left) zero* of S if $sz = z$ $(zs = z)$ for each $s \in S$, and a *zero* if it is both a left and a right zero of S. A semigroup whose every element is a right zero is a *right zero band*. For an ideal T of a semigroup S a congruence relation ϱ_T on S, called the *Rees congruence* on S determined by T, is defined by: for $s, t \in S$ we say that $(s, t) \in \varrho_T$ if and only if either $s = t$ or $s, t \in T$. The factor semigroup S/ϱ_T, usually denoted by S/T, is the *Rees factor* of S determined by T. A semigroup S is an *ideal extension* of a semigroup T by a semigroup Q with zero if T is an ideal of S and the Rees factor S/T is isomorphic to Q. If, in addition, Q is a nilpotent semigroup, i.e., if $Q^k = \{0\}$ for some $k \in \mathbb{N}$, where 0 is the zero of Q, then S is a *nilpotent extension* of T. Note that throughout the paper, $\mathbb{N}$ denotes the set of positive natural numbers.

Next we present some notions and notation from automata theory. Automata considered throughout this paper will be automata without outputs in the sense of the definition from [7]. The state set and the input set of an automaton will not be necessarily finite. It is well known that automata without outputs, with the input alphabet X, called *X-automata*, can be considered as unary algebras of type indexed by X, and they are also known as *X-algebras*. Hence notions such as a *congruence*, *homomorphism*, *generating set*, *direct product*, *subdirect product* etc., will have their usual algebraic meanings (see [4]). In order to simplify notation, an automaton with a state set A is also denoted by the same letter A. For an alphabet X the free monoid over X is denoted by X^*, its neutral element is denoted by e, and the free semigroup over X is denoted by X^+. An X-automaton A goes from a state a into the state denoted by au, or au^A if it is necessary to distinguish the automaton A, under the action of an input word $u \in X^*$.

For an alphabet X, by $X_@^*$ we denote the X-automaton whose state set is X^* and the transitions are defined by $(u, x) \mapsto ux$. By $X_@^+$ the subautomaton of $X_@^*$ with the state set X^+ is denoted. For a congruence θ of X^+, the congruence θ^* defined on X^* by $\theta^* = \theta \cup \{(e, e)\}$ is a congruence on the automaton $X_@^*$ and the *Myhill automaton of the congruence* θ, in notation $M(\theta)$, is the factor automaton of $X_@^*$ with respect to the congruence θ^*, i.e., $M(\theta) = X_@^*/\theta^*$.

For a given word $u \in X^*$, an automaton A is *u-directable* if $au = bu$ for all $a, b \in A$, and u is a *directing word* of A. Furthermore, A is a *directable* automaton if A is u-directable for some word $u \in X^*$. By $DW(A)$ the set of all directable words of a directable automaton A is denoted. A state a_0 of an automaton A is a *trap* if $a_0 x = a_0$ holds for every $x \in X$. A directable automaton having a trap is *trap-directable*. A directable automaton A satisfying $X^{\geq n} \subseteq DW(A)$ is *n-definite*, where $n \in \mathbb{N}$, $X^{\geq n} = \{u \mid |u| \geq n\}$ and $|u|$ is the length of the word u. An automaton A is *definite* if it is n-definite for some $n \in \mathbb{N}$. An n-definite automaton, for $n \in \mathbb{N}$, having a trap is *n-nilpotent*, and it is *nilpotent* if it is n-nilpotent for some $n \in \mathbb{N}$.

Let G be a non-empty set whose elements are called *variables*. By a *term of type X over G* we mean any word over the set $G \cup X$ of the form gu, with $g \in G$ and $u \in X^*$. The set of all terms of type X over G, denoted by $T(G)$, is an X-automaton with transitions defined by $(gu)x = g(ux)$ for $gu \in T(G)$ and $x \in X$. This automaton is a *term automaton* over G. Any pair (s, t) of terms from $T(G)$, usually written as the formal equality $s = t$, is an *(automaton) identity* over G or over $T(G)$. If terms s and t contain the same variable, i.e., if the identity $s = t$ has the form $gu = gv$, with $g \in G$ and $u, v \in X^*$, then it is a *regular identity*; otherwise, it is an *irregular identity*. An automaton A *satisfies the identity $s = t$ over $T(G)$*, in notation $A \models s = t$, if the pair (s, t) of terms belongs to the kernel of any homomorphism from $T(G)$ into A, i.e., if $s\varphi = t\varphi$ for any homomorphism $\varphi : T(G) \to A$. An automaton satisfying some irregular identity is *irregular*, otherwise it is *regular*. It is not hard to notice that an automaton is irregular if and only if it is directable. Evidently, when working with identities satisfied on an automaton, it is enough to deal with the two-element set $G = \{g, h\}$ of variables, and this will be done throughout the paper.

A *variety of semigroups (automata)* is a class of semigroups (automata) closed under subsemigroups (subautomata), homomorphic images and direct products, or equivalently, a class closed under homomorphic images and subdirect products, whereas a class of finite semigroups (automata) is a *pseudovariety* if it is closed under subsemigroups (subautomata), homomorphic images and finite direct products. A *generalized variety of semigroups (automata)* is a class of semigroups (automata) closed under subsemigroups (subautomata), homomorphic images, finite direct products and arbitrary direct powers. A family $\{a_i\}_{i \in I}$ of elements of a partially ordered set is *directed* if for every $i, j \in I$ there exists $k \in I$ such that $a_i, a_j \leq a_k$. It is known (see [2]) that a class of semigroups (automata) is a generalized variety if and only if it can be represented as the union of a directed family of varieties, and a class of finite semigroups (automata) is a pseudovariety if and only if it is the class of all finite semigroups (automata) from some generalized variety. It is known that a class of semigroups (automata) is a variety if and only if it is the set of all semigroups (automata) satisfying certain set of identities. Then a variety is *regular* if it is determined by a set of regular identities, whereas it is *irregular* if there are irregular identities in the set of identities characterizing it.

For undefined notions and notation we refer to the books [4] and [7].

2   Characteristic Semigroups of Directable Automata and Their Properties

The *transition semigroup* $S(A)$ of an automaton A, in some sources called the *characteristic semigroup* of A, can be defined in two equivalent ways. The first one is to define $S(A)$ as the subsemigroup of the full transformation semigroup on A consisting of all *transition mappings* on A, i.e., $S(A) = \{u^A \mid u \in X^+\}$, where $u^A : a \mapsto au$. Another way is to define $S(A)$ as the factor semigroup of the input semigroup X^+ with respect to the *Myhill congruence* μ_A on X^+ defined by: $(u, v) \in \mu_A$ if and only if $au = av$ for every $a \in A$. Note that $(u, v) \in \mu_A$ if and only if $u^A = v^A$. But, we can observe that the Myhill congruence μ_A of A is determined by regular identities satisfied on A, i.e., for $u, v \in X^+$ we have that

$$(u, v) \in \mu_A \iff A \models gu = gv.$$

Therefore, if an automaton A satisfies some irregular identity, i.e., if it is directable, then the transition semigroup $S(A)$ of A does not contain enough information about A. Motivated by this fact, we introduce the notion of a characteristic semigroup of a directable automaton defined in terms of irregular identities satisfied on it. Namely, if A is a directable automaton, then the relation θ_A on X^+ defined by

$$(u, v) \in \theta_A \iff u = v \text{ or } A \models gu = hv$$

is a congruence relation on X^+, and the factor semigroup X^+/θ_A, denoted by $C(A)$, is the *characteristic semigroup* of A. An equivalent definition of the congruence θ_A can be given using the following lemma.

In a u-directable automaton A the word u directs the states of A into a single state called a *u-neck*, denoted by d_u. Then it is not hard to notice that the following lemma holds.

Lemma 1. *An irregular identity $gu = hv$ is satisfied on an automaton A if and only if $u, v \in DW(A)$ and $d_u = d_v$.*

Therefore, the congruence θ_A can be defined by $(u, v) \in \theta_A$ if and only if $u = v$ or $u, v \in DW(A)$ and $d_u = d_v$.

For a directable automaton A let ϱ_A denote the Rees congruence on X^+ that corresponds to the ideal $DW(A)$ of X^+. Then the congruence θ_A can be expressed through ϱ_A and the Myhill congruence μ_A as follows.

Theorem 1. *Let A be a directable automaton. Then $\theta_A = \varrho_A \cap \mu_A$.*

By the previous theorem it follows immediately that the characteristic semigroup $C(A)$ is a subdirect product of the semigroup X^+/ϱ_A and the transition semigroup $S(A)$ of A. Also if A is a regular automaton then $C(A)$ is equal to X^+, and, hence, it is not interesting for consideration.

In the next theorem we give some properties of characteristic semigroups of automata similar to those known for transition semigroups (see [7]).

Theorem 2. (a) *If an automaton B is a homomorphic image (subautomaton) of the automaton A then $C(B)$ is a homomorphic image of $C(A)$.*

(b) *If an automaton A is a subdirect product of automata A_i, $i \in I$, then $C(A)$ is a subdirect product of the semigroups $C(A_i)$, $i \in I$.*

Proof. (a) It can be easily proved that in both cases $\theta_A \subseteq \theta_B$ holds.

(b) It is not hard to prove that $\theta_A = \bigcap_{i \in I} \theta_{A_i}$, what implies the claim.

3 The Structure of Characteristic Semigroups

As known, every semigroup is isomorphic to the transition semigroup of some automaton, and it is interesting to consider the following question: Is every semigroup the characteristic semigroup of some directable automaton? We shall show that the answer is negative and determine necessary and sufficient conditions for a semigroup to be the characteristic semigroup of some directable automaton. In order to describe such semigroups, we define a semigroup S generated by a set Y with the zero 0 to be a *0-free semigroup* over Y if every nonzero element of S can be uniquely represented as the product of elements from Y.

Theorem 3. *For a semigroup S the following conditions are equivalent:*

(i) *S is the characteristic semigroup of some directable automaton;*

(ii) *S is a factor semigroup of some free semigroup X^+ with respect to a congruence θ on X^+ such that each nontrivial θ-class is a left ideal of X^+;*

(iii) *S is an ideal extension of a right zero band by a 0-free semigroup.*

Proof. (i)$\Rightarrow$(ii). Suppose that $S \cong C(A)$, for a directable X-automaton A, where X is an alphabet. We are proving that θ_A satisfies described conditions. Assume a nontrivial class $p\theta_A$ and words $u \in p\theta_A$ and $v \in X^+$. Since $|p\theta_A| > 1$, it follows that there exists $w \in X^+$ such that $u \neq w$ and $A \models gu = hw$. Then also $A \models gvu = hw$. So, $(vu, w) \in \theta_A$, and hence, by transitivity of θ_A, follows $(vu, u) \in \theta_A$, i.e., $vu \in u\theta_A = p\theta_A$, what has to be proved.

(ii)$\Rightarrow$(iii). Assume that $S \cong X^+/\theta$ and every nontrivial θ-class is a left ideal in X^+. We are going to prove that $R = \{u\theta \,|\, |u\theta| \neq 1\}$ is a set of right zeroes from S and S/R is a 0-free semigroup.

Assume two elements $u\theta \in R$ and $v\theta \in S$. Then $vu \in u\theta$, since $u\theta$ is a left ideal, what implies that $u\theta$ is a right zero in S. It is not hard to notice that R is an ideal in S, and hence S is an ideal extension of a right zero band R.

Now we are proving that S/R is a 0-free semigroup with generating set $X\theta \backslash R$. It is enough to prove that every element from $S \backslash R$ has the unique representation trough the elements from $X\theta \backslash R$. Choose arbitrarily $u\theta \in S \backslash R$ and assume that u has the form $u = x_1 x_2 \ldots x_n$ for some $x_1, x_2, \ldots, x_n \in X$. Suppose that $u\theta$ does not have the unique representation over the set $X\theta \backslash R$, i.e., $(x_1 x_2 \ldots x_n)\theta = (x_1' x_2' \ldots x_m')\theta$ holds. Since $u\theta \notin R$ we have that $|u\theta| = 1$, what means that $x_1 x_2 \ldots x_n = x_1' x_2' \ldots x_m'$ in X^+, so $n = m$ and $x_i = x_i'$ for every $i \in \{1, 2, \ldots, n\}$, what proves the claim.

(iii)$\Rightarrow$(i). Suppose that S is an extension of a right zero band R by a 0-free semigroup with generating set X. Let us first prove that the elements from R are right zeroes in S. Indeed, choose $z \in R$ and $s \in S$ arbitrarily. Then for arbitrary $z' \in R$ is $z's, sz \in R$. Using the fact that R is a right zero band we have $sz = z'(sz) = (z's)z = z$.

Consider the homomorphism $\phi : X^+ \to S$ generated by $x\phi = x$ for every $x \in X$. Let us prove that $\theta_A = \ker\phi$, where $A = M(\ker\phi)$. Suppose that $(u, v) \in \theta_A$ and $u \neq v$, for some $u, v \in X^+$. Then $A \models gu = hv$, what for $g = h = e \ker\phi$ gives $u \ker\phi = v \ker\phi$, i.e., $(u, v) \in \ker\phi$. On the other hand, assume $(u, v) \in \ker\phi$. There are two different cases. In the first case $u\phi = v\phi \in S \setminus R$. Since S/R is a 0-free semigroup with the generating set X, it follows that equality $u = v$ holds in X^+ what implies $(u, v) \in \theta_A$. The second case is $u\phi = v\phi \in R$. Since R contains right zeroes from S, then for arbitrary $p \ker\phi, q \ker\phi \in A$ is $(p \ker\phi)u = (pu) \ker\phi = u \ker\phi = v \ker\phi = (qv) \ker\phi = (q \ker\phi)v$. Therefore $A \models gu = hv$, what implies $(u, v) \in \theta_A$. So, $S \cong X^+/\ker\phi = X^+/\theta_A \cong C(A)$.

Next we characterize semigroups that are characteristic semigroups of trap-directable, definite and nilpotent automata.

Theorem 4. *For a semigroup S the following conditions are equivalent:*

(i) S *is the characteristic semigroup of some trap-directable automaton;*
(ii) S *is a Rees factor semigroup of some free semigroup;*
(iii) S *is a 0-free semigroup.*

Proof. (i)$\Rightarrow$(ii). Suppose that $S = C(A)$ for some trap-directable automaton A with the trap a_0. Then $S \cong X^+/\theta_A$ and, according to Theorem 3, S is an extension of a right zero band $R = \{u\theta_A \mid |u\theta_A| \neq 1\}$ by a 0-free semigroup. It is enough to prove that $|R| = 1$. Assume $u\theta_A, v\theta_A \in R$. Then $u, v \in DW(A)$ according to Lemma 1. Clearly $d_u = d_v = a_0$, what again by Lemma 1 gives $A \models gu = hv$, i.e., $(u, v) \in \theta_A$.

(ii)$\Rightarrow$(iii). This is obvious.

(iii)$\Rightarrow$(i). Since S is a 0-free semigroup, it is an extension of a trivial right zero band by a 0-free semigroup and hence, according to Theorem 3, $S \cong C(A)$ where $A = M(\ker\phi)$ for the mapping ϕ defined in the part (iii)$\Rightarrow$(i) of the proof of Theorem 3. It remains to prove that A has a trap. For a word $u \in X^+$ such that $u\phi = 0$ and an arbitrary $v \in X^*$ we have $(uv)\phi = (u\phi)(v\phi) = 0 = u\phi$, i.e., $(uv, u) \in \ker\phi$, what means that A has a trap $u \ker\phi$.

The equivalence (ii)$\Leftrightarrow$(iii) of Theorem 4 gives another characterization of 0-free semigroups which was proved in [1].

Theorem 5. *A semigroup S is the characteristic semigroup of some definite automaton if and only if S is an ideal extension of a right zero band by a nilpotent 0-free semigroup.*

Proof. Assume first that $S \cong C(A)$, where A is an n-definite automaton, for some $n \in \mathbb{N}$. According to Theorem 3, the semigroup S is an extension of a right zero band $R = \{u\theta_A \mid |u\theta_A| \neq 1\}$ by a 0-free semigroup. We are going to prove that S/R is an n-nilpotent semigroup. Assume arbitrary elements $u_1\theta_A, u_2\theta_A, \ldots, u_n\theta_A \in S \setminus R$. Then $|u_1u_2\cdots u_n| \geq n$, and so $au_1u_2\cdots u_n = bu_1u_2\cdots u_n = bu_1^2u_2\cdots u_n$ hold for all $a, b \in A$. So, $(u_1u_2\cdots u_n, u_1^2u_2\cdots u_n) \in \theta_A$ what gives $(u_1u_2\cdots u_n)\theta_A \in R$.

On the other hand, suppose that S is an extension of a right zero band R by an n-nilpotent 0-free semigroup with generating set X. Then, by Theorem 3, $S \cong C(A)$, where $A = M(\ker\phi)$ and the mapping $\phi : X^+ \to S$ is defined in the part (iii)$\Rightarrow$(i) of the proof of Theorem 3. It remains to prove that A is an n-definite automaton. Assume arbitrarily $w \ker\phi \in A$ and $u \in X^{\geq n}$. Since $|u| \geq n$, it follows that $u\phi \in R$, and, as it is proved in the part (iii)$\Rightarrow$(i) of the proof of Theorem 3, $u\phi$ is a right zero, i.e., $u\phi = (wu)\phi$. So, $(w \ker\phi)u = (wu)\ker\phi = u\ker\phi$, i.e., A is a definite automaton.

Theorem 6. *A semigroup S is the characteristic semigroup of some nilpotent automaton if and only if S is a nilpotent 0-free semigroup.*

Proof. Follows by Theorems 4 and 5.

4   The Variety Theorem

Finally, we can establish correspondences between certain varieties, generalized varieties and pseudovarieties of irregular automata and semigroups. Here we follow the method used in [14] for regular case.

A congruence θ on a semigroup S satisfying the condition (ii) of Theorem 3, i.e., whose every nontrivial class is a left ideal in S, is called a C-*congruence*. A C-*homomorphism* (C-*epimorphism*) is a homomorphism (epimorphism) $\phi : S \to T$ such that $\ker\phi$ is a C-congruence. On the other hand, a semigroup which is a factor of a free semigroup trough some C-congruence, i.e., which is a C-homomorphic image of a free semigroup, is a C-*semigroup*. By Theorem 3, a semigroup is the characteristic semigroup of some directable automaton if and only if it is a C-semigroup.

Since the class of C-semigroups is not closed for homomorphic images, we have been encouraged to introduce the notion of a C-homomorphism and a C-homomorphic image. Naturally, it leads us to some new "varieties" of semigroups in the following way. Namely, for an arbitrary cardinal κ by a κ_C-*subdirect product* of κ-generated C-semigroups S_i, $i \in I$, we mean a κ-generated C-semigroup S such that there are C-epimorphisms $\phi_i : S \to S_i$, $i \in I$, such that $\bigcap_{i \in I} \ker\phi_i = \Delta_S$. Finally, a κ_C-*variety* of semigroups is a class of κ-generated C-semigroups closed for C-homomorphic images and κ_C-subdirect products. The set of all κ-generated semigroups is denoted by $\mathcal{S}_\kappa$.

Now we are going to prove some auxiliary results.

Lemma 2. *Let S, T and R be semigroups and $\phi : S \to T$ and $\psi : T \to R$ epimorphisms. Then the following hold:*

(a) *if ϕ and ψ are C-epimorphisms, then $\phi\psi$ is also a C-epimorphism;*
(b) *if ϕ and $\phi\psi$ are C-epimorphisms, then ψ is also a C-epimorphism.*

Proof. (a) Assume that $|s\ker(\phi\psi)| > 1$ for some $s \in S$, and $s' \in S$ is an arbitrary element. Clearly $|s\ker\phi| \leq |s\ker(\phi\psi)|$. If $|s\ker\phi| > 1$ then $s\ker\phi$ is a left ideal in S and hence, $(s, s's) \in \ker\phi \subseteq \ker(\phi\psi)$, i.e., $s\ker(\phi\psi)$ is also a left ideal in S. In the other case, if $|s\ker\phi| = 1$ then for any $p \in s\ker(\phi\psi)$ with $p \neq s$ is $p\phi \neq s\phi$ and $p\phi \in (s\phi)\ker\psi$, i.e., $|(s\phi)\ker\psi| > 1$. Since ψ is a C-epimorphism $(s\phi, (s'\phi)(s\phi)) \in \ker\psi$ holds, and so $(s, s's) \in \ker(\phi\psi)$. Hence $\phi\psi$ is a C-epimorphism.

(b) Assume $|t\ker\psi| > 1$ for $t \in T$ and $t' \in T$ is an arbitrary element. Then $t = s\phi$ and $t' = s'\phi$ for some $s, s' \in S$. So $|(s\phi)\ker\psi| > 1$, and hence $|s\ker(\phi\psi)| > 1$. Since $\phi\psi$ is a C-homomorphism, it follows that $s\ker(\phi\psi) = (s's)\ker(\phi\psi)$. Hence $s\phi\psi = (s's)\phi\psi = ((s'\phi)(s\phi))\psi$, i.e., $t\psi = (t't)\psi$. Therefore, $(t, t't) \in \ker\psi$, what means that ψ is a C-homomorphism.

Lemma 3. *Let X be an alphabet and ρ a C-congruence on X^+. Then $C(M(\rho)) \cong X^+/\rho$.*

Proof. The equality $\theta_{M(\rho)} = \rho$ can be proved similarly as in the part (iii)$\Rightarrow$(i) of the proof of Theorem 3.

Similarly to the case when we deal with regular varieties of automata and transition semigroups where a σ-variety of X-automata, for some alphabet X, is defined, here we also define the notion of a σ_C-*variety of X-automata* as an irregular variety V satisfying the following condition

$$(\forall A, B \in V)\ C(A) \cong C(B) \Rightarrow (A \in V \Leftrightarrow B \in V).$$

Then the following result can be stated:

Theorem 7. *Let κ be a cardinal and X an alphabet of cardinality κ. Then to every κ_C-variety of semigroups K corresponds the σ_C-variety of X-automata*

$$\mathcal{V}(K) = \{A \mid A \text{ is an } X\text{-automaton and } C(A) \in K\},$$

and vice versa, to every σ_C-variety of X-automata V corresponds the κ_C-variety of semigroups

$$\mathcal{K}(V) = \{S \in \mathcal{S}_\kappa \mid S \cong C(A) \text{ for some } A \in V\}.$$

Proof. Let K be a κ_C-variety of semigroups. Assume that $A \in \mathcal{V}(K)$ and B is a homomorphic image of A. Then, by Theorem 2, $\phi : C(A) \to C(B)$ defined by $(u\theta_A)\phi = u\theta_B$ is an epimorphism. Besides that, since the mappings $u \mapsto u\theta_A$ and $u \mapsto u\theta_B$, for $u \in X^*$, are C-epimorphisms, it follows, by Lemma 2, that ϕ is also a C-homomorphism. Hence from $C(A) \in K$ follows $C(B) \in K$, i.e., $B \in \mathcal{V}(K)$. The case when an automaton B is a subautomaton of an automaton $A \in \mathcal{V}(K)$ can be treated similarly.

Assume now that an automaton A is a subdirect product of automata $A_i \in \mathcal{V}(K)$, $i \in I$. As noted in Theorem 2, $\theta_A = \bigcap_{i \in I} \theta_{A_i}$, what implies that the mappings $\phi_i : C(A) \to C(A_i)$ defined by $(u\theta_A)\phi_i = u\theta_{A_i}$, for any $i \in I$, are C-epimorphisms, according to Lemma 2, such that $\bigcap_{i \in I} \ker \phi_i = \Delta_{C(A)}$. So, $C(A) \in K$ as a κ_C-subdirect product of $C(A_i) \in K$, what gives $A \in \mathcal{V}(K)$.

Therefore, $\mathcal{V}(K)$ is a variety of X-automata. Its σ_C-closure is obvious.

Suppose now that V is a σ_C-variety of X-automata. Let us prove that $\mathcal{K}(V)$ is a κ_C-variety of semigroups.

Let $S \in \mathcal{K}(V)$ be an arbitrary semigroup and $\phi : S \to T$ is a C-epimorphism. Since $S \in \mathcal{K}(V)$ then there exists an automaton $A \in V$ such that $S \cong C(A) = X^+/\theta_A$. Let us denote by $\pi : X^+ \to S$ the natural epimorphism of the congruence θ_A. Obviously, π is a C-epimorphism. Then the mapping $\psi = \pi\phi : X^+ \to T$ is, by Lemma 2, a C-epimorphism and $T = C(M(\ker \psi))$ according to Lemma 3. Since $\theta_A = \ker \pi \subseteq \ker \psi$ it follows that $M(\ker \psi)$ is a homomorphic image of the automaton $M(\theta_A)$. Now we have that $A \in V$, by assumption, and $C(A) = S \cong C(M(\theta_A))$. Since V is σ_C-closed, it follows that $M(\theta_A) \in V$, and also $M(\ker \psi) \in V$ as a homomorphic image of $M(\theta_A)$. Therefore, $T = C(M(\ker \psi)) \in \mathcal{K}(V)$.

Suppose that a semigroup S is a κ_C-subdirect product of semigroups $S_i \in \mathcal{K}(V)$, $i \in I$. Then there exist automata $A_i \in V$, $i \in I$, such that $S_i = C(A_i)$ for every $i \in I$. Also there exist C-epimorphisms $\phi_i : S \to S_i$, for every $i \in I$, such that $\bigcap_{i \in I} \ker \phi_i = \Delta_S$. Since S is a κ-generated C-semigroup there exists a C-epimorphism $\psi : X^+ \to S$. Denote by $\pi_i = \psi\phi_i : X^+ \to S_i$, for every $i \in I$. By Lemma 2, π_i are C-epimorphisms. It can easily be proved that $\bigcap_{i \in I} \ker \pi_i = \ker \psi$. Then $M(\ker \psi)$ is a subdirect product of the automata $M(\ker \pi_i)$, $i \in I$. Now, since $C(M(\ker \pi_i)) \cong C(A_i)$ and V is σ_C-closed, it follows that $M(\ker \pi_i) \in V$ and $M(\ker \psi) \in V$ as a subdirect product of automata from V. Then $S = C(M(\ker \psi)) \in \mathcal{K}(V)$.

Since the concepts of varieties of automata and semigroups that we consider here are rather different from the usual ones, that will also be the case with generalized varieties and pseudovarieties. Namely, we follow the way of construction of these classes proposed in [2]. Let κ be an arbitrary cardinal. A class K of κ-generated C-semigroups is a *generalized κ_C-variety* if it can be represented as the union of a directed family of κ_C-varieties. If K is the class of all finite semigroups belonging to some generalized κ_C-variety of semigroups, then it is a *pseudo-κ_C-variety*. Similarly, for an alphabet X, a class of X-automata V is a *generalized σ_C-variety* if it can be represented as the union of a directed family of σ_C-varieties of X-automata, whereas V is a *pseudo-σ_C-variety* if it is the class of all finite members of some generalized σ_C-variety of X-automata.

Theorem 8. *Let κ be a cardinal and X an alphabet of cardinality κ. Then to every generalized κ_C-variety of semigroups K corresponds the generalized σ_C-variety of X-automata $\mathcal{V}(K) = \{ A \mid A$ is an X-automaton and $C(A) \in K \}$, and vice versa, to every generalized σ_C-variety of X-automata V corresponds the generalized κ_C-variety of semigroups $\mathcal{K}(V) = \{ S \in \mathcal{S}_\kappa \mid S \cong C(A)$ for some $A \in V \}$.*

Proof. It is straightforward by using the fact that the operators $\mathcal{V}$ and $\mathcal{K}$ go trough directed unions and Theorem 7.

Theorem 9. *Let κ be a cardinal and X an alphabet of cardinality κ. Then to every pseudo-κ_C-variety of semigroups K corresponds the pseudo-σ_C-variety of X-automata $\mathcal{V}(K) = \{A \mid A$ is a finite X-automaton and $C(A) \in K\}$, and vice versa, to every pseudo-σ_C-variety of X-automata V corresponds the pseudo-κ_C-variety of semigroups $\mathcal{K}(V) = \{S \in \mathcal{S}_\kappa \mid S \cong C(A)$ for some $A \in V\}$.*

Proof. Follows from Theorems 8 and 7.

Finally, we can prove the following theorem.

Theorem 10. *Let κ be a cardinal and X an alphabet of cardinality κ. Then the mappings $K \mapsto \mathcal{V}(K)$ and $V \mapsto \mathcal{K}(V)$ defined in Theorem 7 (Theorem 8, Theorem 9) are mutually inverse isomorphisms of the complete lattice of all κ_C-varieties (generalized κ_C-varieties, pseudo-κ_C-varieties) of semigroups and the complete lattice of all σ_C-varieties (generalized σ_C-varieties, pseudo-σ_C-varieties) of X-automata.*

Proof. It is not hard to check that all mentioned lattices are complete. We are proving the result concerning varieties, while the others can be proved similarly.

Having known the result of Theorem 7, it remains to notice that for κ_C-varieties of semigroups K_1 and K_2 from $K_1 \subseteq K_2$ follows $\mathcal{V}(K_1) \subseteq \mathcal{V}(K_2)$, and for σ_C-varieties of automata V_1 and V_2 such that $V_1 \subseteq V_2$ the inclusion $\mathcal{K}(V_1) \subseteq \mathcal{K}(V_2)$ holds. Besides that, it is easy to notice that $V = \mathcal{V}(\mathcal{K}(V))$ and $K = \mathcal{K}(\mathcal{V}(K))$, for any σ_C-variety of automata V and any κ_C-variety of semigroups K, what ends the proof.

Example 1. For a fixed word $u \in X^*$ the set $\mathbf{Dir}_u$ of all u-directable automata is a variety, but it is not σ_C-closed. Indeed, consider $X = \{x, y\}$ and $A \in \mathbf{Dir}_{xx}$. Then for the automaton $B = (A, X)$ with transitions given by $ax^B = ay^A$ and $ay^B = ax^A$ for any $a \in A$, we have that $C(A) \cong C(B)$ and $B \notin \mathbf{Dir}_{xx}$. Similarly the case of trap-directable automata can be discussed.

For $n \in \mathbb{N}$ the varieties $\mathbf{Def}_n$ of all n-definite automata and $\mathbf{Nilp}_n$ of all n-nilpotent automata are σ_C-varieties. The corresponding κ_C-varieties are, according to Theorems 5 and 6, formed of all κ-generated ideal extensions of right zero bands by n-nilpotent 0-free semigroups, and all 0-free n-nilpotent semigroups, respectively. To the generalized σ_C-varieties $\mathbf{Def} = \bigcup_{n \in \mathbb{N}} \mathbf{Def}_n$ and $\mathbf{Nilp} = \bigcup_{n \in N} \mathbf{Nilp}_n$ correspond generalized κ_C-varieties of all nilpotent 0-free extensions of right zero bands and all nilpotent 0-free semigroups, respectively. To the pseudo-σ_C-varieties of all finite definite and finite nilpotent automata correspond pseudo-κ_C-varieties of all finite semigroups that are nilpotent 0-free extensions of right zero bands and all finite nilpotent 0-free semigroups, respectively.

References

1. B. D. Ardent and C. J. Stuth, On partial homomorphisms of semigroups, Pacific J. Math. **35**, No. 1, (1970), 7–9.

2. C. J. Ash, Pseudovarieties, generalized varieties and similarly described classes, J. Algebra **92** (1985), 104–115.

3. S. Bogdanović, B. Imreh, M. Ćirić and T. Petković, Directable automata and their generalizations (A survey), in: S. Crvenković and I. Dolinka (eds.), Proc. VIII Int. Conf. "Algebra and Logic" (Novi Sad, 1998), Novi Sad J. Math. **29** (2) (1999), 31-74.

4. S. Burris and H. P. Sankappanavar, A course in universal algebra, Springer-Verlag, New York, 1981.

5. J. Černý, Poznámka k homogénnym experimentom s konečnými automatmi, Mat.-fyz. čas. SAV **14** (1964), 208–216.

6. D. H. Dao, Az irányítható automaták félcsoportjairól, Automataelméleti füzetek **1** (1978), 1-28, Eötvös Loránd University of Science, Budapest, 1978.

7. F. Gécseg and I. Peák, Algebraic Theory of Automata, Akadémiai Kiadó, Budapest, 1972.

8. V. M. Glushkov, Abstract theory of automata, Uspehi matem. nauk **16:5 (101)** (1961), 3–62 (in Russian).

9. V. M. Glushkov, Abstract automata and partitions of free semigroups, DAN SSSR **136** (1961), 765–767 (in Russian).

10. S. C. Kleene, Representation of events in nerve nets and finite automata, Automata Studies, Princeton University Press, Princeton, N.J., 1956, 3–41.

11. M. Perles, M. O. Rabin and E. Shamir, The theory of definite automata, IEEE Trans. Electronic Computers **EC-12** (1963), 233–243.

12. T. Petković, Varieties of automata and semigroups, Ph. D. Thesis, University of Niš, 1998, (in Serbian).

13. T. Petković, M. Ćirić and S. Bogdanović, Decompositions of automata and transition semigroups, Acta Cybernetica (Szeged) **13** (1998), 385–403.

14. T. Petković, M. Ćirić and S. Bogdanović, Correspondences between X-algebras, semigroups and congruences on free semigroups, Theoretical Computer Science (to appear).

15. T. Petković, M. Ćirić and S. Bogdanović, Theorems of Eilenberg's Type for Automata, (to appear).

16. L. N. Shevrin, About some classes of abstract automata, Uspehi matem. nauk **17:6 108** (1962), p. 219 (in Russian).

Author Index

Lecture Notes in Computer Science

For information about Vols. 1–2625
please contact your bookseller or Springer-Verlag

Vol. 2666: C. Guerra, S. Istrail (Eds.), Mathematical Methods for Protein Structure Analysis and Design. Proceedings, 2000. XI, 157 pages. 2003. (Subseries LNBI).

Vol. 2667: V. Kumar, M.L. Gavrilova, C.J.K. Tan, P. L'Ecuyer (Eds.), Computational Science and Its Applications – ICCSA 2003. Proceedings, Part I. 2003. XXXIV, 1060 pages. 2003.

Vol. 2668: V. Kumar, M.L. Gavrilova, C.J.K. Tan, P. L'Ecuyer (Eds.), Computational Science and Its Applications – ICCSA 2003. Proceedings, Part II. 2003. XXXIV, 942 pages. 2003.

Vol. 2669: V. Kumar, M.L. Gavrilova, C.J.K. Tan, P. L'Ecuyer (Eds.), Computational Science and Its Applications – ICCSA 2003. Proceedings, Part III. 2003. XXXIV, 948 pages. 2003.

Vol. 2670: R. Peña, T. Arts (Eds.), Implementation of Functional Languages. Proceedings, 2002. X, 249 pages. 2003.

Vol. 2671: Y. Xiang, B. Chaib-draa (Eds.), Advances in Artificial Intelligence. Proceedings, 2003. XIV, 642 pages. 2003. (Subseries LNAI).

Vol. 2672: M. Endler, D. Schmidt (Eds.), Middleware 2003. Proceedings, 2003. XIII, 513 pages. 2003.

Vol. 2673: N. Ayache, H. Delingette (Eds.), Surgery Simulation and Soft Tissue Modeling. Proceedings, 2003. XII, 386 pages. 2003.

Vol. 2674: I.E. Magnin, J. Montagnat, P. Clarysse, J. Nenonen, T. Katila (Eds.), Functional Imaging and Modeling of the Heart. Proceedings, 2003. XI, 308 pages. 2003.

Vol. 2675: M. Marchesi, G. Succi (Eds.), Extreme Programming and Agile Processes in Software Engineering. Proceedings, 2003. XV, 464 pages. 2003.

Vol. 2676: R. Baeza-Yates, E. Chávez, M. Crochemore (Eds.), Combinatorial Pattern Matching. Proceedings, 2003. XI, 403 pages. 2003.

Vol. 2678: W. van der Aalst, A. ter Hofstede, M. Weske (Eds.), Business Process Management. Proceedings, 2003. XI, 391 pages. 2003.

Vol. 2679: W. van der Aalst, E. Best (Eds.), Applications and Theory of Petri Nets 2003. Proceedings, 2003. XI, 508 pages. 2003.

Vol. 2680: P. Blackburn, C. Ghidini, R.M. Turner, F. Giunchiglia (Eds.), Modeling and Using Context. Proceedings, 2003. XII, 525 pages. 2003. (Subseries LNAI).

Vol. 2681: J. Eder, M. Missikoff (Eds.), Advanced Information Systems Engineering. Proceedings, 2003. XV, 740 pages. 2003.

Vol. 2685: C. Freksa, W. Brauer, C. Habel, K.F. Wender (Eds.), Spatial Cognition III. X, 415 pages. 2003. (Subseries LNAI).

Vol. 2686: J. Mira, J.R. Álvarez (Eds.), Computational Methods in Neural Modeling. Proceedings, Part I. 2003. XXVII, 764 pages. 2003.

Vol. 2687: J. Mira, J.R. Álvarez (Eds.), Artificial Neural Nets Problem Solving Methods. Proceedings, Part II. 2003. XXVII, 820 pages. 2003.

Vol. 2688: J. Kittler, M.S. Nixon (Eds.), Audio- and Video-Based Biometric Person Authentication. Proceedings, 2003. XVII, 978 pages. 2003.

Vol. 2689: K.D. Ashley, D.G. Bridge (Eds.), Case-Based Reasoning Research and Development. Proceedings, 2003. XV, 734 pages. 2003. (Subseries LNAI).

Vol. 2691: V. Mařík, J. Müller, M. Pěchouček (Eds.), Multi-Agent Systems and Applications III. Proceedings, 2003. XIV, 660 pages. 2003. (Subseries LNAI).

Vol. 2692: P. Nixon, S. Terzis (Eds.), Trust Management. Proceedings, 2003. X, 349 pages. 2003.

Vol. 2694: R. Cousot (Ed.), Static Analysis. Proceedings, 2003. XIV, 505 pages. 2003.

Vol. 2695: L.D. Griffin, M. Lillholm (Eds.), Scale Space Methods in Computer Vision. Proceedings, 2003. XII, 816 pages. 2003.

Vol. 2701: M. Hofmann (Ed.), Typed Lambda Calculi and Applications. Proceedings, 2003. VIII, 317 pages. 2003.

Vol. 2702: P. Brusilovsky, A. Corbett, F. de Rosis (Eds.), User Modeling 2003. Proceedings, 2003. XIV, 436 pages. 2003. (Subseries LNAI).

Vol. 2704: S.-T. Huang, T. Herman (Eds.), Self-Stabilizing Systems. Proceedings, 2003. X, 215 pages. 2003.

Vol. 2706: R. Nieuwenhuis (Ed.), Rewriting Techniques and Applications. Proceedings, 2003. XI, 515 pages. 2003.

Vol. 2707: K. Jeffay, I. Stoica, K. Wehrle (Eds.), Quality of Service – IWQoS 2003. Proceedings, 2003. XI, 517 pages. 2003.

Vol. 2709: T. Windeatt, F. Roli (Eds.), Multiple Classifier Systems. Proceedings, 2003. X, 406 pages. 2003.

Vol. 2710: Z. Ésik, Z, Fülöp (Eds.), Developments in Language Theory. Proceedings, 2003. XI, 437 pages. 2003.

Vol. 2711: T.D. Nielsen, N.L. Zhang (Eds.), Symbolic and Quantitative Approaches to Reasoning with Uncertainty. Proceedings, 2003. XII, 608 pages. 2003. (Subseries LNAI).

Vol. 2713: C.-W. Chung, C.-K. Kim, W. Kim, T.-W. Ling, K.-H. Song (Eds.), Web and Communication Technologies and Internet-Related Social Issues – HSI 2003. Proceedings, 2003. XXII, 773 pages. 2003.

Vol. 2714: O. Kaynak, E. Alpaydin, E. Oja, L. Xu (Eds.), Artificial Neural Networks and Neural Information Processing – ICANN/ICONIP 2003. Proceedings, 2003. XXII, 1188 pages. 2003.

Vol. 2715: T. Bilgiç, B. De Baets, O. Kaynak (Eds.), Fuzzy Sets and Systems – IFSA 2003. Proceedings, 2003. XV, 735 pages. 2003. (Subseries LNAI).

Vol. 2716: M.J. Voss (Ed.), OpenMP Shared Memory Parallel Programming. Proceedings, 2003. VIII, 271 pages. 2003.

Vol. 2718: P. W. H. Chung, C. Hinde, M. Ali (Eds.), Developments in Applied Artificial Intelligence. Proceedings, 2003. XIV, 817 pages. 2003. (Subseries LNAI).

Vol. 2719: J.C.M. Baeten, J.K. Lenstra, J. Parrow, G.J. Woeginger (Eds.), Automata, Languages and Programming. Proceedings, 2003. XVIII, 1199 pages. 2003.

Vol. 2721: N.J. Mamede, J. Baptista, I. Trancoso, M. das Graças Volpe Nunes (Eds.), Computational Processing of the Portuguese Language. Proceedings, 2003. XIV, 268 pages. 2003. (Subseries LNAI).

Vol. 2726: E. Hancock, M. Vento (Eds.), Graph Based Representations in Pattern Recognition. Proceedings, 2003. VIII, 271 pages. 2003.

Vol. 2734: P. Perner, A. Rosenfeld (Eds.), Machine Learning and Data Mining in Pattern Recognition. Proceedings, 2003. XII, 440 pages. 2003. (Subseries LNAI).